STUDENTS SOLUTIONS MANUAL

JUDITH A. PENNA
Indiana University Purdue University Indianapolis

COLLEGE ALGEBRA
THIRD EDITION

Judith A. Beecher
Indiana University Purdue University Indianapolis

Judith A. Penna
Indiana University Purdue University Indianapolis

Marvin L. Bittinger
Indiana University Purdue University Indianapolis

PEARSON

Addison
Wesley

Boston San Francisco New York
London Toronto Sydney Tokyo Singapore Madrid
Mexico City Munich Paris Cape Town Hong Kong Montreal

Reproduced by Pearson Addison-Wesley from QuarkXPress® files.

Copyright © 2008 Pearson Education, Inc.
Publishing as Pearson Addison-Wesley, 75 Arlington Street, Boston, MA 02116.

ISBN-13: 978-0-321-45918-3
ISBN-10: 0-321-45918-0

2 3 4 5 6 BB 10 09 08 07

Contents

Chapter R

Basic Concepts of Algebra

Exercise Set R.1

1. Whole numbers: $\sqrt[3]{8}$, 0, 9, $\sqrt{25}$ ($\sqrt[3]{8} = 2$, $\sqrt{25} = 5$)

3. Irrational numbers: $\sqrt{7}$, 5.242242224..., $-\sqrt{14}$, $\sqrt[5]{5}$, $\sqrt[3]{4}$

(Although there is a pattern in 5.242242224..., there is no repeating block of digits.)

5. Rational numbers: -12, $5.\overline{3}$, $-\dfrac{7}{3}$, $\sqrt[3]{8}$, 0, -1.96, 9, $4\dfrac{2}{3}$, $\sqrt{25}$, $\dfrac{5}{7}$

7. Rational numbers but not integers: $5.\overline{3}$, $-\dfrac{7}{3}$, -1.96, $4\dfrac{2}{3}$, $\dfrac{5}{7}$

9. Integers but not rational numbers: -12, 0

11. This is a closed interval, so we use brackets. Interval notation is $[-3, 3]$.

13. This is a half-open interval. We use a bracket on the left and a parenthesis on the right. Interval notation is $[-4, -1)$.

15. This interval is of unlimited extent in the negative direction, and the endpoint -2 is included. Interval notation is $(-\infty, -2]$.

17. This interval is of unlimited extent in the positive direction, and the endpoint 3.8 is not included. Interval notation is $(3.8, \infty)$.

19. $\{x | 7 < x\}$, or $\{x | x > 7\}$.

This interval is of unlimited extent in the positive direction and the endpoint 7 is not included. Interval notation is $(7, \infty)$.

21. The endpoints 0 and 5 are not included in the interval, so we use parentheses. Interval notation is $(0, 5)$.

23. The endpoint -9 is included in the interval, so we use a bracket before the -9. The endpoint -4 is not included, so we use a parenthesis after the -4. Interval notation is $[-9, -4)$.

25. Both endpoints are included in the interval, so we use brackets. Interval notation is $[x, x + h]$.

27. The endpoint p is not included in the interval, so we use a parenthesis before the p. The interval is of unlimited extent in the positive direction, so we use the infinity symbol ∞. Interval notation is (p, ∞).

29. Since 6 is an element of the set of natural numbers, the statement is true.

31. Since 3.2 is not an element of the set of integers, the statement is false.

33. Since $-\dfrac{11}{5}$ is an element of the set of rational numbers, the statement is true.

35. Since $\sqrt{11}$ is an element of the set of real numbers, the statement is false.

37. Since 24 is an element of the set of whole numbers, the statement is false.

39. Since 1.089 is not an element of the set of irrational numbers, the statement is true.

41. Since every whole number is an integer, the statement is true.

43. Since every rational number is a real number, the statement is true.

45. Since there are real numbers that are not integers, the statement is false.

47. The sentence $6 \cdot x = x \cdot 6$ illustrates the commutative property of multiplication.

49. The sentence $-3 \cdot 1 = -3$ illustrates the multiplicative identity property.

51. The sentence $5(ab) = (5a)b$ illustrates the associative property of multiplication.

53. The sentence $2(a+b) = (a+b)2$ illustrates the commutative property of multiplication.

55. The sentence $-6(m+n) = -6(n+m)$ illustrates the commutative property of addition.

57. The sentence $8 \cdot \dfrac{1}{8} = 1$ illustrates the multiplicative inverse property.

59. The distance of -7.1 from 0 is 7.1, so $|-7.1| = 7.1$.

61. The distance of 347 from 0 is 347, so $|347| = 347$.

63. The distance of $-\sqrt{97}$ from 0 is $\sqrt{97}$, so $|-\sqrt{97}| = \sqrt{97}$.

65. The distance of 0 from 0 is 0, so $|0| = 0$.

67. The distance of $\frac{5}{4}$ from 0 is $\frac{5}{4}$, so $\left|\frac{5}{4}\right| = \frac{5}{4}$.

69. $|-5-6| = |-11| = 11$, or
$|6-(-5)| = |6+5| = |11| = 11$

71. $|-2-(-8)| = |-2+8| = |6| = 6$, or
$|-8-(-2)| = |-8+2| = |-6| = 6$

73. $|12.1-6.7| = |5.4| = 5.4$, or
$|6.7-12.1| = |-5.4| = 5.4$

75. $\left|-\frac{3}{4}-\frac{15}{8}\right| = \left|-\frac{6}{8}-\frac{15}{8}\right| = \left|-\frac{21}{8}\right| = \frac{21}{8}$, or

$\left|\frac{15}{8}-\left(-\frac{3}{4}\right)\right| = \left|\frac{15}{8}+\frac{3}{4}\right| = \left|\frac{15}{8}+\frac{6}{8}\right| = \left|\frac{21}{8}\right| = \frac{21}{8}$

77. $|-7-0| = |-7| = 7$, or
$|0-(-7)| = |0+7| = |7| = 7$

79. Discussion and Writing

81. Answers may vary. One such number is
$0.124124412444\ldots$.

83. Answers may vary. Since $-\frac{1}{101} = 0.\overline{0099}$ and
$-\frac{1}{100} = -0.01$, one such number is -0.00999.

85. Since $1^2 + 3^2 = 10$, the hypotenuse of a right triangle with legs of lengths 1 unit and 3 units has a length of $\sqrt{10}$ units.

$$c^2 = 1^2 + 3^2$$
$$c^2 = 10$$
$$c = \sqrt{10}$$

Exercise Set R.2

1. $3^{-7} = \dfrac{1}{3^7}$ $\left(a^{-m} = \dfrac{1}{a^m},\ a \neq 0\right)$

3. Observe that each exponent is negative. We move each factor to the other side of the fraction bar and change the sign of each exponent.
$$\frac{x^{-5}}{y^{-4}} = \frac{y^4}{x^5}$$

5. Observe that each exponent is negative. We move each factor to the other side of the fraction bar and change the sign of each exponent.
$$\frac{m^{-1}n^{-12}}{t^{-6}} = \frac{t^6}{m^1 n^{12}}, \text{ or } \frac{t^6}{mn^{12}}$$

7. $18° = 1$ (For any nonzero real number, $a^0 = 1$.)

9. $x^9 \cdot x^0 = x^{9+0} = x^9$

11. $5^8 \cdot 5^{-6} = 5^{8+(-6)} = 5^2$, or 25

13. $m^{-5} \cdot m^5 = m^{-5+5} = m^0 = 1$

15. $y^3 \cdot y^{-7} = y^{3+(-7)} = y^{-4}$, or $\dfrac{1}{y^4}$

17. $7^3 \cdot 7^{-5} \cdot 7 = 7^{3+(-5)+1} = 7^{-1}$, or $\dfrac{1}{7}$

19. $2x^3 \cdot 3x^2 = 2 \cdot 3 \cdot x^{3+2} = 6x^5$

21. $(-3a^{-5})(5a^{-7}) = -3 \cdot 5 \cdot a^{-5+(-7)} = -15a^{-12}$, or
$-\dfrac{15}{a^{12}}$

23. $(5a^2b)(3a^{-3}b^4) = 5 \cdot 3 \cdot a^{2+(-3)} \cdot b^{1+4} = 15a^{-1}b^5$, or
$\dfrac{15b^5}{a}$

25. $(6x^{-3}y^5)(-7x^2y^{-9}) = 6(-7)x^{-3+2}y^{5+(-9)} =$
$-42x^{-1}y^{-4}$, or $-\dfrac{42}{xy^4}$

27. $(2x)^3(3x)^2 = 2^3x^3 \cdot 3^2x^2 = 8 \cdot 9 \cdot x^{3+2} = 72x^5$

29. $(-2n)^3(5n)^2 = (-2)^3n^3 \cdot 5^2n^2 = -8 \cdot 25 \cdot n^{3+2} =$
$-200n^5$

31. $\dfrac{b^{40}}{b^{37}} = b^{40-37} = b^3$

33. $\dfrac{x^{-5}}{x^{16}} = x^{-5-16} = x^{-21}$, or $\dfrac{1}{x^{21}}$

35. $\dfrac{x^2y^{-2}}{x^{-1}y} = x^{2-(-1)}y^{-2-1} = x^3y^{-3}$, or $\dfrac{x^3}{y^3}$

37. $\dfrac{32x^{-4}y^3}{4x^{-5}y^8} = \dfrac{32}{4}x^{-4-(-5)}y^{3-8} = 8xy^{-5}$, or $\dfrac{8x}{y^5}$

39. $(2ab^2)^3 = 2^3a^3(b^2)^3 = 2^3a^3b^{2\cdot3} = 8a^3b^6$

41. $(-2x^3)^5 = (-2)^5(x^3)^5 = (-2)^5x^{3\cdot5} = -32x^{15}$

43. $(-5c^{-1}d^{-2})^{-2} = (-5)^{-2}c^{-1(-2)}d^{-2(-2)} =$
$\dfrac{c^2d^4}{(-5)^2} = \dfrac{c^2d^4}{25}$

45. $(3m^4)^3(2m^{-5})^4 = 3^3m^{12} \cdot 2^4m^{-20} =$
$27 \cdot 16m^{12+(-20)} = 432m^{-8}$, or $\dfrac{432}{m^8}$

47. $\left(\dfrac{2x^{-3}y^7}{z^{-1}}\right)^3 = \dfrac{(2x^{-3}y^7)^3}{(z^{-1})^3} = \dfrac{2^3x^{-9}y^{21}}{z^{-3}} =$
$\dfrac{8x^{-9}y^{21}}{z^{-3}}$, or $\dfrac{8y^{21}z^3}{x^9}$

49. $\left(\dfrac{24a^{10}b^{-8}c^7}{12a^6b^{-3}c^5}\right)^{-5} = (2a^4b^{-5}c^2)^{-5} = 2^{-5}a^{-20}b^{25}c^{-10}$,

or $\dfrac{b^{25}}{32a^{20}c^{10}}$

51. Convert 405,000 to scientific notation.

We want the decimal point to be positioned between the 4 and the first 0, so we move it 5 places to the left. Since 405,000 is greater than 10, the exponent must be positive.

$$405,000 = 4.05 \times 10^5$$

53. Convert 0.00000039 to scientific notation.

We want the decimal point to be positioned between the 3 and the 9, so we move it 7 places to the right. Since 0.00000039 is a number between 0 and 1, the exponent must be negative.

$$0.00000039 = 3.9 \times 10^{-7}$$

55. Convert 234,600,000,000 to scientific notation. We want the decimal point to be positioned between the 2 and the 3, so we move it 11 places to the left. Since 234,600,000,000 is greater than 10, the exponent must be positive.

$$234,600,000,000 = 2.346 \times 10^{11}$$

57. Convert 0.00104 to scientific notation. We want the decimal point to be positioned between the 1 and the last 0, so we move it 3 places to the right. Since 0.00104 is a number between 0 and 1, the exponent must be negative.

$$0.00104 = 1.04 \times 10^{-3}$$

59. Convert 0.000016 to scientific notation.

We want the decimal point to be positioned between the 1 and the 6, so we move it 5 places to the right. Since 0.000016 is a number between 0 and 1, the exponent must be negative.

$$0.000016 = 1.6 \times 10^{-5}$$

61. Convert 8.3×10^{-5} to decimal notation.

The exponent is negative, so the number is between 0 and 1. We move the decimal point 5 places to the left.

$$8.3 \times 10^{-5} = 0.000083$$

63. Convert 2.07×10^7 to decimal notation.

The exponent is positive, so the number is greater than 10. We move the decimal point 7 places to the right.

$$2.07 \times 10^7 = 20,700,000$$

65. Convert 3.496×10^{10} to decimal notation.

The exponent is positive, so the number is greater than 10. We move the decimal point 10 places to the right.

$$3.496 \times 10^{10} = 34,960,000,000$$

67. Convert 5.41×10^{-8} to decimal notation.

The exponent is negative, so the number is between 0 and 1. We move the decimal point 8 places to the left.

$$5.41 \times 10^{-8} = 0.0000000541$$

69. Convert 2.319×10^8 to decimal notation.

The exponent is positive, so the number is greater than 10. We move the decimal point 8 places to the right.

$$2.319 \times 10^8 = 231,900,000$$

71. $(3.1 \times 10^5)(4.5 \times 10^{-3})$

$= (3.1 \times 4.5) \times (10^5 \times 10^{-3})$

$= 13.95 \times 10^2$ This is not scientific notation.

$= (1.395 \times 10) \times 10^2$

$= 1.395 \times 10^3$ Writing scientific notation

73. $(2.6 \times 10^{-18})(8.5 \times 10^7)$

$= (2.6 \times 8.5) \times (10^{-18} \times 10^7)$

$= 22.1 \times 10^{-11}$ This is not scientific notation.

$= (2.21 \times 10) \times 10^{-11}$

$= 2.21 \times 10^{-10}$

75. $\dfrac{6.4 \times 10^{-7}}{8.0 \times 10^6} = \dfrac{6.4}{8.0} \times \dfrac{10^{-7}}{10^6}$

$= 0.8 \times 10^{-13}$ This is not scientific notation.

$= (8 \times 10^{-1}) \times 10^{-13}$

$= 8 \times 10^{-14}$ Writing scientific notation

77. $\dfrac{1.8 \times 10^{-3}}{7.2 \times 10^{-9}}$

$= \dfrac{1.8}{7.2} \times \dfrac{10^{-3}}{10^{-9}}$

$= 0.25 \times 10^6$ This is not scientific notation.

$= (2.5 \times 10^{-1}) \times 10^6$

$= 2.5 \times 10^5$

79. We multiply the number of AUs from Earth to Pluto by the number of miles in 1 AU.

39×93 million

$= 39 \times 93,000,000$

$= (3.9 \times 10) \times (9.3 \times 10^7)$ Writing scientific notation

$= (3.9 \times 9.3) \times (10 \times 10^7)$

$= 36.27 \times 10^8$

$= (3.627 \times 10) \times 10^8$

$= 3.627 \times 10^9$

The distance from Earth to Pluto is about 3.627×10^9 mi.

81. We multiply the diameter, in nanometers, by the number of meters in 1 nanometer.

360×0.000000001

$= (3.6 \times 10^2) \times 10^{-9}$

$= 3.6 \times (10^2 \times 10^{-9})$

$= 3.6 \times 10^{-7}$

The diameter of the wire is 3.6×10^{-7} m.

83. The average cost per mile is the total cost divided by the number of miles.

$$\frac{\$210 \times 10^6}{17.6}$$

$$= \frac{\$210 \times 10^6}{1.76 \times 10}$$

$$\approx \$119 \times 10^5$$

$$\approx (\$1.19 \times 10^2) \times 10^5$$

$$\approx \$1.19 \times 10^7$$

The average cost per mile was about $\$1.19 \times 10^7$.

85. First find the number of seconds in 1 hour:

$$1 \text{ hour} = 1 \text{ hr} \times \frac{60 \text{ min}}{1 \text{ hr}} \times \frac{60 \text{ sec}}{1 \text{ min}} = 3600 \text{ sec}$$

The number of disintegrations produced in 1 hour is the number of disintegrations per second times the number of seconds in 1 hour.

$$37 \text{ billion} \times 3600$$

$$= 37,000,000,000 \times 3600$$

$$= 3.7 \times 10^{10} \times 3.6 \times 10^3 \quad \text{Writing scientific}$$
$$\text{notation}$$

$$= (3.7 \times 3.6) \times (10^{10} \times 10^3)$$

$$= 13.32 \times 10^{13} \quad \text{Multiplying}$$

$$= (1.332 \times 10) \times 10^{13}$$

$$= 1.332 \times 10^{14}$$

One gram of radium produces 1.332×10^{14} disintegrations in 1 hour.

87. $\quad 3 \cdot 2 - 4 \cdot 2^2 + 6(3-1)$

$$= 3 \cdot 2 - 4 \cdot 2^2 + 6 \cdot 2 \quad \text{Working inside}$$
$$\text{parentheses}$$

$$= 3 \cdot 2 - 4 \cdot 4 + 6 \cdot 2 \quad \text{Evaluating } 2^2$$

$$= 6 - 16 + 12 \quad \text{Multiplying}$$

$$= -10 + 12 \quad \text{Adding in order}$$

$$= 2 \quad \text{from left to right}$$

89. $\quad 16 \div 4 \cdot 4 \div 2 \cdot 256$

$$= 4 \cdot 4 \div 2 \cdot 256 \quad \text{Multiplying and dividing}$$
$$\text{in order from left to right}$$

$$= 16 \div 2 \cdot 256$$

$$= 8 \cdot 256$$

$$= 2048$$

91. $\quad \dfrac{4(8-6)^2 - 4 \cdot 3 + 2 \cdot 8}{3^1 + 19^0}$

$$= \frac{4 \cdot 2^2 - 4 \cdot 3 + 2 \cdot 8}{3 + 1} \quad \text{Calculating in the}$$
$$\text{numerator and in}$$
$$\text{the denominator}$$

$$= \frac{4 \cdot 4 - 4 \cdot 3 + 2 \cdot 8}{4}$$

$$= \frac{16 - 12 + 16}{4}$$

$$= \frac{4 + 16}{4}$$

$$= \frac{20}{4}$$

$$= 5$$

93. Since interest is compounded semiannually, $n = 2$. Substitute $\$2125$ for P, 6.2% or 0.062 for i, 2 for n, and 5 for t in the compound interest formula.

$$A = P\left(1 + \frac{i}{n}\right)^{nt}$$

$$= \$2125\left(1 + \frac{0.062}{2}\right)^{2 \cdot 5} \quad \text{Substituting}$$

$$= \$2125(1 + 0.031)^{2 \cdot 5} \quad \text{Dividing}$$

$$= \$2125(1.031)^{2 \cdot 5} \quad \text{Adding}$$

$$= \$2125(1.031)^{10} \quad \text{Multiplying 2 and 5}$$

$$\approx \$2125(1.357021264) \quad \text{Evaluating the}$$
$$\text{exponential expression}$$

$$\approx \$2883.670185 \quad \text{Multiplying}$$

$$\approx \$2883.67 \quad \text{Rounding to the nearest cent}$$

95. Since interest is compounded quarterly, $n = 4$. Substitute $\$6700$ for P, 4.5% or 0.045 for i, 4 for n, and 6 for t in the compound interest formula.

$$A = P\left(1 + \frac{i}{n}\right)^{nt}$$

$$= \$6700\left(1 + \frac{0.045}{4}\right)^{4 \cdot 6} \quad \text{Substituting}$$

$$= \$6700(1 + 0.01125)^{4 \cdot 6} \quad \text{Dividing}$$

$$= \$6700(1.01125)^{4 \cdot 6} \quad \text{Adding}$$

$$= \$6700(1.01125)^{24} \quad \text{Multiplying 4 and 6}$$

$$\approx \$6700(1.307991226) \quad \text{Evaluating the}$$
$$\text{exponential expression}$$

$$\approx \$8763.541217 \quad \text{Multiplying}$$

$$\approx \$8763.54 \quad \text{Rounding to the nearest cent}$$

97. Discussion and Writing

99. Substitute $\$250$ for P, 0.05 for r and 27 for t and perform the resulting computation.

$$S = P\left[\frac{\left(1 + \dfrac{r}{12}\right)^{12 \cdot t} - 1}{\dfrac{r}{12}}\right]$$

$$= \$250\left[\frac{\left(1 + \dfrac{0.05}{12}\right)^{12 \cdot 27} - 1}{\dfrac{0.05}{12}}\right]$$

$$\approx \$170,797.30$$

101. Substitute $120,000 for S, 0.06 for r, and 18 for t and solve for P.

$$S = P\left[\frac{\left(1+\frac{r}{12}\right)^{12\cdot t} - 1}{\frac{r}{12}}\right]$$

$$\$120,000 = P\left[\frac{\left(1+\frac{0.06}{12}\right)^{12\cdot 18} - 1}{\frac{0.06}{12}}\right]$$

$$\$120,000 = P\left[\frac{(1.005)^{216} - 1}{0.05}\right]$$

$$\$120,000 \approx P(387.3532)$$

$$\$309.79 \approx P$$

103. $(x^t \cdot x^{3t})^2 = (x^{4t})^2 = x^{4t\cdot 2} = x^{8t}$

105. $(t^{a+x} \cdot t^{x-a})^4 = (t^{2x})^4 = t^{2x\cdot 4} = t^{8x}$

107. $\left[\frac{(3x^a y^b)^3}{(-3x^a y^b)^2}\right]^2 = \left[\frac{27x^{3a}y^{3b}}{9x^{2a}y^{2b}}\right]^2$

$$= \left[3x^a y^b\right]^2$$

$$= 9x^{2a}y^{2b}$$

Exercise Set R.3

1. $-5y^4 + 3y^3 + 7y^2 - y - 4 =$
$-5y^4 + 3y^3 + 7y^2 + (-y) + (-4)$

Terms: $-5y^4$, $3y^3$, $7y^2$, $-y$, -4

The degree of the term of highest degree, $-5y^4$, is 4. Thus, the degree of the polynomial is 4.

3. $3a^4 b - 7a^3 b^3 + 5ab - 2 = 3a^4 b + (-7a^3 b^3) + 5ab + (-2)$

Terms: $3a^4 b$, $-7a^3 b^3$, $5ab$, -2

The degrees of the terms are 5, 6, 2, and, 0, respectively, so the degree of the polynomial is 6.

5. $(5x^2 y - 2xy^2 + 3xy - 5) +$
$\qquad (-2x^2 y - 3xy^2 + 4xy + 7)$
$= (5-2)x^2 y + (-2-3)xy^2 + (3+4)xy +$
$\qquad (-5+7)$
$= 3x^2 y - 5xy^2 + 7xy + 2$

7. $(2x + 3y + z - 7) + (4x - 2y - z + 8) +$
$\qquad (-3x + y - 2z - 4)$
$= (2+4-3)x + (3-2+1)y + (1-1-2)z +$
$\qquad (-7+8-4)$
$= 3x + 2y - 2z - 3$

9. $(3x^2 - 2x - x^3 + 2) - (5x^2 - 8x - x^3 + 4)$
$= (3x^2 - 2x - x^3 + 2) + (-5x^2 + 8x + x^3 - 4)$
$= (3-5)x^2 + (-2+8)x + (-1+1)x^3 + (2-4)$
$= -2x^2 + 6x - 2$

11. $(x^4 - 3x^2 + 4x) - (3x^3 + x^2 - 5x + 3)$
$= (x^4 - 3x^2 + 4x) + (-3x^3 - x^2 + 5x - 3)$
$= x^4 - 3x^3 + (-3-1)x^2 + (4+5)x - 3$
$= x^4 - 3x^3 - 4x^2 + 9x - 3$

13. $(a - b)(2a^3 - ab + 3b^2)$
$= (a-b)(2a^3) + (a-b)(-ab) + (a-b)(3b^2)$
$\qquad$ Using the distributive property
$= 2a^4 - 2a^3 b - a^2 b + ab^2 + 3ab^2 - 3b^3$
$\qquad$ Using the distributive property three more times
$= 2a^4 - 2a^3 b - a^2 b + 4ab^2 - 3b^3$ Collecting like terms

15. $(x + 5)(x - 3)$
$= x^2 - 3x + 5x - 15 \qquad$ Using FOIL
$= x^2 + 2x - 15 \qquad$ Collecting like terms

17. $(x + 6)(x + 4)$
$= x^2 + 4x + 6x + 24 \qquad$ Using FOIL
$= x^2 + 10x + 24 \qquad$ Collecting like terms

19. $(2a + 3)(a + 5)$
$= 2a^2 + 10a + 3a + 15 \qquad$ Using FOIL
$= 2a^2 + 13a + 15 \qquad$ Collecting like terms

21. $(2x + 3y)(2x + y)$
$= 4x^2 + 2xy + 6xy + 3y^2 \qquad$ Using FOIL
$= 4x^2 + 8xy + 3y^2$

23. $(y + 5)^2$
$= y^2 + 2\cdot y\cdot 5 + 5^2$
$\qquad [(A+B)^2 = A^2 + 2AB + B^2]$
$= y^2 + 10y + 25$

25. $(x - 4)^2$
$= x^2 - 2\cdot x\cdot 4 + 4^2$
$\qquad [(A-B)^2 = A^2 - 2AB + B^2]$
$= x^2 - 8x + 16$

27. $(5x - 3)^2$
$= (5x)^2 - 2\cdot 5x\cdot 3 + 3^2$
$\qquad [(A-B)^2 = A^2 - 2AB + B^2]$
$= 25x^2 - 30x + 9$

29. $(2x + 3y)^2$
$= (2x)^2 + 2(2x)(3y) + (3y)^2$
$\qquad [(A+B)^2 = A^2 + 2AB + B^2]$
$= 4x^2 + 12xy + 9y^2$

31. $(2x^2 - 3y)^2$
$= (2x^2)^2 - 2(2x^2)(3y) + (3y)^2$
$\qquad [(A-B)^2 = A^2 - 2AB + B^2]$
$= 4x^4 - 12x^2 y + 9y^2$

33. $(a+3)(a-3)$

$= a^2 - 3^2$ $\quad [(A+B)(A-B) = A^2 - B^2]$

$= a^2 - 9$

35. $(2x-5)(2x+5)$

$= (2x)^2 - 5^2$ $\quad [(A+B)(A-B) = A^2 - B^2]$

$= 4x^2 - 25$

37. $(3x-2y)(3x+2y)$

$= (3x)^2 - (2y)^2$ $\quad [(A-B)(A+B) = A^2 - B^2]$

$= 9x^2 - 4y^2$

39. $(2x+3y+4)(2x+3y-4)$

$= [(2x+3y)+4][(2x+3y)-4]$

$= (2x+3y)^2 - 4^2$

$= 4x^2 + 12xy + 9y^2 - 16$

41. $(x+1)(x-1)(x^2+1)$

$= (x^2-1)(x^2+1)$

$= x^4 - 1$

43. Discussion and Writing

45. $(a^n + b^n)(a^n - b^n) = (a^n)^2 - (b^n)^2$

$ = a^{2n} - b^{2n}$

47. $(a^n + b^n)^2 = (a^n)^2 + 2 \cdot a^n \cdot b^n + (b^n)^2$

$ = a^{2n} + 2a^n b^n + b^{2n}$

49. $(x-1)(x^2+x+1)(x^3+1)$

$= [(x-1)x^2 + (x-1)x + (x-1)\cdot 1](x^3+1)$

$= (x^3 - x^2 + x^2 - x + x - 1)(x^3 + 1)$

$= (x^3 - 1)(x^3 + 1)$

$= (x^3)^2 - 1^2$

$= x^6 - 1$

51. $(x^{a-b})^{a+b}$

$= x^{(a-b)(a+b)}$

$= x^{a^2 - b^2}$

53. $(a+b+c)^2$

$= (a+b+c)(a+b+c)$

$= (a+b+c)(a) + (a+b+c)(b) + (a+b+c)(c)$

$= a^2 + ab + ac + ab + b^2 + bc + ac + bc + c^2$

$= a^2 + b^2 + c^2 + 2ab + 2ac + 2bc$

Exercise Set R.4

1. $2x - 10 = 2 \cdot x - 2 \cdot 5 = 2(x - 5)$

3. $3x^4 - 9x^2 = 3x^2 \cdot x^2 - 3x^2 \cdot 3 = 3x^2(x^2 - 3)$

5. $4a^2 - 12a + 16 = 4 \cdot a^2 - 4 \cdot 3a + 4 \cdot 4 = 4(a^2 - 3a + 4)$

7. $a(b-2) + c(b-2) = (b-2)(a+c)$

9. $x^3 + 3x^2 + 6x + 18$

$= x^2(x+3) + 6(x+3)$

$= (x+3)(x^2+6)$

11. $y^3 - y^2 + 3y - 3$

$= y^2(y-1) + 3(y-1)$

$= (y-1)(y^2+3)$

13. $24x^3 - 36x^2 + 72x - 108$

$= 12(2x^3 - 3x^2 + 6x - 9)$

$= 12[x^2(2x-3) + 3(2x-3)]$

$= 12(2x-3)(x^2+3)$

15. $a^3 - 3a^2 - 2a + 6$

$= a^2(a-3) - 2(a-3)$

$= (a-3)(a^2-2)$

17. $x^3 - x^2 - 5x + 5$

$= x^2(x-1) - 5(x-1)$

$= (x-1)(x^2-5)$

19. $p^2 + 6p + 8$

We look for two numbers with a product of 8 and a sum of 6. By trial, we determine that they are 2 and 4.

$$p^2 + 6p + 8 = (p+2)(p+4)$$

21. $x^2 - 8x + 12$

We look for two numbers with a product of 12 and a sum of -8. By trial, we determine that they are -2 and -6.

$$x^2 - 8x + 12 = (x-2)(x-6)$$

23. $t^2 + 8t + 15$

We look for two numbers with a product of 15 and a sum of 8. By trial, we determine that they are 3 and 5.

$$t^2 + 8t + 15 = (t+3)(t+5)$$

25. $x^2 - 6xy - 27y^2$

We look for two numbers with a product of -27 and a sum of -6. By trial, we determine that they are 3 and -9.

$$x^2 - 6xy - 27y^2 = (x+3y)(x-9y)$$

27. $2n^2 - 20n - 48 = 2(n^2 - 10n - 24)$

Now factor $n^2 - 10n - 24$. We look for two numbers with a product of -24 and a sum of -10. By trial, we determine that they are 2 and -12. Then $n^2 - 10n - 24 = (n+2)(n-12)$. We must include the common factor, 2, to have a factorization of the original trinomial.

$$2n^2 - 20n - 48 = 2(n+2)(n-12)$$

29. $y^4 - 4y^2 - 21 = (y^2)^2 - 4y^2 - 21$

We look for two numbers with a product of -21 and a sum of -4. By trial, we determine that they are 3 and -7.

$$y^4 - 4y^2 - 21 = (y^2+3)(y^2-7)$$

31. $y^4 + 9y^3 + 14y^2 = y^2(y^2 + 9y + 14)$

Now factor $y^2 + 9y + 14$. Look for two numbers with a product of 14 and a sum of 9. The numbers are 2 and 7. Then $y^2 + 9y + 14 = (y + 2)(y + 7)$. We must include the common factor, y^2, in order to have a factorization of the original trinomial.

$$y^4 + 9y^3 + 14y^2 = y^2(y + 2)(y + 7)$$

33. $2x^3 - 2x^2y - 24xy^2 = 2x(x^2 - xy - 12y^2)$

Now factor $x^2 - xy - 12y^2$. Look for two numbers with a product of -12 and a sum of -1. The numbers are -4 and 3. Then $x^2 - xy - 12y^2 = (x - 4y)(x + 3y)$. We must include the common factor, $2x$, in order to have a factorization of the original trinomial.

$$2x^3 - 2x^2y - 24xy^2 = 2x(x - 4y)(x + 3y)$$

35. $2n^2 + 9n - 56$

We use the FOIL method.

1. There is no common factor other than 1 or -1.

2. The factorization must be of the form $(2n+\quad)(n+\quad)$.

3. Factor the constant term, -56. The possibilities are $-1 \cdot 56$, $1(-56)$, $-2 \cdot 28$, $2(-28)$, $-4 \cdot 16$, $4(-16)$, $-7 \cdot 8$, and $7(-8)$. The factors can be written in the opposite order as well: $56(-1)$, $-56 \cdot 1$, $28(-2)$, $-28 \cdot 2$, $16(-4)$, $-16 \cdot 4$, $8(-7)$, and $-8 \cdot 7$.

4. Find a pair of factors for which the sum of the outside and the inside products is the middle term, $9n$. By trial, we determine that the factorization is $(2n - 7)(n + 8)$.

37. $12x^2 + 11x + 2$

We use the grouping method.

1. There is no common factor other than 1 or -1.

2. Multiply the leading coefficient and the constant: $12 \cdot 2 = 24$.

3. Try to factor 24 so that the sum of the factors is the coefficient of the middle term, 11. The factors we want are 3 and 8.

4. Split the middle term using the numbers found in step (3):
$$11x = 3x + 8x$$

5. Factor by grouping.
$$12x^2 + 11x + 2 = 12x^2 + 3x + 8x + 2$$
$$= 3x(4x + 1) + 2(4x + 1)$$
$$= (4x + 1)(3x + 2)$$

39. $4x^2 + 15x + 9$

We use the FOIL method.

1. There is no common factor other than 1 or -1.

2. The factorization must be of the form $(4x+\quad)(x+\quad)$ or $(2x+\quad)(2x+\quad)$.

3. Factor the constant term, 9. The possibilities are $1 \cdot 9$, $-1(-9)$, $3 \cdot 3$, and $-3(-3)$. The first two pairs of factors can be written in the opposite order as well: $9 \cdot 1$, $-9(-1)$.

4. Find a pair of factors for which the sum of the outside and the inside products is the middle term, $15x$. By trial, we determine that the factorization is $(4x + 3)(x + 3)$.

41. $2y^2 + y - 6$

We use the grouping method.

1. There is no common factor other than 1 or -1.

2. Multiply the leading coefficient and the constant: $2(-6) = -12$.

3. Try to factor -12 so that the sum of the factors is the coefficient of the middle term, 1. The factors we want are 4 and -3.

4. Split the middle term using the numbers found in step (3):
$$y = 4y - 3y$$

5. Factor by grouping.
$$2y^2 + y - 6 = 2y^2 + 4y - 3y - 6$$
$$= 2y(y + 2) - 3(y + 2)$$
$$= (y + 2)(2y - 3)$$

43. $6a^2 - 29ab + 28b^2$

We use the FOIL method.

1. There is no common factor other than 1 or -1.

2. The factorization must be of the form $(6x+\quad)(x+\quad)$ or $(3x+\quad)(2x+\quad)$.

3. Factor the coefficient of the last term, 28. The possibilities are $1 \cdot 28$, $-1(-28)$, $2 \cdot 14$, $-2(-14)$, $4 \cdot 7$, and $-4(-7)$. The factors can be written in the opposite order as well: $28 \cdot 1$, $-28(-1)$, $14 \cdot 2$, $-14(-2)$, $7 \cdot 4$, and $-7(-4)$.

4. Find a pair of factors for which the sum of the outside and the inside products is the middle term, -29. Observe that the second term of each binomial factor will contain a factor of b. By trial, we determine that the factorization is $(3a - 4b)(2a - 7b)$.

45. $12a^2 - 4a - 16$

We will use the grouping method.

1. Factor out the common factor, 4.
$$12a^2 - 4a - 16 = 4(3a^2 - a - 4)$$

2. Now consider $3a^2 - a - 4$. Multiply the leading coefficient and the constant: $3(-4) = -12$.

3. Try to factor -12 so that the sum of the factors is the coefficient of the middle term, -1. The factors we want are -4 and 3.

4. Split the middle term using the numbers found in step (3):
$$-a = -4a + 3a$$

5. Factor by grouping.
$$3a^2 - a - 4 = 3a^2 - 4a + 3a - 4$$
$$= a(3a - 4) + (3a - 4)$$
$$= (3a - 4)(a + 1)$$

We must include the common factor to get a factorization of the original trinomial.
$$12a^2 - 4a - 16 = 4(3a - 4)(a + 1)$$

47. $m^2 - 4 = m^2 - 2^2$
$$= (m + 2)(m - 2)$$

49. $4z^2 - 81 = (2z)^2 - 9^2$
$$= (2z + 9)(2z - 9)$$

51. $6x^2 - 6y^2 = 6(x^2 - y^2) = 6(x + y)(x - y)$

53. $4xy^4 - 4xz^2 = 4x(y^4 - z^2)$
$$= 4x[(y^2)^2 - z^2]$$
$$= 4x(y^2 + z)(y^2 - z)$$

55. $7pq^4 - 7py^4 = 7p(q^4 - y^4)$
$$= 7p[(q^2)^2 - (y^2)^2]$$
$$= 7p(q^2 + y^2)(q^2 - y^2)$$
$$= 7p(q^2 + y^2)(q + y)(q - y)$$

57. $y^2 - 6y + 9 = y^2 - 2 \cdot y \cdot 3 + 3^2$
$$= (y - 3)^2$$

59. $4z^2 + 12z + 9 = (2z)^2 + 2 \cdot 2z \cdot 3 + 3^2$
$$= (2z + 3)^2$$

61. $1 - 8x + 16x^2 = 1^2 - 2 \cdot 1 \cdot 4x + (4x)^2$
$$= (1 - 4x)^2$$

63. $a^3 + 24a^2 + 144a$
$$= a(a^2 + 24a + 144)$$
$$= a(a^2 + 2 \cdot a \cdot 12 + 12^2)$$
$$= a(a + 12)^2$$

65. $4p^2 - 8pq + 4q^2$
$$= 4(p^2 - 2pq + q^2)$$
$$= 4(p - q)^2$$

67. $x^3 + 8 = x^3 + 2^3$
$$= (x + 2)(x^2 - 2x + 4)$$

69. $m^3 - 1 = m^3 - 1^3$
$$= (m - 1)(m^2 + m + 1)$$

71. $2y^3 - 128 = 2(y^3 - 64)$
$$= 2(y^3 - 4^3)$$
$$= 2(y - 4)(y^2 + 4y + 16)$$

73. $3a^5 - 24a^2 = 3a^2(a^3 - 8)$
$$= 3a^2(a^3 - 2^3)$$
$$= 3a^2(a - 2)(a^2 + 2a + 4)$$

75. $t^6 + 1 = (t^2)^3 + 1^3$
$$= (t^2 + 1)(t^4 - t^2 + 1)$$

77. $18a^2b - 15ab^2 = 3ab \cdot 6a - 3ab \cdot 5b$
$$= 3ab(6a - 5b)$$

79. $x^3 - 4x^2 + 5x - 20 = x^2(x - 4) + 5(x - 4)$
$$= (x - 4)(x^2 + 5)$$

81. $8x^2 - 32 = 8(x^2 - 4)$
$$= 8(x + 2)(x - 2)$$

83. $4y^2 - 5$

There are no common factors. We might try to factor this polynomial as a difference of squares, but there is no integer which yields 5 when squared. Thus, the polynomial is prime.

85. $m^2 - 9n^2 = m^2 - (3n)^2$
$$= (m + 3n)(m - 3n)$$

87. $x^2 + 9x + 20$

We look for two numbers with a product of 20 and a sum of 9. They are 4 and 5.
$$x^2 + 9x + 20 = (x + 4)(x + 5)$$

89. $y^2 - 6y + 5$

We look for two numbers with a product of 5 and a sum of -6. They are -5 and -1.
$$y^2 - 6y + 5 = (y - 5)(y - 1)$$

91. $2a^2 + 9a + 4$

We use the FOIL method.

1. There is no common factor other than 1 or -1.

2. The factorization must be of the form
   $(2a + \quad)(a + \quad)$.

3. Factor the constant term, 4. The possibilities are
   $1 \cdot 4$, $-1(-4)$, and $2 \cdot 2$. The first two pairs of factors can be written in the opposite order as well: $4 \cdot 1$, $-4(-1)$.

4. Find a pair of factors for which the sum of the outside and the inside products is the middle term, $9a$. By trial, we determine that the factorization is $(2a + 1)(a + 4)$.

93. $6x^2 + 7x - 3$

We use the grouping method.

1. There is no common factor other than 1 or -1.

2. Multiply the leading coefficient and the constant: $6(-3) = -18$.

3. Try to factor -18 so that the sum of the factors is the coefficient of the middle term, 7. The factors we want are 9 and -2.

4. Split the middle term using the numbers found in step (3):
   $$7x = 9x - 2x$$

5. Factor by grouping.
$$6x^2 + 7x - 3 = 6x^2 + 9x - 2x - 3$$
$$= 3x(2x + 3) - (2x + 3)$$
$$= (2x + 3)(3x - 1)$$

95. $y^2 - 18y + 81 = y^2 - 2 \cdot y \cdot 9 + 9^2$
$$= (y - 9)^2$$

97. $9z^2 - 24z + 16 = (3z)^2 - 2 \cdot 3z \cdot 4 + 4^2$
$$= (3z - 4)^2$$

99. $x^2 y^2 - 14xy + 49 = (xy)^2 - 2 \cdot xy \cdot 7 + 7^2$
$$= (xy - 7)^2$$

101. $4ax^2 + 20ax - 56a = 4a(x^2 + 5x - 14)$
$$= 4a(x + 7)(x - 2)$$

103. $3z^3 - 24 = 3(z^3 - 8)$
$$= 3(z^3 - 2^3)$$
$$= 3(z - 2)(z^2 + 2z + 4)$$

105. $16a^7 b + 54ab^7$
$$= 2ab(8a^6 + 27b^6)$$
$$= 2ab[(2a^2)^3 + (3b^2)^3]$$
$$= 2ab(2a^2 + 3b^2)(4a^4 - 6a^2 b^2 + 9b^4)$$

107. $y^3 - 3y^2 - 4y + 12$
$$= y^2(y - 3) - 4(y - 3)$$
$$= (y - 3)(y^2 - 4)$$
$$= (y - 3)(y + 2)(y - 2)$$

109. $x^3 - x^2 + x - 1$
$$= x^2(x - 1) + (x - 1)$$
$$= (x - 1)(x^2 + 1)$$

111. $5m^4 - 20 = 5(m^4 - 4)$
$$= 5(m^2 + 2)(m^2 - 2)$$

113. $2x^3 + 6x^2 - 8x - 24$
$$= 2(x^3 + 3x^2 - 4x - 12)$$
$$= 2[x^2(x + 3) - 4(x + 3)]$$
$$= 2(x + 3)(x^2 - 4)$$
$$= 2(x + 3)(x + 2)(x - 2)$$

115. $4c^2 - 4cd - d^2 = (2c)^2 - 2 \cdot 2c \cdot d - d^2$
$$= (2c - d)^2$$

117. $m^6 + 8m^3 - 20 = (m^3)^2 + 8m^3 - 20$

We look for two numbers with a product of -20 and a sum of 8. They are 10 and -2.
$$m^6 + 8m^3 - 20 = (m^3 + 10)(m^3 - 2)$$

119. $p - 64p^4 = p(1 - 64p^3)$
$$= p[1^3 - (4p)^3]$$
$$= p(1 - 4p)(1 + 4p + 16p^2)$$

121. Discussion and Writing

123. $y^4 - 84 + 5y^2$
$$= y^4 + 5y^2 - 84$$
$$= u^2 + 5u - 84 \qquad \text{Substituting } u \text{ for } y^2$$
$$= (u + 12)(u - 7)$$
$$= (y^2 + 12)(y^2 - 7) \quad \text{Substituting } y^2 \text{ for } u$$

125. $y^2 - \dfrac{8}{49} + \dfrac{2}{7}y = y^2 + \dfrac{2}{7}y - \dfrac{8}{49}$
$$= \left(y + \frac{4}{7}\right)\left(y - \frac{2}{7}\right)$$

127. $x^2 + 3x + \dfrac{9}{4} = x^2 + 2 \cdot x \cdot \dfrac{3}{2} + \left(\dfrac{3}{2}\right)^2$
$$= \left(x + \frac{3}{2}\right)^2$$

129. $x^2 - x + \dfrac{1}{4} = x^2 - 2 \cdot x \cdot \dfrac{1}{2} + \left(\dfrac{1}{2}\right)^2$
$$= \left(x - \frac{1}{2}\right)^2$$

131. $(x + h)^3 - x^3$
$$= [(x + h) - x][(x + h)^2 + x(x + h) + x^2]$$
$$= (x + h - x)(x^2 + 2xh + h^2 + x^2 + xh + x^2)$$
$$= h(3x^2 + 3xh + h^2)$$

133. $(y - 4)^2 + 5(y - 4) - 24$
$$= u^2 + 5u - 24 \qquad \text{Substituting } u \text{ for } y - 4$$
$$= (u + 8)(u - 3)$$
$$= (y - 4 + 8)(y - 4 - 3) \quad \text{Substituting } y - 4$$
$$\text{for } u$$
$$= (y + 4)(y - 7)$$

135. $x^{2n} + 5x^n - 24 = (x^n)^2 + 5x^n - 24$
$$= (x^n + 8)(x^n - 3)$$

137. $x^2 + ax + bx + ab = x(x + a) + b(x + a)$
$$= (x + a)(x + b)$$

139. $25y^{2m} - (x^{2n} - 2x^n + 1)$
$$= (5y^m)^2 - (x^n - 1)^2$$
$$= [5y^m + (x^n - 1)][5y^m - (x^n - 1)]$$
$$= (5y^m + x^n - 1)(5y^m - x^n + 1)$$

141. $(y - 1)^4 - (y - 1)^2$
$$= (y - 1)^2[(y - 1)^2 - 1]$$
$$= (y - 1)^2[y^2 - 2y + 1 - 1]$$
$$= (y - 1)^2(y^2 - 2y)$$
$$= y(y - 1)^2(y - 2)$$

Exercise Set R.5

1. Since $-\dfrac{3}{4}$ is defined for all real numbers, the domain is $\{x|x$ is a real number$\}$.

3. $\dfrac{3x-3}{x(x-1)}$

The denominator is 0 when the factor $x = 0$ and also when $x - 1 = 0$, or $x = 1$. The domain is $\{x|x$ is a real number $and\ x \neq 0\ and\ x \neq 1\}$.

5. $\dfrac{x+5}{x^2+4x-5} = \dfrac{x+5}{(x+5)(x-1)}$

We see that $x + 5 = 0$ when $x = -5$ and $x - 1 = 0$ when $x = 1$. Thus, the domain is $\{x|x$ is a real number $and\ x \neq -5\ and\ x \neq 1\}$.

7. We first factor the denominator completely.

$$\dfrac{7x^2-28x+28}{(x^2-4)(x^2+3x-10)} = \dfrac{7x^2-28x+28}{(x+2)(x-2)(x+5)(x-2)}$$

We see that $x + 2 = 0$ when $x = -2$, $x - 2 = 0$ when $x = 2$, and $x + 5 = 0$ when $x = -5$. Thus, the domain is $\{x|x$ is a real number $and\ x \neq -2\ and\ x \neq 2\ and\ x \neq -5\}$.

9. $\dfrac{x^2-4}{x^2-4x+4} = \dfrac{(x+2)(x-2)}{(x-2)(x-2)} = \dfrac{x+2}{x-2}$

11. $\dfrac{x^3-6x^2+9x}{x^3-3x^2} = \dfrac{x(x^2-6x+9)}{x^2(x-3)}$

$$= \dfrac{x(x-3)(x-3)}{x \cdot x(x-3)}$$

$$= \dfrac{x-3}{x}$$

13. $\dfrac{6y^2+12y-48}{3y^2-9y+6} = \dfrac{6(y^2+2y-8)}{3(y^2-3y+2)}$

$$= \dfrac{2 \cdot 3 \cdot (y+4)(y-2)}{3(y-1)(y-2)}$$

$$= \dfrac{2(y+4)}{y-1}$$

15. $\dfrac{4-x}{x^2+4x-32} = \dfrac{-1(x-4)}{(x-4)(x+8)}$

$$= \dfrac{-1}{x+8}, \text{ or } -\dfrac{1}{x+8}$$

17. $\dfrac{x^2-y^2}{(x-y)^2} \cdot \dfrac{1}{x+y}$

$$= \dfrac{(x^2-y^2) \cdot 1}{(x-y)^2(x+y)}$$

$$= \dfrac{(x+y)(x-y) \cdot 1}{(x-y)(x-y)(x+y)}$$

$$= \dfrac{1}{x-y}$$

19. $\dfrac{x^2-2x-35}{2x^3-3x^2} \cdot \dfrac{4x^3-9x}{7x-49}$

$$= \dfrac{(x-7)(x+5)(x)(2x+3)(2x-3)}{x \cdot x(2x-3)(7)(x-7)}$$

$$= \dfrac{(x+5)(2x+3)}{7x}$$

21. $\dfrac{a^2-a-6}{a^2-7a+12} \cdot \dfrac{a^2-2a-8}{a^2-3a-10}$

$$= \dfrac{(a-3)(a+2)(a-4)(a+2)}{(a-4)(a-3)(a-5)(a+2)}$$

$$= \dfrac{a+2}{a-5}$$

23. $\dfrac{m^2-n^2}{r+s} \div \dfrac{m-n}{r+s}$

$$= \dfrac{m^2-n^2}{r+s} \cdot \dfrac{r+s}{m-n}$$

$$= \dfrac{(m+n)(m-n)(r+s)}{(r+s)(m-n)}$$

$$= m+n$$

25. $\dfrac{3x+12}{2x-8} \div \dfrac{(x+4)^2}{(x-4)^2}$

$$= \dfrac{3x+12}{2x-8} \cdot \dfrac{(x-4)^2}{(x+4)^2}$$

$$= \dfrac{3(x+4)(x-4)(x-4)}{2(x-4)(x+4)(x+4)}$$

$$= \dfrac{3(x-4)}{2(x+4)}$$

27. $\dfrac{x^2-y^2}{x^3-y^3} \cdot \dfrac{x^2+xy+y^2}{x^2+2xy+y^2}$

$$= \dfrac{(x+y)(x-y)(x^2+xy+y^2)}{(x-y)(x^2+xy+y^2)(x+y)(x+y)}$$

$$= \dfrac{1}{x+y} \cdot \dfrac{(x+y)(x-y)(x^2+xy+y^2)}{(x+y)(x-y)(x^2+xy+y^2)}$$

$$= \dfrac{1}{x+y} \cdot 1 \quad \text{Removing a factor of 1}$$

$$= \dfrac{1}{x+y}$$

29. $\dfrac{(x-y)^2-z^2}{(x+y)^2-z^2} \div \dfrac{x-y+z}{x+y-z}$

$$= \dfrac{(x-y)^2-z^2}{(x+y)^2-z^2} \cdot \dfrac{x+y-z}{x-y+z}$$

$$= \dfrac{(x-y+z)(x-y-z)(x+y-z)}{(x+y+z)(x+y-z)(x-y+z)}$$

$$= \dfrac{(x-y+z)(x+y-z)}{(x-y+z)(x+y-z)} \cdot \dfrac{x-y-z}{x+y+z}$$

$$= 1 \cdot \dfrac{x-y-z}{x+y+z} \quad \text{Removing a factor of 1}$$

$$= \dfrac{x-y-z}{x+y+z}$$

31. $\dfrac{5}{2x} + \dfrac{1}{2x} = \dfrac{5+1}{2x}$

$\qquad = \dfrac{6}{2x}$

$\qquad = \dfrac{\cancel{2} \cdot 3}{\cancel{2} \cdot x}$

$\qquad = \dfrac{3}{x}$

33. $\dfrac{3}{2a+3} + \dfrac{2a}{2a+3}$

$= \dfrac{3+2a}{2a+3}$

$= 1$

35. $\dfrac{5}{4z} - \dfrac{3}{8z}$, LCD is $8z$

$= \dfrac{5}{4z} \cdot \dfrac{2}{2} - \dfrac{3}{8z}$

$= \dfrac{10}{8z} - \dfrac{3}{8z}$

$= \dfrac{7}{8z}$

37. $\dfrac{3}{x+2} + \dfrac{2}{x^2-4}$

$= \dfrac{3}{x+2} + \dfrac{2}{(x+2)(x-2)}$, LCD is $(x+2)(x-2)$

$= \dfrac{3}{x+2} \cdot \dfrac{x-2}{x-2} + \dfrac{2}{(x+2)(x-2)}$

$= \dfrac{3x-6}{(x+2)(x-2)} + \dfrac{2}{(x+2)(x-2)}$

$= \dfrac{3x-4}{(x+2)(x-2)}$

39. $\dfrac{y}{y^2-y-20} - \dfrac{2}{y+4}$

$= \dfrac{y}{(y+4)(y-5)} - \dfrac{2}{y+4}$, LCD is $(y+4)(y-5)$

$= \dfrac{y}{(y+4)(y-5)} - \dfrac{2}{y+4} \cdot \dfrac{y-5}{y-5}$

$= \dfrac{y}{(y+4)(y-5)} - \dfrac{2y-10}{(y+4)(y-5)}$

$= \dfrac{y-(2y-10)}{(y+4)(y-5)}$

$= \dfrac{y-2y+10}{(y+4)(y-5)}$

$= \dfrac{-y+10}{(y+4)(y-5)}$

41. $\dfrac{3}{x+y} + \dfrac{x-5y}{x^2-y^2}$

$= \dfrac{3}{x+y} + \dfrac{x-5y}{(x+y)(x-y)}$, LCD is $(x+y)(x-y)$

$= \dfrac{3}{x+y} \cdot \dfrac{x-y}{x-y} + \dfrac{x-5y}{(x+y)(x-y)}$

$= \dfrac{3x-3y}{(x+y)(x-y)} + \dfrac{x-5y}{(x+y)(x-y)}$

$= \dfrac{4x-8y}{(x+y)(x-y)}$

43. $\dfrac{y}{y-1} + \dfrac{2}{1-y}$

$= \dfrac{y}{y-1} + \dfrac{-1}{-1} \cdot \dfrac{2}{1-y}$

$= \dfrac{y}{y-1} + \dfrac{-2}{y-1}$

$= \dfrac{y-2}{y-1}$

45. $\dfrac{x}{2x-3y} - \dfrac{y}{3y-2x}$

$= \dfrac{x}{2x-3y} - \dfrac{-1}{-1} \cdot \dfrac{y}{3y-2x}$

$= \dfrac{x}{2x-3y} - \dfrac{-y}{2x-3y}$

$= \dfrac{x+y}{2x-3y} \qquad [x-(-y)=x+y]$

47. $\dfrac{9x+2}{3x^2-2x-8} + \dfrac{7}{3x^2+x-4}$

$= \dfrac{9x+2}{(3x+4)(x-2)} + \dfrac{7}{(3x+4)(x-1)}$,

$\qquad$ LCD is $(3x+4)(x-2)(x-1)$

$= \dfrac{9x+2}{(3x+4)(x-2)} \cdot \dfrac{x-1}{x-1} + \dfrac{7}{(3x+4)(x-1)} \cdot \dfrac{x-2}{x-2}$

$= \dfrac{9x^2-7x-2}{(3x+4)(x-2)(x-1)} + \dfrac{7x-14}{(3x+4)(x-1)(x-2)}$

$= \dfrac{9x^2-16}{(3x+4)(x-2)(x-1)}$

$= \dfrac{\cancel{(3x+4)}(3x-4)}{\cancel{(3x+4)}(x-2)(x-1)}$

$= \dfrac{3x-4}{(x-2)(x-1)}$

49. $\dfrac{5a}{a-b} + \dfrac{ab}{a^2-b^2} + \dfrac{4b}{a+b}$

$= \dfrac{5a}{a-b} + \dfrac{ab}{(a+b)(a-b)} + \dfrac{4b}{a+b}$,

$\qquad$ LCD is $(a+b)(a-b)$

$= \dfrac{5a}{a-b} \cdot \dfrac{a+b}{a+b} + \dfrac{ab}{(a+b)(a-b)} + \dfrac{4b}{a+b} \cdot \dfrac{a-b}{a-b}$

$= \dfrac{5a^2+5ab}{(a+b)(a-b)} + \dfrac{ab}{(a+b)(a-b)} + \dfrac{4ab-4b^2}{(a+b)(a-b)}$

$= \dfrac{5a^2+10ab-4b^2}{(a+b)(a-b)}$

51.

$$\frac{7}{x+2} - \frac{x+8}{4-x^2} + \frac{3x-2}{4-4x+x^2}$$

$$= \frac{7}{x+2} - \frac{x+8}{(2+x)(2-x)} + \frac{3x-2}{(2-x)^2},$$
$$\text{LCD is } (2+x)(2-x)^2$$

$$= \frac{7}{2+x} \cdot \frac{(2-x)^2}{(2-x)^2} - \frac{x+8}{(2+x)(2-x)} \cdot \frac{2-x}{2-x} +$$
$$\frac{3x-2}{(2-x)^2} \cdot \frac{2+x}{2+x}$$

$$= \frac{28-28x+7x^2-(16-6x-x^2)+3x^2+4x-4}{(2+x)(2-x)^2}$$

$$= \frac{28-28x+7x^2-16+6x+x^2+3x^2+4x-4}{(2+x)(2-x)^2}$$

$$= \frac{11x^2-18x+8}{(2+x)(2-x)^2}, \text{ or } \frac{11x^2-18x+8}{(x+2)(x-2)^2}$$

53.

$$\frac{1}{x+1} + \frac{x}{2-x} + \frac{x^2+2}{x^2-x-2}$$

$$= \frac{1}{x+1} + \frac{x}{2-x} + \frac{x^2+2}{(x+1)(x-2)}$$

$$= \frac{1}{x+1} + \frac{-1}{-1} \cdot \frac{x}{2-x} + \frac{x^2+2}{(x+1)(x-2)}$$

$$= \frac{1}{x+1} + \frac{-x}{x-2} + \frac{x^2+2}{(x+1)(x-2)},$$
$$\text{LCD is } (x+1)(x-2)$$

$$= \frac{1}{x+1} \cdot \frac{x-2}{x-2} + \frac{-x}{x-2} \cdot \frac{x+1}{x+1} + \frac{x^2+2}{(x+1)(x-2)}$$

$$= \frac{x-2}{(x+1)(x-2)} + \frac{-x^2-x}{(x+1)(x-2)} + \frac{x^2+2}{(x+1)(x-2)}$$

$$= \frac{x-2-x^2-x+x^2+2}{(x+1)(x-2)}$$

$$= \frac{0}{(x+1)(x-2)}$$

$$= 0$$

55.

$$\frac{\dfrac{x^2-y^2}{xy}}{\dfrac{x-y}{y}} = \frac{x^2-y^2}{xy} \cdot \frac{y}{x-y}$$

$$= \frac{(x+y)(x-y)y}{xy(x-y)}$$

$$= \frac{x+y}{x}$$

57.

$$\frac{\dfrac{x}{y}-\dfrac{y}{x}}{\dfrac{1}{y}+\dfrac{1}{x}} = \frac{\dfrac{x}{y}-\dfrac{y}{x}}{\dfrac{1}{y}+\dfrac{1}{x}} \cdot \frac{xy}{xy}, \text{ LCM is } xy$$

$$= \frac{\left(\dfrac{x}{y}-\dfrac{y}{x}\right)(xy)}{\left(\dfrac{1}{y}+\dfrac{1}{x}\right)(xy)}$$

$$= \frac{x^2-y^2}{x+y}$$

$$= \frac{(x+y)(x-y)}{(x+y) \cdot 1}$$

$$= x-y$$

59.

$$\frac{c+\dfrac{8}{c^2}}{1+\dfrac{2}{c}} = \frac{c \cdot \dfrac{c^2}{c^2}+\dfrac{8}{c^2}}{1 \cdot \dfrac{c}{c}+\dfrac{2}{c}}$$

$$= \frac{\dfrac{c^3+8}{c^2}}{\dfrac{c+2}{c}}$$

$$= \frac{c^3+8}{c^2} \cdot \frac{c}{c+2}$$

$$= \frac{(c+2)(c^2-2c+4)c}{c \cdot c(c+2)}$$

$$= \frac{c^2-2c+4}{c}$$

61.

$$\frac{x^2+xy+y^2}{\dfrac{x^2}{y}-\dfrac{y^2}{x}} = \frac{x^2+xy+y^2}{\dfrac{x^2}{y} \cdot \dfrac{x}{x}-\dfrac{y^2}{x} \cdot \dfrac{y}{y}}$$

$$= \frac{x^2+xy+y^2}{\dfrac{x^3-y^3}{xy}}$$

$$= (x^2+xy+y^2) \cdot \frac{xy}{x^3-y^3}$$

$$= \frac{(x^2+xy+y^2)(xy)}{(x-y)(x^2+xy+y^2)}$$

$$= \frac{x^2+xy+y^2}{x^2+xy+y^2} \cdot \frac{xy}{x-y}$$

$$= 1 \cdot \frac{xy}{x-y}$$

$$= \frac{xy}{x-y}$$

63. $\dfrac{a - a^{-1}}{a + a^{-1}} = \dfrac{a - \dfrac{1}{a}}{a + \dfrac{1}{a}} = \dfrac{a \cdot \dfrac{a}{a} - \dfrac{1}{a}}{a \cdot \dfrac{a}{a} + \dfrac{1}{a}}$

$= \dfrac{\dfrac{a^2 - 1}{a}}{\dfrac{a^2 + 1}{a}}$

$= \dfrac{a^2 - 1}{a} \cdot \dfrac{a}{a^2 + 1}$

$= \dfrac{a^2 - 1}{a^2 + 1}$

65. $\dfrac{\dfrac{1}{x - 3} + \dfrac{2}{x + 3}}{\dfrac{3}{x - 1} - \dfrac{4}{x + 2}} = \dfrac{\dfrac{1}{x - 3} \cdot \dfrac{x + 3}{x + 3} + \dfrac{2}{x + 3} \cdot \dfrac{x - 3}{x - 3}}{\dfrac{3}{x - 1} \cdot \dfrac{x + 2}{x + 2} - \dfrac{4}{x + 2} \cdot \dfrac{x - 1}{x - 1}}$

$= \dfrac{\dfrac{x + 3 + 2(x - 3)}{(x - 3)(x + 3)}}{\dfrac{3(x + 2) - 4(x - 1)}{(x - 1)(x + 2)}}$

$= \dfrac{\dfrac{x + 3 + 2x - 6}{(x - 3)(x + 3)}}{\dfrac{3x + 6 - 4x + 4}{(x - 1)(x + 2)}}$

$= \dfrac{\dfrac{3x - 3}{(x - 3)(x + 3)}}{\dfrac{-x + 10}{(x - 1)(x + 2)}}$

$= \dfrac{3x - 3}{(x - 3)(x + 3)} \cdot \dfrac{(x - 1)(x + 2)}{-x + 10}$

$= \dfrac{(3x - 3)(x - 1)(x + 2)}{(x - 3)(x + 3)(-x + 10)}$, or

$\dfrac{3(x - 1)^2(x + 2)}{(x - 3)(x + 3)(-x + 10)}$

67. $\dfrac{\dfrac{a}{1 - a} + \dfrac{1 + a}{a}}{\dfrac{1 - a}{a} + \dfrac{a}{1 + a}} = \dfrac{\dfrac{a}{1 - a} \cdot \dfrac{a}{a} + \dfrac{1 + a}{a} \cdot \dfrac{1 - a}{1 - a}}{\dfrac{1 - a}{a} \cdot \dfrac{1 + a}{1 + a} + \dfrac{a}{1 + a} \cdot \dfrac{a}{a}}$

$= \dfrac{\dfrac{a^2 + (1 - a^2)}{a(1 - a)}}{\dfrac{(1 - a^2) + a^2}{a(1 + a)}}$

$= \dfrac{1}{a(1 - a)} \cdot \dfrac{a(1 + a)}{1}$

$= \dfrac{1 + a}{1 - a}$

69. $\dfrac{\dfrac{1}{a^2} + \dfrac{2}{ab} + \dfrac{1}{b^2}}{\dfrac{1}{a^2} - \dfrac{1}{b^2}} = \dfrac{\dfrac{1}{a^2} + \dfrac{2}{ab} + \dfrac{1}{b^2}}{\dfrac{1}{a^2} - \dfrac{1}{b^2}} \cdot \dfrac{a^2 b^2}{a^2 b^2}$,

LCM is $a^2 b^2$

$= \dfrac{b^2 + 2ab + a^2}{b^2 - a^2}$

$= \dfrac{(b + a)(b + a)}{(b + a)(b - a)}$

$= \dfrac{b + a}{b - a}$

71. Discussion and Writing

73. $\dfrac{(x + h)^2 - x^2}{h} = \dfrac{x^2 + 2xh + h^2 - x^2}{h}$

$= \dfrac{2xh + h^2}{h}$

$= \dfrac{h(2x + h)}{h \cdot 1}$

$= 2x + h$

75. $\dfrac{(x + h)^3 - x^3}{h} = \dfrac{x^3 + 3x^2 h + 3xh^2 + h^3 - x^3}{h}$

$= \dfrac{3x^2 h + 3xh^2 + h^3}{h}$

$= \dfrac{h(3x^2 + 3xh + h^2)}{h \cdot 1}$

$= 3x^2 + 3xh + h^2$

77. $\left[\dfrac{\dfrac{x + 1}{x - 1} + 1}{\dfrac{x + 1}{x - 1} - 1} \right]^5 = \left[\dfrac{\dfrac{(x + 1) + (x - 1)}{x - 1}}{\dfrac{(x + 1) - (x - 1)}{x - 1}} \right]^5$

$= \left[\dfrac{2x}{x - 1} \cdot \dfrac{x - 1}{2} \right]^5$

$= \left[\dfrac{2x(x - 1)}{1 \cdot 2(x - 1)} \right]^5$

$= x^5$

79. $\dfrac{n(n + 1)(n + 2)}{2 \cdot 3} + \dfrac{(n + 1)(n + 2)}{2}$

$= \dfrac{n(n + 1)(n + 2)}{2 \cdot 3} + \dfrac{(n + 1)(n + 2)}{2} \cdot \dfrac{3}{3}$,

LCD is $2 \cdot 3$

$= \dfrac{n(n + 1)(n + 2) + 3(n + 1)(n + 2)}{2 \cdot 3}$

$= \dfrac{(n + 1)(n + 2)(n + 3)}{2 \cdot 3}$ Factoring the numerator by grouping

81.
$$\frac{x^2 - 9}{x^3 + 27} \cdot \frac{5x^2 - 15x + 45}{x^2 - 2x - 3} + \frac{x^2 + x}{4 + 2x}$$

$$= \frac{(x+3)(x-3)(5)(x^2 - 3x + 9)}{(x+3)(x^2 - 3x + 9)(x-3)(x+1)} + \frac{x^2 + x}{4 + 2x}$$

$$= \frac{(x+3)(x-3)(x^2 - 3x + 9)}{(x+3)(x-3)(x^2 - 3x + 9)} \cdot \frac{5}{x+1} + \frac{x^2 + x}{4 + 2x}$$

$$= 1 \cdot \frac{5}{x+1} + \frac{x^2 + x}{4 + 2x}$$

$$= \frac{5}{x+1} + \frac{x^2 + x}{2(2+x)}$$

$$= \frac{5 \cdot 2(2+x) + (x^2 + x)(x+1)}{2(x+1)(2+x)}$$

$$= \frac{20 + 10x + x^3 + 2x^2 + x}{2(x+1)(2+x)}$$

$$= \frac{x^3 + 2x^2 + 11x + 20}{2(x+1)(2+x)}$$

Exercise Set R.6

1. $\sqrt{(-11)^2} = |-11| = 11$

3. $\sqrt{16y^2} = \sqrt{(4y)^2} = |4y| = 4|y|$

5. $\sqrt{(b+1)^2} = |b+1|$

7. $\sqrt[3]{-27x^3} = \sqrt[3]{(-3x)^3} = -3x$

9. $\sqrt[4]{81x^8} = \sqrt[4]{(3x^2)^4} = |3x^2| = 3x^2$

11. $\sqrt[5]{32} = \sqrt[5]{2^5} = 2$

13. $\sqrt{180} = \sqrt{36 \cdot 5} = \sqrt{36} \cdot \sqrt{5} = 6\sqrt{5}$

15. $\sqrt{72} = \sqrt{36 \cdot 2} = \sqrt{36} \cdot \sqrt{2} = 6\sqrt{2}$

17. $\sqrt[3]{54} = \sqrt[3]{27 \cdot 2} = \sqrt[3]{27} \cdot \sqrt[3]{2} = 3\sqrt[3]{2}$

19. $\sqrt{128c^2d^4} = \sqrt{64c^2d^4 \cdot 2} = |8cd^2|\sqrt{2} = 8\sqrt{2}\,|c|d^2$

21. $\sqrt[4]{48x^6y^4} = \sqrt[4]{16x^4y^4 \cdot 3x^2} = |2xy|\sqrt[4]{3x^2} = 2|x||y|\sqrt[4]{3x^2}$

23. $\sqrt{x^2 - 4x + 4} = \sqrt{(x-2)^2} = |x-2|$

25. $\sqrt{10}\sqrt{30} = \sqrt{10 \cdot 30} = \sqrt{300} = \sqrt{100 \cdot 3} = \sqrt{100} \cdot \sqrt{3} = 10\sqrt{3}$

27. $\sqrt{12}\sqrt{33} = \sqrt{12 \cdot 33} = \sqrt{3 \cdot 4 \cdot 3 \cdot 11} = \sqrt{3^2 \cdot 4 \cdot 11} = 3 \cdot 2\sqrt{11} = 6\sqrt{11}$

29. $\sqrt{2x^3y}\sqrt{12xy} = \sqrt{24x^4y^2} = \sqrt{4x^4y^2 \cdot 6} = 2x^2y\sqrt{6}$

31. $\sqrt[3]{3x^2y}\sqrt[3]{36x} = \sqrt[3]{108x^3y} = \sqrt[3]{27x^3 \cdot 4y} = 3x\sqrt[3]{4y}$

33. $\sqrt[3]{2(x+4)}\sqrt[3]{4(x+4)^4} = \sqrt[3]{8(x+4)^5}$
$$= \sqrt[3]{8(x+4)^3 \cdot (x+4)^2}$$
$$= 2(x+4)\sqrt[3]{(x+4)^2}$$

35. $\sqrt[6]{\dfrac{m^{12}n^{24}}{64}} = \sqrt[6]{\left(\dfrac{m^2n^4}{2}\right)^6} = \dfrac{m^2n^4}{2}$

37. $\dfrac{\sqrt[3]{40m}}{\sqrt[3]{5m}} = \sqrt[3]{\dfrac{40m}{5m}} = \sqrt[3]{8} = 2$

39. $\dfrac{\sqrt[3]{3x^2}}{\sqrt[3]{24x^5}} = \sqrt[3]{\dfrac{3x^2}{24x^5}} = \sqrt[3]{\dfrac{1}{8x^3}} = \dfrac{1}{2x}$

41. $\sqrt[3]{\dfrac{64a^4}{27b^3}} = \sqrt[3]{\dfrac{64 \cdot a^3 \cdot a}{27 \cdot b^3}}$
$$= \dfrac{\sqrt[3]{64a^3}\sqrt[3]{a}}{\sqrt[3]{27b^3}}$$
$$= \dfrac{4a\sqrt[3]{a}}{3b}$$

43. $\sqrt{\dfrac{7x^3}{36y^6}} = \sqrt{\dfrac{7 \cdot x^2 \cdot x}{36 \cdot y^6}}$
$$= \dfrac{\sqrt{x^2}\sqrt{7x}}{\sqrt{36y^6}}$$
$$= \dfrac{x\sqrt{7x}}{6y^3}$$

45. $9\sqrt{50} + 6\sqrt{2} = 9\sqrt{25 \cdot 2} + 6\sqrt{2}$
$$= 9 \cdot 5\sqrt{2} + 6\sqrt{2}$$
$$= 45\sqrt{2} + 6\sqrt{2}$$
$$= (45 + 6)\sqrt{2}$$
$$= 51\sqrt{2}$$

47. $6\sqrt{20} - 4\sqrt{45} + \sqrt{80} = 6\sqrt{4 \cdot 5} - 4\sqrt{9 \cdot 5} + \sqrt{16 \cdot 5}$
$$= 6 \cdot 2\sqrt{5} - 4 \cdot 3\sqrt{5} + 4\sqrt{5}$$
$$= 12\sqrt{5} - 12\sqrt{5} + 4\sqrt{5}$$
$$= (12 - 12 + 4)\sqrt{5}$$
$$= 4\sqrt{5}$$

49. $8\sqrt{2x^2} - 6\sqrt{20x} - 5\sqrt{8x^2}$
$$= 8x\sqrt{2} - 6\sqrt{4 \cdot 5x} - 5\sqrt{4x^2 \cdot 2}$$
$$= 8x\sqrt{2} - 6 \cdot 2\sqrt{5x} - 5 \cdot 2x\sqrt{2}$$
$$= 8x\sqrt{2} - 12\sqrt{5x} - 10x\sqrt{2}$$
$$= -2x\sqrt{2} - 12\sqrt{5x}$$

51. $(\sqrt{3} - \sqrt{2})(\sqrt{3} + \sqrt{2})$
$$= (\sqrt{3})^2 - (\sqrt{2})^2$$
$$= 3 - 2$$
$$= 1$$

53. $(2\sqrt{3} + \sqrt{5})(\sqrt{3} - 3\sqrt{5})$
$$= 2\sqrt{3} \cdot \sqrt{3} - 2\sqrt{3} \cdot 3\sqrt{5} + \sqrt{5} \cdot \sqrt{3} - \sqrt{5} \cdot 3\sqrt{5}$$
$$= 2 \cdot 3 - 6\sqrt{15} + \sqrt{15} - 3 \cdot 5$$
$$= 6 - 6\sqrt{15} + \sqrt{15} - 15$$
$$= -9 - 5\sqrt{15}$$

55. $(1 + \sqrt{3})^2 = 1^2 + 2 \cdot 1 \cdot \sqrt{3} + (\sqrt{3})^2$
$$= 1 + 2\sqrt{3} + 3$$
$$= 4 + 2\sqrt{3}$$

57. $(\sqrt{5} - \sqrt{6})^2 = (\sqrt{5})^2 - 2\sqrt{5} \cdot \sqrt{6} + (\sqrt{6})^2$
$$= 5 - 2\sqrt{30} + 6$$
$$= 11 - 2\sqrt{30}$$

59. We use the Pythagorean theorem to find b, the airplane's horizontal distance from the airport. We have $a = 3700$ and $c = 14,200$.
$$c^2 = a^2 + b^2$$
$$14,200^2 = 3700^2 + b^2$$
$$201,640,000 = 13,690,000 + b^2$$
$$187,950,000 = b^2$$
$$13,709.5 \approx b$$

The airplane is about $13,709.5$ ft horizontally from the airport.

61. a) $h^2 + \left(\dfrac{a}{2}\right)^2 = a^2$ Pythagorean theorem

$$h^2 + \frac{a^2}{4} = a^2$$
$$h^2 = \frac{3a^2}{4}$$
$$h = \sqrt{\frac{3a^2}{4}}$$
$$h = \frac{a}{2}\sqrt{3}$$

b) Using the result of part (a) we have

$$A = \frac{1}{2} \cdot \text{base} \cdot \text{height}$$
$$A = \frac{1}{2}a \cdot \frac{a}{2}\sqrt{3} \quad \left(\frac{a}{2} + \frac{a}{2} = a\right)$$
$$A = \frac{a^2}{4}\sqrt{3}$$

63.

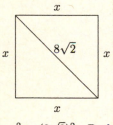

$$x^2 + x^2 = (8\sqrt{2})^2 \quad \text{Pythagorean theorem}$$
$$2x^2 = 128$$
$$x^2 = 64$$
$$x = 8$$

65. $\sqrt{\dfrac{2}{3}} = \sqrt{\dfrac{2}{3} \cdot \dfrac{3}{3}} = \sqrt{\dfrac{6}{9}} = \dfrac{\sqrt{6}}{\sqrt{9}} = \dfrac{\sqrt{6}}{3}$

67. $\dfrac{\sqrt[3]{5}}{\sqrt[3]{4}} = \dfrac{\sqrt[3]{5}}{\sqrt[3]{4}} \cdot \dfrac{\sqrt[3]{2}}{\sqrt[3]{2}} = \dfrac{\sqrt[3]{10}}{\sqrt[3]{8}} = \dfrac{\sqrt[3]{10}}{2}$

69. $\sqrt[3]{\dfrac{16}{9}} = \sqrt[3]{\dfrac{16}{9} \cdot \dfrac{3}{3}} = \sqrt[3]{\dfrac{48}{27}} = \dfrac{\sqrt[3]{48}}{\sqrt[3]{27}} =$

$$\dfrac{\sqrt[3]{8 \cdot 6}}{3} = \dfrac{2\sqrt[3]{6}}{3}$$

71. $\dfrac{6}{3 + \sqrt{5}} = \dfrac{6}{3 + \sqrt{5}} \cdot \dfrac{3 - \sqrt{5}}{3 - \sqrt{5}}$

$$= \dfrac{6(3 - \sqrt{5})}{9 - 5}$$
$$= \dfrac{6(3 - \sqrt{5})}{4}$$
$$= \dfrac{3(3 - \sqrt{5})}{2} = \dfrac{9 - 3\sqrt{5}}{2}$$

73. $\dfrac{1 - \sqrt{2}}{2\sqrt{3} - \sqrt{6}} = \dfrac{1 - \sqrt{2}}{2\sqrt{3} - \sqrt{6}} \cdot \dfrac{2\sqrt{3} + \sqrt{6}}{2\sqrt{3} + \sqrt{6}}$

$$= \dfrac{2\sqrt{3} + \sqrt{6} - 2\sqrt{6} - \sqrt{12}}{4 \cdot 3 - 6}$$
$$= \dfrac{2\sqrt{3} + \sqrt{6} - 2\sqrt{6} - 2\sqrt{3}}{12 - 6}$$
$$= \dfrac{-\sqrt{6}}{6}, \text{ or } -\dfrac{\sqrt{6}}{6}$$

75. $\dfrac{6}{\sqrt{m} - \sqrt{n}} = \dfrac{6}{\sqrt{m} - \sqrt{n}} \cdot \dfrac{\sqrt{m} + \sqrt{n}}{\sqrt{m} + \sqrt{n}}$

$$= \dfrac{6(\sqrt{m} + \sqrt{n})}{(\sqrt{m})^2 - (\sqrt{n})^2}$$
$$= \dfrac{6\sqrt{m} + 6\sqrt{n}}{m - n}$$

77. $\dfrac{\sqrt{12}}{5} = \dfrac{\sqrt{12}}{5} \cdot \dfrac{\sqrt{3}}{\sqrt{3}} = \dfrac{\sqrt{36}}{5\sqrt{3}} = \dfrac{6}{5\sqrt{3}}$

79. $\sqrt[3]{\dfrac{7}{2}} = \sqrt[3]{\dfrac{7}{2} \cdot \dfrac{49}{49}} = \sqrt[3]{\dfrac{343}{98}} = \dfrac{\sqrt[3]{343}}{\sqrt[3]{98}} = \dfrac{7}{\sqrt[3]{98}}$

81. $\dfrac{\sqrt{11}}{\sqrt{3}} = \dfrac{\sqrt{11}}{\sqrt{3}} \cdot \dfrac{\sqrt{11}}{\sqrt{11}} = \dfrac{\sqrt{121}}{\sqrt{33}} = \dfrac{11}{\sqrt{33}}$

83. $\dfrac{9 - \sqrt{5}}{3 - \sqrt{3}} = \dfrac{9 - \sqrt{5}}{3 - \sqrt{3}} \cdot \dfrac{9 + \sqrt{5}}{9 + \sqrt{5}}$

$$= \dfrac{9^2 - (\sqrt{5})^2}{27 + 3\sqrt{5} - 9\sqrt{3} - \sqrt{15}}$$
$$= \dfrac{81 - 5}{27 + 3\sqrt{5} - 9\sqrt{3} - \sqrt{15}}$$
$$= \dfrac{76}{27 + 3\sqrt{5} - 9\sqrt{3} - \sqrt{15}}$$

85. $\dfrac{\sqrt{a} + \sqrt{b}}{3a} = \dfrac{\sqrt{a} + \sqrt{b}}{3a} \cdot \dfrac{\sqrt{a} - \sqrt{b}}{\sqrt{a} - \sqrt{b}}$

$$= \dfrac{(\sqrt{a})^2 - (\sqrt{b})^2}{3a(\sqrt{a} - \sqrt{b})}$$
$$= \dfrac{a - b}{3a\sqrt{a} - 3a\sqrt{b}}$$

87. $x^{3/4} = \sqrt[4]{x^3}$

89. $16^{3/4} = (16^{1/4})^3 = \left(\sqrt[4]{16}\right)^3 = 2^3 = 8$

91. $125^{-1/3} = \dfrac{1}{125^{1/3}} = \dfrac{1}{\sqrt[3]{125}} = \dfrac{1}{5}$

93. $a^{5/4}b^{-3/4} = \dfrac{a^{5/4}}{b^{3/4}} = \dfrac{\sqrt[4]{a^5}}{\sqrt[4]{b^3}} = \dfrac{a\sqrt[4]{a}}{\sqrt[4]{b^3}}$, or $a\sqrt[4]{\dfrac{a}{b^3}}$

95. $m^{5/3}n^{7/3} = \sqrt[3]{m^5}\sqrt[3]{n^7} = \sqrt[3]{m^5n^7} = mn^2\sqrt[3]{m^2n}$

97. $\left(\sqrt[4]{13}\right)^5 = \sqrt[4]{13^5} = 13^{5/4}$

99. $\sqrt[3]{20^2} = 20^{2/3}$

101. $\sqrt[3]{\sqrt{11}} = \left(\sqrt{11}\right)^{1/3} = (11^{1/2})^{1/3} = 11^{1/6}$

103. $\sqrt{5}\sqrt[3]{5} = 5^{1/2}\cdot 5^{1/3} = 5^{1/2+1/3} = 5^{5/6}$

105. $\sqrt[5]{32^2} = 32^{2/5} = (32^{1/5})^2 = 2^2 = 4$

107. $(2a^{3/2})(4a^{1/2}) = 8a^{3/2+1/2} = 8a^2$

109. $\left(\dfrac{x^6}{9b^{-4}}\right)^{1/2} = \left(\dfrac{x^6}{3^2b^{-4}}\right)^{1/2} = \dfrac{x^3}{3b^{-2}}$, or $\dfrac{x^3b^2}{3}$

111. $\dfrac{x^{2/3}y^{5/6}}{x^{-1/3}y^{1/2}} = x^{2/3-(-1/3)}y^{5/6-1/2} = xy^{1/3} = x\sqrt[3]{y}$

113. $(m^{1/2}n^{5/2})^{2/3} = m^{\frac{1}{2}\cdot\frac{2}{3}}n^{\frac{5}{2}\cdot\frac{2}{3}} = m^{1/3}n^{5/3} =$
$\sqrt[3]{m}\sqrt[3]{n^5} = \sqrt[3]{mn^5} = n\sqrt[3]{mn^2}$

115. $a^{3/4}(a^{2/3} + a^{4/3}) = a^{3/4+2/3} + a^{3/4+4/3} =$
$a^{17/12} + a^{25/12} = \sqrt[12]{a^{17}} + \sqrt[12]{a^{25}} =$
$a\sqrt[12]{a^5} + a^2\sqrt[12]{a}$

117. $\sqrt[3]{6}\sqrt{2} = 6^{1/3}2^{1/2} = 6^{2/6}2^{3/6}$
$\qquad\qquad\qquad = (6^2 2^3)^{1/6}$
$\qquad\qquad\qquad = \sqrt[6]{36\cdot 8}$
$\qquad\qquad\qquad = \sqrt[6]{288}$

119. $\sqrt[4]{xy}\sqrt[3]{x^2y} = (xy)^{1/4}(x^2y)^{1/3} = (xy)^{3/12}(x^2y)^{4/12}$
$\qquad\qquad\qquad = \left[(xy)^3(x^2y)^4\right]^{1/12}$
$\qquad\qquad\qquad = \left[x^3y^3x^8y^4\right]^{1/12}$
$\qquad\qquad\qquad = \sqrt[12]{x^{11}y^7}$

121. $\sqrt[3]{a^4\sqrt{a^3}} = \left(a^4\sqrt{a^3}\right)^{1/3} = (a^4a^{3/2})^{1/3}$
$\qquad\qquad\qquad = (a^{11/2})^{1/3}$
$\qquad\qquad\qquad = a^{11/6}$
$\qquad\qquad\qquad = \sqrt[6]{a^{11}}$
$\qquad\qquad\qquad = a\sqrt[6]{a^5}$

123. $\dfrac{\sqrt{(a+x)^3}\sqrt[3]{(a+x)^2}}{\sqrt[4]{a+x}} = \dfrac{(a+x)^{3/2}(a+x)^{2/3}}{(a+x)^{1/4}}$
$\qquad\qquad\qquad = \dfrac{(a+x)^{26/12}}{(a+x)^{3/12}}$
$\qquad\qquad\qquad = (a+x)^{23/12}$
$\qquad\qquad\qquad = \sqrt[12]{(a+x)^{23}}$
$\qquad\qquad\qquad = (a+x)\sqrt[12]{(a+x)^{11}}$

125. Discussion and Writing

127. $\sqrt{1+x^2} + \dfrac{1}{\sqrt{1+x^2}}$

$= \sqrt{1+x^2}\cdot\dfrac{1+x^2}{1+x^2} + \dfrac{1}{\sqrt{1+x^2}}\cdot\dfrac{\sqrt{1+x^2}}{\sqrt{1+x^2}}$

$= \dfrac{(1+x^2)\sqrt{1+x^2}}{1+x^2} + \dfrac{\sqrt{1+x^2}}{1+x^2}$

$= \dfrac{(2+x^2)\sqrt{1+x^2}}{1+x^2}$

129. $\left(\sqrt{a^{\sqrt{a}}}\right)^{\sqrt{a}} = \left(a^{\sqrt{a}/2}\right)^{\sqrt{a}} = a^{a/2}$

Exercise Set R.7

1. $6x - 15 = 45$
$\qquad 6x = 60 \quad$ Adding 15
$\qquad\ x = 10 \quad$ Dividing by 6
The solution is 10.

3. $5x - 10 = 45$
$\qquad 5x = 55 \quad$ Adding 10
$\qquad\ x = 11 \quad$ Dividing by 5
The solution is 11.

5. $9t + 4 = -5$
$\qquad 9t = -9 \quad$ Subtracting 4
$\qquad\ t = -1 \quad$ Dividing by 9
The solution is -1.

7. $8x + 48 = 3x - 12$
$\qquad 5x + 48 = -12 \qquad$ Subtracting $3x$
$\qquad\quad 5x = -60 \qquad$ Subtracting 48
$\qquad\quad\ x = -12 \qquad$ Dividing by 5
The solution is -12.

9. $7y - 1 = 23 - 5y$
$\quad 12y - 1 = 23 \qquad$ Adding $5y$
$\quad 12y = 24 \qquad$ Adding 1
$\qquad y = 2 \qquad$ Dividing by 12
The solution is 2.

11. $3x - 4 = 5 + 12x$
$\quad -9x - 4 = 5 \qquad$ Subtracting $12x$
$\quad -9x = 9 \qquad$ Adding 4
$\qquad\ x = -1 \qquad$ Dividing by -9
The solution is -1.

13. $5 - 4a = a - 13$
$\quad 5 - 5a = -13 \qquad$ Subtracting a
$\qquad -5a = -18 \qquad$ Subtracting 5
$\qquad\quad a = \dfrac{18}{5} \qquad$ Dividing by -5

The solution is $\dfrac{18}{5}$.

15. $3m - 7 = -13 + m$

$\quad 2m - 7 = -13$ Subtracting m

$\quad\quad 2m = -6$ Adding 7

$\quad\quad\; m = -3$ Dividing by 2

The solution is -3.

17. $11 - 3x = 5x + 3$

$\quad 11 - 8x = 3$ Subtracting $5x$

$\quad\quad -8x = -8$ Subtracting 11

$\quad\quad\quad x = 1$

The solution is 1.

19. $2(x + 7) = 5x + 14$

$\quad 2x + 14 = 5x + 14$

$\quad -3x + 14 = 14$ Subtracting $5x$

$\quad\quad -3x = 0$ Subtracting 14

$\quad\quad\quad x = 0$

The solution is 0.

21. $24 = 5(2t + 5)$

$\quad 24 = 10t + 25$

$\quad -1 = 10t$ Subtracting 25

$\quad -\dfrac{1}{10} = t$ Dividing by 10

The solution is $-\dfrac{1}{10}$.

23. $5y - (2y - 10) = 25$

$\quad 5y - 2y + 10 = 25$

$\quad\quad 3y + 10 = 25$ Collecting like terms

$\quad\quad\quad 3y = 15$ Subtracting 10

$\quad\quad\quad\; y = 5$ Dividing by 3

The solution is 5.

25. $7(3x + 6) = 11 - (x + 2)$

$\quad 21x + 42 = 11 - x - 2$

$\quad 21x + 42 = 9 - x$ Collecting like terms

$\quad 22x + 42 = 9$ Adding x

$\quad\quad 22x = -33$ Subtracting 42

$\quad\quad\quad x = -\dfrac{3}{2}$ Dividing by 22

The solution is $-\dfrac{3}{2}$.

27. $4(3y - 1) - 6 = 5(y + 2)$

$\quad 12y - 4 - 6 = 5y + 10$

$\quad 12y - 10 = 5y + 10$ Collecting like terms

$\quad\quad 7y - 10 = 10$ Subtracting $5y$

$\quad\quad\quad 7y = 20$ Adding 10

$\quad\quad\quad\; y = \dfrac{20}{7}$ Dividing by 7

The solution is $\dfrac{20}{7}$.

29. $x^2 + 3x - 28 = 0$

$\quad (x + 7)(x - 4) = 0$ Factoring

$\quad x + 7 = 0 \;\; or \;\; x - 4 = 0$ Principle of zero products

$\quad\quad x = -7 \; or \quad\quad x = 4$

The solutions are -7 and 4.

31. $x^2 - 8x = 0$

$\quad x(x - 8) = 0$ Factoring

$\quad x = 0 \; or \; x - 8 = 0$ Principle of zero products

$\quad x = 0 \; or \quad\quad x = 8$

The solutions are 0 and 8.

33. $y^2 + 6y + 9 = 0$

$\quad (y + 3)(y + 3) = 0$

$\quad y + 3 = 0 \;\; or \;\; y + 3 = 0$

$\quad\quad y = -3 \; or \quad\quad y = -3$

The solution is -3.

35. $x^2 + 100 = 20x$

$\quad x^2 - 20x + 100 = 0$ Subtracting $20x$

$\quad (x - 10)(x - 10) = 0$

$\quad x - 10 = 0 \;\; or \;\; x - 10 = 0$

$\quad\quad x = 10 \; or \quad\quad x = 10$

The solution is 10.

37. $x^2 - 4x - 32 = 0$

$\quad (x - 8)(x + 4) = 0$

$\quad x - 8 = 0 \; or \; x + 4 = 0$

$\quad\quad x = 8 \; or \quad\quad x = -4$

The solutions are 8 and -4.

39. $3y^2 + 8y + 4 = 0$

$\quad (3y + 2)(y + 2) = 0$

$\quad 3y + 2 = 0 \;\; or \;\; y + 2 = 0$

$\quad\quad 3y = -2 \; or \quad\quad y = -2$

$\quad\quad\; y = -\dfrac{2}{3} \; or \quad\quad y = -2$

The solutions are $-\dfrac{2}{3}$ and -2.

41. $12z^2 + z = 6$

$\quad 12z^2 + z - 6 = 0$

$\quad (4z + 3)(3z - 2) = 0$

$\quad 4z + 3 = 0 \;\; or \;\; 3z - 2 = 0$

$\quad\quad 4z = -3 \; or \quad\quad 3z = 2$

$\quad\quad\; z = -\dfrac{3}{4} \; or \quad\quad z = \dfrac{2}{3}$

The solutions are $-\dfrac{3}{4}$ and $\dfrac{2}{3}$.

43.
$$12a^2 - 28 = 5a$$
$$12a^2 - 5a - 28 = 0$$
$$(3a + 4)(4a - 7) = 0$$
$$3a + 4 = 0 \quad or \quad 4a - 7 = 0$$
$$3a = -4 \quad or \quad 4a = 7$$
$$a = -\frac{4}{3} \quad or \quad a = \frac{7}{4}$$
The solutions are $-\frac{4}{3}$ and $\frac{7}{4}$.

45.
$$14 = x(x - 5)$$
$$14 = x^2 - 5x$$
$$0 = x^2 - 5x - 14$$
$$0 = (x - 7)(x + 2)$$
$$x - 7 = 0 \ or \ x + 2 = 0$$
$$x = 7 \ or \quad x = -2$$
The solutions are 7 and -2.

47.
$$x^2 - 36 = 0$$
$$(x + 6)(x - 6) = 0$$
$$x + 6 = 0 \quad or \quad x - 6 = 0$$
$$x = -6 \ or \quad x = 6$$
The solutions are -6 and 6.

49.
$$z^2 = 144$$
$$z^2 - 144 = 0$$
$$(z + 12)(z - 12) = 0$$
$$z + 12 = 0 \quad or \quad z - 12 = 0$$
$$z = -12 \ or \quad z = 12$$
The solutions are -12 and 12.

51.
$$2x^2 - 20 = 0$$
$$2x^2 = 0$$
$$x^2 = 10$$
$$x = \sqrt{10} \ or \ x = -\sqrt{10} \quad \text{Principle of square roots}$$
The solutions are $\sqrt{10}$ and $-\sqrt{10}$, or $\pm\sqrt{10}$.

53.
$$6z^2 - 18 = 0$$
$$6z^2 = 18$$
$$z^2 = 3$$
$$z = \sqrt{3} \ or \ z = -\sqrt{3}$$
The solutions are $\sqrt{3}$ and $-\sqrt{3}$, or $\pm\sqrt{3}$.

55. Discussion and Writing

57.
$$3[5 - 3(4 - t)] - 2 = 5[3(5t - 4) + 8] - 26$$
$$3[5 - 12 + 3t] - 2 = 5[15t - 12 + 8] - 26$$
$$3[-7 + 3t] - 2 = 5[15t - 4] - 26$$
$$-21 + 9t - 2 = 75t - 20 - 26$$
$$9t - 23 = 75t - 46$$
$$-66t - 23 = -46$$
$$-66t = -23$$
$$t = \frac{23}{66}$$
The solution is $\frac{23}{66}$.

59.
$$x - \{3x - [2x - (5x - (7x - 1))]\} = x + 7$$
$$x - \{3x - [2x - (5x - 7x + 1)]\} = x + 7$$
$$x - \{3x - [2x - (-2x + 1)]\} = x + 7$$
$$x - \{3x - [2x + 2x - 1]\} = x + 7$$
$$x - \{3x - [4x - 1]\} = x + 7$$
$$x - \{3x - 4x + 1\} = x + 7$$
$$x - \{-x + 1\} = x + 7$$
$$x + x - 1 = x + 7$$
$$2x - 1 = x + 7$$
$$x - 1 = 7$$
$$x = 8$$
The solution is 8.

61.
$$(5x^2 + 6x)(12x^2 - 5x - 2) = 0$$
$$x(5x + 6)(4x + 1)(3x - 2) = 0$$
$$x = 0 \ or \ 5x + 6 = 0 \quad or \ 4x + 1 = 0 \quad or \ 3x - 2 = 0$$
$$x = 0 \ or \quad 5x = -6 \ or \quad 4x = -1 \ or \quad 3x = 2$$
$$x = 0 \ or \quad x = -\frac{6}{5} \ or \quad x = -\frac{1}{4} \ or \quad x = \frac{2}{3}$$
The solutions are 0, $-\frac{6}{5}$, $-\frac{1}{4}$, and $\frac{2}{3}$.

63.
$$3x^3 + 6x^2 - 27x - 54 = 0$$
$$3(x^3 + 2x^2 - 9x - 18) = 0$$
$$3[x^2(x + 2) - 9(x + 2)] = 0 \quad \text{Factoring by grouping}$$
$$3(x + 2)(x^2 - 9) = 0$$
$$3(x + 2)(x + 3)(x - 3) = 0$$
$$x + 2 = 0 \quad or \ x + 3 = 0 \quad or \ x - 3 = 0$$
$$x = -2 \ or \quad x = -3 \ or \quad x = 3$$
The solutions are -2, -3, and 3.

Chapter R Review Exercises

1. True

3. True

5. Integers: $12, -3, -1, -19, 31, 0$

7. Rational numbers:

$$-43.89, 12, -3, -\frac{1}{5}, -1, -\frac{4}{3}, 7\frac{2}{3}, -19, 31, 0$$

9. Irrational numbers: $\sqrt{7}, \sqrt[3]{10}$

11. $[-3, 5)$

13. $|16| = 16$

15.
$$5^3 - [2(4^2 - 3^2 - 6)]^3$$
$$= 5^3 - [2(16 - 9 - 6)]^3$$
$$= 5^3 - [2(1)]^3$$
$$= 5^3 - [2]^3$$
$$= 125 - 8$$
$$= 117$$

17. The exponent is positive, so the number is greater than 10. We move the decimal point 6 places to the right.
$$3.261 \times 10^6 = 3,261,000$$

19. Position the decimal point 2 places to the right, between the 1 and the 4. Since 0.01432 is a number between 0 and 1, the exponent must be negative.
$$0.01432 = 1.432 \times 10^{-2}$$

21.
$$\frac{2.5 \times 10^{-8}}{3.2 \times 10^{13}}$$
$$= \frac{2.5}{3.2} \times \frac{10^{-8}}{10^{13}}$$
$$= 0.78125 \times 10^{-21}$$
$$= (7.8125 \times 10^{-1}) \times 10^{-21}$$
$$= 7.8125 \times 10^{-22}$$

23.
$$(7a^2b^4)(-2a^{-4}b^3)$$
$$= 7(-2)a^{2+(-4)}b^{4+3}$$
$$= -14a^{-2}b^7, \text{ or } -\frac{14b^7}{a^2}$$

25. $\sqrt[4]{81} = \sqrt[4]{3^4} = 3$

27.
$$\frac{b - a^{-1}}{a - b^{-1}} = \frac{b - \dfrac{1}{a}}{a - \dfrac{1}{b}}$$
$$= \frac{b \cdot \dfrac{a}{a} - \dfrac{1}{a}}{a \cdot \dfrac{b}{b} - \dfrac{1}{b}}$$
$$= \frac{\dfrac{ab}{a} - \dfrac{1}{a}}{\dfrac{ab}{b} - \dfrac{1}{b}}$$
$$= \frac{\dfrac{ab - 1}{a}}{\dfrac{ab - 1}{b}}$$
$$= \frac{ab - 1}{a} \cdot \frac{b}{ab - 1}$$
$$= \frac{(ab - 1)b}{a(ab - 1)}$$
$$= \frac{b}{a}$$

29.
$$(\sqrt{3} - \sqrt{7})(\sqrt{3} + \sqrt{7}) = (\sqrt{3})^2 - (\sqrt{7})^2$$
$$= 3 - 7$$
$$= -4$$

31.
$$8\sqrt{5} + \frac{25}{\sqrt{5}} = 8\sqrt{5} + \frac{25}{\sqrt{5}} \cdot \frac{\sqrt{5}}{\sqrt{5}}$$
$$= 8\sqrt{5} + \frac{25\sqrt{5}}{5}$$
$$= 8\sqrt{5} + 5\sqrt{5}$$
$$= 13\sqrt{5}$$

33.
$$(5a + 4b)(2a - 3b)$$
$$= 10a^2 - 15ab + 8ab - 12b^2$$
$$= 10a^2 - 7ab - 12b^2$$

35.
$$x^3 + 2x^2 - 3x - 6$$
$$= x^2(x + 2) - 3(x + 2)$$
$$= (x + 2)(x^2 - 3)$$

37.
$$24x + 144 + x^2$$
$$= x^2 + 24x + 144$$
$$= (x + 12)^2$$

39.
$$8x^3 - 1$$
$$= (2x)^3 - 1^3$$
$$= (2x - 1)(4x^2 + 2x + 1)$$

41.
$$6x^3 + 48$$
$$= 6(x^3 + 8)$$
$$= 6(x + 2)(x^2 - 2x + 4)$$

43. $9x^2 - 30x + 25 = (3x - 5)^2$

45. $a^2b^2 - ab - 6 = (ab - 3)(ab + 2)$

47. $\dfrac{x}{x^2 + 9x + 20} - \dfrac{4}{x^2 + 7x + 12}$

$= \dfrac{x}{(x+5)(x+4)} - \dfrac{4}{(x+4)(x+3)}$

$\quad$ LCD is $(x+5)(x+4)(x+3)$

$= \dfrac{x}{(x+5)(x+4)} \cdot \dfrac{x+3}{x+3} - \dfrac{x}{(x+4)(x+3)} \cdot \dfrac{x+5}{x+5}$

$= \dfrac{x(x+3) - 4(x+5)}{(x+5)(x+4)(x+3)}$

$= \dfrac{x^2 + 3x - 4x - 20}{(x+5)(x+4)(x+3)}$

$= \dfrac{x^2 - x - 20}{(x+5)(x+4)(x+3)}$

$= \dfrac{(x-5)(\cancel{x+4})}{(x+5)(\cancel{x+4})(x+3)}$

$= \dfrac{x-5}{(x+5)(x+3)}$

49. $\dfrac{\sqrt{(a+b)^3}\,\sqrt[3]{a+b}}{\sqrt[6]{(a+b)^7}}$

$= \dfrac{(a+b)^{3/2}(a+b)^{1/3}}{(a+b)^{7/6}}$

$= (a+b)^{3/2 + 1/3 - 7/6}$

$= (a+b)^{9/6 + 2/6 - 7/6}$

$= (a+b)^{2/3}$

$= \sqrt[3]{(a+b)^2}$

51. $\sqrt[8]{\dfrac{m^{32}n^{16}}{3^8}} = \left(\dfrac{m^{32}n^{16}}{3^8}\right)^{1/8} = \dfrac{m^4 n^2}{3}$

53. $a = 8$ and $b = 17$. Find c.

$c^2 = a^2 + b^2$

$c^2 = 8^2 + 17^2$

$c^2 = 64 + 289$

$c^2 = 353$

$c \approx 18.8$

The guy wire is about 18.8 ft long.

55. $5x - 7 = 3x - 9$

$2x - 7 = -9$

$2x = -2$

$x = -1$

The solution is -1.

57. $6(2x - 1) = 3 - (x + 10)$

$12x - 6 = 3 - x - 10$

$12x - 6 = -x - 7$

$13x - 6 = -7$

$13x = -1$

$x = -\dfrac{1}{13}$

The solution is $-\dfrac{1}{13}$.

59. $x^2 - x = 20$

$x^2 - x - 20 = 0$

$(x - 5)(x + 4) = 0$

$x - 5 = 0 \quad or \quad x + 4 = 0$

$x = 5 \quad or \quad\quad x = -4$

The solutions are 5 and -4.

61. $x(x - 2) = 3$

$x^2 - 2x = 3$

$x^2 - 2x - 3 = 0$

$(x + 1)(x - 3) = 0$

$x + 1 = 0 \quad or \quad x - 3 = 0$

$x = -1 \quad or \quad\quad x = 3$

The solutions are -1 and 3.

63. $n^2 - 7 = 0$

$n^2 = 7$

$n = \sqrt{7} \quad or \quad n = -\sqrt{7}$

The solutions are $\sqrt{7}$ and $-\sqrt{7}$, or $\pm\sqrt{7}$.

65. $9x^2 - 36y^2 = 9(x^2 - 4y^2)$

$\quad\quad\quad\quad\quad\quad = 9[x^2 - (2y)^2]$

$\quad\quad\quad\quad\quad\quad = 9(x + 2y)(x - 2y)$

Answer C is correct.

67. When the number 4 is raised to a positive integer power, the last digit of the result is 4 or 6. Since the calculator returns $4.398046511 \times 10^{12}$, or $4{,}398{,}046{,}511{,}000$, we can conclude that this result is an approximation.

69. Substitute $\$124{,}000 - \$20{,}000$, or $\$104{,}000$ for P, 0.0575 for r, and $12 \cdot 30$, or 360, for n and perform the resulting computation.

$M = P\left[\dfrac{\dfrac{r}{12}\left(1 + \dfrac{r}{12}\right)^n}{\left(1 + \dfrac{r}{12}\right)^n - 1}\right]$

$= \$104{,}000\left[\dfrac{\dfrac{0.0575}{12}\left(1 + \dfrac{0.0575}{12}\right)^{360}}{\left(1 + \dfrac{0.0575}{12}\right)^{360} - 1}\right]$

$\approx \$606.92$

71. $P = \$151{,}000 - \$21{,}000 = \$130{,}000$, $r = 0.0625$, $n = 12 \cdot 25 = 300$.

$M = \$130{,}000\left[\dfrac{\dfrac{0.0625}{12}\left(1 + \dfrac{0.0625}{12}\right)^{300}}{\left(1 + \dfrac{0.0625}{12}\right)^{300} - 1}\right]$

$\approx \$857.57$

73. $(t^a + t^{-a})^2 = (t^a)^2 + 2 \cdot t^a \cdot t^{-a} + (t^{-a})^2$

$\quad\quad\quad\quad\quad\quad = t^{2a} + 2 + t^{-2a}$

75. $(a^n - b^n)^3 = (a^n - b^n)(a^n - b^n)^2$

$\qquad = (a^n - b^n)(a^{2n} - 2a^n b^n + b^{2n})$

$\qquad = a^{3n} - 2a^{2n}b^n + a^n b^{2n} - a^{2n}b^n + 2a^n b^{2n} - b^{3n}$

$\qquad = a^{3n} - 3a^{2n}b^n + 3a^n b^{2n} - b^{3n}$

77. $x^{2t} - 3x^t - 28 = (x^t)^2 - 3x^t - 28$

$\qquad = (x^t - 7)(x^t + 4)$

Chapter R Test

1. a) Integers: $-8, 0, 36$

b) Rational numbers: $-8, \dfrac{11}{3}, 0, -5.49, 36, 10\dfrac{1}{6}$

c) Rational numbers but not integers: $\dfrac{11}{3}, -5.49, 10\dfrac{1}{6}$

d) Integers but not natural numbers: $-8, 0$

2. $\left| -\dfrac{14}{5} \right| = \dfrac{14}{5}$

3. $|19.4| = 19.4$

4. $|-1.2xy| = |-1.2||x||y| = 1.2|x||y|$

5. $(-3, 6]$

6. $|-7 - 5| = |-12| = 12$, or

$|5 - (-7)| = |5 + 7| = |12| = 12$

7. $\quad 32 \div 2^3 - 12 \div 4 \cdot 3$

$= 32 \div 8 - 12 \div 4 \cdot 3$

$= 4 - 12 \div 4 \cdot 3$

$= 4 - 3 \cdot 3$

$= 4 - 9$

$= -5$

8. Position the decimal point 5 places to the right, between the 3 and the 6. Since 0.0000367 is a number between 0 and 1, the exponent must be negative.

$0.0000367 = 3.67 \times 10^{-5}$

9. The exponent is positive, so the number is greater than 10. We move the decimal point 6 places to the right.

$4.51 \times 10^6 = 4,510,000$

10. $\dfrac{2.7 \times 10^4}{3.6 \times 10^{-3}} = 0.75 \times 10^7$

$\qquad = (7.5 \times 10^{-1}) \times 10^7$

$\qquad = 7.5 \times 10^6$

11. $x^{-8} \cdot x^5 = x^{-8+5} = x^{-3}$, or $\dfrac{1}{x^3}$

12. $(2y^2)^3(3y^4)^2 = 2^3 y^6 \cdot 3^2 y^8 = 8 \cdot 9 \cdot y^{6+8} = 72y^{14}$

13. $\quad (-3a^5 b^{-4})(5a^{-1} b^3)$

$= -3 \cdot 5 \cdot a^{5+(-1)} \cdot b^{-4+3}$

$= -15a^4 b^{-1}$, or $-\dfrac{15a^4}{b}$

14. $\quad (3x^4 - 2x^2 + 6x) - (5x^3 - 3x^2 + x)$

$= 3x^4 - 2x^2 + 6x - 5x^3 + 3x^2 - x$

$= 3x^4 - 5x^3 + x^2 + 5x$

15. $(x + 3)(2x - 5) = 2x^2 - 5x + 6x - 15 = 2x^2 + x - 15$

16. $(2y - 1)^2 = (2y)^2 - 2 \cdot 2y \cdot 1 + (1)^2 = 4y^2 - 4y + 1$

17. $\dfrac{\dfrac{x}{y} - \dfrac{y}{x}}{x + y} = \dfrac{\dfrac{x}{y} \cdot \dfrac{x}{x} - \dfrac{y}{x} \cdot \dfrac{y}{y}}{x + y}$

$= \dfrac{\dfrac{x^2}{xy} - \dfrac{y^2}{xy}}{x + y}$

$= \dfrac{\dfrac{x^2 - y^2}{xy}}{x + y}$

$= \dfrac{x^2 - y^2}{xy} \cdot \dfrac{1}{x + y}$

$= \dfrac{(x + y)(x - y)}{xy(x + y)}$

$= \dfrac{x - y}{xy}$

18. $\sqrt{54} = \sqrt{9 \cdot 6} = \sqrt{9}\sqrt{6} = 3\sqrt{6}$

19. $\sqrt[3]{40} = \sqrt[3]{8 \cdot 5} = \sqrt[3]{8}\sqrt[3]{5} = 2\sqrt[3]{5}$

20. $3\sqrt{75} + 2\sqrt{27} = 3\sqrt{25 \cdot 3} + 2\sqrt{9 \cdot 3}$

$\qquad = 3 \cdot 5\sqrt{3} + 2 \cdot 3\sqrt{3}$

$\qquad = 15\sqrt{3} + 6\sqrt{3}$

$\qquad = 21\sqrt{3}$

21. $\sqrt{18}\sqrt{10} = \sqrt{18 \cdot 10} = \sqrt{2 \cdot 3 \cdot 3 \cdot 2 \cdot 5} = 2 \cdot 3\sqrt{5} = 6\sqrt{5}$

22. $\quad (2 + \sqrt{3})(5 - 2\sqrt{3})$

$= 2 \cdot 5 - 4\sqrt{3} + 5\sqrt{3} - 2 \cdot 3$

$= 10 - 4\sqrt{3} + 5\sqrt{3} - 6$

$= 4 + \sqrt{3}$

23. $y^2 - 3y - 18 = (y + 3)(y - 6)$

24. $x^3 + 10x^2 + 25x = x(x^2 + 10x + 25) = x(x + 5)^2$

25. $2n^2 + 5n - 12 = (2n - 3)(n + 4)$

26. $8x^2 - 18 = 2(4x^2 - 9) = 2(2x + 3)(2x - 3)$

27. $m^3 - 8 = (m - 2)(m^2 + 2m + 4)$

28.
$$\frac{x^2 + x - 6}{x^2 + 8x + 15} \cdot \frac{x^2 - 25}{x^2 - 4x + 4}$$
$$= \frac{(x^2 + x - 6)(x^2 - 25)}{(x^2 + 8x + 15)(x^2 - 4x + 4)}$$
$$= \frac{(x+3)(x-2)(x+5)(x-5)}{(x+3)(x+5)(x-2)(x-2)}$$
$$= \frac{x-5}{x-2}$$

29.
$$\frac{x}{x^2 - 1} - \frac{3}{x^2 + 4x - 5}$$
$$= \frac{x}{(x+1)(x-1)} - \frac{3}{(x-1)(x+5)}$$
$$\text{LCD is } (x+1)(x-1)(x+5)$$
$$= \frac{x}{(x+1)(x-1)} \cdot \frac{x+5}{x+5} - \frac{3}{(x-1)(x+5)} \cdot \frac{x+1}{x+1}$$
$$= \frac{x(x+5) - 3(x+1)}{(x+1)(x-1)(x+5)}$$
$$= \frac{x^2 + 5x - 3x - 3}{(x+1)(x-1)(x+5)}$$
$$= \frac{x^2 + 2x - 3}{(x+1)(x-1)(x+5)}$$
$$= \frac{(x+3)(x-1)}{(x+1)(x-1)(x+5)}$$
$$= \frac{x+3}{(x+1)(x+5)}$$

30. $\dfrac{5}{7 - \sqrt{3}} = \dfrac{5}{7 - \sqrt{3}} \cdot \dfrac{7 + \sqrt{3}}{7 + \sqrt{3}} = \dfrac{35 + 5\sqrt{3}}{49 - 3} = $
$$\frac{35 + 5\sqrt{3}}{46}$$

31. $t^{5/7} = \sqrt[7]{t^5}$

32. $(\sqrt[5]{7})^3 = (7^{1/5})^3 = 7^{3/5}$

33. $a = 5$ and $b = 12$. Find c.
$$c^2 = a^2 + b^2$$
$$c^2 = 5^2 + 12^2$$
$$c^2 = 25 + 144$$
$$c^2 = 169$$
$$c = 13$$
The guy wire is 13 ft long.

34. $7x - 4 = 24$
$$7x = 28$$
$$x = 4$$
The solution is 4.

35. $3(y - 5) + 6 = 8 - (y + 2)$
$$3y - 15 + 6 = 8 - y - 2$$
$$3y - 9 = -y + 6$$
$$4y - 9 = 6$$
$$4y = 15$$
$$y = \frac{15}{4}$$
The solution is $\dfrac{15}{4}$.

36. $2x^2 + 5x + 3 = 0$
$$(2x + 3)(x + 1) = 0$$
$$2x + 3 = 0 \quad or \quad x + 1 = 0$$
$$2x = -3 \quad or \quad \quad x = -1$$
$$x = -\frac{3}{2} \quad or \quad \quad x = -1$$
The solutions are $-\dfrac{3}{2}$ and -1.

37. $z^2 - 11 = 0$
$$z^2 = 11$$
$$z = \sqrt{11} \quad or \quad z = -\sqrt{11}$$
The solutions are $\sqrt{11}$ and $-\sqrt{11}$, or $\pm\sqrt{11}$.

38. $(x - y - 1)^2$
$$= [(x - y) - 1]^2$$
$$= (x - y)^2 - 2(x - y)(1) + 1^2$$
$$= x^2 - 2xy + y^2 - 2x + 2y + 1$$

Chapter 1

Graphs, Functions, and Models

Exercise Set 1.1

1. To graph $(4, 0)$ we move from the origin 4 units to the right of the y-axis. Since the second coordinate is 0, we do not move up or down from the x-axis.

   To graph $(-3, -5)$ we move from the origin 3 units to the left of the y-axis. Then we move 5 units down from the x-axis.

   To graph $(-1, 4)$ we move from the origin 1 unit to the left of the y-axis. Then we move 4 units up from the x-axis.

   To graph $(0, 2)$ we do not move to the right or the left of the y-axis since the first coordinate is 0. From the origin we move 2 units up.

   To graph $(2, -2)$ we move from the origin 2 units to the right of the y-axis. Then we move 2 units down from the x-axis.

3. To graph $(-5, 1)$ we move from the origin 5 units to the left of the y-axis. Then we move 1 unit up from the x-axis.

   To graph $(5, 1)$ we move from the origin 5 units to the right of the y-axis. Then we move 1 unit up from the x-axis.

   To graph $(2, 3)$ we move from the origin 2 units to the right of the y-axis. Then we move 3 units up from the x-axis.

   To graph $(2, -1)$ we move from the origin 2 units to the right of the y-axis. Then we move 1 unit down from the x-axis.

   To graph $(0, 1)$ we do not move to the right or the left of the y-axis since the first coordinate is 0. From the origin we move 1 unit up.

5. The first coordinate represents the year and the corresponding second coordinate represents total advertisement spending, in millions of dollars. The ordered pairs are $(2000, 310.6)$, $(2001, 311.2)$, $(2002, 348.2)$, $(2003, 361.6)$, $(2004, 435.8)$, and $(2005, 467.7)$.

7. To determine whether $(1, -1)$ is a solution, substitute 1 for x and -1 for y.

$$\begin{array}{c|c} \multicolumn{2}{c}{y = 2x - 3} \\ \hline -1 \ ? \ 2 \cdot 1 - 3 & \\ \qquad 2 - 3 & \\ -1 \ \big| \ -1 & \text{TRUE} \end{array}$$

   The equation $-1 = -1$ is true, so $(1, -1)$ is a solution.

   To determine whether $(0, 3)$ is a solution, substitute 0 for x and -3 for y.

$$\begin{array}{c|c} \multicolumn{2}{c}{y = 2x - 3} \\ \hline 3 \ ? \ 2 \cdot 0 - 3 & \\ \qquad 0 - 3 & \\ 3 \ \big| \ -3 & \text{FALSE} \end{array}$$

   The equation $3 = -3$ is false, so $(0, 3)$ is not a solution.

9. To determine whether $\left(\dfrac{2}{3}, \dfrac{3}{4}\right)$ is a solution, substitute $\dfrac{2}{3}$ for x and $\dfrac{3}{4}$ for y.

$$\begin{array}{c|c} \multicolumn{2}{c}{6x - 4y = 1} \\ \hline 6 \cdot \dfrac{2}{3} - 4 \cdot \dfrac{3}{4} \ ? \ 1 & \\ 4 - 3 & \\ 1 \ \big| \ 1 & \text{TRUE} \end{array}$$

   The equation $1 = 1$ is true, so $\left(\dfrac{2}{3}, \dfrac{3}{4}\right)$ is a solution.

   To determine whether $\left(1, \dfrac{3}{2}\right)$ is a solution, substitute 1 for x and $\dfrac{3}{2}$ for y.

$$\begin{array}{c|c} \multicolumn{2}{c}{6x - 4y = 1} \\ \hline 6 \cdot 1 - 4 \cdot \dfrac{3}{2} \ ? \ 1 & \\ 6 - 6 & \\ 0 \ \big| \ 1 & \text{FALSE} \end{array}$$

   The equation $0 = 1$ is false, so $\left(1, \dfrac{3}{2}\right)$ is not a solution.

11. To determine whether $\left(-\frac{1}{2}, -\frac{4}{5}\right)$ is a solution, substitute $-\frac{1}{2}$ for a and $-\frac{4}{5}$ for b.

$$
\begin{array}{c|c}
\multicolumn{2}{c}{2a + 5b = 3} \\
\hline
2\left(-\frac{1}{2}\right) + 5\left(-\frac{4}{5}\right) \ ? \ 3 & \\
-1 - 4 & \\
-5 & 3 \quad \text{FALSE}
\end{array}
$$

The equation $-5 = 3$ is false, so $\left(-\frac{1}{2}, -\frac{4}{5}\right)$ is not a solution.

To determine whether $\left(0, \frac{3}{5}\right)$ is a solution, substitute 0 for a and $\frac{3}{5}$ for b.

$$
\begin{array}{c|c}
\multicolumn{2}{c}{2a + 5b = 3} \\
\hline
2 \cdot 0 + 5 \cdot \frac{3}{5} \ ? \ 3 & \\
0 + 3 & \\
3 & 3 \quad \text{TRUE}
\end{array}
$$

The equation $3 = 3$ is true, so $\left(0, \frac{3}{5}\right)$ is a solution.

13. To determine whether $(-0.75, 2.75)$ is a solution, substitute -0.75 for x and 2.75 for y.

$$
\begin{array}{c|c}
\multicolumn{2}{c}{x^2 - y^2 = 3} \\
\hline
(-0.75)^2 - (2.75)^2 \ ? \ 3 & \\
0.5625 - 7.5625 & \\
-7 & 3 \quad \text{FALSE}
\end{array}
$$

The equation $-7 = 3$ is false, so $(-0.75, 2.75)$ is not a solution.

To determine whether $(2, -1)$ is a solution, substitute 2 for x and -1 for y.

$$
\begin{array}{c|c}
\multicolumn{2}{c}{x^2 - y^2 = 3} \\
\hline
2^2 - (-1)^2 \ ? \ 3 & \\
4 - 1 & \\
3 & 3 \quad \text{TRUE}
\end{array}
$$

The equation $3 = 3$ is true, so $(2, -1)$ is a solution.

15. Graph $5x - 3y = -15$.

To find the x-intercept we replace y with 0 and solve for x.

$$
\begin{aligned}
5x - 3 \cdot 0 &= -15 \\
5x &= -15 \\
x &= -3
\end{aligned}
$$

The x-intercept is $(-3, 0)$.

To find the y-intercept we replace x with 0 and solve for y.

$$
\begin{aligned}
5 \cdot 0 - 3y &= -15 \\
-3y &= -15 \\
y &= 5
\end{aligned}
$$

The y-intercept is $(0, 5)$.

We plot the intercepts and draw the line that contains them. We could find a third point as a check that the intercepts were found correctly.

17. Graph $2x + y = 4$.

To find the x-intercept we replace y with 0 and solve for x.

$$
\begin{aligned}
2x + 0 &= 4 \\
2x &= 4 \\
x &= 2
\end{aligned}
$$

The x-intercept is $(2, 0)$.

To find the y-intercept we replace x with 0 and solve for y.

$$
\begin{aligned}
2 \cdot 0 + y &= 4 \\
y &= 4
\end{aligned}
$$

The y-intercept is $(0, 4)$.

We plot the intercepts and draw the line that contains them. We could find a third point as a check that the intercepts were found correctly.

19. Graph $4y - 3x = 12$.

To find the x-intercept we replace y with 0 and solve for x.

$$
\begin{aligned}
4 \cdot 0 - 3x &= 12 \\
-3x &= 12 \\
x &= -4
\end{aligned}
$$

The x-intercept is $(-4, 0)$.

To find the y-intercept we replace x with 0 and solve for y.

$$
\begin{aligned}
4y - 3 \cdot 0 &= 12 \\
4y &= 12 \\
y &= 3
\end{aligned}
$$

The y-intercept is $(0, 3)$.

We plot the intercepts and draw the line that contains them. We could find a third point as a check that the intercepts were found correctly.

$$4y - 3x = 12$$

21. Graph $y = 3x + 5$.

We choose some values for x and find the corresponding y-values.

When $x = -3$, $y = 3x + 5 = 3(-3) + 5 = -9 + 5 = -4$.

When $x = -1$, $y = 3x + 5 = 3(-1) + 5 = -3 + 5 = 2$.

When $x = 0$, $y = 3x + 5 = 3 \cdot 0 + 5 = 0 + 5 = 5$

We list these points in a table, plot them, and draw the graph.

x	y	(x, y)
-3	-4	$(-3, -4)$
-1	2	$(-1, 2)$
0	5	$(0, 5)$

23. Graph $x - y = 3$.

Make a table of values, plot the points in the table, and draw the graph.

x	y	(x, y)
-2	-5	$(-2, -5)$
0	-3	$(0, -3)$
3	0	$(3, 0)$

25. Graph $y = -\dfrac{3}{4}x + 3$.

By choosing multiples of 4 for x, we can avoid fraction values for y. Make a table of values, plot the points in the table, and draw the graph.

x	y	(x, y)
-4	6	$(-4, 6)$
0	3	$(0, 3)$
4	0	$(4, 0)$

27. Graph $5x - 2y = 8$.

We could solve for y first.

$$5x - 2y = 8$$

$-2y = -5x + 8$ Subtracting $5x$ on both sides

$y = \dfrac{5}{2}x - 4$ Multiplying by $-\dfrac{1}{2}$ on both sides

By choosing multiples of 2 for x we can avoid fraction values for y. Make a table of values, plot the points in the table, and draw the graph.

x	y	(x, y)
0	-4	$(0, -4)$
2	1	$(2, 1)$
4	6	$(4, 6)$

29. Graph $x - 4y = 5$.

Make a table of values, plot the points in the table, and draw the graph.

x	y	(x, y)
-3	-2	$(-3, -2)$
1	-1	$(1, -1)$
5	0	$(5, 0)$

31. Graph $2x + 5y = -10$.

In this case, it is convenient to find the intercepts along with a third point on the graph. Make a table of values, plot the points in the table, and draw the graph.

x	y	(x, y)
-5	0	$(-5, 0)$
0	-2	$(0, -2)$
5	-4	$(5, -4)$

33. Graph $y = -x^2$.

Make a table of values, plot the points in the table, and draw the graph.

x	y	(x, y)
-2	-4	$(-2, -4)$
-1	-1	$(-1, -1)$
0	0	$(0, 0)$
1	-1	$(1, -1)$
2	-4	$(2, -4)$

35. Graph $y = x^2 - 3$.

Make a table of values, plot the points in the table, and draw the graph.

x	y	(x, y)
-3	6	$(-3, 6)$
-1	-2	$(-1, -2)$
0	-3	$(0, -3)$
1	-2	$(1, -2)$
3	6	$(3, 6)$

37. Graph $y = -x^2 + 2x + 3$.

Make a table of values, plot the points in the table, and draw the graph.

x	y	(x, y)
-2	-5	$(-2, -5)$
-1	0	$(-1, 0)$
0	3	$(0, 3)$
1	4	$(1, 4)$
2	3	$(2, 3)$
3	0	$(3, 0)$
4	-5	$(4, -5)$

39. Either point can be considered as (x_1, y_1).
$$d = \sqrt{(4 - 5)^2 + (6 - 9)^2}$$
$$= \sqrt{(-1)^2 + (-3)^2} = \sqrt{10} \approx 3.162$$

41. Either point can be considered as (x_1, y_1).
$$d = \sqrt{(6 - 9)^2 + (-1 - 5)^2}$$
$$= \sqrt{(-3)^2 + (-6)^2} = \sqrt{45} \approx 6.708$$

43. Either point can be considered as (x_1, y_1).
$$d = \sqrt{(-4.2 - 2.1)^2 + [3 - (-6.4)]^2}$$
$$= \sqrt{(-6.3)^2 + (9.4)^2} = \sqrt{128.05} \approx 11.316$$

45. Either point can be considered as (x_1, y_1).

$$d = \sqrt{\left(-\frac{1}{2} - \frac{5}{2}\right)^2 + (4 - 4)^2}$$
$$= \sqrt{(-3)^2 + 0^2} = \sqrt{9} = 3$$

47. Either point can be considered as (x_1, y_1).

$$d = \sqrt{(-\sqrt{6} - \sqrt{3})^2 + (0 - (-\sqrt{5}))^2}$$
$$= \sqrt{6 + 2\sqrt{18} + 3 + 5} = \sqrt{14 + 2\sqrt{9 \cdot 2}}$$
$$= \sqrt{14 + 2 \cdot 3\sqrt{2}} = \sqrt{14 + 6\sqrt{2}} \approx 4.742$$

49. Either point can be considered as (x_1, y_1).

$$d = \sqrt{(0 - a)^2 + (0 - b)^2} = \sqrt{a^2 + b^2}$$

51. First we find the length of the diameter:

$$d = \sqrt{(-3 - 9)^2 + (-1 - 4)^2}$$
$$= \sqrt{(-12)^2 + (-5)^2} = \sqrt{169} = 13$$

The length of the radius is one-half the length of the diameter, or $\frac{1}{2}(13)$, or 6.5.

53. First we find the distance between each pair of points.

For $(-4, 5)$ and $(6, 1)$:

$$d = \sqrt{(-4 - 6)^2 + (5 - 1)^2}$$
$$= \sqrt{(-10)^2 + 4^2} = \sqrt{116}$$

For $(-4, 5)$ and $(-8, -5)$:

$$d = \sqrt{(-4 - (-8))^2 + (5 - (-5))^2}$$
$$= \sqrt{4^2 + 10^2} = \sqrt{116}$$

For $(6, 1)$ and $(-8, -5)$:

$$d = \sqrt{(6 - (-8))^2 + (1 - (-5))^2}$$
$$= \sqrt{14^2 + 6^2} = \sqrt{232}$$

Since $(\sqrt{116})^2 + (\sqrt{116})^2 = (\sqrt{232})^2$, the points could be the vertices of a right triangle.

55. First we find the distance between each pair of points.

For $(-4, 3)$ and $(0, 5)$:

$$d = \sqrt{(-4 - 0)^2 + (3 - 5)^2}$$
$$= \sqrt{(-4)^2 + (-2)^2} = \sqrt{20}$$

For $(-4, 3)$ and $(3, -4)$:

$$d = \sqrt{(-4 - 3)^2 + [3 - (-4)]^2}$$
$$= \sqrt{(-7)^2 + 7^2} = \sqrt{98}$$

For $(0, 5)$ and $(3, -4)$:

$$d = \sqrt{(0 - 3)^2 + [5 - (-4)]^2}$$
$$= \sqrt{(-3)^2 + 9^2} = \sqrt{90}$$

The greatest distance is $\sqrt{98}$, so if the points are the vertices of a right triangle, then it is the hypotenuse. But $(\sqrt{20})^2 + (\sqrt{90})^2 \neq (\sqrt{98})^2$, so the points are not the vertices of a right triangle.

57. We use the midpoint formula.

$$\left(\frac{4 + (-12)}{2}, \frac{-9 + (-3)}{2}\right) = \left(-\frac{8}{2}, -\frac{12}{2}\right) = (-4, -6)$$

59. We use the midpoint formula.

$$\left(\frac{6.1 + 3.8}{2}, \frac{-3.8 + (-6.1)}{2}\right) = \left(\frac{9.9}{2}, -\frac{9.9}{2}\right) =$$
$$(4.95, -4.95)$$

61. We use the midpoint formula.

$$\left(\frac{-6 + (-6)}{2}, \frac{5 + 8}{2}\right) = \left(-\frac{12}{2}, \frac{13}{2}\right) = \left(-6, \frac{13}{2}\right)$$

63. We use the midpoint formula.

$$\left(\frac{-\frac{1}{6} + \left(-\frac{2}{3}\right)}{2}, \frac{-\frac{3}{5} + \frac{5}{4}}{2}\right) = \left(\frac{-\frac{5}{6}}{2}, \frac{\frac{13}{20}}{2}\right) =$$
$$\left(-\frac{5}{12}, \frac{13}{40}\right)$$

65. We use the midpoint formula.

$$\left(\frac{\sqrt{3} + 3\sqrt{3}}{2}, \frac{-1 + 4}{2}\right) = \left(\frac{4\sqrt{3}}{2}, \frac{3}{2}\right) = \left(2\sqrt{3}, \frac{3}{2}\right)$$

67.

For the side with vertices $(-3, 4)$ and $(2, -1)$:

$$\left(\frac{-3 + 2}{2}, \frac{4 + (-1)}{2}\right) = \left(-\frac{1}{2}, \frac{3}{2}\right)$$

For the side with vertices $(2, -1)$ and $(5, 2)$:

$$\left(\frac{2 + 5}{2}, \frac{-1 + 2}{2}\right) = \left(\frac{7}{2}, \frac{1}{2}\right)$$

For the side with vertices $(5, 2)$ and $(0, 7)$:

$$\left(\frac{5 + 0}{2}, \frac{2 + 7}{2}\right) = \left(\frac{5}{2}, \frac{9}{2}\right)$$

For the side with vertices $(0, 7)$ and $(-3, 4)$:

$$\left(\frac{0 + (-3)}{2}, \frac{7 + 4}{2}\right) = \left(-\frac{3}{2}, \frac{11}{2}\right)$$

For the quadrilateral whose vertices are the points found above, the diagonals have endpoints

$$\left(-\frac{1}{2}, \frac{3}{2}\right), \left(\frac{5}{2}, \frac{9}{2}\right) \text{ and } \left(\frac{7}{2}, \frac{1}{2}\right), \left(-\frac{3}{2}, \frac{11}{2}\right).$$

We find the length of each of these diagonals.

For $\left(-\frac{1}{2}, \frac{3}{2}\right), \left(\frac{5}{2}, \frac{9}{2}\right)$:

$$d = \sqrt{\left(-\frac{1}{2} - \frac{5}{2}\right)^2 + \left(\frac{3}{2} - \frac{9}{2}\right)^2}$$
$$= \sqrt{(-3)^2 + (-3)^2} = \sqrt{18}$$

For $\left(\frac{7}{2}, \frac{1}{2}\right)$, $\left(-\frac{3}{2}, \frac{11}{2}\right)$:

$$d = \sqrt{\left(\frac{7}{2} - \left(-\frac{3}{2}\right)\right)^2 + \left(\frac{1}{2} - \frac{11}{2}\right)^2}$$
$$= \sqrt{5^2 + (-5)^2} = \sqrt{50}$$

Since the diagonals do not have the same lengths, the midpoints are not vertices of a rectangle.

69. We use the midpoint formula.

$$\left(\frac{\sqrt{7} + \sqrt{2}}{2}, \frac{-4 + 3}{2}\right) = \left(\frac{\sqrt{7} + \sqrt{2}}{2}, -\frac{1}{2}\right)$$

71. $(x - h)^2 + (y - k)^2 = r^2$

$(x - 2)^2 + (y - 3)^2 = \left(\frac{5}{3}\right)^2$ Substituting

$(x - 2)^2 + (y - 3)^2 = \frac{25}{9}$

73. The length of a radius is the distance between $(-1, 4)$ and $(3, 7)$:

$r = \sqrt{(-1 - 3)^2 + (4 - 7)^2}$
$= \sqrt{(-4)^2 + (-3)^2} = \sqrt{25} = 5$

$(x - h)^2 + (y - k)^2 = r^2$
$[x - (-1)]^2 + (y - 4)^2 = 5^2$
$(x + 1)^2 + (y - 4)^2 = 25$

75. The center is the midpoint of the diameter:

$$\left(\frac{7 + (-3)}{2}, \frac{13 + (-11)}{2}\right) = (2, 1)$$

Use the center and either endpoint of the diameter to find the length of a radius. We use the point $(7, 13)$:

$r = \sqrt{(7 - 2)^2 + (13 - 1)^2}$
$= \sqrt{5^2 + 12^2} = \sqrt{169} = 13$

$(x - h)^2 + (y - k)^2 = r^2$
$(x - 2)^2 + (y - 1)^2 = 13^2$
$(x - 2)^2 + (y - 1)^2 = 169$

77. Since the center is 2 units to the left of the y-axis and the circle is tangent to the y-axis, the length of a radius is 2.

$(x - h)^2 + (y - k)^2 = r^2$
$[x - (-2)]^2 + (y - 3)^2 = 2^2$
$(x + 2)^2 + (y - 3)^2 = 4$

79. $x^2 + y^2 = 4$
$(x - 0)^2 + (y - 0)^2 = 2^2$

Center: $(0, 0)$; radius: 2

81. $x^2 + (y - 3)^2 = 16$
$(x - 0)^2 + (y - 3)^2 = 4^2$

Center: $(0, 3)$; radius: 4

83. $(x - 1)^2 + (y - 5)^2 = 36$
$(x - 1)^2 + (y - 5)^2 = 6^2$

Center: $(1, 5)$; radius: 6

85. $(x + 4)^2 + (y + 5)^2 = 9$
$[x - (-4)]^2 + [y - (-5)]^2 = 3^2$

Center: $(-4, -5)$; radius: 3

87. From the graph we see that the center of the circle is $(-2, 1)$ and the radius is 3. The equation of the circle is $[x - (-2)]^2 + (y - 1)^2 = 3^2$, or $(x + 2)^2 + (y - 1)^2 = 3^2$.

89. From the graph we see that the center of the circle is $(5, -5)$ and the radius is 15. The equation of the circle is $(x - 5)^2 + [y - (-5)]^2 = 15^2$, or $(x - 5)^2 + (y + 5)^2 = 15^2$.

91. Graph (b) is the graph of $y = 3 - x$.

93. Graph (a) is the graph of $y = x^2 + 2x + 1$.

95. First solve the equation for y: $y = -4x + 7$. Enter the equation in this form, select the standard window, and graph the equation.

$4x + y = 7$

97. Enter the equation, select the standard window, and graph the equation.

$y = \frac{1}{3}x + 2$

99. Enter the equation, select the standard window, and graph the equation.

$y = x^2 + 6$

101. Enter the equation, select the standard window, and graph the equation.

$y = 2 - x^2$

103. Standard window:

$[-4, 4, -4, 4]$

We see that the standard window is a better choice for this graph.

105. Standard window:

$[-1, 1, -0.3, 0.3]$, Xscl $= 0.1$, Yscl $= 0.1$

We see that $[-1, 1, -0.3, 0.3]$ is a better choice for this graph.

107. Square the viewing window. For the graph shown, one possibility is $[-12, 9, -4, 10]$.

109.

$x^2 + y^2 = 4$

$y_1 = \sqrt{4 - x^2}, \ y_2 = -\sqrt{4 - x^2}$

111.

$$(x-1)^2 + (y-5)^2 = 36$$

$$y_1 = 5 + \sqrt{36 - (x-1)^2}, \quad y_2 = 5 - \sqrt{36 - (x-1)^2}$$

113. Discussion and Writing

115. If the point (p,q) is in the fourth quadrant, then $p > 0$ and $q < 0$. If $p > 0$, then $-p < 0$ so both coordinates of the point $(q, -p)$ are negative and $(q, -p)$ is in the third quadrant.

117. Use the distance formula. Either point can be considered as (x_1, y_1).

$$d = \sqrt{(a+h-a)^2 + (\sqrt{a+h} - \sqrt{a})^2}$$
$$= \sqrt{h^2 + a + h - 2\sqrt{a^2 + ah} + a}$$
$$= \sqrt{h^2 + 2a + h - 2\sqrt{a^2 + ah}}$$

Next we use the midpoint formula.

$$\left(\frac{a+a+h}{2}, \frac{\sqrt{a} + \sqrt{a+h}}{2}\right) = \left(\frac{2a+h}{2}, \frac{\sqrt{a} + \sqrt{a+h}}{2}\right)$$

119. First use the formula for the area of a circle to find r^2:

$$A = \pi r^2$$
$$36\pi = \pi r^2$$
$$36 = r^2$$

Then we have:

$$(x-h)^2 + (y-k)^2 = r^2$$
$$(x-2)^2 + [y - (-7)]^2 = 36$$
$$(x-2)^2 + (y+7)^2 = 36$$

121. Let $(0, y)$ be the required point. We set the distance from $(-2, 0)$ to $(0, y)$ equal to the distance from $(4, 6)$ to $(0, y)$ and solve for y.

$$\sqrt{[0-(-2)]^2 + (y-0)^2} = \sqrt{(0-4)^2 + (y-6)^2}$$
$$\sqrt{4 + y^2} = \sqrt{16 + y^2 - 12y + 36}$$
$$4 + y^2 = 16 + y^2 - 12y + 36$$
$$\text{Squaring both sides}$$
$$-48 = -12y$$
$$4 = y$$

The point is $(0, 4)$.

123. Label the drawing with additional information and lettering.

Find b using the Pythagorean theorem.

$$b^2 + 10^2 = 20^2$$
$$b^2 + 100 = 400$$
$$b^2 = 300$$
$$b = 10\sqrt{3}$$
$$b \approx 17.3$$

Find a_1:

$$a_1 = 20 - b \approx 20 - 17.3 \approx 2.7 \text{ ft}$$

Find a_2:

$$a_2 = 2b + a_1 \approx 2(17.3) + 2.7 \approx 37.3 \text{ ft}$$

125.

$$x^2 + y^2 = 1$$
$$\left(\frac{\sqrt{3}}{2}\right)^2 + \left(-\frac{1}{2}\right)^2 \ ?\ 1$$
$$\frac{3}{4} + \frac{1}{4} \ \bigg| $$
$$1 \ \bigg|\ 1 \quad \text{TRUE}$$

$\left(\dfrac{\sqrt{3}}{2}, -\dfrac{1}{2}\right)$ lies on the unit circle.

127.

$$x^2 + y^2 = 1$$
$$\left(\frac{\sqrt{2}}{2}\right)^2 + \left(\frac{\sqrt{2}}{2}\right)^2 \ ?\ 1$$
$$\frac{2}{4} + \frac{2}{4} \ \bigg| $$
$$1 \ \bigg|\ 1 \quad \text{TRUE}$$

$\left(\dfrac{\sqrt{2}}{2}, \dfrac{\sqrt{2}}{2}\right)$ lies on the unit circle.

129. a), b) See the answer section in the text.

Exercise Set 1.2

1. This correspondence is a function, because each member of the domain corresponds to exactly one member of the range.

3. This correspondence is a function, because each member of the domain corresponds to exactly one member of the range.

5. This correspondence is not a function, because there is a member of the domain (m) that corresponds to more than one member of the range (A and B).

7. This correspondence is a function, because each member of the domain corresponds to exactly one member of the range.

9. This correspondence is a function, because each car has exactly one license number.

11. This correspondence is a function, because each member of the family has exactly one eye color.

13. This correspondence is not a function, because at least one student will have more than one neighboring seat occupied by another student.

15. The relation is a function, because no two ordered pairs have the same first coordinate and different second coordinates.

   The domain is the set of all first coordinates: $\{2, 3, 4\}$.

   The range is the set of all second coordinates: $\{10, 15, 20\}$.

17. The relation is not a function, because the ordered pairs $(-2, 1)$ and $(-2, 4)$ have the same first coordinate and different second coordinates.

   The domain is the set of all first coordinates: $\{-7, -2, 0\}$.

   The range is the set of all second coordinates: $\{3, 1, 4, 7\}$.

19. The relation is a function, because no two ordered pairs have the same first coordinate and different second coordinates.

   The domain is the set of all first coordinates: $\{-2, 0, 2, 4, -3\}$.

   The range is the set of all second coordinates: $\{1\}$.

21. From the graph we see that, when the input is 1, the output is -2, so $h(1) = -2$. When the input is 3, the output is 2, so $h(3) = 2$. When the input is 4, the output is 1, so $h(4) = 1$.

23. From the graph we see that, when the input is -4, the output is 3, so $s(-4) = 3$. When the input is -2, the output is 0, so $s(-2) = 0$. When the input is 0, the output is -3, so $s(0) = -3$.

25. From the graph we see that, when the input is -1, the output is 2, so $f(-1) = 2$. When the input is 0, the output is 0, so $f(0) = 0$. When the input is 1, the output is -2, so $f(1) = -2$.

27. $g(x) = 3x^2 - 2x + 1$

   a) $g(0) = 3 \cdot 0^2 - 2 \cdot 0 + 1 = 1$

   b) $g(-1) = 3(-1)^2 - 2(-1) + 1 = 6$

   c) $g(3) = 3 \cdot 3^2 - 2 \cdot 3 + 1 = 22$

   d) $g(-x) = 3(-x)^2 - 2(-x) + 1 = 3x^2 + 2x + 1$

   e) $g(1 - t) = 3(1 - t)^2 - 2(1 - t) + 1 =$
      $3(1 - 2t + t^2) - 2(1 - t) + 1 = 3 - 6t + 3t^2 - 2 + 2t + 1 =$
      $3t^2 - 4t + 2$

29. $g(x) = x^3$

   a) $g(2) = 2^3 = 8$

   b) $g(-2) = (-2)^3 = -8$

   c) $g(-x) = (-x)^3 = -x^3$

   d) $g(3y) = (3y)^3 = 27y^3$

   e) $g(2 + h) = (2 + h)^3 = 8 + 12h + 6h^2 + h^3$

31. $g(x) = \dfrac{x - 4}{x + 3}$

   a) $g(5) = \dfrac{5 - 4}{5 + 3} = \dfrac{1}{8}$

   b) $g(4) = \dfrac{4 - 4}{4 + 7} = 0$

   c) $g(-3) = \dfrac{-3 - 4}{-3 + 3} = \dfrac{-7}{0}$

   Since division by 0 is not defined, $g(-3)$ does not exist.

   d) $g(-16.25) = \dfrac{-16.25 - 4}{-16.25 + 3} = \dfrac{-20.25}{-13.25} = \dfrac{81}{53}$

   e) $g(x + h) = \dfrac{x + h - 4}{x + h + 3}$

33. $g(x) = \dfrac{x}{\sqrt{1 - x^2}}$

   $g(0) = \dfrac{0}{\sqrt{1 - 0^2}} = \dfrac{0}{\sqrt{1}} = \dfrac{0}{1} = 0$

   $g(-1) = \dfrac{-1}{\sqrt{1 - (-1)^2}} = \dfrac{-1}{\sqrt{1 - 1}} = \dfrac{-1}{\sqrt{0}} = \dfrac{-1}{0}$

   Since division by 0 is not defined, $g(-1)$ does not exist.

   $g(5) = \dfrac{5}{\sqrt{1 - 5^2}} = \dfrac{5}{\sqrt{1 - 25}} = \dfrac{5}{\sqrt{-24}}$

   Since $\sqrt{-24}$ is not defined as a real number, $g(5)$ does not exist as a real number.

   $g\left(\dfrac{1}{2}\right) = \dfrac{\dfrac{1}{2}}{\sqrt{1 - \left(\dfrac{1}{2}\right)^2}} = \dfrac{\dfrac{1}{2}}{\sqrt{1 - \dfrac{1}{4}}} = \dfrac{\dfrac{1}{2}}{\sqrt{\dfrac{3}{4}}} =$

   $\dfrac{\dfrac{1}{2}}{\dfrac{\sqrt{3}}{2}} = \dfrac{1}{2} \cdot \dfrac{2}{\sqrt{3}} = \dfrac{1 \cdot 2}{2\sqrt{3}} = \dfrac{1}{\sqrt{3}}, \text{ or } \dfrac{\sqrt{3}}{3}$

35. This is not the graph of a function, because we can find a vertical line that crosses the graph more than once.

37. This is the graph of a function, because there is no vertical line that crosses the graph more than once.

39. This is the graph of a function, because there is no vertical line that crosses the graph more than once.

41. This is not the graph of a function, because we can find a vertical line that crosses the graph more than once.

43. We can substitute any real number for x. Thus, the domain is the set of all real numbers, or $(-\infty, \infty)$.

45. The input 0 results in a denominator of 0. Thus, the domain is $\{x | x \neq 0\}$, or $(-\infty, 0) \cup (0, \infty)$.

47. We can substitute any real number in the numerator, but we must avoid inputs that make the denominator 0. We find these inputs.

$$2 - x = 0$$
$$2 = x$$

The domain is $\{x | x \neq 2\}$, or $(-\infty, 2) \cup (2, \infty)$.

49. We find the inputs that make the denominator 0:

$$x^2 - 4x - 5 = 0$$
$$(x - 5)(x + 1) = 0$$
$$x - 5 = 0 \ \ or \ \ x + 1 = 0$$
$$x = 5 \ \ or \ \ \ \ \ \ x = -1$$

The domain is $\{x | x \neq 5 \ and \ x \neq -1\}$, or $(-\infty, -1) \cup (-1, 5) \cup (5, \infty)$.

51. We can substitute any real number for which the radicand is nonnegative. We see that $8 - x \geq 0$ for $x \leq 8$, so the domain is $\{x | x \leq 8\}$, or $(-\infty, 8]$.

53. We can substitute any real number for x. Thus, the domain is the set of all real numbers, or $(-\infty, \infty)$.

55. The inputs on the x-axis that correspond to points on the graph extend from 0 to 5, inclusive. Thus, the domain is $\{x | 0 \leq x \leq 5\}$, or $[0, 5]$.

The outputs on the y-axis extend from 0 to 3, inclusive. Thus, the range is $\{y | 0 \leq y \leq 3\}$, or $[0, 3]$.

57. The inputs on the x-axis that correspond to points on the graph extend from -2π to 2π inclusive. Thus, the domain is $\{x | -2\pi \leq x \leq 2\pi\}$, or $[-2\pi, 2\pi]$.

The outputs on the y-axis extend from -1 to 1, inclusive. Thus, the range is $\{y | -1 \leq y \leq 1\}$, or $[-1, 1]$.

59. The graph extends to the left and to the right without bound. Thus, the domain is the set of all real numbers, or $(-\infty, \infty)$.

The only output is -3, so the range is $\{-3\}$.

61. The inputs on the x-axis extend from -5 to 3, inclusive. Thus, the domain is $[-5, 3]$.

The outputs on the y-axis extend from -2 to 2, inclusive. Thus, the range is $[-2, 2]$.

63.

To find the domain we look for the inputs on the x-axis that correspond to a point on the graph. We see that each point on the x-axis corresponds to a point on the graph so the domain is the set of all real numbers, or $(-\infty, \infty)$.

To find the range we look for outputs on the y-axis. The number 0 is the smallest output, and every number greater than 0 is also an output. Thus, the range is $[0, \infty)$.

65.

The inputs on the x-axis extend from -3 to 3, so the domain is $[-3, 3]$.

The outputs on the y-axis extend from 0 to 3, so the range is $[0, 3]$.

67.

Each point on the x-axis corresponds to a point on the

graph, so the domain is the set of all real numbers, or $(-\infty, \infty)$.

Each point on the y-axis also corresponds to a point on the graph, so the range is the set of all real numbers, $(-\infty, \infty)$.

69.

The largest input on the x-axis is 7 and every number less than 7 is also an input. Thus, the domain is $(-\infty, 7]$.

The number 0 is the smallest output, and every number greater than 0 is also an output. Thus, the range is $[0, \infty)$.

71.

Each point on the x-axis corresponds to a point on the graph, so the domain is the set of all real numbers, or $(-\infty, \infty)$.

The largest output is 3 and every number less than 3 is also an output. Thus, the range is $(-\infty, 3]$.

73. $E(t) = 1000(100 - t) + 580(100 - t)^2$

  a) $E(99.5) = 1000(100 - 99.5) + 580(100 - 99.5)^2$

        $= 1000(0.5) + 580(0.5)^2$

        $= 500 + 580(0.25) = 500 + 145$

        $= 645$ m above sea level

  b) $E(100) = 1000(100 - 100) + 580(100 - 100)^2$

        $= 1000 \cdot 0 + 580(0)^2 = 0 + 0$

        $= 0$ m above sea level, or at sea level

75. $T(0.5) = 0.5^{1.31} \approx 0.4$ acre

    $T(10) = 10^{1.31} \approx 20.4$ acres

    $T(20) = 20^{1.31} \approx 50.6$ acres

    $T(100) = 100^{1.31} \approx 416.9$ acres

    $T(200) = 200^{1.31} \approx 1033.6$ acres

77.

Rounding to the nearest tenth, we see that $g(-2.1) \approx -21.8$, $g(5.08) \approx -130.4$, and $g(10.003) \approx -468.3$.

79. Discussion and Writing

81. To determine whether $(0, -7)$ is a solution, substitute 0 for x and -7 for y.

$$y = 0.5x + 7$$

-7 ? $0.5(0) + 7$	
$0 + 7$	
-7 $\vert$ 7	FALSE

The equation $-7 = 7$ is false, so $(0, -7)$ is not a solution.

To determine whether $(8, 11)$ is a solution, substitute 8 for x and 11 for y.

$$y = 0.5x + 7$$

11 ? $0.5(8) + 7$	
$4 + 7$	
11 $\vert$ 11	TRUE

The equation $11 = 11$ is true, so $(8, 11)$ is a solution.

83. Graph $y = (x - 1)^2$.

Make a table of values, plot the points in the table, and draw the graph.

x	y	(x, y)
-1	4	$(-1, 4)$
0	1	$(0, 1)$
1	0	$(1, 0)$
2	1	$(2, 1)$
3	4	$(3, 4)$

85. Graph $-2x - 5y = 10$.

Make a table of values, plot the points in the table, and draw the graph.

x	y	(x, y)
-5	0	$(-5, 0)$
0	-2	$(0, -2)$
5	-4	$(5, -4)$

87. Answers may vary. Two possibilities are $f(x) = x$, $g(x) = x + 1$ and $f(x) = x^2$, $g(x) = x^2 - 4$.

89.

91. First find the value of x for which $x + 3 = -1$.

$$x + 3 = -1$$
$$x = -4$$

Then we have:

$$g(x+3) = 2x + 1$$
$$g(-1) = g(-4+3) = 2(-4) + 1 = -8 + 1 = -7$$

93. $f(x) = |x| + |x - 1|$

a) If x is in the interval $(-\infty, 0)$, then $x < 0$ and $x - 1 < 0$. We have:

$$f(x) = |x| + |x - 1|$$
$$= -x - (x - 1)$$
$$= -x - x + 1$$
$$= -2x + 1$$

b) If x is in the interval $[0, 1)$, then $x \geq 0$ and $x - 1 < 0$. We have:

$$f(x) = |x| + |x - 1|$$
$$= x - (x - 1)$$
$$= x - x + 1$$
$$= 1$$

c) If x is in the interval $[1, \infty)$, then $x > 0$ and $x - 1 \geq 0$. We have:

$$f(x) = |x| + |x - 1|$$
$$= x + x - 1$$
$$= 2x - 1$$

Exercise Set 1.3

1. a) Yes. Each input is 1 more than the one that precedes it.

b) Yes. Each output is 3 more than the one that precedes it.

c) Yes. Constant changes in inputs result in constant changes in outputs.

3. a) Yes. Each input is 15 more than the one that precedes it.

b) No. The change in the outputs varies.

c) No. Constant changes in inputs do not result in constant changes in outputs.

5. Two points on the line are $(0, 3)$ and $(5, 0)$.

$$m = \frac{y_2 - y_1}{x_2 - x_1} = \frac{0 - 3}{5 - 0} = \frac{-3}{5}, \text{ or } -\frac{3}{5}$$

7. $m = \dfrac{y_2 - y_1}{x_2 - x_1} = \dfrac{3 - 3}{3 - 0} = \dfrac{0}{3} = 0$

9. $m = \dfrac{y_2 - y_1}{x_2 - x_1} = \dfrac{2 - 4}{-1 - 9} = \dfrac{-2}{-10} = \dfrac{1}{5}$

11. $m = \dfrac{y_2 - y_1}{x_2 - x_1} = \dfrac{6 - (-9)}{-5 - 4} = \dfrac{15}{-9} = -\dfrac{5}{3}$

13. $m = \dfrac{y_2 - y_1}{x_2 - x_1} = \dfrac{-0.4 - (-0.1)}{-0.3 - 0.7} = \dfrac{-0.3}{-1} = 0.3$

15. $m = \dfrac{y_2 - y_1}{x_2 - x_1} = \dfrac{-2 - (-2)}{4 - 2} = \dfrac{0}{2} = 0$

17. $m = \dfrac{y_2 - y_1}{x_2 - x_1} = \dfrac{\dfrac{3}{5} - \left(-\dfrac{3}{5}\right)}{-\dfrac{1}{2} - \dfrac{1}{2}} = \dfrac{\dfrac{6}{5}}{-1} = -\dfrac{6}{5}$

19. $m = \dfrac{y_2 - y_1}{x_2 - x_1} = \dfrac{-5 - (-13)}{-8 - 16} = \dfrac{8}{-24} = -\dfrac{1}{3}$

21. $m = \dfrac{y_2 - y_1}{x_2 - x_1} = \dfrac{2 - (-3)}{\pi - \pi} = \dfrac{5}{0}$

Since division by 0 is not defined, the slope is not defined.

23. We have the points $(4, 3)$ and $(-2, 15)$.

$$m = \frac{y_2 - y_1}{x_2 - x_1} = \frac{15 - 3}{-2 - 4} = \frac{12}{-6} = -2$$

25. We have the points $\left(\dfrac{1}{5}, \dfrac{1}{2}\right)$ and $\left(-1, -\dfrac{11}{2}\right)$.

$$m = \frac{y_2 - y_1}{x_2 - x_1} = \frac{-\dfrac{11}{2} - \dfrac{1}{2}}{-1 - \dfrac{1}{5}} = \frac{-6}{-\dfrac{6}{5}} = -6 \cdot \left(-\dfrac{5}{6}\right) = 5$$

27. $y = -\dfrac{1}{2}x + 3$

We find and plot two ordered pairs on the line and connect the points.

When $x = 0$, $y = -\dfrac{1}{2} \cdot 0 + 3 = 3$.

When $x = 2$, $y = -\dfrac{1}{2} \cdot 2 + 3 = -1 + 3 = 2$.

Since the equation is in the form $y = mx + b$ with $m = -\dfrac{1}{2}$, we know that the slope is $-\dfrac{1}{2}$.

29. $2y - 3x = -6$

We find and plot two ordered pairs on the line and connect the points.

First let $y = 0$ and solve for x.

$$2 \cdot 0 - 3x = -6$$
$$-3x = -6$$
$$x = 2$$

The point $(2, 0)$ is on the graph.

Next let $x = 0$ and solve for y.

$$2y - 3 \cdot 0 = -6$$
$$2y = -6$$
$$y = -3$$

The point $(0, -3)$ is on the graph.

Plot the points $(2, 0)$ and $(0, -3)$ and connect them.

Now find the slope.

$$m = \frac{y_2 - y_1}{x_2 - x_1} = \frac{-3 - 0}{0 - 2} = \frac{-3}{-2} = \frac{3}{2}$$

31. $5x + 2y = 10$

We find and plot two ordered pairs on the line and connect the points.

First let $y = 0$ and solve for x.

$$5x + 2 \cdot 0 = 10$$
$$5x = 10$$
$$x = 2$$

The point $(2, 0)$ is on the graph.

Next let $x = 0$ and solve for y.

$$5 \cdot 0 + 2y = 10$$
$$2y = 10$$
$$y = 5$$

The point $(0, 5)$ is on the graph.

Plot the points $(2, 0)$ and $(0, 5)$ and connect them.

Now find the slope

$$m = \frac{5 - 0}{0 - 2} = \frac{5}{-2} = -\frac{5}{2}$$

33. $y = -\dfrac{2}{3}$

This is the equation of a horizontal line whose graph is $\dfrac{2}{3}$ units below the x-axis.

The slope of a horizontal line is 0.

35. $y = 1.3x - 5$ is in the form $y = mx + b$ with $m = 1.3$, so the slope is 1.3.

37. The graph of $x = -2$ is a vertical line, so the slope is not defined.

39. First solve the equation for y.

$$10y + x = 9$$
$$10y = -x + 9$$
$$y = -\frac{1}{10}x + \frac{9}{10}$$

The equation $y = -\dfrac{1}{10}x + \dfrac{9}{10}$ is in the form $y = mx + b$ with $m = -\dfrac{1}{10}$, so the slope is $-\dfrac{1}{10}$.

41. We have the points $(1998, 1.1)$ and $(2003, 1.3)$. We find the average rate of change, or slope.

$$m = \frac{1.3 - 1.1}{2003 - 1998} = \frac{0.2}{5} = 0.04$$

The average rate of change over the 5-yr period was 0.04 billion, or 40 million, visits per year.

43. The change in attendance was $107,097 - 59,368 = 47,729$. We divide this number by the number of races, 29.

$$\frac{47,729}{29} \approx 1646$$

The average rate of change in attendance for the 29 races was about 1646 per race.

45. We have the points $(1994, 43)$ and $(2004, 18.1)$. We find the average rate of change, or slope.

$$m = \frac{18.1 - 43}{2004 - 1994} = \frac{-24.9}{10} = -2.49$$

The average rate of change over the 10-yr period was -2.49 per 1000 women per year.

47. First we convert the times to minutes.

$$10 \text{ sec} = 10 \text{ sec} \cdot \frac{1 \text{ min}}{60 \text{ sec}} = \frac{10}{60} \cdot \frac{\text{sec}}{\text{sec}} \cdot \text{min} = \frac{1}{6} \text{ min}$$

Thus, 31 min 10 sec $= 31 \text{ min} + \dfrac{1}{6} \text{ min} = 31\dfrac{1}{6} \text{ min}$, or $\dfrac{187}{6}$ min.

1 hr = 60 min

$38 \sec = 38 \sec \cdot \dfrac{1 \text{ min}}{60 \sec} = \dfrac{38}{60} \cdot \dfrac{\sec}{\sec} \cdot \text{min} = \dfrac{19}{30} \text{ min}$

Thus, $1 \text{ hr } 1 \text{ min } 38 \sec = 60 \text{ min} + 1 \text{ min} + \dfrac{19}{30} \text{ min} =$

$61\dfrac{19}{30} \text{ min, or } \dfrac{1849}{30} \text{ min.}$

Now we find the speed.

$\dfrac{10-5}{\dfrac{1849}{30} - \dfrac{187}{6}} = \dfrac{5}{\dfrac{1849}{30} - \dfrac{935}{30}} = \dfrac{5}{\dfrac{914}{30}} = 5 \cdot \dfrac{30}{914} =$

$\dfrac{150}{914} = \dfrac{\cancel{2} \cdot 75}{\cancel{2} \cdot 457} = \dfrac{75}{457} \approx \dfrac{1}{6}$

Lucie's speed was $\dfrac{75}{457}$ mi per minute, or about $\dfrac{1}{6}$ mi per minute.

49. a) $W(h) = 4h - 130$

b) $W(62) = 4 \cdot 62 - 130 = 248 - 130 = 118 \text{ lb}$

c) Both the height and weight must be positive. Solving $h > 0$ *and* $4h - 130 > 0$, we find that the domain of the function is $\{h | h > 32.5\}$, or $(32.5, \infty)$.

51. $D(F) = 2F + 115$

a) $D(0) = 2 \cdot 0 + 115 = 115 \text{ ft}$

$D(-20) = 2(-20) + 115 = -40 + 115 = 75 \text{ ft}$

$D(10) = 2 \cdot 10 + 115 = 20 + 115 = 135 \text{ ft}$

$D(32) = 2 \cdot 32 + 115 = 64 + 115 = 179 \text{ ft}$

b) Below $-57.5°$, stopping distance is negative; above $32°$, ice doesn't form.

53. a) $D(r) = \dfrac{11r + 5}{10} = \dfrac{11}{10}r + \dfrac{5}{10}$

The slope is $\dfrac{11}{10}$.

For each mph faster the car travels, it takes $\dfrac{11}{10}$ ft longer to stop.

b) $D(5) = \dfrac{11 \cdot 5 + 5}{10} = \dfrac{60}{10} = 6 \text{ ft}$

$D(10) = \dfrac{11 \cdot 10 + 5}{10} = \dfrac{115}{10} = 11.5 \text{ ft}$

$D(20) = \dfrac{11 \cdot 20 + 5}{10} = \dfrac{225}{10} = 22.5 \text{ ft}$

$D(50) = \dfrac{11 \cdot 50 + 5}{10} = \dfrac{555}{10} = 55.5 \text{ ft}$

$D(65) = \dfrac{11 \cdot 65 + 5}{10} = \dfrac{720}{10} = 72 \text{ ft}$

c) The speed cannot be negative. $D(0) = \dfrac{1}{2}$ which says that a stopped car travels $\dfrac{1}{2}$ ft before stopping. Thus, 0 is not in the domain. The speed can be positive, so the domain is $\{r | r > 0\}$, or $(0, \infty)$.

55. $C(t) = 60 + 29t$

$C(6) = 60 + 29 \cdot 6 = \$234$

57. Let $x =$ the number of shirts produced.

$C(x) = 800 + 3x$

$C(75) = 800 + 3 \cdot 75 = \$1025$

59. Left to the student.

61. Discussion and Writing

63. $f(x) = x^2 - 3x$

$f(-5) = (-5)^2 - 3(-5) = 25 + 15 = 40$

65. $f(x) = x^2 - 3x$

$f(a+h) = (a+h)^2 - 3(a+h) = a^2 + 2ah + h^2 - 3a - 3h$

67. $m = \dfrac{y_2 - y_1}{x_2 - x_1} = \dfrac{-2d - (-d)}{9c - (-c)} = \dfrac{-2d + d}{9c + c} = \dfrac{-d}{10c} = -\dfrac{d}{10c}$

69. $m = \dfrac{y_2 - y_1}{x_2 - x_1} = \dfrac{z - z}{z - q - (z + q)} = \dfrac{0}{z - q - z - q} =$

$\dfrac{0}{-2q} = 0$

71. $m = \dfrac{y_2 - y_1}{x_2 - x_1} = \dfrac{(a+h)^2 - a^2}{a + h - a} = \dfrac{a^2 + 2ah + h^2 - a^2}{h} =$

$\dfrac{2ah + h^2}{h} = \dfrac{h(2a + h)}{h} = 2a + h$

73. False. For example, let $f(x) = x + 1$. Then $f(cd) = cd + 1$, but $f(c)f(d) = (c+1)(d+1) = cd + c + d + 1 \neq cd + 1$ for $c \neq -d$.

75. False. For example, let $f(x) = x + 1$. Then $f(c - d) = c - d + 1$, but $f(c) - f(d) = c + 1 - (d + 1) = c - d$.

77. $f(x) = mx + b$

$f(x + 2) = f(x) + 2$

$m(x + 2) + b = mx + b + 2$

$mx + 2m + b = mx + b + 2$

$2m = 2$

$m = 1$

Thus, $f(x) = 1 \cdot x + b$, or $f(x) = x + b$.

Exercise Set 1.4

1. $y = \dfrac{3}{5}x - 7$

The equation is in the form $y = mx + b$ where $m = \dfrac{3}{5}$ and $b = -7$. Thus, the slope is $\dfrac{3}{5}$, and the y-intercept is $(0, -7)$.

3. $x = -\dfrac{2}{5}$

This is the equation of a vertical line $\dfrac{2}{5}$ unit to the left of the y-axis. The slope is not defined, and there is no y-intercept.

5. $f(x) = 5 - \frac{1}{2}x$, or $f(x) = -\frac{1}{2}x + 5$

The second equation is in the form $y = mx + b$ where $m = -\frac{1}{2}$ and $b = 5$. Thus, the slope is $-\frac{1}{2}$ and the y-intercept is $(0, 5)$.

7. Solve the equation for y.
$$3x + 2y = 10$$
$$2y = -3x + 10$$
$$y = -\frac{3}{2}x + 5$$

Slope: $-\frac{3}{2}$; y-intercept: $(0, 5)$

9. $y = -6 = 0 \cdot x - 6$

Slope: 0; y-intercept: $(0, -6)$

11. Solve the equation for y.
$$5y - 4x = 8$$
$$5y = 4x + 8$$
$$y = \frac{4}{5}x + \frac{8}{5}$$

Slope: $\frac{4}{5}$; y-intercept: $\left(0, \frac{8}{5}\right)$

13. We see that the y-intercept is $(0, -2)$. Another point on the graph is $(1, 2)$. Use these points to find the slope.
$$m = \frac{y_2 - y_1}{x_2 - x_1} = \frac{2 - (-2)}{1 - 0} = \frac{4}{1} = 4$$
We have $m = 4$ and $b = -2$, so the equation is $y = 4x - 2$.

15. We see that the y-intercept is $(0, 0)$. Another point on the graph is $(3, -3)$. Use these points to find the slope.
$$m = \frac{y_2 - y_1}{x_2 - x_1} = \frac{-3 - 0}{3 - 0} = \frac{-3}{3} = -1$$
We have $m = -1$ and $b = 0$, so the equation is $y = -1 \cdot x + 0$, or $y = -x$.

17. We see that the y-intercept is $(0, -3)$. This is a horizontal line, so the slope is 0. We have $m = 0$ and $b = -3$, so the equation is $y = 0 \cdot x - 3$, or $y = -3$.

19. We substitute $\frac{2}{9}$ for m and 4 for b in the slope-intercept equation.
$$y = mx + b$$
$$y = \frac{2}{9}x + 4$$

21. We substitute -4 for m and -7 for b in the slope-intercept equation.
$$y = mx + b$$
$$y = -4x - 7$$

23. We substitute -4.2 for m and $\frac{3}{4}$ for b in the slope-intercept equation.
$$y = mx + b$$
$$y = -4.2x + \frac{3}{4}$$

25. Using the point-slope equation:
$$y - y_1 = m(x - x_1)$$
$$y - 7 = \frac{2}{9}(x - 3) \quad \text{Substituting}$$
$$y - 7 = \frac{2}{9}x - \frac{2}{3}$$
$$y = \frac{2}{9}x + \frac{19}{3} \quad \text{Slope-intercept equation}$$

Using the slope-intercept equation:

Substitute $\frac{2}{9}$ for m, 3 for x, and 7 for y in the slope-intercept equation and solve for b.
$$y = mx + b$$
$$7 = \frac{2}{9} \cdot 3 + b$$
$$7 = \frac{2}{3} + b$$
$$\frac{19}{3} = b$$

Now substitute $\frac{2}{9}$ for m and $\frac{19}{3}$ for b in $y = mx + b$.
$$y = \frac{2}{9}x + \frac{19}{3}$$

27. Using the point-slope equation:
$$y - y_1 = m(x - x_1)$$
$$y - (-2) = 3(x - 1)$$
$$y + 2 = 3x - 3$$
$$y = 3x - 5 \quad \text{Slope-intercept equation}$$

Using the slope-intercept equation:
$$y = mx + b$$
$$-2 = 3 \cdot 1 + b$$
$$-2 = 3 + b$$
$$-5 = b$$

Then we have $y = 3x - 5$.

29. Using the point-slope equation:
$$y - y_1 = m(x - x_1)$$
$$y - (-1) = -\frac{3}{5}(x - (-4))$$
$$y + 1 = -\frac{3}{5}(x + 4)$$
$$y + 1 = -\frac{3}{5}x - \frac{12}{5}$$
$$y = -\frac{3}{5}x - \frac{17}{5} \quad \text{Slope-intercept equation}$$

Using the slope-intercept equation:
$$y = mx + b$$
$$-1 = -\frac{3}{5}(-4) + b$$
$$-1 = \frac{12}{5} + b$$
$$-\frac{17}{5} = b$$

Then we have $y = -\frac{3}{5}x - \frac{17}{5}$.

31. First we find the slope.

$$m = \frac{-4 - 5}{2 - (-1)} = \frac{-9}{3} = -3$$

Using the point-slope equation:

Using the point $(-1, 5)$, we get

$$y - 5 = -3(x - (-1)), \text{ or } y - 5 = -3(x + 1).$$

Using the point $(2, -4)$, we get

$$y - (-4) = -3(x - 2), \text{ or } y + 4 = -3(x - 2).$$

In either case, the slope-intercept equation is $y = -3x + 2$.

Using the slope-intercept equation and the point $(-1, 5)$:

$$y = mx + b$$
$$5 = -3(-1) + b$$
$$5 = 3 + b$$
$$2 = b$$

Then we have $y = -3x + 2$.

33. First we find the slope.

$$m = \frac{4 - 0}{-1 - 7} = \frac{4}{-8} = -\frac{1}{2}$$

Using the point-slope equation:

Using the point $(7, 0)$, we get

$$y - 0 = -\frac{1}{2}(x - 7).$$

Using the point $(-1, 4)$, we get

$$y - 4 = -\frac{1}{2}(x - (-1)), \text{ or }$$
$$y - 4 = -\frac{1}{2}(x + 1).$$

In either case, the slope-intercept equation is $y = -\frac{1}{2}x + \frac{7}{2}$.

Using the slope-intercept equation and the point $(7, 0)$:

$$0 = -\frac{1}{2} \cdot 7 + b$$
$$\frac{7}{2} = b$$

Then we have $y = -\frac{1}{2}x + \frac{7}{2}$.

35. First we find the slope.

$$m = \frac{-4 - (-6)}{3 - 0} = \frac{2}{3}$$

We know the y-intercept is $(0, -6)$, so we substitute in the slope-intercept equation.

$$y = mx + b$$
$$y = \frac{2}{3}x - 6$$

37. Graph $y = -\frac{1}{2}x - 3$.

Plot the y-intercept, $(0, -3)$. We can think of the slope as $\frac{-1}{2}$. Start at $(0, -3)$ and find another point by moving down 1 unit and right 2 units. We have the point $(2, -4)$.

We could also think of the slope as $\frac{1}{-2}$. Then we can start at $(0, -3)$ and get another point by moving up 1 unit and left 2 units. We have the point $(-2, -2)$. Connect the three points to draw the graph.

39. Graph $f(x) = 3x - 1$.

Plot the y-intercept, $(0, -1)$. We can think of the slope as $\frac{3}{1}$. Start at $(0, -1)$ and find another point by moving up 3 units and right 1 unit. We have the point $(1, 2)$. We can move from the point $(1, 2)$ in a similar manner to get a third point, $(2, 5)$. Connect the three points to draw the graph.

41. First solve the equation for y.

$$3x - 4y = 20$$
$$-4y = -3x + 20$$
$$y = \frac{3}{4}x - 5$$

Plot the y-intercept, $(0, -5)$. Then using the slope, $\frac{3}{4}$, start at $(0, -5)$ and find another point by moving up 3 units and right 4 units. We have the point $(4, -2)$. We can move from the point $(4, -2)$ in a similar manner to get a third point, $(8, 1)$. Connect the three points to draw the graph.

43. First solve the equation for y.

$$x + 3y = 18$$
$$3y = -x + 18$$
$$y = -\frac{1}{3}x + 6$$

Plot the y-intercept, $(0, 6)$. We can think of the slope as $\frac{-1}{3}$. Start at $(0, 6)$ and find another point by moving down 1 unit and right 3 units. We have the point $(3, 5)$. We can move from the point $(3, 5)$ in a similar manner to get a third point, $(6, 4)$. Connect the three points and draw the graph.

45. We have the points $(1, 4)$ and $(-2, 13)$. First we find the slope.

$$m = \frac{13 - 4}{-2 - 1} = \frac{9}{-3} = -3$$

We will use the point-slope equation, choosing $(1, 4)$ for the given point.

$$y - 4 = -3(x - 1)$$
$$y - 4 = -3x + 3$$
$$y = -3x + 7, \text{ or}$$
$$h(x) = -3x + 7$$

47. We have the points $(5, 1)$ and $(-5, -3)$. First we find the slope.

$$m = \frac{-3 - 1}{-5 - 5} = \frac{-4}{-10} = \frac{2}{5}$$

We will use the slope-intercept equation, choosing $(5, 1)$ for the given point.

$$y = mx + b$$
$$1 = \frac{2}{5} \cdot 5 + b$$
$$1 = 2 + b$$
$$-1 = b$$

Then we have $f(x) = \frac{2}{5}x - 1$.

49. The equation of the horizontal line through $(0, -3)$ is of the form $y = b$ where b is -3. We have $y = -3$.

The equation of the vertical line through $(0, -3)$ is of the form $x = a$ where a is 0. We have $x = 0$.

51. The equation of the horizontal line through $\left(\frac{2}{11}, -1\right)$ is of the form $y = b$ where b is -1. We have $y = -1$.

The equation of the vertical line through $\left(\frac{2}{11}, -1\right)$ is of the form $x = a$ where a is $\frac{2}{11}$. We have $x = \frac{2}{11}$.

53. The slopes are $\frac{26}{3}$ and $-\frac{3}{26}$. Their product is -1, so the lines are perpendicular.

55. The slopes are $\frac{2}{5}$ and $-\frac{2}{5}$. The slopes are not the same and their product is not -1, so the lines are neither parallel nor perpendicular.

57. We solve each equation for y.

$$x + 2y = 5 \qquad\qquad 2x + 4y = 8$$
$$y = -\frac{1}{2}x + \frac{5}{2} \qquad\qquad y = -\frac{1}{2}x + 2$$

We see that $m_1 = -\frac{1}{2}$ and $m_2 = -\frac{1}{2}$. Since the slopes are the same and the y-intercepts, $\frac{5}{2}$ and 2, are different, the lines are parallel.

59. We solve each equation for y.

$$y = 4x - 5 \qquad 4y = 8 - x$$
$$y = -\frac{1}{4}x + 2$$

We see that $m_1 = 4$ and $m_2 = -\frac{1}{4}$. Since $m_1 m_2 = 4\left(-\frac{1}{4}\right) = -1$, the lines are perpendicular.

61. $y = \frac{2}{7}x + 1; \ m = \frac{2}{7}$

The line parallel to the given line will have slope $\frac{2}{7}$. We use the point-slope equation for a line with slope $\frac{2}{7}$ and containing the point $(3, 5)$:

$$y - y_1 = m(x - x_1)$$
$$y - 5 = \frac{2}{7}(x - 3)$$
$$y - 5 = \frac{2}{7}x - \frac{6}{7}$$
$$y = \frac{2}{7}x + \frac{29}{7} \quad \text{Slope-intercept form}$$

The slope of the line perpendicular to the given line is the opposite of the reciprocal of $\frac{2}{7}$, or $-\frac{7}{2}$. We use the point-slope equation for a line with slope $-\frac{7}{2}$ and containing the point $(3, 5)$:

$$y - y_1 = m(x - x_1)$$
$$y - 5 = -\frac{7}{2}(x - 3)$$
$$y - 5 = -\frac{7}{2}x + \frac{21}{2}$$
$$y = -\frac{7}{2}x + \frac{31}{2} \quad \text{Slope-intercept form}$$

63. $y = -0.3x + 4.3; \ m = -0.3$

The line parallel to the given line will have slope -0.3. We use the point-slope equation for a line with slope -0.3 and containing the point $(-7, 0)$:

$$y - y_1 = m(x - x_1)$$
$$y - 0 = -0.3(x - (-7))$$
$$y = -0.3x - 2.1 \quad \text{Slope-intercept form}$$

The slope of the line perpendicular to the given line is the opposite of the reciprocal of -0.3, or $\dfrac{1}{0.3} = \dfrac{10}{3}$.

We use the point-slope equation for a line with slope $\dfrac{10}{3}$ and containing the point $(-7, 0)$:

$$y - y_1 = m(x - x_1)$$
$$y - 0 = \frac{10}{3}(x - (-7))$$
$$y = \frac{10}{3}x + \frac{70}{3} \quad \text{Slope-intercept form}$$

65. $3x + 4y = 5$
$$4y = -3x + 5$$
$$y = -\frac{3}{4}x + \frac{5}{4}; \quad m = -\frac{3}{4}$$

The line parallel to the given line will have slope $-\dfrac{3}{4}$. We use the point-slope equation for a line with slope $-\dfrac{3}{4}$ and containing the point $(3, -2)$:

$$y - y_1 = m(x - x_1)$$
$$y - (-2) = -\frac{3}{4}(x - 3)$$
$$y + 2 = -\frac{3}{4}x + \frac{9}{4}$$
$$y = -\frac{3}{4}x + \frac{1}{4} \quad \text{Slope-intercept form}$$

The slope of the line perpendicular to the given line is the opposite of the reciprocal of $-\dfrac{3}{4}$, or $\dfrac{4}{3}$. We use the point-slope equation for a line with slope $\dfrac{4}{3}$ and containing the point $(3, -2)$:

$$y - y_1 = m(x - x_1)$$
$$y - (-2) = \frac{4}{3}(x - 3)$$
$$y + 2 = \frac{4}{3}x - 4$$
$$y = \frac{4}{3}x - 6 \quad \text{Slope-intercept form}$$

67. $x = -1$ is the equation of a vertical line. The line parallel to the given line is a vertical line containing the point $(3, -3)$, or $x = 3$.

The line perpendicular to the given line is a horizontal line containing the point $(3, -3)$, or $y = -3$.

69. $x = -3$ is a vertical line and $y = 5$ is a horizontal line, so it is true that the lines are perpendicular.

71. The lines have the same slope, $\dfrac{2}{5}$, and different y-intercepts, $(0, 4)$ and $(0, -4)$, so it is true that the lines are parallel.

73. $x = -1$ and $x = 1$ are both vertical lines, so it is false that they are perpendicular.

75. a) Answers will vary depending on the data points used. We will use $(0, 96.7)$ and $(6, 121.2)$.
$$m = \frac{121.2 - 96.7}{6 - 0} = \frac{24.5}{6} \approx 4.08$$
We know the y-intercept is $(0, 96.7)$, so we use the slope-intercept equation.
$$y = mx + b$$
$$y = 4.08x - 96.7$$
Since we have
$$\frac{24.5}{6} = \frac{245}{60} = \frac{\cancel{5} \cdot 49}{\cancel{5} \cdot 12} = \frac{49}{12} \text{ and } 96.7 = \frac{967}{10},$$
we can also write the equation as $y = \dfrac{49}{12}x + \dfrac{967}{10}$.

b) In 2007, $x = 2007 - 1995 = 12$.
$$y = 4.08(12) + 96.7 = 145.66$$
The number of twin births in 2007 will be about 145.66 thousand, or 145,660.

In 2010, $x = 2010 - 1995 = 15$.
$$y = 4.08(15) + 96.7 = 157.9$$
The number of twin births in 2010 will be about 157.9 thousand, or 157,900.

77. Answers will vary depending on the data points used. We will use $(1, 47.37)$ and $(4, 50.64)$.
$$m = \frac{50.64 - 47.37}{4 - 1} = \frac{3.27}{3} = 1.09$$
Using the point-slope equation and the point $(1, 47.37)$:
$$y - 47.37 = 1.09(x - 1)$$
$$y - 47.37 = 1.09x - 1.09$$
$$y = 1.09x + 46.28$$
In 2007, $x = 2007 - 2000 = 7$.
$$y = 1.09(7) + 46.28 = \$53.91$$
In 2009, $x = 2009 - 2000 = 9$
$$y = 1.09(9) + 46.28 = \$56.09$$

79. Answers will vary depending on the data points used. We will use $(10, 321.10)$ and $(30, 844.60)$.
$$m = \frac{844.60 - 321.10}{30 - 10} = \frac{523.5}{20} = 26.175$$
Using the slope-intercept equation and the point $(10, 321.10)$:
$$y = mx + b$$
$$321.10 = 26.175(10) + b$$
$$321.10 = 261.75 + b$$
$$59.35 = b$$
The equation is $y = 26.175x + 59.35$.

In 2005, $x = 2005 - 1970 = 35$.
$$y = 26.175(35) + 59.35 \approx \$975.48$$
In 2010, $x = 2010 - 1970 = 40$.
$$y = 26.175(40) + 59.35 = \$1106.35$$
In 2020, $x = 2020 - 1970 = 50$.
$$y = 26.175(50) + 59.35 = \$1368.10$$

81. a) Using the linear regression feature on a graphing calculator, we get $M = 0.2H + 156$.

b) For $H = 40$: $M = 0.2(40) + 156 = 164$ beats per minute

For $H = 65$: $M = 0.2(65) + 156 = 169$ beats per minute

For $H = 76$: $M = 0.2(76) + 156 \approx 171$ beats per minute

For $H = 84$: $M = 0.2(84) + 156 \approx 173$ beats per minute

c) $r = 1$; all the data points are on the regression line so it should be a good predictor.

83. a) Using the linear regression feature on a graphing calculator, we get $y = 0.02x + 1.77$, where x is the number of years after 1999 and y is in billions of dollars.

b) In 2005, $x = 2005 - 1999 = 6$, and $y = 0.02(6) + 1.77 = \$1.89$ billion.

In 2010, $x = 2010 - 1999 = 11$, and $y = 0.02(11) + 1.77 = \$1.99$ billion.

c) $r \approx 0.3162$; since there is a low correlation, the regression line does not fit the data closely.

85. a) Using the linear regression feature on a graphing calculator, we get $y = 4.055x + 96.78$, where x is the number of years after 1995 and y is in thousands.

b) In 2010, $x = 2010 - 1995 = 15$.

$y = 4.055(15) + 96.78 = 157.605$ thousand, or 157,605

This value is only 295 less than the value found in Exercise 75.

c) $r \approx 0.9980$; the line fits the data well.

87. Discussion and Writing

89. $m = \dfrac{y_2 - y_1}{x_2 - x_1}$

$= \dfrac{-1 - (-8)}{-5 - 2} = \dfrac{-1 + 8}{-7}$

$= \dfrac{7}{-7} = -1$

91. $(x - h)^2 + (y - k)^2 = r^2$

$[x - (-7)]^2 + [y - (-1)]^2 = \left(\dfrac{9}{5}\right)^2$

$(x + 7)^2 + (y + 1)^2 = \dfrac{81}{25}$

93. The slope of the line containing $(-3, k)$ and $(4, 8)$ is

$\dfrac{8 - k}{4 - (-3)} = \dfrac{8 - k}{7}$.

The slope of the line containing $(5, 3)$ and $(1, -6)$ is

$\dfrac{-6 - 3}{1 - 5} = \dfrac{-9}{-4} = \dfrac{9}{4}$.

The slopes must be equal in order for the lines to be parallel:

$\dfrac{8 - k}{7} = \dfrac{9}{4}$

$32 - 4k = 63$ Multiplying by 28

$-4k = 31$

$k = -\dfrac{31}{4}$, or -7.75

Exercise Set 1.5

1. a) For x-values from -5 to 1, the y-values increase from -3 to 3. Thus the function is increasing on the interval $(-5, 1)$.

b) For x-values from 3 to 5, the y-values decrease from 3 to 1. Thus the function is decreasing on the interval $(3, 5)$.

c) For x-values from 1 to 3, y is 3. Thus the function is constant on $(1, 3)$.

3. a) For x-values from -3 to -1, the y-values increase from -4 to 4. Also, for x-values from 3 to 5, the y-values increase from 2 to 6. Thus the function is increasing on $(-3, -1)$ and on $(3, 5)$.

b) For x-values from 1 to 3, the y-values decrease from 3 to 2. Thus the function is decreasing on the interval $(1, 3)$.

c) For x-values from -5 to -3, y is 1. Thus the function is constant on $(-5, -3)$.

5. a) For x-values from $-\infty$ to -8, the y-values increase from $-\infty$ to 2. Also, for x-values from -3 to -2, the y-values increase from -2 to 3. Thus the function is increasing on $(-\infty, -8)$ and on $(-3, -2)$.

b) For x-values from -8 to -6, the y-values decrease from 2 to -2. Thus the function is decreasing on the interval $(-8, -6)$.

c) For x-values from -6 to -3, y is -2. Also, for x-values from -2 to ∞, y is 3. Thus the function is constant on $(-6, -3)$ and on $(-2, \infty)$.

7. The x-values extend from -5 to 5, so the domain is $[-5, 5]$.

The y-values extend from -3 to 3, so the range is $[-3, 3]$.

9. The x-values extend from -5 to -1 and from 1 to 5, so the domain is $[-5, -1] \cup [1, 5]$.

The y-values extend from -4 to 6, so the range is $[-4, 6]$.

11. The x-values extend from $-\infty$ to ∞, so the domain is $(-\infty, \infty)$.

The y-values extend from $-\infty$ to 3, so the range is $(-\infty, 3]$.

13. From the graph we see that a relative maximum value of the function is 3.25. It occurs at $x = 2.5$. There is no relative minimum value.

The graph starts rising, or increasing, from the left and stops increasing at the relative maximum. From this point, the graph decreases. Thus the function is increasing on $(-\infty, 2.5)$ and is decreasing on $(2.5, \infty)$.

15. From the graph we see that a relative maximum value of the function is 2.370. It occurs at $x = -0.667$. We also see that a relative minimum value of 0 occurs at $x = 2$.

The graph starts rising, or increasing, from the left and stops increasing at the relative maximum. From this point it decreases to the relative minimum and then increases again. Thus the function is increasing on $(-\infty, -0.667)$ and on $(2, \infty)$. It is decreasing on $(-0.667, 2)$.

17.

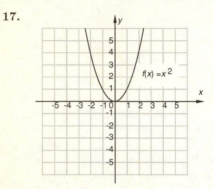

The function is increasing on $(0, \infty)$ and decreasing on $(-\infty, 0)$. We estimate that the minimum is 0 at $x = 0$. There are no maxima.

19.

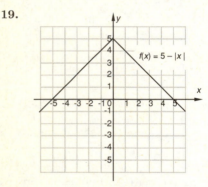

The function is increasing on $(-\infty, 0)$ and decreasing on $(0, \infty)$. We estimate that the maximum is 5 at $x = 0$. There are no minima.

21.

The function is decreasing on $(-\infty, 3)$ and increasing on $(3, \infty)$. We estimate that the minimum is 1 at $x = 3$. There are no maxima.

23. If $x =$ the length of the rectangle, in meters, then the width is $\dfrac{60 - 2x}{2}$, or $30 - x$. We use the formula Area = length × width:

$$A(x) = x(30 - x)$$
$$A(x) = 30x - x^2$$

25. After t minutes, the balloon has risen $120t$ ft. We use the Pythagorean theorem.

$$[d(t)]^2 = (120t)^2 + (400)^2$$
$$d(t) = \sqrt{(120t)^2 + (400)^2}$$

We only considered the positive square root since distance must be nonnegative.

27. Let $w =$ the width of the rectangle. Then the length $= \dfrac{40 - 2w}{2}$, or $20 - w$. Divide the rectangle into quadrants as shown below.

In each quadrant there are two congruent triangles. One triangle is part of the rhombus and both are part of the rectangle. Thus, in each quadrant the area of the rhombus is one-half the area of the rectangle. Then, in total, the area of the rhombus is one-half the area of the rectangle.

$$A(w) = \frac{1}{2}(20 - w)(w)$$

$$A(w) = 10w - \frac{w^2}{2}$$

29. We will use similar triangles, expressing all distances in feet. $\left(6 \text{ in.} = \dfrac{1}{2} \text{ ft, } s \text{ in.} = \dfrac{s}{12} \text{ ft, and } d \text{ yd} = 3d \text{ ft}\right)$ We have

$$\frac{3d}{7} = \frac{\frac{1}{2}}{\frac{s}{12}}$$

$$\frac{s}{12} \cdot 3d = 7 \cdot \frac{1}{2}$$

$$\frac{sd}{4} = \frac{7}{2}$$

$$d = \frac{4}{s} \cdot \frac{7}{2}, \text{ so}$$

$$d(s) = \frac{14}{s}.$$

31. a) If the length $= x$ feet, then the width $= 30 - x$ feet.

$$A(x) = x(30 - x)$$
$$A(x) = 30x - x^2$$

b) The length of the rectangle must be positive and less than 30 ft, so the domain of the function is $\{x | 0 < x < 30\}$, or $(0, 30)$.

c) We see from the graph that the maximum value of the area function on the interval $(0, 30)$ appears to be 225 when $x = 15$. Then the dimensions that yield the maximum area are length = 15 ft and width $= 30 - 15$, or 15 ft.

33. a) When a square with sides of length x are cut from each corner, the length of each of the remaining sides of the piece of cardboard is $12 - 2x$. Then the dimensions of the box are x by $12 - 2x$ by $12 - 2x$. We use the formula Volume = length × width × height to find the volume of the box:

$$V(x) = (12 - 2x)(12 - 2x)(x)$$
$$V(x) = (144 - 48x + 4x^2)(x)$$
$$V(x) = 144x - 48x^2 + 4x^3$$

This can also be expressed as $V(x) = 4x(x - 6)^2$, or $V(x) = 4x(6 - x)^2$.

b) The length of the sides of the square corners that are cut out must be positive and less than half the length of a side of the piece of cardboard. Thus, the domain of the function is $\{x | 0 < x < 6\}$, or $(0, 6)$.

c) We see from the graph that the maximum value of the area function on the interval $(0, 6)$ appears to be 128 when $x = 2$. When $x = 2$, then $12 - 2x = 12 - 2 \cdot 2 = 8$, so the dimensions that yield the maximum volume are 8 cm by 8 cm by 2 cm.

35. $g(x) = \begin{cases} x + 4, & \text{for } x \le 1, \\ 8 - x, & \text{for } x > 1 \end{cases}$

Since $-4 \le 1$, $g(-4) = -4 + 4 = 0$.

Since $0 \le 1$, $g(0) = 0 + 4 = 4$.

Since $1 \le 1$, $g(1) = 1 + 4 = 5$.

Since $3 > 1$, $g(3) = 8 - 3 = 5$.

37. $h(x) = \begin{cases} -3x - 18, & \text{for } x < -5, \\ 1, & \text{for } -5 \le x < 1, \\ x + 2, & \text{for } x \ge 1 \end{cases}$

Since -5 is in the interval $[-5, 1)$, $h(-5) = 1$.

Since 0 is in the interval $[-5, 1)$, $h(0) = 1$.

Since $1 \ge 1$, $h(1) = 1 + 2 = 3$.

Since $4 \ge 1$, $h(4) = 4 + 2 = 6$.

39. $f(x) = \begin{cases} \dfrac{1}{2}x, & \text{for } x < 0, \\ x + 3, & \text{for } x \ge 0 \end{cases}$

We create the graph in two parts. Graph $f(x) = \frac{1}{2}x$ for inputs x less than 0. Then graph $f(x) = x + 3$ for inputs x greater than or equal to 0.

41. $f(x) = \begin{cases} -\dfrac{3}{4}x + 2, & \text{for } x < 4, \\ -1 + x, & \text{for } x \ge 4 \end{cases}$

We create the graph in two parts. Graph $f(x) = -\frac{3}{4}x + 2$ for inputs x less than 4. Then graph $f(x) = -1 + x$ for inputs x greater than or equal to 4.

43. $f(x) = \begin{cases} x + 1, & \text{for } x \le -3, \\ -1, & \text{for } -3 < x < 4 \\ \dfrac{1}{2}x, & \text{for } x \ge 4 \end{cases}$

We create the graph in three parts. Graph $f(x) = x + 1$ for inputs x less than or equal to -3. Graph $f(x) = -1$ for inputs greater than -3 and less than 4. Then graph $f(x) = \frac{1}{2}x$ for inputs greater than or equal to 4.

45. $g(x) = \begin{cases} \dfrac{1}{2}x - 1, & \text{for } x < 0, \\ 3, & \text{for } 0 \le x \le 1 \\ -2x, & \text{for } x > 1 \end{cases}$

We create the graph in three parts. Graph $g(x) = \frac{1}{2}x - 1$ for inputs less than 0. Graph $g(x) = 3$ for inputs greater than or equal to 0 and less than or equal to 1. Then graph $g(x) = -2x$ for inputs greater than 1.

47. $f(x) = \begin{cases} 2, & \text{for } x = 5, \\ \dfrac{x^2 - 25}{x - 5}, & \text{for } x \neq 5 \end{cases}$

When $x \neq 5$, the denominator of $(x^2 - 25)/(x - 5)$ is nonzero so we can simplify:

$$\frac{x^2 - 25}{x - 5} = \frac{(x + 5)(x - 5)}{x - 5} = x + 5.$$

Thus, $f(x) = x + 5$, for $x \neq 5$.

The graph of this part of the function consists of a line with a "hole" at the point $(5, 10)$, indicated by an open dot. At $x = 5$, we have $f(5) = 2$, so the point $(5, 2)$ is plotted below the open dot.

49. $f(x) = \text{int}(x)$

See Example 8.

51. $f(x) = 1 + \text{int}(x)$

This function can be defined by a piecewise function with an infinite number of statements:

$$f(x) = \begin{cases} \vdots \\ -1, & \text{for } -2 \leq x < -1, \\ 0, & \text{for } -1 \leq x < 0, \\ 1, & \text{for } 0 \leq x < 1, \\ 2, & \text{for } 1 \leq x < 2, \\ \vdots \end{cases}$$

$g(x) = 1 + [\![x]\!]$

53. From the graph we see that the domain is $(-\infty, \infty)$ and the range is $(-\infty, 0) \cup [3, \infty)$.

55. From the graph we see that the domain is $(-\infty, \infty)$ and the range is $[-1, \infty)$.

57. From the graph we see that the domain is $(-\infty, \infty)$ and the range is $(-\infty, -2] \cup \{-1\} \cup [2, \infty)$, or $\{y | y \leq -2 \text{ or } y = -1 \text{ or } y \geq 2\}$.

59. From the graph we see that the domain is $(-\infty, \infty)$ and the range is $\{-5, -2, 4\}$. An equation for the function is:

$$f(x) = \begin{cases} -2, & \text{for } x < 2, \\ -5, & \text{for } x = 2, \\ 4, & \text{for } x > 2 \end{cases}$$

61. From the graph we see that the domain is $(-\infty, \infty)$ and the range is $(-\infty, -1] \cup [2, \infty)$. Finding the slope of each segment and using the slope-intercept or point-slope formula, we find that an equation for the function is:

$$g(x) = \begin{cases} x, & \text{for } x \leq -1, \\ 2, & \text{for } -1 < x \leq 2, \\ x, & \text{for } x > 2 \end{cases}$$

This can also be expressed as follows:

$$g(x) = \begin{cases} x, & \text{for } x \leq -1, \\ 2, & \text{for } -1 < x < 2, \\ x, & \text{for } x \geq 2 \end{cases}$$

63. From the graph we see that the domain is $[-5, 3]$ and the range is $(-3, 5)$. Finding the slope of each segment and using the slope-intercept or point-slope formula, we find that an equation for the function is:

$$h(x) = \begin{cases} x + 8, & \text{for } -5 \leq x < -3, \\ 3, & \text{for } -3 \leq x \leq 1, \\ 3x - 6, & \text{for } 1 < x \leq 3 \end{cases}$$

65.

Beginning at the left side of the window, the graph first drops as we move to the right. We see that the function is decreasing on $(-\infty, 1)$. We then find that the function is increasing on $(1, 3)$ and decreasing again on $(3, \infty)$. The MAXIMUM and MINIMUM features also show that the relative maximum is -4 at $x = 3$ and the relative minimum is -8 at $x = 1$.

67.

We find that the function is increasing on $(-1.552, 0)$ and on $(1.552, \infty)$ and decreasing on $(-\infty, -1.552)$ and on

$(0, 1.552)$. The relative maximum is 4.07 at $x = 0$ and the relative minima are -2.314 at $x = -1.552$ and -2.314 at $x = 1.552$.

69. a)

$y = -0.1x^2 + 1.2x + 98.6$

b) Using the MAXIMUM feature we find that the relative maximum is 102.2 at $t = 6$. Thus, we know that the patient's temperature was the highest at $t = 6$, or 6 days after the onset of the illness and that the highest temperature was 102.2°F.

71. Graph $y = \dfrac{8x}{x^2 + 1}$.

Increasing: $(-1, 1)$

Decreasing: $(-\infty, -1), (1, \infty)$

73. Graph $y = x\sqrt{4 - x^2}$, for $-2 \le x \le 2$.

Increasing: $(-1.414, 1.414)$

Decreasing: $(-2, -1.414), (1.414, 2)$

75. a) The length of a diameter of the circle (and a diagonal of the rectangle) is $2 \cdot 8$, or 16 ft. Let $l =$ the length of the rectangle. Use the Pythagorean theorem to write l as a function of x.

$$x^2 + l^2 = 16^2$$
$$x^2 + l^2 = 256$$
$$l^2 = 256 - x^2$$
$$l = \sqrt{256 - x^2}$$

Since the length must be positive, we considered only the positive square root.

Use the formula Area = length × width to find the area of the rectangle:

$$A(x) = x\sqrt{256 - x^2}$$

b) The width of the rectangle must be positive and less than the diameter of the circle. Thus, the domain of the function is $\{x | 0 < x < 16\}$, or $(0, 16)$.

c)

$y = x\sqrt{256 - x^2}$

d) Using the MAXIMUM feature, we find that the maximum area occurs when x is about 11.314. When $x \approx 11.314$, $\sqrt{256 - x^2} \approx \sqrt{256 - (11.314)^2} \approx 11.313$. Thus, the dimensions that maximize the area are about 11.314 ft by 11.313 ft. (Answers may vary slightly due to rounding differences.)

77. Discussion and Writing

79. Function; domain; range; domain; exactly one; range

81. x-intercept

83. If $[\text{int}(x)]^2 = 25$, then $\text{int}(x) = -5$ or $\text{int}(x) = 5$. For $-5 \le x < -4$, $\text{int}(x) = -5$. For $5 \le x < 6$, $\text{int}(x) = 5$. Thus, the possible inputs for x are $\{x | -5 \le x < -4 \text{ or } 5 \le x < 6\}$.

85. a) We add labels to the drawing in the text.

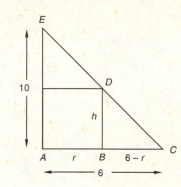

We write a proportion involving the lengths of the sides of the similar triangles BCD and ACE. Then we solve it for h.

$$\frac{h}{6 - r} = \frac{10}{6}$$
$$h = \frac{10}{6}(6 - r) = \frac{5}{3}(6 - r)$$
$$h = \frac{30 - 5r}{3}$$

Thus, $h(r) = \dfrac{30 - 5r}{3}$.

b)
$$V = \pi r^2 h$$
$$V(r) = \pi r^2 \left(\frac{30 - 5r}{3}\right) \quad \text{Substituting for } h$$

c) We first express r in terms of h.

$$h = \frac{30 - 5r}{3}$$
$$3h = 30 - 5r$$
$$5r = 30 - 3h$$
$$r = \frac{30 - 3h}{5}$$
$$V = \pi r^2 h$$
$$V(h) = \pi \left(\frac{30 - 3h}{5}\right)^2 h$$

Substituting for r

We can also write $V(h) = \pi h \left(\dfrac{30 - 3h}{5}\right)^2$.

Exercise Set 1.6

1. $(f + g)(5) = f(5) + g(5)$
$$= (5^2 - 3) + (2 \cdot 5 + 1)$$
$$= 25 - 3 + 10 + 1$$
$$= 33$$

3. $(f - g)(-1) = f(-1) - g(-1)$
$$= ((-1)^2 - 3) - (2(-1) + 1)$$
$$= -2 - (-1) = -2 + 1$$
$$= -1$$

5. $(f/g)\left(-\dfrac{1}{2}\right) = \dfrac{f\left(-\dfrac{1}{2}\right)}{g\left(-\dfrac{1}{2}\right)}$

$$= \dfrac{\left(-\dfrac{1}{2}\right)^2 - 3}{2\left(-\dfrac{1}{2}\right) + 1}$$

$$= \dfrac{\dfrac{1}{4} - 3}{-1 + 1}$$

$$= \dfrac{-\dfrac{11}{4}}{0}$$

Since division by 0 is not defined, $(f/g)\left(-\dfrac{1}{2}\right)$ does not exist.

7. $(fg)\left(-\dfrac{1}{2}\right) = f\left(-\dfrac{1}{2}\right) \cdot g\left(-\dfrac{1}{2}\right)$

$$= \left[\left(-\dfrac{1}{2}\right)^2 - 3\right]\left[2\left(-\dfrac{1}{2}\right) + 1\right]$$

$$= -\dfrac{11}{4} \cdot 0 = 0$$

9. $(g - f)(-1) = g(-1) - f(-1)$
$$= [2(-1) + 1] - [(-1)^2 - 3]$$
$$= (-2 + 1) - (1 - 3)$$
$$= -1 - (-2)$$
$$= -1 + 2$$
$$= 1$$

11. $(h - g)(-4) = h(-4) - g(-4)$
$$= (-4 + 4) - \sqrt{-4 - 1}$$
$$= 0 - \sqrt{-5}$$

Since $\sqrt{-5}$ is not a real number, $(h-g)(-4)$ does not exist.

13. $(g/h)(1) = \dfrac{g(1)}{h(1)}$

$$= \dfrac{\sqrt{1 - 1}}{1 + 4}$$

$$= \dfrac{\sqrt{0}}{5}$$

$$= \dfrac{0}{5} = 0$$

15. $(g + h)(1) = g(1) + h(1)$
$$= \sqrt{1 - 1} + (1 + 4)$$
$$= \sqrt{0} + 5$$
$$= 0 + 5 = 5$$

17. $f(x) = 2x + 3$, $g(x) = 3 - 5x$

a) The domain of f and of g is the set of all real numbers, or $(-\infty, \infty)$. Then the domain of $f + g$, $f - g$, ff, and fg is also $(-\infty, \infty)$. For f/g we must exclude $\dfrac{3}{5}$ since $g\left(\dfrac{3}{5}\right) = 0$. Then the domain of f/g is $\left(-\infty, \dfrac{3}{5}\right) \cup \left(\dfrac{3}{5}, \infty\right)$. For g/f we must exclude $-\dfrac{3}{2}$ since $f\left(-\dfrac{3}{2}\right) = 0$. The domain of g/f is $\left(-\infty, -\dfrac{3}{2}\right) \cup \left(-\dfrac{3}{2}, \infty\right)$.

b) $(f + g)(x) = f(x) + g(x) = (2x + 3) + (3 - 5x) = -3x + 6$

$(f - g)(x) = f(x) - g(x) = (2x + 3) - (3 - 5x) = 2x + 3 - 3 + 5x = 7x$

$(fg)(x) = f(x) \cdot g(x) = (2x + 3)(3 - 5x) = 6x - 10x^2 + 9 - 15x = -10x^2 - 9x + 9$

$(ff)(x) = f(x) \cdot f(x) = (2x + 3)(2x + 3) = 4x^2 + 12x + 9$

$(f/g)(x) = \dfrac{f(x)}{g(x)} = \dfrac{2x + 3}{3 - 5x}$

$(g/f)(x) = \dfrac{g(x)}{f(x)} = \dfrac{3 - 5x}{2x + 3}$

19. $f(x) = x - 3$, $g(x) = \sqrt{x + 4}$

a) Any number can be an input in f, so the domain of f is the set of all real numbers, or $(-\infty, \infty)$.

The inputs of g must be nonnegative, so we have $x + 4 \geq 0$, or $x \geq -4$. Thus, the domain of g is $[-4, \infty)$.

The domain of $f + g$, $f - g$, and fg is the set of all numbers in the domains of both f and g. This is $[-4, \infty)$.

The domain of ff is the domain of f, or $(-\infty, \infty)$.

The domain of f/g is the set of all numbers in the domains of f and g, excluding those for which $g(x) = 0$. Since $g(-4) = 0$, the domain of f/g is $(-4, \infty)$.

The domain of g/f is the set of all numbers in the domains of g and f, excluding those for which $f(x) = 0$. Since $f(3) = 0$, the domain of g/f is $[-4, 3) \cup (3, \infty)$.

b) $(f + g)(x) = f(x) + g(x) = x - 3 + \sqrt{x + 4}$

$(f - g)(x) = f(x) - g(x) = x - 3 - \sqrt{x + 4}$

$(fg)(x) = f(x) \cdot g(x) = (x - 3)\sqrt{x + 4}$

$(ff)(x) = \left[f(x)\right]^2 = (x - 3)^2 = x^2 - 6x + 9$

$(f/g)(x) = \dfrac{f(x)}{g(x)} = \dfrac{x - 3}{\sqrt{x + 4}}$

$(g/f)(x) = \dfrac{g(x)}{f(x)} = \dfrac{\sqrt{x + 4}}{x - 3}$

21. $f(x) = 2x - 1$, $g(x) = -2x^2$

a) The domain of f and of g is $(-\infty, \infty)$. Then the domain of $f + g$, $f - g$, fg, and ff is $(-\infty, \infty)$. For f/g, we must exclude 0 since $g(0) = 0$. The domain of f/g is $(-\infty, 0) \cup (0, \infty)$. For g/f, we must exclude $\dfrac{1}{2}$ since $f\left(\dfrac{1}{2}\right) = 0$. The domain of g/f is $\left(-\infty, \dfrac{1}{2}\right) \cup \left(\dfrac{1}{2}, \infty\right)$.

b) $(f + g)(x) = f(x) + g(x) = (2x - 1) + (-2x^2) = -2x^2 + 2x - 1$

$(f - g)(x) = f(x) - g(x) = (2x - 1) - (-2x^2) = 2x^2 + 2x - 1$

$(fg)(x) = f(x) \cdot g(x) = (2x - 1)(-2x^2) = -4x^3 + 2x^2$

$(ff)(x) = f(x) \cdot f(x) = (2x - 1)(2x - 1) = 4x^2 - 4x + 1$

$(f/g)(x) = \dfrac{f(x)}{g(x)} = \dfrac{2x - 1}{-2x^2}$

$(g/f)(x) = \dfrac{g(x)}{f(x)} = \dfrac{-2x^2}{2x - 1}$

23. $f(x) = \sqrt{x - 3}$, $g(x) = \sqrt{x + 3}$

a) Since $f(x)$ is nonnegative for values of x in $[3, \infty)$, this is the domain of f. Since $g(x)$ is nonnegative for values of x in $[-3, \infty)$, this is the domain of g. The domain of $f + g$, $f - g$, and fg is the intersection of the domains of f and g, or $[3, \infty)$. The domain of ff is the same as the domain of f, or $[3, \infty)$. For f/g, we must exclude -3 since $g(-3) = 0$. This is not in $[3, \infty)$, so the domain of f/g is $[3, \infty)$. For g/f, we must exclude 3 since $f(3) = 0$. The domain of g/f is $(3, \infty)$.

b) $(f + g)(x) = f(x) + g(x) = \sqrt{x - 3} + \sqrt{x + 3}$

$(f - g)(x) = f(x) - g(x) = \sqrt{x - 3} - \sqrt{x + 3}$

$(fg)(x) = f(x) \cdot g(x) = \sqrt{x - 3} \cdot \sqrt{x + 3} = \sqrt{x^2 - 9}$

$(ff)(x) = f(x) \cdot f(x) = \sqrt{x - 3} \cdot \sqrt{x - 3} = |x - 3|$

$(f/g)(x) = \dfrac{\sqrt{x - 3}}{\sqrt{x + 3}}$

$(g/f)(x) = \dfrac{\sqrt{x + 3}}{\sqrt{x - 3}}$

25. $f(x) = x + 1$, $g(x) = |x|$

a) The domain of f and of g is $(-\infty, \infty)$. Then the domain of $f + g$, $f - g$, fg, and ff is $(-\infty, \infty)$. For f/g, we must exclude 0 since $g(0) = 0$. The domain of f/g is $(-\infty, 0) \cup (0, \infty)$. For g/f, we must exclude -1 since $f(-1) = 0$. The domain of g/f is $(-\infty, -1) \cup (-1, \infty)$.

b) $(f + g)(x) = f(x) + g(x) = x + 1 + |x|$

$(f - g)(x) = f(x) - g(x) = x + 1 - |x|$

$(fg)(x) = f(x) \cdot g(x) = (x + 1)|x|$

$(ff)(x) = f(x) \cdot f(x) = (x + 1)(x + 1) = x^2 + 2x + 1$

$(f/g)(x) = \dfrac{x + 1}{|x|}$

$(g/f)(x) = \dfrac{|x|}{x + 1}$

27. $f(x) = x^3$, $g(x) = 2x^2 + 5x - 3$

a) Since any number can be an input for either f or g, the domain of f, g, $f + g$, $f - g$, fg, and ff is the set of all real numbers, or $(-\infty, \infty)$.

Since $g(-3) = 0$ and $g\left(\dfrac{1}{2}\right) = 0$, the domain of f/g is $(-\infty, -3) \cup \left(-3, \dfrac{1}{2}\right) \cup \left(\dfrac{1}{2}, \infty\right)$.

Since $f(0) = 0$, the domain of g/f is $(-\infty, 0) \cup (0, \infty)$.

b) $(f + g)(x) = f(x) + g(x) = x^3 + 2x^2 + 5x - 3$

$(f - g)(x) = f(x) - g(x) = x^3 - (2x^2 + 5x - 3) = x^3 - 2x^2 - 5x + 3$

$(fg)(x) = f(x) \cdot g(x) = x^3(2x^2 + 5x - 3) = 2x^5 + 5x^4 - 3x^3$

$(ff)(x) = f(x) \cdot f(x) = x^3 \cdot x^3 = x^6$

$(f/g)(x) = \dfrac{f(x)}{g(x)} = \dfrac{x^3}{2x^2 + 5x - 3}$

$(g/f)(x) = \dfrac{g(x)}{f(x)} = \dfrac{2x^2 + 5x - 3}{x^3}$

29. $f(x) = \dfrac{4}{x + 1}$, $g(x) = \dfrac{1}{6 - x}$

a) Since $x + 1 = 0$ when $x = -1$, we must exclude -1 from the domain of f. It is $(-\infty, -1) \cup (-1, \infty)$. Since $6 - x = 0$ when $x = 6$, we must exclude 6 from the domain of g. It is $(-\infty, 6) \cup (6, \infty)$. The domain of $f + g$, $f - g$, and fg is the intersection of the domains of f and g, or $(-\infty, -1) \cup (-1, 6) \cup (6, \infty)$. The domain of ff is the same as the domain of f, or $(-\infty, -1) \cup (-1, \infty)$. Since there are no values of x for which $g(x) = 0$ or $f(x) = 0$, the domain of f/g and g/f is $(-\infty, -1) \cup (-1, 6) \cup (6, \infty)$.

b) $(f + g)(x) = f(x) + g(x) = \dfrac{4}{x + 1} + \dfrac{1}{6 - x}$

$(f - g)(x) = f(x) - g(x) = \dfrac{4}{x + 1} - \dfrac{1}{6 - x}$

$(fg)(x) = f(x) \cdot g(x) = \dfrac{4}{x + 1} \cdot \dfrac{1}{6 - x} = \dfrac{4}{(x + 1)(6 - x)}$

$(ff)(x) = f(x) \cdot f(x) = \dfrac{4}{x + 1} \cdot \dfrac{4}{x + 1} = \dfrac{16}{(x + 1)^2}$, or $\dfrac{16}{x^2 + 2x + 1}$

$(f/g)(x) = \dfrac{\dfrac{4}{x + 1}}{\dfrac{1}{6 - x}} = \dfrac{4}{x + 1} \cdot \dfrac{6 - x}{1} = \dfrac{4(6 - x)}{x + 1}$

$(g/f)(x) = \dfrac{\dfrac{1}{6 - x}}{\dfrac{4}{x + 1}} = \dfrac{1}{6 - x} \cdot \dfrac{x + 1}{4} = \dfrac{x + 1}{4(6 - x)}$

31. $f(x) = \dfrac{1}{x}$, $g(x) = x - 3$

a) Since $f(0)$ is not defined, the domain of f is $(-\infty, 0) \cup (0, \infty)$. The domain of g is $(-\infty, \infty)$. Then the domain of $f + g$, $f - g$, fg, and ff is $(-\infty, 0) \cup (0, \infty)$. Since $g(3) = 0$, the domain of f/g is $(-\infty, 0) \cup (0, 3) \cup (3, \infty)$. There are no values of x for which $f(x) = 0$, so the domain of g/f is $(-\infty, 0) \cup (0, \infty)$.

b) $(f + g)(x) = f(x) + g(x) = \dfrac{1}{x} + x - 3$

$(f - g)(x) = f(x) - g(x) = \dfrac{1}{x} - (x - 3) = \dfrac{1}{x} - x + 3$

$(fg)(x) = f(x) \cdot g(x) = \dfrac{1}{x} \cdot (x - 3) = \dfrac{x - 3}{x}$, or $1 - \dfrac{3}{x}$

$(ff)(x) = f(x) \cdot f(x) = \dfrac{1}{x} \cdot \dfrac{1}{x} = \dfrac{1}{x^2}$

$(f/g)(x) = \dfrac{f(x)}{g(x)} = \dfrac{\frac{1}{x}}{x - 3} = \dfrac{1}{x} \cdot \dfrac{1}{x - 3} = \dfrac{1}{x(x - 3)}$

$(g/f)(x) = \dfrac{g(x)}{f(x)} = \dfrac{x - 3}{\frac{1}{x}} = (x - 3) \cdot \dfrac{x}{1} = x(x - 3)$, or

$x^2 - 3x$

33. From the graph we see that the domain of F is $[2, 11]$ and the domain of G is $[1, 9]$. The domain of $F + G$ is the set of numbers in the domains of both F and G. This is $[2, 9]$.

35. The domain of G/F is the set of numbers in the domains of both F and G (See Exercise 33.), excluding those for which $F = 0$. Since $F(3) = 0$, the domain of G/F is $[2, 3) \cup (3, 9]$.

37.

39. From the graph, we see that the domain of F is $[0, 9]$ and the domain of G is $[3, 10]$. The domain of $F + G$ is the set of numbers in the domains of both F and G. This is $[3, 9]$.

41. The domain of G/F is the set of numbers in the domains of both F and G (See Exercise 39.), excluding those for which $F = 0$. Since $F(6) = 0$ and $F(8) = 0$, the domain of G/F is $[3, 6) \cup (6, 8) \cup (8, 9]$.

43.

45. a) $P(x) = R(x) - C(x) = 60x - 0.4x^2 - (3x + 13) =$
$60x - 0.4x^2 - 3x - 13 = -0.4x^2 + 57x - 13$

b) $R(100) = 60 \cdot 100 - 0.4(100)^2 = 6000 - 0.4(10,000) =$
$6000 - 4000 = 2000$

$C(100) = 3 \cdot 100 + 13 = 300 + 13 = 313$

$P(100) = R(100) - C(100) = 2000 - 313 = 1687$

47. $f(x) = -\dfrac{3}{5}x + 10$

$f(x + h) = -\dfrac{3}{5}(x + h) + 10 = -\dfrac{3}{5}x - \dfrac{3}{5}h + 10$

$\dfrac{f(x + h) - f(x)}{h} = \dfrac{-\frac{3}{5}x - \frac{3}{5}h + 10 - \left(-\frac{3}{5}x + 10\right)}{h}$

$= \dfrac{-\frac{3}{5}x - \frac{3}{5}h + 10 + \frac{3}{5}x - 10}{h}$

$= \dfrac{-\frac{3}{5}h}{h} = -\dfrac{3}{5}$

49. $f(x) = x^2 + 1$

$f(x + h) = (x + h)^2 + 1 = x^2 + 2xh + h^2 + 1$

$\dfrac{f(x + h) - f(x)}{h} = \dfrac{x^2 + 2xh + h^2 + 1 - (x^2 + 1)}{h}$

$= \dfrac{x^2 + 2xh + h^2 + 1 - x^2 - 1}{h}$

$= \dfrac{2xh + h^2}{h}$

$= \dfrac{h(2x + h)}{h}$

$= \dfrac{h}{h} \cdot \dfrac{2x + h}{1}$

$= 2x + h$

51. $f(x) = 3x - 5$

$f(x + h) = 3(x + h) - 5 = 3x + 3h - 5$

$\dfrac{f(x + h) - f(x)}{h} = \dfrac{3x + 3h - 5 - (3x - 5)}{h}$

$= \dfrac{3x + 3h - 5 - 3x + 5}{h}$

$= \dfrac{3h}{h} = 3$

53. $f(x) = 3x^2 - 2x + 1$

$f(x + h) = 3(x + h)^2 - 2(x + h) + 1 =$
$3(x^2 + 2xh + h^2) - 2(x + h) + 1 =$
$3x^2 + 6xh + 3h^2 - 2x - 2h + 1$

$f(x) = 3x^2 - 2x + 1$

$\dfrac{f(x + h) - f(x)}{h} =$

$\dfrac{(3x^2 + 6xh + 3h^2 - 2x - 2h + 1) - (3x^2 - 2x + 1)}{h} =$

$\dfrac{3x^2 + 6xh + 3h^2 - 2x - 2h + 1 - 3x^2 + 2x - 1}{h} =$

$$\frac{6xh + 3h^2 - 2h}{h} = \frac{h(6x + 3h - 2)}{h \cdot 1} =$$

$$\frac{h}{h} \cdot \frac{6x + 3h - 2}{1} = 6x + 3h - 2$$

55. $f(x) = 4 + 5|x|$

$f(x + h) = 4 + 5|x + h|$

$$\frac{f(x+h) - f(x)}{h} = \frac{4 + 5|x + h| - (4 + 5|x|)}{h}$$

$$= \frac{4 + 5|x + h| - 4 - 5|x|}{h}$$

$$= \frac{5|x + h| - 5|x|}{h}$$

57. $f(x) = x^3$

$f(x + h) = (x + h)^3 = x^3 + 3x^2h + 3xh^2 + h^3$

$f(x) = x^3$

$$\frac{f(x+h) - f(x)}{h} = \frac{x^3 + 3x^2h + 3xh^2 + h^3 - x^3}{h} =$$

$$\frac{3x^2h + 3xh^2 + h^3}{h} = \frac{h(3x^2 + 3xh + h^2)}{h \cdot 1} =$$

$$\frac{h}{h} \cdot \frac{3x^2 + 3xh + h^2}{1} = 3x^2 + 3xh + h^2$$

59. $f(x) = \dfrac{x - 4}{x + 3}$

$$\frac{f(x+h) - f(x)}{h} = \frac{\dfrac{x+h-4}{x+h+3} - \dfrac{x-4}{x+3}}{h} =$$

$$\frac{\dfrac{x+h-4}{x+h+3} - \dfrac{x-4}{x+3}}{h} \cdot \frac{(x+h+3)(x+3)}{(x+h+3)(x+3)} =$$

$$\frac{(x + h - 4)(x + 3) - (x - 4)(x + h + 3)}{h(x + h + 3)(x + 3)} =$$

$$\frac{x^2 + hx - 4x + 3x + 3h - 12 - (x^2 + hx + 3x - 4x - 4h - 12)}{h(x + h + 3)(x + 3)} =$$

$$\frac{x^2 + hx - x + 3h - 12 - x^2 - hx + x + 4h + 12}{h(x + h + 3)(x + 3)} =$$

$$\frac{7h}{h(x + h + 3)(x + 3)} = \frac{h}{h} \cdot \frac{7}{(x + h + 3)(x + 3)} =$$

$$\frac{7}{(x + h + 3)(x + 3)}$$

61. $(f \circ g)(-1) = f(g(-1)) = f((-1)^2 - 2(-1) - 6) =$
$f(1 + 2 - 6) = f(-3) = 3(-3) + 1 = -9 + 1 = -8$

63. $(h \circ f)(1) = h(f(1)) = h(3 \cdot 1 + 1) = h(3 + 1) =$
$h(4) = 4^3 = 64$

65. $(g \circ f)(5) = g(f(5)) = g(3 \cdot 5 + 1) = g(15 + 1) =$
$g(16) = 16^2 - 2 \cdot 16 - 6 = 218$

67. $(f \circ h)(-3) = f(h(-3)) = f((-3)^3) = f(-27) =$
$3(-27) + 1 = -81 + 1 = -80$

69. $(f \circ g)(x) = f(g(x)) = f(x - 3) = x - 3 + 3 = x$

$(g \circ f)(x) = g(f(x)) = g(x + 3) = x + 3 - 3 = x$

The domain of f and of g is $(-\infty, \infty)$, so the domain of $f \circ g$ and of $g \circ f$ is $(-\infty, \infty)$.

71. $(f \circ g)(x) = f(g(x)) = f(3x^2 - 2x - 1) = 3x^2 - 2x - 1 + 1 = 3x^2 - 2x$

$(g \circ f)(x) = g(f(x)) = g(x + 1) = 3(x + 1)^2 - 2(x + 1) - 1 =$
$3(x^2 + 2x + 1) - 2(x + 1) - 1 = 3x^2 + 6x + 3 - 2x - 2 - 1 =$
$3x^2 + 4x$

The domain of f and of g is $(-\infty, \infty)$, so the domain of $f \circ g$ and of $g \circ f$ is $(-\infty, \infty)$.

73. $(f \circ g)(x) = f(g(x)) = f(4x - 3) = (4x - 3)^2 - 3 =$
$16x^2 - 24x + 9 - 3 = 16x^2 - 24x + 6$

$(g \circ f)(x) = g(f(x)) = g(x^2 - 3) = 4(x^2 - 3) - 3 =$
$4x^2 - 12 - 3 = 4x^2 - 15$

The domain of f and of g is $(-\infty, \infty)$, so the domain of $f \circ g$ and of $g \circ f$ is $(-\infty, \infty)$.

75. $(f \circ g)(x) = f(g(x)) = f\left(\dfrac{1}{x}\right) = \dfrac{4}{1 - 5 \cdot \dfrac{1}{x}} = \dfrac{4}{1 - \dfrac{5}{x}} =$

$$\frac{4}{\dfrac{x - 5}{x}} = 4 \cdot \frac{x}{x - 5} = \frac{4x}{x - 5}$$

$(g \circ f)(x) = g(f(x)) = g\left(\dfrac{4}{1 - 5x}\right) = \dfrac{1}{\dfrac{4}{1 - 5x}} =$

$$1 \cdot \frac{1 - 5x}{4} = \frac{1 - 5x}{4}$$

The domain of f is $\left\{x \,\middle|\, x \neq \dfrac{1}{5}\right\}$ and the domain of g is $\{x | x \neq 0\}$. Consider the domain of $f \circ g$. Since 0 is not in the domain of g, 0 is not in the domain of $f \circ g$. Since $\dfrac{1}{5}$ is not in the domain of f, we know that $g(x)$ cannot be $\dfrac{1}{5}$. We find the value(s) of x for which $g(x) = \dfrac{1}{5}$.

$$\frac{1}{x} = \frac{1}{5}$$
$$5 = x \qquad \text{Multiplying by } 5x$$

Thus 5 is also not in the domain of $f \circ g$. Then the domain of $f \circ g$ is $\{x | x \neq 0 \ and \ x \neq 5\}$, or $(-\infty, 0) \cup (0, 5) \cup (5, \infty)$.

Now consider the domain of $g \circ f$. Recall that $\dfrac{1}{5}$ is not in the domain of f, so it is not in the domain of $g \circ f$. Now 0 is not in the domain of g but $f(x)$ is never 0, so the domain of $g \circ f$ is $\left\{x \,\middle|\, x \neq \dfrac{1}{5}\right\}$, or $\left(-\infty, \dfrac{1}{5}\right) \cup \left(\dfrac{1}{5}, \infty\right)$.

77. $(f \circ g)(x) = f(g(x)) = f\left(\dfrac{x + 7}{3}\right) =$

$$3\left(\frac{x + 7}{3}\right) - 7 = x + 7 - 7 = x$$

$(g \circ f)(x) = g(f(x)) = g(3x - 7) = \dfrac{(3x - 7) + 7}{3} =$

$$\frac{3x}{3} = x$$

The domain of f and of g is $(-\infty, \infty)$, so the domain of $f \circ g$ and of $g \circ f$ is $(-\infty, \infty)$.

79. $(f \circ g)(x) = f(g(x)) = f(\sqrt{x}) = 2\sqrt{x} + 1$

$(g \circ f)(x) = g(f(x)) = g(2x + 1) = \sqrt{2x + 1}$

The domain of f is $(-\infty, \infty)$ and the domain of g is $\{x | x \geq 0\}$. Thus the domain of $f \circ g$ is $\{x | x \geq 0\}$, or $[0, \infty)$.

Now consider the domain of $g \circ f$. There are no restrictions on the domain of f, but the domain of g is $\{x | x \geq 0\}$. Since $f(x) \geq 0$ for $x \geq -\frac{1}{2}$, the domain of $g \circ f$ is $\left\{x | x \geq -\frac{1}{2}\right\}$, or $\left[-\frac{1}{2}, \infty\right)$.

81. $(f \circ g)(x) = f(g(x)) = f(0.05) = 20$

$(g \circ f)(x) = g(f(x)) = g(20) = 0.05$

The domain of f and of g is $(-\infty, \infty)$, so the domain of $f \circ g$ and of $g \circ f$ is $(-\infty, \infty)$.

83. $(f \circ g)(x) = f(g(x)) = f(x^2 - 5) =$
$$\sqrt{x^2 - 5 + 5} = \sqrt{x^2} = |x|$$

$(g \circ f)(x) = g(f(x)) = g(\sqrt{x + 5}) =$
$$(\sqrt{x + 5})^2 - 5 = x + 5 - 5 = x$$

The domain of f is $\{x | x \geq -5\}$ and the domain of g is $(-\infty, \infty)$. Since $x^2 \geq 0$ for all values of x, then $x^2 - 5 \geq -5$ for all values of x and the domain of $g \circ f$ is $(-\infty, \infty)$.

Now consider the domain of $f \circ g$. There are no restrictions on the domain of g, so the domain of $f \circ g$ is the same as the domain of f, $\{x | x \geq -5\}$, or $[-5, \infty)$.

85. $(f \circ g)(x) = f(g(x)) = f(\sqrt{3 - x}) = (\sqrt{3 - x})^2 + 2 =$
$$3 - x + 2 = 5 - x$$

$(g \circ f)(x) = g(f(x)) = g(x^2 + 2) = \sqrt{3 - (x^2 + 2)} =$
$$\sqrt{3 - x^2 - 2} = \sqrt{1 - x^2}$$

The domain of f is $(-\infty, \infty)$ and the domain of g is $\{x | x \leq 3\}$, so the domain of $f \circ g$ is $\{x | x \leq 3\}$, or $(-\infty, 3]$.

Now consider the domain of $g \circ f$. There are no restrictions on the domain of f and the domain of g is $\{x | x \leq 3\}$, so we find the values of x for which $f(x) \leq 3$. We see that $x^2 + 2 \leq 3$ for $-1 \leq x \leq 1$, so the domain of $g \circ f$ is $\{x | -1 \leq x \leq 1\}$, or $[-1, 1]$.

87. $(f \circ g)(x) = f(g(x)) = f\left(\dfrac{1}{1 + x}\right) =$

$$\frac{1 - \left(\dfrac{1}{1 + x}\right)}{\dfrac{1}{1 + x}} = \frac{\dfrac{1 + x - 1}{1 + x}}{\dfrac{1}{1 + x}} =$$

$$\frac{x}{1 + x} \cdot \frac{1 + x}{1} = x$$

$(g \circ f)(x) = g(f(x)) = g\left(\dfrac{1 - x}{x}\right) =$

$$\frac{1}{1 + \left(\dfrac{1 - x}{x}\right)} = \frac{1}{\dfrac{x + 1 - x}{x}} =$$

$$\frac{1}{\dfrac{1}{x}} = 1 \cdot \frac{x}{1} = x$$

The domain of f is $\{x | x \neq 0\}$ and the domain of g is $\{x | x \neq -1\}$, so we know that -1 is not in the domain of $f \circ g$. Since 0 is not in the domain of f, values of x for which $g(x) = 0$ are not in the domain of $f \circ g$. But $g(x)$ is never 0, so the domain of $f \circ g$ is $\{x | x \neq -1\}$, or $(-\infty, -1) \cup (-1, \infty)$.

Now consider the domain of $g \circ f$. Recall that 0 is not in the domain of f. Since -1 is not in the domain of g, we know that $g(x)$ cannot be -1. We find the value(s) of x for which $f(x) = -1$.

$$\frac{1 - x}{x} = -1$$

$$1 - x = -x \quad \text{Multiplying by } x$$

$$1 = 0 \quad \text{False equation}$$

We see that there are no values of x for which $f(x) = -1$, so the domain of $g \circ f$ is $\{x | x \neq 0\}$, or $(-\infty, 0) \cup (0, \infty)$.

89. $(f \circ g)(x) = f(g(x)) = f(x + 1) =$
$(x + 1)^3 - 5(x + 1)^2 + 3(x + 1) + 7 =$
$x^3 + 3x^2 + 3x + 1 - 5x^2 - 10x - 5 + 3x + 3 + 7 =$
$x^3 - 2x^2 - 4x + 6$

$(g \circ f)(x) = g(f(x)) = g(x^3 - 5x^2 + 3x + 7) =$
$x^3 - 5x^2 + 3x + 7 + 1 = x^3 - 5x^2 + 3x + 8$

The domain of f and of g is $(-\infty, \infty)$, so the domain of $f \circ g$ and of $g \circ f$ is $(-\infty, \infty)$.

91. $h(x) = (4 + 3x)^5$

This is $4 + 3x$ to the 5th power. The most obvious answer is $f(x) = x^5$ and $g(x) = 4 + 3x$.

93. $h(x) = \dfrac{1}{(x - 2)^4}$

This is 1 divided by $(x - 2)$ to the 4th power. One obvious answer is $f(x) = \dfrac{1}{x^4}$ and $g(x) = x - 2$. Another possibility is $f(x) = \dfrac{1}{x}$ and $g(x) = (x - 2)^4$.

95. $f(x) = \dfrac{x - 1}{x + 1}$, $g(x) = x^3$

97. $f(x) = x^6$, $g(x) = \dfrac{2 + x^3}{2 - x^3}$

99. $f(x) = \sqrt{x}$, $g(x) = \dfrac{x - 5}{x + 2}$

101. $f(x) = x^3 - 5x^2 + 3x - 1$, $g(x) = x + 2$

103. $f(x) = (t \circ s)(x) = t(s(x)) = t(x - 3) = x - 3 + 4 = x + 1$

We have $f(x) = x + 1$.

105.

107. Discussion and Writing

109. Equations $(a) - (f)$ are in the form $y = mx + b$, so we can read the y-intercepts directly from the equations. Equations (g) and (h) can be written in this form as $y = \frac{2}{3}x - 2$ and $y = -2x + 3$, respectively. We see that only equation (c) has y-intercept $(0, 1)$.

111. If a line slopes down from left to right, its slope is negative. The equations $y = mx + b$ for which m is negative are (b), (d), (f), and (h). (See Exercise 109.)

113. The only equation that has $(0, 0)$ as a solution is (a).

115. Only equations (c) and (g) have the same slope and different y-intercepts. They represent parallel lines.

117. Answers may vary. One example is $f(x) = 2x + 5$ and $g(x) = \frac{x - 5}{2}$. Other examples are found in Exercises 69, 70, 77, 78, 84, and 87.

119. The domain of $h(x)$ is $\left\{ x \big| x \neq \frac{7}{3} \right\}$, and the domain of $g(x)$ is $\{x | x \neq 3\}$, so $\frac{7}{3}$ and 3 are not in the domain of $(h/g)(x)$. We must also exclude the value of x for which $g(x) = 0$.

$$\frac{x^4 - 1}{5x - 15} = 0$$
$$x^4 - 1 = 0 \quad \text{Multiplying by } 5x - 15$$
$$x^4 = 1$$
$$x = \pm 1$$

Then the domain of $(h/g)(x)$ is
$\left\{ x \big| x \neq \frac{7}{3} \text{ and } x \neq 3 \text{ and } x \neq -1 \text{ and } x \neq 1 \right\}$, or
$(-\infty, -1) \cup (-1, 1) \cup \left(1, \frac{7}{3}\right) \cup \left(\frac{7}{3}, 3\right) \cup (3, \infty)$.

Exercise Set 1.7

1. If the graph were folded on the x-axis, the parts above and below the x-axis would not coincide, so the graph is not symmetric with respect to the x-axis.

If the graph were folded on the y-axis, the parts to the left and right of the y-axis would coincide, so the graph is symmetric with respect to the y-axis.

If the graph were rotated $180°$, the resulting graph would not coincide with the original graph, so it is not symmetric with respect to the origin.

3. If the graph were folded on the x-axis, the parts above and below the x-axis would coincide, so the graph is symmetric with respect to the x-axis.

If the graph were folded on the y-axis, the parts to the left and right of the y-axis would not coincide, so the graph is not symmetric with respect to the y-axis.

If the graph were rotated $180°$, the resulting graph would not coincide with the original graph, so it is not symmetric with respect to the origin.

5. If the graph were folded on the x-axis, the parts above and below the x-axis would not coincide, so the graph is not symmetric with respect to the x-axis.

If the graph were folded on the y-axis, the parts to the left and right of the y-axis would not coincide, so the graph is not symmetric with respect to the y-axis.

If the graph were rotated $180°$, the resulting graph would coincide with the original graph, so it is symmetric with respect to the origin.

7.

The graph is symmetric with respect to the y-axis. It is not symmetric with respect to the x-axis or the origin.

Test algebraically for symmetry with respect to the x-axis:

$$y = |x| - 2 \quad \text{Original equation}$$
$$-y = |x| - 2 \quad \text{Replacing } y \text{ by } -y$$
$$y = -|x| + 2 \quad \text{Simplifying}$$

The last equation is not equivalent to the original equation, so the graph is not symmetric with respect to the x-axis.

Test algebraically for symmetry with respect to the y-axis:

$$y = |x| - 2 \quad \text{Original equation}$$
$$y = |-x| - 2 \quad \text{Replacing } x \text{ by } -x$$
$$y = |x| - 2 \quad \text{Simplifying}$$

The last equation is equivalent to the original equation, so the graph is symmetric with respect to the y-axis.

Test algebraically for symmetry with respect to the origin:

$$y = |x| - 2 \quad \text{Original equation}$$
$$-y = |-x| - 2 \quad \text{Replacing } x \text{ by } -x \text{ and } y \text{ by } -y$$
$$-y = |x| - 2 \quad \text{Simplifying}$$
$$y = -|x| + 2$$

The last equation is not equivalent to the original equation, so the graph is not symmetric with respect to the origin.

9.

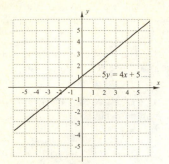

The graph is not symmetric with respect to the x-axis, the y-axis, or the origin.

Test algebraically for symmetry with respect to the x-axis:

$$5y = 4x + 5 \qquad \text{Original equation}$$
$$5(-y) = 4x + 5 \qquad \text{Replacing } y \text{ by } -y$$
$$-5y = 4x + 5 \qquad \text{Simplifying}$$
$$5y = -4x - 5$$

The last equation is not equivalent to the original equation, so the graph is not symmetric with respect to the x-axis.

Test algebraically for symmetry with respect to the y-axis:

$$5y = 4x + 5 \qquad \text{Original equation}$$
$$5y = 4(-x) + 5 \qquad \text{Replacing } x \text{ by } -x$$
$$5y = -4x + 5 \qquad \text{Simplifying}$$

The last equation is not equivalent to the original equation, so the graph is not symmetric with respect to the y-axis.

Test algebraically for symmetry with respect to the origin:

$$5y = 4x + 5 \qquad \text{Original equation}$$
$$5(-y) = 4(-x) + 5 \qquad \text{Replacing } x \text{ by } -x \text{ and } y \text{ by } -y$$
$$-5y = -4x + 5 \qquad \text{Simplifying}$$
$$5y = 4x - 5$$

The last equation is not equivalent to the original equation, so the graph is not symmetric with respect to the origin.

11.

The graph is symmetric with respect to the y-axis. It is not symmetric with respect to the x-axis or the origin.

Test algebraically for symmetry with respect to the x-axis:

$$5y = 2x^2 - 3 \qquad \text{Original equation}$$
$$5(-y) = 2x^2 - 3 \qquad \text{Replacing } y \text{ by } -y$$
$$-5y = 2x^2 - 3 \qquad \text{Simplifying}$$
$$5y = -2x^2 + 3$$

The last equation is not equivalent to the original equation, so the graph is not symmetric with respect to the x-axis.

Test algebraically for symmetry with respect to the y-axis:

$$5y = 2x^2 - 3 \qquad \text{Original equation}$$
$$5y = 2(-x)^2 - 3 \qquad \text{Replacing } x \text{ by } -x$$
$$5y = 2x^2 - 3$$

The last equation is equivalent to the original equation, so the graph is symmetric with respect to the y-axis.

Test algebraically for symmetry with respect to the origin:

$$5y = 2x^2 - 3 \qquad \text{Original equation}$$
$$5(-y) = 2(-x)^2 - 3 \qquad \text{Replacing } x \text{ by } -x \text{ and } y \text{ by } -y$$
$$-5y = 2x^2 - 3 \qquad \text{Simplifying}$$
$$5y = -2x^2 + 3$$

The last equation is not equivalent to the original equation, so the graph is not symmetric with respect to the origin.

13.

The graph is not symmetric with respect to the x-axis or the y-axis. It is symmetric with respect to the origin.

Test algebraically for symmetry with respect to the x-axis:

$$y = \frac{1}{x} \qquad \text{Original equation}$$
$$-y = \frac{1}{x} \qquad \text{Replacing } y \text{ by } -y$$
$$y = -\frac{1}{x} \qquad \text{Simplifying}$$

The last equation is not equivalent to the original equation, so the graph is not symmetric with respect to the x-axis.

Test algebraically for symmetry with respect to the y-axis:

$$y = \frac{1}{x} \qquad \text{Original equation}$$
$$y = \frac{1}{-x} \qquad \text{Replacing } x \text{ by } -x$$
$$y = -\frac{1}{x} \qquad \text{Simplifying}$$

The last equation is not equivalent to the original equation, so the graph is not symmetric with respect to the y-axis.

Test algebraically for symmetry with respect to the origin:

$$y = \frac{1}{x} \qquad \text{Original equation}$$
$$-y = \frac{1}{-x} \qquad \text{Replacing } x \text{ by } -x \text{ and } y \text{ by } -y$$
$$y = \frac{1}{x} \qquad \text{Simplifying}$$

The last equation is equivalent to the original equation, so the graph is symmetric with respect to the origin.

15. Test for symmetry with respect to the x-axis:

$$5x - 5y = 0 \quad \text{Original equation}$$
$$5x - 5(-y) = 0 \quad \text{Replacing } y \text{ by } -y$$
$$5x + 5y = 0 \quad \text{Simplifying}$$

The last equation is not equivalent to the original equation, so the graph is not symmetric with respect to the x-axis.

Test for symmetry with respect to the y-axis:

$$5x - 5y = 0 \quad \text{Original equation}$$
$$5(-x) - 5y = 0 \quad \text{Replacing } x \text{ by } -x$$
$$-5x - 5y = 0 \quad \text{Simplifying}$$
$$5x + 5y = 0$$

The last equation is not equivalent to the original equation, so the graph is not symmetric with respect to the y-axis.

Test for symmetry with respect to the origin:

$$5x - 5y = 0 \quad \text{Original equation}$$
$$5(-x) - 5(-y) = 0 \quad \text{Replacing } x \text{ by } -x \text{ and } y \text{ by } -y$$
$$-5x + 5y = 0 \quad \text{Simplifying}$$
$$5x - 5y = 0$$

The last equation is equivalent to the original equation, so the graph is symmetric with respect to the origin.

17. Test for symmetry with respect to the x-axis:

$$3x^2 - 2y^2 = 3 \quad \text{Original equation}$$
$$3x^2 - 2(-y)^2 = 3 \quad \text{Replacing } y \text{ by } -y$$
$$3x^2 - 2y^2 = 3 \quad \text{Simplifying}$$

The last equation is equivalent to the original equation, so the graph is symmetric with respect to the x-axis.

Test for symmetry with respect to the y-axis:

$$3x^2 - 2y^2 = 3 \quad \text{Original equation}$$
$$3(-x)^2 - 2y^2 = 3 \quad \text{Replacing } x \text{ by } -x$$
$$3x^2 - 2y^2 = 3 \quad \text{Simplifying}$$

The last equation is equivalent to the original equation, so the graph is symmetric with respect to the y-axis.

Test for symmetry with respect to the origin:

$$3x^2 - 2y^2 = 3 \quad \text{Original equation}$$
$$3(-x)^2 - 2(-y)^2 = 3 \quad \text{Replacing } x \text{ by } -x \text{ and } y \text{ by } -y$$
$$3x^2 - 2y^2 = 3 \quad \text{Simplifying}$$

The last equation is equivalent to the original equation, so the graph is symmetric with respect to the origin.

19. Test for symmetry with respect to the x-axis:

$$y = |2x| \quad \text{Original equation}$$
$$-y = |2x| \quad \text{Replacing } y \text{ by } -y$$
$$y = -|2x| \quad \text{Simplifying}$$

The last equation is not equivalent to the original equation, so the graph is not symmetric with respect to the x-axis.

Test for symmetry with respect to the y-axis:

$$y = |2x| \quad \text{Original equation}$$
$$y = |2(-x)| \quad \text{Replacing } x \text{ by } -x$$
$$y = |-2x| \quad \text{Simplifying}$$
$$y = |2x|$$

The last equation is equivalent to the original equation, so the graph is symmetric with respect to the y-axis.

Test for symmetry with respect to the origin:

$$y = |2x| \quad \text{Original equation}$$
$$-y = |2(-x)| \quad \text{Replacing } x \text{ by } -x \text{ and } y \text{ by } -y$$
$$-y = |-2x| \quad \text{Simplifying}$$
$$-y = |2x|$$
$$y = -|2x|$$

The last equation is not equivalent to the original equation, so the graph is not symmetric with respect to the origin.

21. Test for symmetry with respect to the x-axis:

$$2x^4 + 3 = y^2 \quad \text{Original equation}$$
$$2x^4 + 3 = (-y)^2 \quad \text{Replacing } y \text{ by } -y$$
$$2x^4 + 3 = y^2 \quad \text{Simplifying}$$

The last equation is equivalent to the original equation, so the graph is symmetric with respect to the x-axis.

Test for symmetry with respect to the y-axis:

$$2x^4 + 3 = y^2 \quad \text{Original equation}$$
$$2(-x)^4 + 3 = y^2 \quad \text{Replacing } x \text{ by } -x$$
$$2x^4 + 3 = y^2 \quad \text{Simplifying}$$

The last equation is equivalent to the original equation, so the graph is symmetric with respect to the y-axis.

Test for symmetry with respect to the origin:

$$2x^4 + 3 = y^2 \quad \text{Original equation}$$
$$2(-x)^4 + 3 = (-y)^2 \quad \text{Replacing } x \text{ by } -x \text{ and } y \text{ by } -y$$
$$2x^4 + 3 = y^2 \quad \text{Simplifying}$$

The last equation is equivalent to the original equation, so the graph is symmetric with respect to the origin.

23. Test for symmetry with respect to the x-axis:

$$3y^3 = 4x^3 + 2 \quad \text{Original equation}$$
$$3(-y)^3 = 4x^3 + 2 \quad \text{Replacing } y \text{ by } -y$$
$$-3y^3 = 4x^3 + 2 \quad \text{Simplifying}$$
$$3y^3 = -4x^3 - 2$$

The last equation is not equivalent to the original equation, so the graph is not symmetric with respect to the x-axis.

Test for symmetry with respect to the y-axis:

$$3y^3 = 4x^3 + 2 \quad \text{Original equation}$$
$$3y^3 = 4(-x)^3 + 2 \quad \text{Replacing } x \text{ by } -x$$
$$3y^3 = -4x^3 + 2 \quad \text{Simplifying}$$

The last equation is not equivalent to the original equation, so the graph is not symmetric with respect to the y-axis.

Test for symmetry with respect to the origin:

$$3y^3 = 4x^3 + 2 \qquad \text{Original equation}$$

$$3(-y)^3 = 4(-x)^3 + 2 \quad \text{Replacing } x \text{ by } -x$$
$$\text{and } y \text{ by } -y$$

$$-3y^3 = -4x^3 + 2 \qquad \text{Simplifying}$$

$$3y^3 = 4x^3 - 2$$

The last equation is not equivalent to the original equation, so the graph is not symmetric with respect to the origin.

25. Test for symmetry with respect to the x-axis:

$$xy = 12 \qquad \text{Original equation}$$

$$x(-y) = 12 \qquad \text{Replacing } y \text{ by } -y$$

$$-xy = 12 \qquad \text{Simplifying}$$

$$xy = -12$$

The last equation is not equivalent to the original equation, so the graph is not symmetric with respect to the x-axis.

Test for symmetry with respect to the y-axis:

$$xy = 12 \qquad \text{Original equation}$$

$$-xy = 12 \qquad \text{Replacing } x \text{ by } -x$$

$$xy = -12 \quad \text{Simplifying}$$

The last equation is not equivalent to the original equation, so the graph is not symmetric with respect to the y-axis.

Test for symmetry with respect to the origin:

$$xy = 12 \qquad \text{Original equation}$$

$$-x(-y) = 12 \quad \text{Replacing } x \text{ by } -x \text{ and } y \text{ by } -y$$

$$xy = 12 \quad \text{Simplifying}$$

The last equation is equivalent to the original equation, so the graph is symmetric with respect to the origin.

27. x-axis: Replace y with $-y$; $(-5, -6)$

y-axis: Replace x with $-x$; $(5, 6)$

Origin: Replace x with $-x$ and y with $-y$; $(5, -6)$

29. x-axis: Replace y with $-y$; $(-10, 7)$

y-axis: Replace x with $-x$; $(10, -7)$

Origin: Replace x with $-x$ and y with $-y$; $(10, 7)$

31. x-axis: Replace y with $-y$; $(0, 4)$

y-axis: Replace x with $-x$; $(0, -4)$

Origin: Replace x with $-x$ and y with $-y$; $(0, 4)$

33. The graph is symmetric with respect to the y-axis, so the function is even.

35. The graph is symmetric with respect to the origin, so the function is odd.

37. The graph is not symmetric with respect to either the y-axis or the origin, so the function is neither even nor odd.

39. $f(x) = -3x^3 + 2x$

$$f(-x) = -3(-x)^3 + 2(-x) = 3x^3 - 2x$$

$$-f(x) = -(-3x^3 + 2x) = 3x^3 - 2x$$

$f(-x) = -f(x)$, so f is odd.

41. $f(x) = 5x^2 + 2x^4 - 1$

$$f(-x) = 5(-x)^2 + 2(-x)^4 - 1 = 5x^2 + 2x^4 - 1$$

$f(x) = f(-x)$, so f is even.

43. $f(x) = x^{17}$

$$f(-x) = (-x)^{17} = -x^{17}$$

$$-f(x) = -x^{17}$$

$f(-x) = -f(x)$, so f is odd.

45. $f(x) = \dfrac{1}{x^2}$

$$f(-x) = \dfrac{1}{(-x)^2} = \dfrac{1}{x^2}$$

$f(x) = f(-x)$, so f is even.

47. $f(x) = 8$

$$f(-x) = 8$$

$f(x) = f(-x)$, so f is even.

49. Shift the graph of $f(x) = x^2$ right 3 units.

51. Shift the graph of $g(x) = x$ down 3 units.

53. Reflect the graph of $h(x) = \sqrt{x}$ across the x-axis.

55. Shift the graph of $h(x) = \dfrac{1}{x}$ up 4 units.

$h(x) = \dfrac{1}{x} + 4$

57. First stretch the graph of $h(x) = x$ vertically by multiplying each y-coordinate by 3. Then reflect it across the x-axis and shift it up 3 units.

$h(x) = -3x + 3$

59. First shrink the graph of $h(x) = |x|$ vertically by multiplying each y-coordinate by $\dfrac{1}{2}$. Then shift it down 2 units.

$h(x) = \dfrac{1}{2}|x| - 2$

61. Shift the graph of $g(x) = x^3$ right 2 units and reflect it across the x-axis.

$g(x) = -(x - 2)^3$

63. Shift the graph of $g(x) = x^2$ left 1 unit and down 1 unit.

$g(x) = (x + 1)^2 - 1$

65. First shrink the graph of $g(x) = x^3$ vertically by multiplying each y-coordinate by $\dfrac{1}{3}$. Then shift it up 2 units.

$g(x) = \dfrac{1}{3}x^3 + 2$

67. Shift the graph of $f(x) = \sqrt{x}$ left 2 units.

$f(x) = \sqrt{x + 2}$

69. Shift the graph of $f(x) = \sqrt[3]{x}$ down 2 units.

$f(x) = \sqrt[3]{x} - 2$

71. Think of the graph of $f(x) = |x|$. Since $g(x) = f(3x)$, the graph of $g(x) = |3x|$ is the graph of $f(x) = |x|$ shrunk horizontally by dividing each x-coordinate by 3 $\left(\text{or multiplying each } x\text{-coordinate by } \dfrac{1}{3}\right)$.

73. Think of the graph of $f(x) = \dfrac{1}{x}$. Since $h(x) = 2f(x)$, the graph of $h(x) = \dfrac{2}{x}$ is the graph of $f(x) = \dfrac{1}{x}$ stretched vertically by multiplying each y-coordinate by 2.

75. Think of the graph of $g(x) = \sqrt{x}$. Since $f(x) = 3g(x) - 5$, the graph of $f(x) = 3\sqrt{x} - 5$ is the graph of $g(x) = \sqrt{x}$ stretched vertically by multiplying each y-coordinate by 3 and then shifted down 5 units.

77. Think of the graph of $f(x) = |x|$. Since $g(x) = f\left(\frac{1}{3}x\right) - 4$, the graph of $g(x) = \left|\frac{1}{3}x\right| - 4$ is the graph of $f(x) = |x|$ stretched horizontally by multiplying each x-coordinate by 3 and then shifted down 4 units.

79. Think of the graph of $g(x) = x^2$. Since $f(x) = -\frac{1}{4}g(x - 5)$, the graph of $f(x) = -\frac{1}{4}(x - 5)^2$ is the graph of $g(x) = x^2$ shifted right 5 units, shrunk vertically by multiplying each y-coordinate by $\frac{1}{4}$, and reflected across the x-axis.

81. Think of the graph of $g(x) = \frac{1}{x}$. Since $f(x) = g(x + 3) + 2$, the graph of $f(x) = \frac{1}{x + 3} + 2$ is the graph of $g(x) = \frac{1}{x}$ shifted left 3 units and up 2 units.

83. Think of the graph of $f(x) = x^2$. Since $h(x) = -f(x-3) + 5$, the graph of $h(x) = -(x-3)^2 + 5$ is the graph of $f(x) = x^2$ shifted right 3 units, reflected across the x-axis, and shifted up 5 units.

85. The graph of $y = g(x)$ is the graph of $y = f(x)$ shrunk vertically by a factor of $\frac{1}{2}$. Multiply the y-coordinate by $\frac{1}{2}$: $(-12, 2)$.

87. The graph of $y = g(x)$ is the graph of $y = f(x)$ reflected across the y-axis, so we reflect the point across the y-axis: $(12, 4)$.

89. The graph of $y = g(x)$ is the graph of $y = f(x)$ shifted down 2 units. Subtract 2 from the y-coordinate: $(-12, 2)$.

91. The graph of $y = g(x)$ is the graph of $y = f(x)$ stretched vertically by a factor of 4. Multiply the y-coordinate by 4: $(-12, 16)$.

93. $g(x) = x^2 + 4$ is the function $f(x) = x^2 + 3$ shifted up 1 unit, so $g(x) = f(x) + 1$. Answer B is correct.

95. If we substitute $x - 2$ for x in f, we get $(x - 2)^3 + 3$, so $g(x) = f(x - 2)$. Answer A is correct.

97. Shape: $h(x) = x^2$

Turn $h(x)$ upside-down (that is, reflect it across the x-axis): $g(x) = -h(x) = -x^2$

Shift $g(x)$ right 8 units: $f(x) = g(x - 8) = -(x - 8)^2$

99. Shape: $h(x) = |x|$

Shift $h(x)$ left 7 units: $g(x) = h(x + 7) = |x + 7|$

Shift $g(x)$ up 2 units: $f(x) = g(x) + 2 = |x + 7| + 2$

101. Shape: $h(x) = \frac{1}{x}$

Shrink $h(x)$ vertically by a factor of $\frac{1}{2}$ $\left(\text{that is, multiply each function value by } \frac{1}{2}\right)$:

$g(x) = \frac{1}{2}h(x) = \frac{1}{2} \cdot \frac{1}{x}$, or $\frac{1}{2x}$

Shift $g(x)$ down 3 units: $f(x) = g(x) - 3 = \frac{1}{2x} - 3$

103. Shape: $m(x) = x^2$

Turn $m(x)$ upside-down (that is, reflect it across the x-axis): $h(x) = -m(x) = -x^2$

Shift $h(x)$ right 3 units: $g(x) = h(x - 3) = -(x - 3)^2$

Shift $g(x)$ up 4 units: $f(x) = g(x) + 4 = -(x - 3)^2 + 4$

105. Shape: $m(x) = \sqrt{x}$

Reflect $m(x)$ across the y-axis: $h(x) = m(-x) = \sqrt{-x}$

Shift $h(x)$ left 2 units: $g(x) = h(x + 2) = \sqrt{-(x + 2)}$

Shift $g(x)$ down 1 unit: $f(x) = g(x) - 1 = \sqrt{-(x + 2)} - 1$

107. Each y-coordinate is multiplied by -2. We plot and connect $(-4, 0)$, $(-3, 4)$, $(-1, 4)$, $(2, -6)$, and $(5, 0)$.

109. The graph is reflected across the y-axis and stretched horizontally by a factor of 2. That is, each x-coordinate is multiplied by -2 $\left(\text{or divided by } -\frac{1}{2}\right)$. We plot and connect $(8, 0)$, $(6, -2)$, $(2, -2)$, $(-4, 3)$, and $(-10, 0)$.

111. The graph is shifted right 1 unit so each x-coordinate is increased by 1. The graph is also reflected across the x-axis, shrunk vertically by a factor of 2, and shifted up 3 units. Thus, each y-coordinate is multiplied by $-\frac{1}{2}$ and then increased by 3. We plot and connect $(-3, 3)$, $(-2, 4)$, $(0, 4)$, $(3, 1.5)$, and $(6, 3)$.

$$y = -\frac{1}{2}f(x-1) + 3$$

113. The graph is reflected across the y-axis so each x-coordinate is replaced by its opposite.

$$g(x) = f(-x)$$

115. The graph is shifted left 2 units so each x-coordinate is decreased by 2. It is also reflected across the x-axis so each y-coordinate is replaced with its opposite. In addition, the graph is shifted up 1 unit, so each y-coordinate is then increased by 1.

$$h(x) = -g(x+2) + 1$$

117. The graph is shrunk horizontally. The x-coordinates of $y = h(x)$ are one-half the corresponding x-coordinates of $y = g(x)$.

$$h(x) = g(2x)$$

119. $g(x) = f(-x) + 3$

The graph of $g(x)$ is the graph of $f(x)$ reflected across the y-axis and shifted up 3 units. This is graph (f). This is graph (h).

121. $g(x) = -f(x) + 3$

The graph of $g(x)$ is the graph of $f(x)$ reflected across the x-axis and shifted up 3 units. This is graph (f).

123. $g(x) = \frac{1}{3}f(x-2)$

The graph of $g(x)$ is the graph of $f(x)$ shrunk vertically by a factor of 3 $\left(\text{that is, each } y\text{-coordinate is multiplied by } \frac{1}{3}\right)$ and then shifted right 2 units. This is graph (d).

125. $g(x) = \frac{1}{3}f(x+2)$

The graph of $g(x)$ is the graph of $f(x)$ shrunk vertically by a factor of 3 $\left(\text{that is, each } y\text{-coordinate is multiplied by } \frac{1}{3}\right)$ and then shifted left 2 units. This is graph (c).

127. $f(-x) = 2(-x)^4 - 35(-x)^3 + 3(-x) - 5 = 2x^4 + 35x^3 - 3x - 5 = g(x)$

129. The graph of $f(x) = x^3 - 3x^2$ is shifted up 2 units. A formula for the transformed function is $g(x) = f(x) + 2$, or $g(x) = x^3 - 3x^2 + 2$.

131. The graph of $f(x) = x^3 - 3x^2$ is shifted left 1 unit. A formula for the transformed function is $k(x) = f(x+1)$, or $k(x) = (x+1)^3 - 3(x+1)^2$.

133. Discussion and Writing

135. $f(x) = 5x^2 - 7$

a) $f(-3) = 5(-3)^2 - 7 = 5 \cdot 9 - 7 = 45 - 7 = 38$

b) $f(3) = 5 \cdot 3^2 - 7 = 5 \cdot 9 - 7 = 45 - 7 = 38$

c) $f(a) = 5a^2 - 7$

d) $f(-a) = 5(-a)^2 - 7 = 5a^2 - 7$

137. First find the slope of the given line.

$$8x - y = 10$$
$$8x = y + 10$$
$$8x - 10 = y$$

The slope of the given line is 8. The slope of a line perpendicular to this line is the opposite of the reciprocal of 8, or $-\frac{1}{8}$.

$$y - y_1 = m(x - x_1)$$
$$y - 1 = -\frac{1}{8}[x - (-1)]$$
$$y - 1 = -\frac{1}{8}(x + 1)$$
$$y - 1 = -\frac{1}{8}x - \frac{1}{8}$$
$$y = -\frac{1}{8}x + \frac{7}{8}$$

139. Each point for which $f(x) < 0$ is reflected across the x-axis.

$$|f(x)|$$

141. The graph of $y = g(|x|)$ consists of the points of $y = g(x)$ for which $x \geq 0$ along with their reflections across the y-axis.

143. $f(x) = x\sqrt{10 - x^2}$

$f(-x) = -x\sqrt{10 - (-x)^2} = -x\sqrt{10 - x^2}$

$-f(x) = -x\sqrt{10 - x^2}$

Since $f(-x) = -f(x)$, f is odd.

145. If the graph were folded on the x-axis, the parts above and below the x-axis would coincide, so the graph is symmetric with respect to the x-axis.

If the graph were folded on the y-axis, the parts to the left and right of the y-axis would not coincide, so the graph is not symmetric with respect to the y-axis.

If the graph were rotated $180°$, the resulting graph would not coincide with the original graph, so it is not symmetric with respect to the origin.

147. If the graph were folded on the x-axis, the parts above and below the x-axis would coincide, so the graph is symmetric with respect to the x-axis.

If the graph were folded on the y-axis, the parts to the left and right of the y-axis would not coincide, so the graph is not symmetric with respect to the y-axis.

If the graph were rotated $180°$, the resulting graph would not coincide with the original graph, so it is not symmetric with respect to the origin.

149. $f(2 - 3) = f(-1) = 5$, so $b = 5$.

(The graph of $y = f(x - 3)$ is the graph of $y = f(x)$ shifted right 3 units, so the point $(-1, 5)$ on $y = f(x)$ is transformed to the point $(-1 + 3, 5)$, or $(2, 5)$ on $y = f(x - 3)$.)

151. Let $f(x)$ and $g(x)$ be even functions. Then by definition, $f(x) = f(-x)$ and $g(x) = g(-x)$. Thus, $(f + g)(x) = f(x) + g(x) = f(-x) + g(-x) = (f + g)(-x)$ and $f + g$ is even. The statement is true.

153. See the answer section in the text.

155. a), b) See the answer section in the text.

Chapter 1 Review Exercises

1. Because the line passes through the origin, its x-intercept is $(0, 0)$. Thus, the statement is false.

3. First we solve each equation for y.

$$ax + y = c \qquad x - by = d$$
$$y = -ax + c \qquad -by = -x + d$$
$$y = \frac{1}{b}x - \frac{d}{b}$$

If the lines are perpendicular, the product of their slopes is -1, so we have $-a \cdot \dfrac{1}{b} = -1$, or $-\dfrac{a}{b} = -1$, or $\dfrac{a}{b} = 1$. The statement is true.

5. For the lines $y = \dfrac{1}{2}$ and $x = -5$, the x-coordinate of the point of intersection is -5 and the y-coordinate is $\dfrac{1}{2}$, so the statement is true.

7. For $\left(3, \dfrac{24}{9}\right)$:

$$2x - 9y = -18$$
$$2 \cdot 3 - 9 \cdot \frac{24}{9} \;?\; -18$$
$$6 - 24$$
$$-18 \;\Big|\; -18 \quad \text{TRUE}$$

$\left(3, \dfrac{24}{9}\right)$ is a solution.

For $(0, -9)$:

$$2x - 9y = -18$$
$$2(0) - 9(-9) \;?\; -18$$
$$0 + 81$$
$$81 \;\Big|\; -18 \quad \text{FALSE}$$

$(0, -9)$ is not a solution.

9.

$y = -\frac{2}{3}x + 1$

11.

$y = 2 - x^2$

13. $m = \left(\dfrac{x_1 + x_2}{2}, \dfrac{y_1 + y_2}{2}\right)$

$ = \left(\dfrac{3 + (-2)}{2}, \dfrac{7 + 4}{2}\right)$

$ = \left(\dfrac{1}{2}, \dfrac{11}{2}\right)$

15. $(x+1)^2 + (y-3)^2 = 16$

$[x-(-1)^2] + (y-3)^2 = 4^2$

Center: $(-1, 3)$; radius: 4

17. The center is the midopint of the diameter:

$$\left(\frac{-3+7}{2}, \frac{5+3}{2}\right) = \left(\frac{4}{2}, \frac{8}{2}\right) = (2, 4)$$

Use the center and either endpoint of the diameter to find the radius. We use the point $(7, 3)$.

$$r = \sqrt{(7-2)^2 + (3-4)^2} = \sqrt{5^2 + (-1)^2} =$$
$$\sqrt{25+1} = \sqrt{26}$$

The equation of the circle is $(x-2)^2 + (y-4)^2 = (\sqrt{26})^2$, or $(x-2)^2 + (y-4)^2 = 26$.

19. The relation is a function, because no two ordered pairs have the same first coordinate and different second co-ordinates. The domain is the set of first coordinates: $\{-2, 0, 1, 2, 7\}$. The range is the set of second coordinates: $\{-7, -4, -2, 2, 7\}$.

21. This is the graph of a function, because there is no vertical line that crosses the graph more than once.

23. This is the graph of a function, because there is no vertical line that crosses the graph more than once.

25. We can substitute any real number for x. Thus, the domain is the set of all real numbers, or $(-\infty, \infty)$.

27. Find the inputs that make the denominator zero:

$$x^2 - 6x + 5 = 0$$
$$(x-1)(x-5) = 0$$
$$x-1 = 0 \ \text{ or } \ x-5 = 0$$
$$x = 1 \ \text{ or } \qquad x = 5$$

The domain is $\{x | x \neq 1 \text{ and } x \neq 5\}$, or $(-\infty, 1) \cup (1, 5) \cup (5, \infty)$.

29.

The inputs on the x axis extend from -4 to 4, so the domain is $[-4, 4]$.

The outputs on the y-axis extend from 0 to 4, so the range is $[0, 4]$.

31. We graph $y = x^3 - 7$ in the window $[-10, 10, -15, 15]$.

Every point on the x-axis corresponds to a point on the graph, so the domain is the set of all real numbers, or $(-\infty, \infty)$.

Each point on the y-axis also corresponds to a point on the graph, so the range is the set of all real numbers, or $(-\infty, \infty)$.

33. $f(x) = x^2 - x - 3$

a) $f(0) = 0^2 - 0 - 3 = -3$

b) $f(-3) = (-3)^2 - (-3) - 3 = 9 + 3 - 3 = 9$

c) $\quad f(a-1) = (a-1)^2 - (a-1) - 3$
$$= a^2 - 2a + 1 - a + 1 - 3$$
$$= a^2 - 3a - 1$$

d) First we find $f(x+h)$.

$$f(x+h) = (x+h)^2 - (x+h) - 3$$
$$= x^2 + 2xh + h^2 - x - h - 3$$

Then

$$\frac{f(x+h) - f(x)}{h}$$
$$= \frac{x^2 + 2xh + h^2 - x - h - 3 - (x^2 - x - 3)}{h}$$
$$= \frac{x^2 + 2xh + h^2 - x - h - 3 - x^2 + x + 3}{h}$$
$$= \frac{2xh + h^2 - h}{h}$$
$$= \frac{h(2x + h - 1)}{h} = 2x + h - 1$$

35. a) Yes. Each input is 10 more than the one that precedes it.

b) Yes. Each output is 12.4 more than the one that precedes it.

c) Yes. Constant changes in inputs result in constant changes in outputs.

37. $m = \dfrac{4-4}{-3-5} = \dfrac{0}{-8} = 0$

39. $\quad -2x - y = 7$

$$-y = 2x + 7$$
$$y = -2x - 7$$

Slope: -2; y-intercept: $(0, -7)$

41. Graph $y = -\dfrac{1}{4}x + 3$.

Plot the y-intercept, $(0, 3)$. We can think of the slope as $\dfrac{-1}{4}$. Start at $(0, 3)$ and find another point by moving down 1 unit and right 4 units. We have the point $(4, 2)$.

We could also think of the slope as $\dfrac{1}{-4}$. Then we can start at $(0, 3)$ and find another point by moving up 1 unit and left 4 units. We have the point $(-4, 4)$. Connect the three points and draw the graph.

43.
$$y - y_1 = m(x - x_1)$$
$$y - (-1) = 3(x - (-2))$$
$$y + 1 = 3(x + 2)$$
$$y + 1 = 3x + 6$$
$$y = 3x + 5$$

45. $\quad 3x - 2y = 8 \qquad\qquad 6x - 4y = 2$
$$y = \frac{3}{2}x - 4 \qquad\qquad y = \frac{3}{2}x - \frac{1}{2}$$

The lines have the same slope, $\dfrac{3}{2}$, and different y-intercepts, $(0, -4)$ and $\left(0, -\dfrac{1}{2}\right)$, so they are parallel.

47. The slope of $y = \dfrac{3}{2}x + 7$ is $\dfrac{3}{2}$ and the slope of $y = -\dfrac{2}{3}x - 4$ is $-\dfrac{2}{3}$. Since $\dfrac{3}{2}\left(-\dfrac{2}{3}\right) = -1$, the lines are perpendicular.

49. $\quad 2x + 3y = 4$
$$3y = -2x + 4$$
$$y = -\frac{2}{3}x + \frac{4}{3}; \ m = -\frac{2}{3}$$

The slope of a line perpendicular to this line is the negative reciprocal of $-\dfrac{2}{3}$, or $\dfrac{3}{2}$.

We use the slope-intercept equation to find the y-intercept.
$$y = mx + b$$
$$-1 = \frac{3}{2} \cdot 1 + b$$
$$-1 = \frac{3}{2} + b$$
$$-\frac{5}{2} = b$$

Then the equation of the desired line is $y = \dfrac{3}{2}x - \dfrac{5}{2}$.

51. a) $T(d) = 10d + 20$
$$T(5) = 10(5) + 20 = 70°C$$
$$T(20) = 10(20) + 20 = 220°C$$
$$T(1000) = 10(1000) + 20 = 10{,}020°C$$

b) 5600 km is the maximum depth. Domain: $[0, 5600]$.

53. $f(x) = \begin{cases} -x, & \text{for } x \leq -4, \\ \dfrac{1}{2}x + 1, & \text{for } x > -4 \end{cases}$

We create the graph in two parts. Graph $f(x) = -x$ for inputs less than or equal to -4. Then graph $f(x) = \dfrac{1}{2}x + 1$ for inputs greater than -4.

55. $f(x) = \begin{cases} \dfrac{x^2 - 1}{x + 1}, & \text{for } x \neq -1, \\ 3, & \text{for } x = -1 \end{cases}$

We create the graph in two parts. Graph $f(x) = \dfrac{x^2 - 1}{x + 1}$ for inputs not equal to -1. Then graph $f(x) = 3$ for $x = -1$.

57. $f(x) = [[x - 3]]$

59. The length of the rectangle is $2x$. The width is the second coordinate of the point (x, y) on the circle. The circle has center $(0, 0)$ and radius 2, so its equation is $x^2 + y^2 = 4$ and $y = \sqrt{4 - x^2}$. Thus the area of the rectangle is given by $A(x) = 2x\sqrt{4 - x^2}$.

61. $(f-g)(6) = f(6) - g(6)$
$ = \sqrt{6-2} - (6^2 - 1)$
$ = \sqrt{4} - (36 - 1)$
$ = 2 - 35$
$ = -33$

63. $(f+g)(-1) = f(-1) + g(-1)$
$ = \sqrt{-1-2} + ((-1)^2 - 1)$
$ = \sqrt{-3} + (1 - 1)$
Since $\sqrt{-3}$ is not a real number, $(f+g)(-1)$ does not exist.

65. a) The domain of f, g, $f+g$, $f-g$, and fg is all real numbers. Since $g\left(\dfrac{1}{2}\right) = 0$, the domain of f/g is
$\left\{x \middle| x \ne \dfrac{1}{2}\right\}$.

b) $(f+g)(x) = (3x^2 + 4x) + (2x - 1) = 3x^2 + 6x - 1$
$(f-g)(x) = (3x^2 + 4x) - (2x - 1) = 3x^2 + 4x - 2x + 1 = 3x^2 + 2x + 1$
$(fg)(x) = (3x^2 + 4x)(2x - 1) = 6x^3 - 3x^2 + 8x^2 - 4x = 6x^3 + 5x^2 - 4x$
$(f/g)(x) = \dfrac{3x^2 + 4x}{2x - 1}$

67. $f(x) = 3 - x^2$
$f(x+h) = 3 - (x+h)^2 = 3 - (x^2 + 2xh + h^2) = 3 - x^2 - 2xh - h^2$
$\dfrac{f(x+h) - f(x)}{h}$
$= \dfrac{3 - x^2 - 2xh - h^2 - (3 - x^2)}{h}$
$= \dfrac{3 - x^2 - 2xh - h^2 - 3 + x^2}{h}$
$= \dfrac{-2xh - h^2}{h}$
$= \dfrac{h(-2x - h)}{h}$
$= -2x - h$

69. a) Domain of f = domain of g = all real numbers, so domain of $f \circ g$ = domain of $g \circ f$ = all real numbers.

b) $f \circ g(x) = f(2x - 1)$
$ = 3(2x - 1)^2 + 4(2x - 1)$
$ = 3(4x^2 - 4x + 1) + 4(2x - 1)$
$ = 12x^2 - 12x + 3 + 8x - 4$
$ = 12x^2 - 4x - 1$
$(g \circ f)(x) = g(3x^2 + 4x)$
$ = 2(3x^2 + 4x) - 1$
$ = 6x^2 + 8x - 1$

71. $f(x) = 4x^2 + 9$, $g(x) = 5x - 1$. Answers may vary.

73. $y^2 = x^2 + 3$

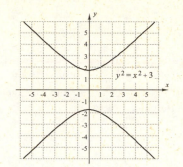

The graph is symmetric with respect to the x-axis, the y-axis, and the origin.

Replace y with $-y$ to test algebraically for symmetry with respect to the x-axis.
$$(-y)^2 = x^2 + 3$$
$$y^2 = x^2 + 3$$
The resulting equation is equivalent to the original equation, so the graph is symmetric with respect to the x-axis.

Replace x with $-x$ to test algebraically for symmetry with respect to the y-axis.
$$y^2 = (-x)^2 + 3$$
$$y^2 = x^2 + 3$$
The resulting equation is equivalent to the original equation, so the graph is symmetric with respect to the y-axis.

Replace x and $-x$ and y with $-y$ to test for symmetry with respect to the origin.
$$(-y)^2 = (-x)^2 + 3$$
$$y^2 = x^2 + 3$$
The resulting equation is equivalent to the original equation, so the graph is symmetric with respect to the origin.

75. $y = x^2$

The graph is symmetric with respect to the y-axis. It is not symmetric with respect to the x-axis or the origin.

Replace y with $-y$ to test algebraically for symmetry with respect to the x-axis.
$$-y = x^2$$
$$y = -x^2$$
The resulting equation is not equivalent to the original equation, so the graph is not symmetric with respect to the x-axis.

Replace x with $-x$ to test algebraically for symmetry with respect to the y-axis.

$$y = (-x)^2$$
$$y = x^2$$

The resulting equation is equivalent to the original equation, so the graph is symmetric with respect to the y-axis.

Replace x and $-x$ and y with $-y$ to test for symmetry with respect to the origin.

$$-y = (-x)^2$$
$$-y = x^2$$
$$y = -x^2$$

The resulting equation is not equivalent to the original equation, so the graph is not symmetric with respect to the origin.

77. $y = x^4 - x^2$

The graph is symmetric with respect to the y-axis. It is not symmetric with respect to the x-axis or the origin.

Replace y with $-y$ to test algebraically for symmetry with respect to the x-axis.

$$-y = x^4 - x^2$$
$$y = -x^4 + x^2$$

The resulting equation is not equivalent to the original equation, so the graph is not symmetric with respect to the x-axis.

Replace x with $-x$ to test algebraically for symmetry with respect to the y-axis.

$$y = (-x)^4 - (-x)^2$$
$$y = x^4 - x^2$$

The resulting equation is equivalent to the original equation, so the graph is symmetric with respect to the y-axis.

Replace x and $-x$ and y with $-y$ to test for symmetry with respect to the origin.

$$-y = (-x)^4 - (-x)^2$$
$$-y = x^4 - x^2$$
$$y = -x^4 + x^2$$

The resulting equation is not equivalent to the original equation, so the graph is not symmetric with respect to the origin.

79. The graph is symmetric with respect to the y-axis, so the function is even.

81. The graph is symmetric with respect to the y-axis, so the function is even.

83. $f(x) = x^3 - 2x + 4$

$$f(-x) = (-x)^3 - 2(-x) + 4 = -x^3 + 2x + 4$$
$$f(x) \neq f(-x), \text{ so } f \text{ is not even.}$$
$$-f(x) = -(x^3 - 2x + 4) = -x^3 + 2x - 4$$
$$f(-x) \neq -f(x), \text{ so } f \text{ is not odd.}$$

Thus, $f(x) = x^3 - 2x + 4$ is neither even or odd.

85. $f(x) = |x|$

$$f(-x) = |-x| = |x|$$
$$f(x) = f(-x), \text{ so } f \text{ is even.}$$

87. $f(x) = \dfrac{10x}{x^2 + 1}$

$$f(-x) = \frac{10(-x)}{(-x)^2 + 1} = -\frac{10x}{x^2 + 1}$$
$$f(x) \neq f(-x), \text{ so } f(x) \text{ is not even.}$$
$$-f(x) = -\frac{10x}{x^2 + 1}$$
$$f(-x) = -f(x), \text{ so } f \text{ is odd.}$$

89. Shape: $t(x) = \sqrt{x}$

Turn $t(x)$ upside down (that is, reflect it across the x-axis): $h(x) = -t(x) = -\sqrt{x}$.

Shift $h(x)$ right 3 units: $g(x) = h(x - 3) = -\sqrt{x - 3}$.

Shift $g(x)$ up 4 units: $f(x) = g(x) + 4 = -\sqrt{x - 3} + 4$.

91. The graph is shifted right 1 unit so each x-coordinate is increased by 1. We plot and connect $(-4, 3)$, $(-2, 0)$, $(1, 1)$ and $(5, -2)$.

93. Each y-coordinate is multiplied by -2. We plot and connect $(-5, -6)$, $(-3, 0)$, $(0, -2)$ and $(4, 4)$.

95. $f(x) = \dfrac{x + 3}{8 - 4x}$

When $x = 2$, the denominator is 0, so 2 is not in the domain of the function. Thus, the domain is $(-\infty, 2) \cup (2, \infty)$ and answer B is correct.

97. a) Enter the data in a graphing calculator and use the linear regression option. We get

$$y = -21.9x + 588.3.$$

In 2009, $x = 2009 - 1995 = 14$. Then we have

$$y = -21.9(14) + 588.3 \approx 282.$$

b) The correlation coefficient is approximately -0.9903. This indicates that the line fits the data well.

99. In the graph of $y = f(cx)$, the constant c stretches or shrinks the graph of $y = f(x)$ horizontally. The constant c in $y = cf(x)$ stretches or shrinks the graph of $y = f(x)$ vertically. For $y = f(cx)$, the x-coordinates of $y = f(x)$ are divided by c; for $y = cf(x)$, the y-coordinates of $y = f(x)$ are multiplied by c.

101. $f(x) = \dfrac{\sqrt{1-x}}{x - |x|}$

We cannot find the square root of a negative number, so $x \leq 1$. Division by zero is undefined, so $x < 0$.

Domain of f is $\{x | x < 0\}$, or $(-\infty, 0)$.

103. Let $f(x)$ and $g(x)$ be odd functions. Then by definition, $f(-x) = -f(x)$, or $f(x) = -f(-x)$, and $g(-x) = -g(x)$, or $g(x) = -g(-x)$. Thus $(f+g)(x) = f(x) + g(x) = -f(-x) + [-g(-x)] = -[f(-x) + g(-x)] = -(f+g)(-x)$ and $f + g$ is odd.

Chapter 1 Test

1.

$$5x - 2y = -10$$

2. $d = \sqrt{(5 - (-1))^2 + (8 - 5)^2} = \sqrt{6^2 + 3^2} = \sqrt{36 + 9} = \sqrt{45} \approx 6.708$

3. $m = \left(\dfrac{-2 + (-4)}{2}, \dfrac{6 + 3}{2}\right) = \left(\dfrac{-6}{2}, \dfrac{9}{2}\right) = \left(-3, \dfrac{9}{2}\right)$

4. $[x - (-1)]^2 + (y - 2)^2 = (\sqrt{5})^2$

$$(x + 1)^2 + (y - 2)^2 = 5$$

5. $(x + 4)^2 + (y - 5)^2 = 36$

$$[x - (-4)]^2 + (y - 5)^2 = 6^2$$

Center: $(-4, 5)$; radius: 6

6. a) The relation is a function, because no two ordered pairs have the same first coordinate and different second coordinates.

b) The domain is the set of first coordinates: $\{-4, 0, 1, 3\}$.

c) The range is the set of second coordinates: $\{0, 5, 7\}$.

7. $f(x) = 2x^2 - x + 5$

a) $f(-1) = 2(-1)^2 - (-1) + 5 = 2 + 1 + 5 = 8$

b) $f(a + 2) = 2(a + 2)^2 - (a + 2) + 5$

$$= 2(a^2 + 4a + 4) - (a + 2) + 5$$
$$= 2a^2 + 8a + 8 - a - 2 + 5$$
$$= 2a^2 + 7a + 11$$

8. a)

$$f(x) = |x - 2| + 3$$

b) Each point on the x-axis corresponds to a point on the graph, so the domain is the set of all real numbers, or $(-\infty, \infty)$.

c) The number 3 is the smallest output on the y-axis and every number greater than 3 is also an output, so the range is $[3, \infty)$.

9. The input 4 results in a denominator of 0. Thus the domain is $\{x | x \neq 4\}$, or $(-\infty, 4) \cup (4, \infty)$.

10. We can substitute any real number for x. Thus the domain is the set of all real numbers, or $(-\infty, \infty)$.

11. We cannot find the square root of a negative number. Thus $25 - x^2 \geq 0$ and the domain is $\{x | -5 \leq x \leq 5\}$, or $[-5, 5]$.

12. a) This is not the graph of a function, because we can find a vertical line that crosses the graph more than once.

b) This is the graph of a function, because there is no vertical line that crosses the graph more than once.

13. $m = \dfrac{5 - \dfrac{2}{3}}{-2 - (-2)} = \dfrac{\dfrac{13}{3}}{0}$

The slope is not defined.

14. $m = \dfrac{12 - (-10)}{-8 - 4} = \dfrac{22}{-12} = -\dfrac{11}{6}$

15. $m = \dfrac{6 - 6}{\dfrac{3}{4} - (-5)} = \dfrac{0}{\dfrac{23}{4}} = 0$

16. We have the points $(1995, 5.3)$ and $(2004, 6.8)$.

$$m = \dfrac{6.8 - 5.3}{2004 - 1995} = \dfrac{1.5}{9} \approx 0.167$$

The average rate of change in weekend attendance from 1995 to 2004 was about 0.167 million per year, or about 167,000 per year.

17. $-3x + 2y = 5$

$$2y = 3x + 5$$
$$y = \dfrac{3}{2}x + \dfrac{5}{2}$$

Slope: $\dfrac{3}{2}$; y-intercept: $\left(0, \dfrac{5}{2}\right)$

18. $y = mx + b$

$$y = -\frac{5}{8}x - 5$$

19. First we find the slope:

$$m = \frac{-2 - 4}{3 - (-5)} = \frac{-6}{8} = -\frac{3}{4}$$

Use the point-slope equation.

Using $(-5, 4)$: $y - 4 = -\frac{3}{4}(x - (-5))$, or

$$y - 4 = -\frac{3}{4}(x + 5)$$

Using $(3, -2)$: $y - (-2) = -\frac{3}{4}(x - 3)$, or

$$y + 2 = -\frac{3}{4}(x - 3)$$

In either case, we have $y = -\frac{3}{4}x + \frac{1}{4}$.

20. First find the slope of the given line.

$$x + 2y = -6$$
$$2y = -x - 6$$
$$y = -\frac{1}{2}x - 3; \quad m = -\frac{1}{2}$$

A line parallel to the given line has slope $-\frac{1}{2}$. We use the point-slope equation.

$$y - 3 = -\frac{1}{2}(x - (-1))$$
$$y - 3 = -\frac{1}{2}(x + 1)$$
$$y - 3 = -\frac{1}{2}x - \frac{1}{2}$$
$$y = -\frac{1}{2}x + \frac{5}{2}$$

21. $\quad 2x + 3y = -12 \qquad\qquad 2y - 3x = 8$

$$y = -\frac{2}{3}x - 4 \qquad\qquad y = \frac{3}{2}x + 4$$

$m_1 = -\frac{2}{3}, m_2 = \frac{3}{2}; m_1 m_2 = -1$.

The lines are perpendicular.

22. Answers will vary depending on the data points used.

We will use $(4, 5.06)$ and $(24, 12.03)$.

$$m = \frac{12.03 - 5.06}{24 - 4} = \frac{6.97}{20} = 0.3485$$

Now we use the point-slope equation with $(4, 5.06)$.

$$y - 5.06 = 0.3485(x - 4)$$
$$y - 5.06 = 0.3485x - 1.394$$
$$y = 0.3485x + 3.666$$

In 2010, $x = 2010 - 1972 = 38$. Then we have
$y = 0.3485(38) + 3.666 \approx \$16.91$.

23. $f(x) = \begin{cases} x^2, & \text{for } x < -1, \\ |x|, & \text{for } -1 \le x \le 1, \\ \sqrt{x - 1}, & \text{for } x > 1 \end{cases}$

24. Since $-1 \le -\frac{7}{8} \le 1$, $f\left(-\frac{7}{8}\right) = \left|-\frac{7}{8}\right| = \frac{7}{8}$.

Since $5 > 1$, $f(5) = \sqrt{5 - 1} = \sqrt{4} = 2$.

Since $-4 < -1$, $f(-4) = (-4)^2 = 16$.

25. $(f - g)(-1) = f(-1) - g(-1) =$
$(-1)^2 - 4(-1) + 3 - \sqrt{3 - (-1)} =$
$1 + 4 + 3 - \sqrt{4} = 1 + 4 + 3 - 2 = 6$

26. a) We can substitute any real number for x, so the domain of f is the set of all real numbers, or $(-\infty, \infty)$.

    b) We must have $x - 3 \ge 0$, or $x \ge 3$, so the domain of g is $[3, \infty)$.

    c) $(f - g)(x) = f(x) - g(x) = x^2 - \sqrt{x - 3}$

    d) $(fg)(x) = f(x) \cdot g(x) = x^2\sqrt{x - 3}$

    e) $g(3) = 0$, so the domain of $(f/g)(x)$ is $(3, \infty)$. (See parts (a) and (b).)

27. Answers may vary. $f(x) = x^4$, $g(x) = 2x - 7$

28. $f(x) = x^2 + 4$

$f(x + h) = (x + h)^2 + 4 = x^2 + 2xh + h^2 + 4$

$$\frac{f(x + h) - f(x)}{h}$$
$$= \frac{x^2 + 2xh + h^2 + 4 - (x^2 + 4)}{h}$$
$$= \frac{x^2 + 2xh + h^2 + 4 - x^2 - 4}{h}$$
$$= \frac{2xh + h^2}{h}$$
$$= \frac{h(2x + h)}{h} = 2x + h$$

29. $(f \circ g)(x) = f(g(x)) = f(x^2 + 1) = \sqrt{x^2 + 1 - 5} = \sqrt{x^2 - 4}$

$(g \circ f)(x) = g(f(x)) = g(\sqrt{x - 5}) = (\sqrt{x - 5})^2 + 1 = x - 5 + 1 = x - 4$

30. The inputs for $f(x)$ must be such that $x - 5 \ge 0$, or $x \ge 5$. Then for $(f \circ g)(x)$ we must have $g(x) \ge 5$, or $x^2 + 1 \ge 5$, or $x^2 \ge 4$. Then the domain of $(f \circ g)(x)$ is $(-\infty, -2] \cup [2, \infty)$.

Since we can substitute any real number for x in g, the domain of $(g \circ f)(x)$ is the same as the domain of $f(x)$, $[5, \infty)$.

31. $y = x^4 - 2x^2$

Replace y with $-y$ to test for symmetry with respect to the x-axis.

$$-y = x^4 - 2x^2$$
$$y = -x^4 + 2x^2$$

The resulting equation is not equivalent to the original equation, so the graph is not symmetric with respect to the x-axis.

Replace x with $-x$ to test for symmetry with respect to the y-axis.

$$y = (-x)^4 - 2(-x)^2$$
$$y = x^4 - 2x^2$$

The resulting equation is equivalent to the original equation, so the graph is symmetric with respect to the y-axis.

Replace x with $-x$ and y with $-y$ to test for symmetry with respect to the origin.

$$-y = (-x)^4 - 2(-x)^2$$
$$-y = x^4 - 2x^2$$
$$y = -x^4 + 2x^2$$

The resulting equation is not equivalent to the original equation, so the graph is not symmetric with respect to the origin.

32. $f(x) = \dfrac{2x}{x^2 + 1}$

$$f(-x) = \frac{2(-x)}{(-x)^2 + 1} = -\frac{2x}{x^2 + 1}$$

$f(x) \neq f(-x)$, so f is not even.

$$-f(x) = -\frac{2x}{x^2 + 1}$$

$f(-x) = -f(x)$, so f is odd.

33. Shape: $h(x) = x^2$

Shift $h(x)$ right 2 units: $g(x) = h(x - 2) = (x - 2)^2$

Shift $g(x)$ down 1 unit: $f(x) = (x - 2)^2 - 1$

34. Shape: $h(x) = x^2$

Shift $h(x)$ left 2 units: $g(x) = h(x + 2) = (x + 2)^2$

Shift $g(x)$ down 3 units: $f(x) = (x + 2)^2 - 3$

35. Each y-coordinate is multiplied by $-\dfrac{1}{2}$. We plot and connect $(-5, 1)$, $(-3, -2)$, $(1, 2)$ and $(4, -1)$.

36. Each x-coordinate on the graph of $y = f(x)$ is divided by 3 on the graph of $y = f(3x)$. Thus the point $\left(\dfrac{-3}{3}, 1\right)$, or $(-1, 1)$ is on the graph of $f(3x)$.

Chapter 2

Functions, Equations, and Inequalities

Exercise Set 2.1

1. $4x + 5 = 21$

    $4x = 16$ Subtracting 5 on both sides

      $x = 4$ Dividing by 4 on both sides

The solution is 4.

3. $4x + 3 = 0$

    $4x = -3$ Subtracting 3 on both sides

     $x = -\dfrac{3}{4}$ Dividing by 4 on both sides

The solution is $-\dfrac{3}{4}$.

5. $3 - x = 12$

    $-x = 9$ Subtracting 3 on both sides

     $x = -9$ Multiplying (or dividing) by -1 on both sides

The solution is -9.

7. $8 = 5x - 3$

  $11 = 5x$ Adding 3 on both sides

  $\dfrac{11}{5} = x$ Dividing by 5 on both sides

The solution is $\dfrac{11}{5}$.

9. $y + 1 = 2y - 7$

    $1 = y - 7$ Subtracting y on both sides

    $8 = y$ Adding 7 on both sides

The solution is 8.

11. $2x + 7 = x + 3$

   $x + 7 = 3$ Subtracting x on both sides

     $x = -4$ Subtracting 7 on both sides

The solution is -4.

13. $3x - 5 = 2x + 1$

   $x - 5 = 1$ Subtracting $2x$ on both sides

     $x = 6$ Adding 5 on both sides

The solution is 6.

15. $4x - 5 = 7x - 2$

    $-5 = 3x - 2$ Subtracting $4x$ on both sides

    $-3 = 3x$ Adding 2 on both sides

    $-1 = x$ Dividing by 3 on both sides

The solution is -1.

17. $5x - 2 + 3x = 2x + 6 - 4x$

   $8x - 2 = 6 - 2x$ Collecting like terms

   $8x + 2x = 6 + 2$ Adding $2x$ and 2 on both sides

     $10x = 8$ Collecting like terms

       $x = \dfrac{8}{10}$ Dividing by 10 on both sides

       $x = \dfrac{4}{5}$ Simplifying

The solution is $\dfrac{4}{5}$.

19. $7(3x + 6) = 11 - (x + 2)$

  $21x + 42 = 11 - x - 2$ Using the distributive property

  $21x + 42 = 9 - x$ Collecting like terms

  $21x + x = 9 - 42$ Adding x and subtracting 42 on both sides

     $22x = -33$ Collecting like terms

       $x = -\dfrac{33}{22}$ Dividing by 22 on both sides

       $x = -\dfrac{3}{2}$ Simplifying

The solution is $-\dfrac{3}{2}$.

21. $3(x + 1) = 5 - 2(3x + 4)$

  $3x + 3 = 5 - 6x - 8$ Removing parentheses

  $3x + 3 = -6x - 3$ Collecting like terms

  $9x + 3 = -3$ Adding $6x$

     $9x = -6$ Subtracting 3

      $x = -\dfrac{2}{3}$ Dividing by 9

The solution is $-\dfrac{2}{3}$.

23. $2(x - 4) = 3 - 5(2x + 1)$

  $2x - 8 = 3 - 10x - 5$ Using the distributive property

  $2x - 8 = -10x - 2$ Collecting like terms

    $12x = 6$ Adding $10x$ and 8 on both sides

      $x = \dfrac{1}{2}$ Dividing by 12 on both sides

The solution is $\dfrac{1}{2}$.

25. *Familiarize.* Let $w =$ the wholesale sales of bottled water in 2001, in billions of dollars. An increase of 45% over this amount is $45\% \cdot w$, or $0.45w$.

Translate.

Wholesale sales in 2001   plus   increase in sales   is   Wholesale sales in 2005.

$$w \quad + \quad 0.45w \quad = \quad 10$$

Carry out. We solve the equation.

$$w + 0.45w = 10$$
$$1.45w = 10$$
$$w = \frac{10}{1.45}$$
$$w \approx 6.9$$

Check. 45% of 6.9 is 0.45(6.9) or about 3.1, and 6.9+3.1 = 10. The answer checks.

State. Wholesale sales of bottled water in 2001 were about $6.9 billion.

27. *Familiarize.* Let $d =$ the average credit card debt per household in 1990.

Translate.

Debt in 1990   plus   additional debt   is   debt in 2004.

$$d \quad + \quad 6346 \quad = \quad 9312$$

Carry out. We solve the equation.

$$d + 6346 = 9312$$
$$d = 2966 \quad \text{Subtracting } 6346$$

Check. $2966 + $6346 = $9312, so the answer checks.

State. The average credit card debt in 1990 was $2966 per household.

29. *Familiarize.* Let $d =$ the number of gigabytes of digital data stored in a typical household in 2004.

Translate.

Data stored in 2004   plus   additional data   is   data stored in 2010.

$$d \quad + \quad 4024 \quad = \quad 4430$$

Carry out. We solve the equation.

$$d + 4024 = 4430$$
$$d = 406 \quad \text{Subtracting } 4024$$

Check. 406 + 4024 = 4430, so the answer checks.

State. In 2004, 406 GB of digital data were stored in a typical household.

31. *Familiarize.* Let $c =$ the average daily calorie requirement for many adults.

Translate.

1560 calories   is $\frac{3}{4}$ of   the average daily requirement for many adults.

$$1560 \quad = \quad \frac{3}{4} \quad \cdot \quad c$$

Carry out. We solve the equation.

$$1560 = \frac{3}{4} \cdot c$$
$$\frac{4}{3} \cdot 1560 = c \quad \text{Multiplying by } \frac{4}{3}$$
$$2080 = c$$

Check. $\frac{3}{4} \cdot 2080 = 1560$, so the answer checks.

State. The average daily calorie requirement for many adults is 2080 calories.

33. *Familiarize.* Let $v =$ the number of ABC viewers, in millions. Then $v + 1.7 =$ the number of CBS viewers and $v - 1.7 =$ the number of NBC viewers.

Translate.

ABC viewers   plus   CBS viewers   plus   NBC viewers   is   total viewers.

$$v \quad + \quad (v+1.7) \quad + \quad (v-1.7) \quad = \quad 29.1$$

Carry out.

$$v + (v + 1.7) + (v - 1.7) = 29.1$$
$$3v = 29.1$$
$$v = 9.7$$

Then $v+1.7 = 9.7+1.7 = 11.4$ and $v-1.7 = 9.7-1.7 = 8.0$.

Check. 9.7 + 11.4 + 8.0 = 29.1, so the answer checks.

State. ABC had 9.7 million viewers, CBS had 11.4 million viewers, and NBC had 8.0 million viewers.

35. *Familiarize.* Let $P =$ the amount Tamisha borrowed. We will use the formula $I = Prt$ to find the interest owed. For $r = 5\%$, or 0.05, and $t = 1$, we have $I = P(0.05)(1)$, or $0.05P$.

Translate.

Amount borrowed   plus   interest   is $1365.

$$P \quad + \quad 0.05P \quad = \quad 1365$$

Carry out. We solve the equation.

$$P + 0.05P = 1365$$
$$1.05P = 1365 \quad \text{Adding}$$
$$P = 1300 \quad \text{Dividing by } 1.05$$

Check. The interest due on a loan of $1300 for 1 year at a rate of 5% is $1300(0.05)(1)$, or $65, and $1300 + $65 = $1365. The answer checks.

State. Tamisha borrowed $1300.

37. *Familiarize.* Let $s =$ Ryan's sales for the month. Then his commission is 8% of s, or $0.08s$.

Translate.

Base salary   plus   commission   is   total pay.

$$1500 \quad + \quad 0.08s \quad = \quad 2284$$

Carry out. We solve the equation.

$$1500 + 0.08s = 2284$$
$$0.08s = 784 \quad \text{Subtracting } 1500$$
$$s = 9800$$

Check. 8% of $9800, or 0.08($9800), is $784 and $1500 + $784 = $2284. The answer checks.

State. Ryan's sales for the month were $9800.

39. Familiarize. Let $d =$ the number of miles Diego traveled in the cab.

Translate.

Carry out. We solve the equation.

$$1.75 + 1.50 \cdot d = 19.75$$
$$1.5d = 18 \quad \text{Subtracting 1.75}$$
$$d = 12 \quad \text{Dividing by 1.5}$$

Check. If Diego travels 12 mi, his fare is $1.75+$1.50·12, or $1.75 + $18, or $19.75. The answer checks.

State. Diego traveled 12 mi in the cab.

41. Familiarize. We make a drawing.

We let $x =$ the measure of angle A. Then $5x =$ the measure of angle B, and $x - 2 =$ the measure of angle C. The sum of the angle measures is 180°.

Translate.

$$x + 5x + x - 2 = 180$$

Carry out. We solve the equation.

$$x + 5x + x - 2 = 180$$
$$7x - 2 = 180$$
$$7x = 182$$
$$x = 26$$

If $x = 26$, then $5x = 5 \cdot 26$, or 130, and $x - 2 = 26 - 2$, or 24.

Check. The measure of angle B, 130°, is five times the measure of angle A, 26°. The measure of angle C, 24°, is 2° less than the measure of angle A, 26°. The sum of the angle measures is $26° + 130° + 24°$, or 180°. The answer checks.

State. The measure of angles A, B, and C are 26°, 130°, and 24°, respectively.

43. Familiarize. Using the labels on the drawing in the text, we let $w =$ the width of the test plot and $w + 25 =$ the length, in meters. Recall that for a rectangle, Perimeter $= 2 \cdot$ length $+ 2 \cdot$ width.

Translate.

$$\underbrace{\text{Perimeter}}_{322} = \underbrace{2 \cdot \text{ length}}_{= 2(w+25)} + \underbrace{2 \cdot \text{ width}}_{2 \cdot w}$$

Carry out. We solve the equation.

$$322 = 2(w + 25) + 2 \cdot w$$
$$322 = 2w + 50 + 2w$$
$$322 = 4w + 50$$
$$272 = 4w$$
$$68 = w$$

When $w = 68$, then $w + 25 = 68 + 25 = 93$.

Check. The length is 25 m more than the width: $93 = 68 + 25$. The perimeter is $2 \cdot 93 + 2 \cdot 68$, or $186 + 136$, or 322 m. The answer checks.

State. The length is 93 m; the width is 68 m.

45. Familiarize. Let $l =$ the length of the soccer field and $l - 35 =$ the width, in yards.

Translate. We use the formula for the perimeter of a rectangle. We substitute 330 for P and $l - 35$ for w.

$$P = 2l + 2w$$
$$330 = 2l + 2(l - 35)$$

Carry out. We solve the equation.

$$330 = 2l + 2(l - 35)$$
$$330 = 2l + 2l - 70$$
$$330 = 4l - 70$$
$$400 = 4l$$
$$100 = l$$

If $l = 100$, then $l - 35 = 100 - 35 = 65$.

Check. The width, 65 yd, is 35 yd less than the length, 100 yd. Also, the perimeter is

$$2 \cdot 100 \text{ yd} + 2 \cdot 65 \text{ yd} = 200 \text{ yd} + 130 \text{ yd} = 330 \text{ yd}.$$

The answer checks.

State. The length of the field is 100 yd, and the width is 65 yd.

47. Familiarize. Let $w =$ the number of pounds of Kimiko's body weight that is water.

Translate.

$$\underset{0.5}{50\%} \underset{\times}{\text{ of }} \underset{135}{\underbrace{\text{body weight}}} \underset{=}{\text{ is }} \underset{w}{\text{ water.}}$$

Carry out. We solve the equation.

$$0.5 \times 135 = w$$
$$67.5 = w$$

Check. Since 50% of 138 is 67.5, the answer checks.

State. 67.5 lb of Kimiko's body weight is water.

49. *Familiarize*. We make a drawing. Let t = the number of hours the passenger train travels before it overtakes the freight train. Then $t+1$ = the number of hours the freight train travels before it is overtaken by the passenger train. Also let d = the distance the trains travel.

Freight train
60 mph $t+1$ hr d

Passenger train
80 mph t hr d

We can also organize the information in a table.

$$d = r \cdot t$$

	Distance	Rate	Time
Freight train	d	60	$t+1$
Passenger train	d	80	t

Translate. Using the formula $d = rt$ in each row of the table, we get two equations.

$$d = 60(t+1) \text{ and } d = 80t.$$

Since the distances are the same, we have the equation

$$60(t+1) = 80t.$$

Carry out. We solve the equation.

$$60(t+1) = 80t$$
$$60t + 60 = 80t$$
$$60 = 20t$$
$$3 = t$$

When $t = 3$, then $t+1 = 3+1 = 4$.

Check. In 4 hr the freight train travels $60 \cdot 4$, or 240 mi. In 3 hr the passenger train travels $80 \cdot 3$, or 240 mi. Since the distances are the same, the answer checks.

State. It will take the passenger train 3 hr to overtake the freight train.

51. *Familiarize*. Let t = the number of hours it takes the kayak to travel 36 mi upstream. The kayak travels upstream at a rate of $12 - 4$, or 8 mph.

Translate. We use the formula $d = rt$.

$$36 = 8 \cdot t$$

Carry out. We solve the equation.

$$36 = 8 \cdot t$$
$$4.5 = t$$

Check. At a rate of 8 mph, in 4.5 hr the kayak travels $8(4.5)$, or 36 mi. The answer checks.

State. It takes the kayak 4.5 hr to travel 36 mi upstream.

53. *Familiarize*. Let t = the number of hours it will take the plane to travel 1050 mi into the wind. The speed into the headwind is $450 - 30$, or 420 mph.

Translate. We use the formula $d = rt$.

$$1050 = 420 \cdot t$$

Carry out. We solve the equation.

$$1050 = 420 \cdot t$$
$$2.5 = t$$

Check. At a rate of 420 mph, in 2.5 hr the plane travels $420(2.5)$, or 1050 mi. The answer checks.

State. It will take the plane 2.5 hr to travel 1050 mi into the wind.

55. *Familiarize*. Let x = the amount invested at 3% interest. Then $5000 - x$ = the amount invested at 4%. We organize the information in a table, keeping in mind the simple interest formula, $I = Prt$.

	Amount invested	Interest rate	Time	Amount of interest
3% investment	x	3%, or 0.03	1 yr	$x(0.03)(1)$, or $0.03x$
4% investment	$5000-x$	4%, or 0.04	1 yr	$(5000-x)(0.04)(1)$, or $0.04(5000-x)$
Total	5000			176

Translate.

Interest on 3% investment	plus	interest on 4% investment	is $176.
$0.03x$	$+$	$0.04(5000-x)$	$=\ 176$

Carry out. We solve the equation.

$$0.03x + 0.04(5000 - x) = 176$$
$$0.03x + 200 - 0.04x = 176$$
$$-0.01x + 200 = 176$$
$$-0.01x = -24$$
$$x = 2400$$

If $x = 2400$, then $5000 - x = 5000 - 2400 = 2600$.

Check. The interest on $2400 at 3% for 1 yr is $2400(0.03)(1) = \$72$. The interest on $2600 at 4% for 1 yr is $2600(0.04)(1) = \$104$. Since $\$72 + \$104 = \$176$, the answer checks.

State. $2400 was invested at 3%, and $2600 was invested at 4%.

57. *Familiarize*. Let c = the calcium content of the cheese, in mg. Then $2c + 4$ = the calcium content of the yogurt.

Translate.

Carry out. We solve the equation.

$$c + (2c + 4) = 676$$
$$3c + 4 = 676$$
$$3c = 672$$
$$c = 224$$

Then $2c + 4 = 2 \cdot 224 + 4 = 448 + 4 = 452$.

Check. $224 + 452 = 676$, so the answer checks.

State. The cheese contains 224 mg of calcium, and the yogurt contains 452 mg.

59. *Familiarize.* Let s = the total holiday spending in 2004, in billions of dollars.

Translate.

$\underbrace{\$29.4 \text{ billion}}$ is 7% of $\underbrace{\text{total spending}}$.
$\downarrow$ $\downarrow$ $\downarrow$ $\downarrow$ $\downarrow$
29.4 $=$ 0.07 $\cdot$ s

Carry out. We solve the equation.

$$29.4 = 0.07 \cdot s$$
$$420 = s \qquad \text{Dividing by } 0.07$$

Check. 7% of 420 is $0.07 \cdot 420$, or 29.4, so the answer checks.

State. Total holiday spending in 2004 was $420 billion.

61. *Familiarize.* Let n = the number of inches the volcano rises in a year. We will express one-half mile in inches:

$$\frac{1}{2} \text{ mi} \times \frac{5280 \text{ ft}}{1 \text{ mi}} \times \frac{12 \text{ in.}}{1 \text{ ft}} = 31,680 \text{ in.}$$

Translate.

$\underbrace{50,000 \text{ yr}}$ times $\underbrace{\substack{\text{number of inches} \\ \text{per year}}}$ is $\underbrace{31,680 \text{ in.}}$
$\downarrow$ $\downarrow$ $\downarrow$ $\downarrow$ $\downarrow$
$50,000$ $\cdot$ n $=$ $31,680$

Carry out. We solve the equation.

$$50,000n = 31,680$$
$$n = 0.6336$$

Check. Rising at a rate of 0.6336 in. per year, in 50,000 yr the volcano will rise $50,000(0.6336)$, or 31,680 in. The answer checks.

State. On average, the volcano rises 0.6336 in. in a year.

63.
$$x + 5 = 0 \qquad \text{Setting } f(x) = 0$$
$$x + 5 - 5 = 0 - 5 \quad \text{Subtracting 5 on both sides}$$
$$x = -5$$
The zero of the function is -5.

65.
$$-x + 18 = 0 \qquad \text{Setting } f(x) = 0$$
$$-x + 18 + x = 0 + x \quad \text{Adding } x \text{ on both sides}$$
$$18 = x$$
The zero of the function is 18.

67.
$$16 - x = 0 \qquad \text{Setting } f(x) = 0$$
$$16 - x + x = 0 + x \quad \text{Adding } x \text{ on both sides}$$
$$16 = x$$
The zero of the function is 16.

69.
$$x + 12 = 0 \qquad \text{Setting } f(x) = 0$$
$$x + 12 - 12 = 0 - 12 \quad \text{Subtracting 12 on}$$
$$\text{both sides}$$
$$x = -12$$
The zero of the function is -12.

71.
$$-x + 6 = 0 \qquad \text{Setting } f(x) = 0$$
$$-x + 6 + x = 0 + x \quad \text{Adding } x \text{ on both sides}$$
$$6 = x$$
The zero of the function is 6.

73.
$$20 - x = 0 \qquad \text{Setting } f(x) = 0$$
$$20 - x + x = 0 + x \quad \text{Adding } x \text{ on both sides}$$
$$20 = x$$
The zero of the function is 20.

75.
$$x - 6 = 0 \quad \text{Setting } f(x) = 0$$
$$x = 6 \quad \text{Adding 6 on both sides}$$
The zero of the function is 6.

77.
$$-x + 15 = 0 \quad \text{Setting } f(x) = 0$$
$$15 = x \quad \text{Adding } x \text{ on both sides}$$
The zero of the function is 15.

79. a) The graph crosses the x-axis at $(4, 0)$. This is the x-intercept.

b) The zero of the function is the first coordinate of the x-intercept. It is 4.

81. a) The graph crosses the x-axis at $(-2, 0)$. This is the x-intercept.

b) The zero of the function is the first coordinate of the x-intercept. It is -2.

83. a) The graph crosses the x-axis at $(-4, 0)$. This is the x-intercept.

b) The zero of the function is the first coordinate of the x-intercept. It is -4.

85.
$$A = \frac{1}{2}bh$$
$$2A = bh \quad \text{Multiplying by 2 on both sides}$$
$$\frac{2A}{h} = b \quad \text{Dividing by } h \text{ on both sides}$$

87.
$$P = 2l + 2w$$
$$P - 2l = 2w \qquad \text{Subtracting } 2l \text{ on both sides}$$
$$\frac{P - 2l}{2} = w \qquad \text{Dividing by 2 on both sides}$$

89.
$$A = \frac{1}{2}h(b_1 + b_2)$$
$$2A = h(b_1 + b_2) \quad \text{Multiplying by 2 on both sides}$$
$$\frac{2A}{b_1 + b_2} = h \quad \text{Dividing by } b_1 + b_2 \text{ on both sides}$$

91.
$$V = \frac{4}{3}\pi r^3$$
$$3V = 4\pi r^3 \quad \text{Multiplying by 3 on both sides}$$
$$\frac{3V}{4r^3} = \pi \quad \text{Dividing by } 4r^3 \text{ on both sides}$$

93.
$$F = \frac{9}{5}C + 32$$
$$F - 32 = \frac{9}{5}C \quad \text{Subtracting 32 on both sides}$$
$$\frac{5}{9}(F - 32) = C \quad \text{Multiplying by } \frac{5}{9} \text{ on both sides}$$

95. $Ax + By = C$
$$Ax = C - By \quad \text{Subtracting } By \text{ on both sides}$$
$$A = \frac{C - By}{x} \quad \text{Dividing by } x \text{ on both sides}$$

97. $2w + 2h + l = p$
$$2h = p - 2w - l \quad \text{Subtracting } 2w \text{ and } l$$
$$h = \frac{p - 2w - l}{2} \quad \text{Dividing by 2}$$

99. $2x - 3y = 6$
$$-3y = 6 - 2x \quad \text{Subtracting } 2x$$
$$y = \frac{6 - 2x}{-3}, \text{ or} \quad \text{Dividing by } -3$$
$$\frac{2x - 6}{3}$$

101. $a = b + bcd$
$$a = b(1 + cd) \quad \text{Factoring}$$
$$\frac{a}{1 + cd} = b \quad \text{Dividing by } 1 + cd$$

103. $z = xy - xy^2$
$$z = x(y - y^2) \quad \text{Factoring}$$
$$\frac{z}{y - y^2} = x \quad \text{Dividing by } y - y^2$$

105. Left to the student

107. Discussion and Writing

109. First find the slope of the given line.
$$3x + 4y = 7$$
$$4y = -3x + 7$$
$$y = -\frac{3}{4}x + \frac{7}{4}$$

The slope is $-\frac{3}{4}$. Now write a slope-intersect equation of the line containing $(-1, 4)$ with slope $-\frac{3}{4}$.
$$y - 4 = -\frac{3}{4}[x - (-1)]$$
$$y - 4 = -\frac{3}{4}(x + 1)$$
$$y - 4 = -\frac{3}{4}x - \frac{3}{4}$$
$$y = -\frac{3}{4}x + \frac{13}{4}$$

111. The domain of f is the set of all real numbers as is the domain of g, so the domain of $f + g$ is the set of all real numbers, or $(-\infty, \infty)$.

113. $(f - g)(x) = f(x) - g(x) = 2x - 1 - (3x + 6) =$
$2x - 1 - 3x - 6 = -x - 7$

115. $f(x) = 7 - \frac{3}{2}x = -\frac{3}{2}x + 7$

The function can be written in the form $y = mx + b$, so it is a linear function.

117. $f(x) = x^2 + 1$ cannot be written in the form $f(x) = mx + b$, so it is not a linear function.

119.
$$2x - \{x - [3x - (6x + 5)]\} = 4x - 1$$
$$2x - \{x - [3x - 6x - 5]\} = 4x - 1$$
$$2x - \{x - [-3x - 5]\} = 4x - 1$$
$$2x - \{x + 3x + 5\} = 4x - 1$$
$$2x - \{4x + 5\} = 4x - 1$$
$$2x - 4x - 5 = 4x - 1$$
$$-2x - 5 = 4x - 1$$
$$-6x - 5 = -1$$
$$-6x = 4$$
$$x = -\frac{2}{3}$$

The solution is $-\frac{2}{3}$.

121. The size of the cup was reduced 8 oz $-$ 6 oz, or 2 oz, and $\frac{2 \text{ oz}}{8 \text{ oz}} = 0.25$, so the size was reduced 25%. The price per ounce of the 8 oz cup was $\frac{89\cent}{8 \text{ oz}}$, or 11.25¢/oz. The price per ounce of the 6 oz cup is $\frac{71\cent}{6 \text{ oz}}$, or 11.83̄¢/oz. Since the price per ounce was not reduced, it is clear that the price per ounce was not reduced by the same percent as the size of the cup. The price was increased by $11.8\overline{3} - 11.125$¢, or 0.708̄3¢ per ounce. This is an increase of $\frac{0.708\overline{3}\cent}{11.8\overline{3}\cent} \approx 0.064$, or about 6.4% per ounce.

123. We use a proportion to determine the number of calories c burned running for 75 minutes, or 1.25 hr.
$$\frac{720}{1} = \frac{c}{1.25}$$
$$720(1.25) = c$$
$$900 = c$$

Next we use a proportion to determine how long the person would have to walk to use 900 calories. Let t represent this time, in hours. We express 90 min as 1.5 hr.
$$\frac{1.5}{480} = \frac{t}{900}$$
$$\frac{900(1.5)}{480} = t$$
$$2.8125 = t$$

Then, at a rate of 4 mph, the person would have to walk $4(2.8125)$, or 11.25 mi.

Exercise Set 2.2

1. $\sqrt{-3} = \sqrt{-1 \cdot 3} = \sqrt{-1} \cdot \sqrt{3} = i\sqrt{3}$, or $\sqrt{3}i$

3. $\sqrt{-25} = \sqrt{-1 \cdot 25} = \sqrt{-1} \cdot \sqrt{25} = i \cdot 5 = 5i$

5. $-\sqrt{-33} = -\sqrt{-1 \cdot 33} = -\sqrt{-1} \cdot \sqrt{33} = -i\sqrt{33}$, or $-\sqrt{33}i$

7. $-\sqrt{-81} = -\sqrt{-1 \cdot 81} = -\sqrt{-1} \cdot \sqrt{81} = -i \cdot 9 = -9i$

9. $\sqrt{-98} = \sqrt{-1 \cdot 98} = \sqrt{-1} \cdot \sqrt{98} = i\sqrt{49 \cdot 2} =$
$i \cdot 7\sqrt{2} = 7i\sqrt{2}$, or $7\sqrt{2}i$

11. $\quad (-5 + 3i) + (7 + 8i)$
$= (-5 + 7) + (3i + 8i)$ Collecting the real parts
                        and the imaginary parts

$= 2 + (3 + 8)i$
$= 2 + 11i$

13. $\quad (4 - 9i) + (1 - 3i)$
$= (4 + 1) + (-9i - 3i)$ Collecting the real parts
                        and the imaginary parts

$= 5 + (-9 - 3)i$
$= 5 - 12i$

15. $\quad (12 + 3i) + (-8 + 5i)$
$= (12 - 8) + (3i + 5i)$
$= 4 + 8i$

17. $\quad (-1 - i) + (-3 - i)$
$= (-1 - 3) + (-i - i)$
$= -4 - 2i$

19. $(3 + \sqrt{-16}) + (2 + \sqrt{-25}) = (3 + 4i) + (2 + 5i)$
$= (3 + 2) + (4i + 5i)$
$= 5 + 9i$

21. $\quad (10 + 7i) - (5 + 3i)$
$= (10 - 5) + (7i - 3i)$ The 5 and the $3i$ are
                        both being subtracted.
$= 5 + 4i$

23. $\quad (13 + 9i) - (8 + 2i)$
$= (13 - 8) + (9i - 2i)$ The 8 and the $2i$ are
                        both being subtracted.
$= 5 + 7i$

25. $\quad (6 - 4i) - (-5 + i)$
$= [6 - (-5)] + (-4i - i)$
$= (6 + 5) + (-4i - i)$
$= 11 - 5i$

27. $\quad (-5 + 2i) - (-4 - 3i)$
$= [-5 - (-4)] + [2i - (-3i)]$
$= (-5 + 4) + (2i + 3i)$
$= -1 + 5i$

29. $\quad (4 - 9i) - (2 + 3i)$
$= (4 - 2) + (-9i - 3i)$
$= 2 - 12i$

31. $\quad 7i(2 - 5i)$
$= 14i - 35i^2$ Using the distributive law
$= 14i + 35$ $i^2 = -1$
$= 35 + 14i$ Writing in the form $a + bi$

33. $\quad -2i(-8 + 3i)$
$= 16i - 6i^2$ Using the distributive law
$= 16i + 6$ $i^2 = -1$
$= 6 + 16i$ Writing in the form $a + bi$

35. $\quad (1 + 3i)(1 - 4i)$
$= 1 - 4i + 3i - 12i^2$ Using FOIL
$= 1 - 4i + 3i - 12(-1)$ $i^2 = -1$
$= 1 - i + 12$
$= 13 - i$

37. $\quad (2 + 3i)(2 + 5i)$
$= 4 + 10i + 6i + 15i^2$ Using FOIL
$= 4 + 10i + 6i - 15$ $i^2 = -1$
$= -11 + 16i$

39. $\quad (-4 + i)(3 - 2i)$
$= -12 + 8i + 3i - 2i^2$ Using FOIL
$= -12 + 8i + 3i + 2$ $i^2 = -1$
$= -10 + 11i$

41. $\quad (8 - 3i)(-2 - 5i)$
$= -16 - 40i + 6i + 15i^2$
$= -16 - 40i + 6i - 15$ $i^2 = -1$
$= -31 - 34i$

43. $\quad (3 + \sqrt{-16})(2 + \sqrt{-25})$
$= (3 + 4i)(2 + 5i)$
$= 6 + 15i + 8i + 20i^2$
$= 6 + 15i + 8i - 20$ $i^2 = -1$
$= -14 + 23i$

45. $(5 - 4i)(5 + 4i) = 5^2 - (4i)^2$
$= 25 - 16i^2$
$= 25 + 16$ $i^2 = -1$
$= 41$

47. $\quad (3 + 2i)(3 - 2i)$
$= 9 - 6i + 6i - 4i^2$
$= 9 - 6i + 6i + 4$ $i^2 = -1$
$= 13$

49. $\quad (7 - 5i)(7 + 5i)$
$= 49 + 35i - 35i - 25i^2$
$= 49 + 35i - 35i + 25$ $i^2 = -1$
$= 74$

51. $(4 + 2i)^2$

$= 16 + 2 \cdot 4 \cdot 2i + (2i)^2$ Recall $(A + B)^2 =$
$\qquad\qquad\qquad\qquad\qquad A^2 + 2AB + B^2$

$= 16 + 16i + 4i^2$

$= 16 + 16i - 4 \qquad\qquad i^2 = -1$

$= 12 + 16i$

53. $(-2 + 7i)^2$

$= (-2)^2 + 2(-2)(7i) + (7i)^2$ Recall $(A + B)^2 =$
$\qquad\qquad\qquad\qquad\qquad\qquad A^2 + 2AB + B^2$

$= 4 - 28i + 49i^2$

$= 4 - 28i - 49 \qquad\qquad i^2 = -1$

$= -45 - 28i$

55. $(1 - 3i)^2$

$= 1^2 - 2 \cdot 1 \cdot (3i) + (3i)^2$

$= 1 - 6i + 9i^2$

$= 1 - 6i - 9 \qquad i^2 = -1$

$= -8 - 6i$

57. $(-1 - i)^2$

$= (-1)^2 - 2(-1)(i) + i^2$

$= 1 + 2i + i^2$

$= 1 + 2i - 1 \qquad i^2 = -1$

$= 2i$

59. $(3 + 4i)^2$

$= 9 + 2 \cdot 3 \cdot 4i + (4i)^2$

$= 9 + 24i + 16i^2$

$= 9 + 24i - 16 \qquad i^2 = -1$

$= -7 + 24i$

61. $\dfrac{3}{5 - 11i}$

$= \dfrac{3}{5 - 11i} \cdot \dfrac{5 + 11i}{5 + 11i}$ $5 - 11i$ is the conjugate
$\qquad\qquad\qquad\qquad\quad$ of $5 + 11i$.

$= \dfrac{3(5 + 11i)}{(5 - 11i)(5 + 11i)}$

$= \dfrac{15 + 33i}{25 - 121i^2}$

$= \dfrac{15 + 33i}{25 + 121} \qquad i^2 = -1$

$= \dfrac{15 + 33i}{146}$

$= \dfrac{15}{146} + \dfrac{33}{146}i$ Writing in the form $a + bi$

63. $\dfrac{5}{2 + 3i}$

$= \dfrac{5}{2 + 3i} \cdot \dfrac{2 - 3i}{2 - 3i}$ $2 - 3i$ is the conjugate
$\qquad\qquad\qquad\qquad\quad$ of $2 + 3i$.

$= \dfrac{5(2 - 3i)}{(2 + 3i)(2 - 3i)}$

$= \dfrac{10 - 15i}{4 - 9i^2}$

$= \dfrac{10 - 15i}{4 + 9} \qquad i^2 = -1$

$= \dfrac{10 - 15i}{13}$

$= \dfrac{10}{13} - \dfrac{15}{13}i$ Writing in the form $a + bi$

65. $\dfrac{4 + i}{-3 - 2i}$

$= \dfrac{4 + i}{-3 - 2i} \cdot \dfrac{-3 + 2i}{-3 + 2i}$ $-3 + 2i$ is the conjugate
$\qquad\qquad\qquad\qquad\qquad$ of the divisor.

$= \dfrac{(4 + i)(-3 + 2i)}{(-3 - 2i)(-3 + 2i)}$

$= \dfrac{-12 + 5i + 2i^2}{9 - 4i^2}$

$= \dfrac{-12 + 5i - 2}{9 + 4} \qquad i^2 = -1$

$= \dfrac{-14 + 5i}{13}$

$= -\dfrac{14}{13} + \dfrac{5}{13}i$ Writing in the form $a + bi$

67. $\dfrac{5 - 3i}{4 + 3i}$

$= \dfrac{5 - 3i}{4 + 3i} \cdot \dfrac{4 - 3i}{4 - 3i}$ $4 - 3i$ is the conjugate
$\qquad\qquad\qquad\qquad\quad$ of $4 + 3i$.

$= \dfrac{(5 - 3i)(4 - 3i)}{(4 + 3i)(4 - 3i)}$

$= \dfrac{20 - 27i + 9i^2}{16 - 9i^2}$

$= \dfrac{20 - 27i - 9}{16 + 9} \qquad i^2 = -1$

$= \dfrac{11 - 27i}{25}$

$= \dfrac{11}{25} - \dfrac{27}{25}i$ Writing in the form $a + bi$

69.
$$\frac{2+\sqrt{3}i}{5-4i}$$

$$= \frac{2+\sqrt{3}i}{5-4i} \cdot \frac{5+4i}{5+4i} \qquad \begin{array}{l} 5+4i \text{ is the conjugate} \\ \text{of the divisor.} \end{array}$$

$$= \frac{(2+\sqrt{3}i)(5+4i)}{(5-4i)(5+4i)}$$

$$= \frac{10+8i+5\sqrt{3}i+4\sqrt{3}i^2}{25-16i^2}$$

$$= \frac{10+8i+5\sqrt{3}i-4\sqrt{3}}{25+16} \qquad i^2 = -1$$

$$= \frac{10-4\sqrt{3}+(8+5\sqrt{3})i}{41}$$

$$= \frac{10-4\sqrt{3}}{41}+\frac{8+5\sqrt{3}}{41}i \qquad \begin{array}{l} \text{Writing in the} \\ \text{form } a+bi \end{array}$$

71.
$$\frac{1+i}{(1-i)^2}$$

$$= \frac{1+i}{1-2i+i^2}$$

$$= \frac{1+i}{1-2i-1} \qquad i^2 = -1$$

$$= \frac{1+i}{-2i}$$

$$= \frac{1+i}{-2i} \cdot \frac{2i}{2i} \qquad \begin{array}{l} 2i \text{ is the conjugate} \\ \text{of } -2i. \end{array}$$

$$= \frac{(1+i)(2i)}{(-2i)(2i)}$$

$$= \frac{2i+2i^2}{-4i^2}$$

$$= \frac{2i-2}{4} \qquad i^2 = -1$$

$$= -\frac{2}{4}+\frac{2}{4}i$$

$$= -\frac{1}{2}+\frac{1}{2}i$$

73.
$$\frac{4-2i}{1+i}+\frac{2-5i}{1+i}$$

$$= \frac{6-7i}{1+i} \qquad \text{Adding}$$

$$= \frac{6-7i}{1+i} \cdot \frac{1-i}{1-i} \qquad \begin{array}{l} 1-i \text{ is the conjugate} \\ \text{of } 1+i. \end{array}$$

$$= \frac{(6-7i)(1-i)}{(1+i)(1-i)}$$

$$= \frac{6-13i+7i^2}{1-i^2}$$

$$= \frac{6-13i-7}{1+1} \qquad i^2 = -1$$

$$= \frac{-1-13i}{2}$$

$$= -\frac{1}{2}-\frac{13}{2}i$$

75. $i^{11} = i^{10} \cdot i = (i^2)^5 \cdot i = (-1)^5 \cdot i = -1 \cdot i = -i$

77. $i^{35} = i^{34} \cdot i = (i^2)^{17} \cdot i = (-1)^{17} \cdot i = -1 \cdot i = -i$

79. $i^{64} = (i^2)^{32} = (-1)^{32} = 1$

81. $(-i)^{71} = (-1 \cdot i)^{71} = (-1)^{71} \cdot i^{71} = -i^{70} \cdot i =$
$-(i^2)^{35} \cdot i = -(-1)^{35} \cdot i = -(-1)i = i$

83. $(5i)^4 = 5^4 \cdot i^4 = 625(i^2)^2 = 625(-1)^2 = 625 \cdot 1 = 625$

85. Left to the student

87. Discussion and Writing

89. First find the slope of the given line.
$$3x - 6y = 7$$
$$-6y = -3x + 7$$
$$y = \frac{1}{2}x - \frac{7}{6}$$

The slope is $\frac{1}{2}$. The slope of the desired line is the opposite of the reciprocal of $\frac{1}{2}$, or -2. Write a slope-intercept equation of the line containing $(3, -5)$ with slope -2.
$$y - (-5) = -2(x - 3)$$
$$y + 5 = -2x + 6$$
$$y = -2x + 1$$

91. The domain of f is the set of all real numbers as is the domain of g. When $x = -\frac{5}{3}$, $g(x) = 0$, so the domain of f/g is $\left(-\infty, -\frac{5}{3}\right) \cup \left(-\frac{5}{3}, \infty\right)$.

93. $(f/g)(2) = \dfrac{f(2)}{g(2)} = \dfrac{2^2+4}{3 \cdot 2+5} = \dfrac{4+4}{6+5} = \dfrac{8}{11}$

95. $(a+bi)+(a-bi) = 2a$, a real number. Thus, the statement is true.

97. $(a+bi)(c+di) = (ac-bd)+(ad+bc)i$. The conjugate of the product is $(ac-bd)-(ad+bc)i =$ $(a-bi)(c-di)$, the product of the conjugates of the individual complex numbers. Thus, the statement is true.

99. $z\bar{z} = (a+bi)(a-bi) = a^2 - b^2i^2 = a^2 + b^2$

Exercise Set 2.3

1. $(2x-3)(3x-2) = 0$

$2x-3 = 0 \quad or \quad 3x-2 = 0 \quad \begin{array}{l} \text{Using the principle} \\ \text{of zero products} \end{array}$

$2x = 3 \quad or \quad 3x = 2$

$x = \dfrac{3}{2} \quad or \quad x = \dfrac{2}{3}$

The solutions are $\dfrac{3}{2}$ and $\dfrac{2}{3}$.

3. $x^2 - 8x - 20 = 0$

$(x - 10)(x + 2) = 0$ Factoring

$x - 10 = 0$ *or* $x + 2 = 0$ Using the principle
                                    of zero products

$\quad\quad x = 10$ *or* $\quad\quad x = -2$

The solutions are 10 and -2.

5. $3x^2 + x - 2 = 0$

$(3x - 2)(x + 1) = 0$ Factoring

$3x - 2 = 0$ *or* $x + 1 = 0$ Using the principle
                                    of zero products

$x = \dfrac{2}{3}$ *or* $\quad\quad x = -1$

The solutions are $\dfrac{2}{3}$ and -1.

7. $4x^2 - 12 = 0$

$\quad 4x^2 = 12$

$\quad\quad x^2 = 3$

$\quad\quad x = \sqrt{3}$ *or* $x = -\sqrt{3}$ Using the principle
                                               of square roots

The solutions are $\sqrt{3}$ and $-\sqrt{3}$.

9. $3x^2 = 21$

$\quad x^2 = 7$

$\quad\; x = \sqrt{7}$ *or* $x = -\sqrt{7}$ Using the principle
                                             of square roots

The solutions are $\sqrt{7}$ and $-\sqrt{7}$.

11. $5x^2 + 10 = 0$

$\quad 5x^2 = -10$

$\quad\; x^2 = -2$

$\quad\quad x = \sqrt{2}i$ *or* $x = -\sqrt{2}i$

The solutions are $\sqrt{2}i$ and $-\sqrt{2}i$.

13. $2x^2 - 34 = 0$

$\quad 2x^2 = 34$

$\quad\; x^2 = 17$

$\quad\quad x = \sqrt{17}$ *or* $x = -\sqrt{17}$

The solutions are $\sqrt{17}$ and $-\sqrt{17}$.

15. $\quad 2x^2 = 6x$

$2x^2 - 6x = 0$ Subtracting $6x$ on both sides

$2x(x - 3) = 0$

$2x = 0$ *or* $x - 3 = 0$

$\; x = 0$ *or* $\quad\; x = 3$

The solutions are 0 and 3.

17. $3y^3 - 5y^2 - 2y = 0$

$y(3y^2 - 5y - 2) = 0$

$y(3y + 1)(y - 2) = 0$

$y = 0$ *or* $3y + 1 = 0$ *or* $y - 2 = 0$

$y = 0$ *or* $\quad\quad y = -\dfrac{1}{3}$ *or* $\quad\; y = 2$

The solutions are $-\dfrac{1}{3}$, 0 and 2.

19. $7x^3 + x^2 - 7x - 1 = 0$

$x^2(7x + 1) - (7x + 1) = 0$

$(x^2 - 1)(7x + 1) = 0$

$(x + 1)(x - 1)(7x + 1) = 0$

$x + 1 = 0$ *or* $x - 1 = 0$ *or* $7x + 1 = 0$

$x = -1$ *or* $\quad\; x = 1$ *or* $\quad\quad x = -\dfrac{1}{7}$

The solutions are -1, $-\dfrac{1}{7}$, and 1.

21. a) The graph crosses the x-axis at $(-4, 0)$ and at $(2, 0)$.
          These are the x-intercepts.

   b) The zeros of the function are the first coordinates of
      the x-intercepts of the graph. They are -4 and 2.

23. a) The graph crosses the x-axis at $(-1, 0)$ and at $(3, 0)$.
          These are the x-intercepts.

   b) The zeros of the function are the first coordinates of
      the x-intercepts of the graph. They are -1 and 3.

25. a) The graph crosses the x-axis at $(-2, 0)$ and at $(2, 0)$.
          These are the x-intercepts.

   b) The zeros of the function are the first coordinates of
      the x-intercepts of the graph. They are -2 and 2.

27. $\quad x^2 + 6x = 7$

$x^2 + 6x + 9 = 7 + 9$ Completing the square:
                          $\frac{1}{2} \cdot 6 = 3$ and $3^2 = 9$

$\quad (x + 3)^2 = 16$ Factoring

$\quad\quad x + 3 = \pm 4$ Using the principle
                              of square roots

$\quad\quad\quad x = -3 \pm 4$

$x = -3 - 4$ *or* $x = -3 + 4$

$x = -7$ $\quad\quad$ *or* $x = 1$

The solutions are -7 and 1.

29. $\quad\quad x^2 = 8x - 9$

$\quad\; x^2 - 8x = -9$ Subtracting $8x$

$x^2 - 8x + 16 = -9 + 16$ Completing the square:

$\quad\quad\quad \frac{1}{2}(-8) = -4$ and $(-4)^2 = 16$

$\quad (x - 4)^2 = 7$ Factoring

$\quad\quad x - 4 = \pm\sqrt{7}$ Using the principle
                                    of square roots

$\quad\quad\quad x = 4 \pm \sqrt{7}$

The solutions are $4 - \sqrt{7}$ and $4 + \sqrt{7}$, or $4 \pm \sqrt{7}$.

31. $x^2 + 8x + 25 = 0$

$\quad x^2 + 8x = -25$ Subtracting 25

$x^2 + 8x + 16 = -25 + 16$ Completing the
                             square:

$\quad\quad\quad \frac{1}{2} \cdot 8 = 4$ and $4^2 = 16$

$\quad (x + 4)^2 = -9$ Factoring

$\quad\quad x + 4 = \pm 3i$ Using the principle
                               of square roots

$\quad\quad\quad x = -4 \pm 3i$

The solutions are $-4 - 3i$ and $-4 + 3i$, or $-4 \pm 3i$.

33. $3x^2 + 5x - 2 = 0$

$3x^2 + 5x = 2$ Adding 2

$x^2 + \dfrac{5}{3}x = \dfrac{2}{3}$ Dividing by 3

$x^2 + \dfrac{5}{3}x + \dfrac{25}{36} = \dfrac{2}{3} + \dfrac{25}{36}$ Completing the square:

$\dfrac{1}{2} \cdot \dfrac{5}{3} = \dfrac{5}{6}$ and $\left(\dfrac{5}{6}\right)^2 = \dfrac{25}{36}$

$\left(x + \dfrac{5}{6}\right)^2 = \dfrac{49}{36}$ Factoring and simplifying

$x + \dfrac{5}{6} = \pm\dfrac{7}{6}$ Using the principle of square roots

$x = -\dfrac{5}{6} \pm \dfrac{7}{6}$

$x = -\dfrac{5}{6} - \dfrac{7}{6}$ $\;or\;$ $x = -\dfrac{5}{6} + \dfrac{7}{6}$

$x = -\dfrac{12}{6}$ $\;or\;$ $x = \dfrac{2}{6}$

$x = -2$ $\;or\;$ $x = \dfrac{1}{3}$

The solutions are -2 and $\dfrac{1}{3}$.

35. $x^2 - 2x = 15$

$x^2 - 2x - 15 = 0$

$(x - 5)(x + 3) = 0$ Factoring

$x - 5 = 0$ $\;or\;$ $x + 3 = 0$

$x = 5$ $\;or\;$ $x = -3$

The solutions are 5 and -3.

37. $5m^2 + 3m = 2$

$5m^2 + 3m - 2 = 0$

$(5m - 2)(m + 1) = 0$ Factoring

$5m - 2 = 0$ $\;or\;$ $m + 1 = 0$

$m = \dfrac{2}{5}$ $\;or\;$ $m = -1$

The solutions are $\dfrac{2}{5}$ and -1.

39. $3x^2 + 6 = 10x$

$3x^2 - 10x + 6 = 0$

We use the quadratic formula. Here $a = 3$, $b = -10$, and $c = 6$.

$x = \dfrac{-b \pm \sqrt{b^2 - 4ac}}{2a}$

$= \dfrac{-(-10) \pm \sqrt{(-10)^2 - 4 \cdot 3 \cdot 6}}{2 \cdot 3}$ Substituting

$= \dfrac{10 \pm \sqrt{28}}{6} = \dfrac{10 \pm 2\sqrt{7}}{6}$

$= \dfrac{2(5 \pm \sqrt{7})}{2 \cdot 3} = \dfrac{5 \pm \sqrt{7}}{3}$

The solutions are $\dfrac{5 - \sqrt{7}}{3}$ and $\dfrac{5 + \sqrt{7}}{3}$, or $\dfrac{5 \pm \sqrt{7}}{3}$.

41. $x^2 + x + 2 = 0$

We use the quadratic formula. Here $a = 1$, $b = 1$, and $c = 2$.

$x = \dfrac{-b \pm \sqrt{b^2 - 4ac}}{2a}$

$= \dfrac{-1 \pm \sqrt{1^2 - 4 \cdot 1 \cdot 2}}{2 \cdot 1}$ Substituting

$= \dfrac{-1 \pm \sqrt{-7}}{2}$

$= \dfrac{-1 \pm \sqrt{7}i}{2} = -\dfrac{1}{2} \pm \dfrac{\sqrt{7}}{2}i$

The solutions are $-\dfrac{1}{2} - \dfrac{\sqrt{7}}{2}i$ and $-\dfrac{1}{2} + \dfrac{\sqrt{7}}{2}i$, or $-\dfrac{1}{2} \pm \dfrac{\sqrt{7}}{2}i$.

43. $5t^2 - 8t = 3$

$5t^2 - 8t - 3 = 0$

We use the quadratic formula. Here $a = 5$, $b = -8$, and $c = -3$.

$t = \dfrac{-b \pm \sqrt{b^2 - 4ac}}{2a}$

$= \dfrac{-(-8) \pm \sqrt{(-8)^2 - 4 \cdot 5(-3)}}{2 \cdot 5}$

$= \dfrac{8 \pm \sqrt{124}}{10} = \dfrac{8 \pm 2\sqrt{31}}{10}$

$= \dfrac{2(4 \pm \sqrt{31})}{2 \cdot 5} = \dfrac{4 \pm \sqrt{31}}{5}$

The solutions are $\dfrac{4 - \sqrt{31}}{5}$ and $\dfrac{4 + \sqrt{31}}{5}$, or $\dfrac{4 \pm \sqrt{31}}{5}$.

45. $3x^2 + 4 = 5x$

$3x^2 - 5x + 4 = 0$

We use the quadratic formula. Here $a = 3$, $b = -5$, and $c = 4$.

$x = \dfrac{-b \pm \sqrt{b^2 - 4ac}}{2a}$

$= \dfrac{-(-5) \pm \sqrt{(-5)^2 - 4 \cdot 3 \cdot 4}}{2 \cdot 3}$

$= \dfrac{5 \pm \sqrt{-23}}{6} = \dfrac{5 \pm \sqrt{23}i}{6}$

$= \dfrac{5}{6} \pm \dfrac{\sqrt{23}}{6}i$

The solutions are $\dfrac{5}{6} - \dfrac{\sqrt{23}}{6}i$ and $\dfrac{5}{6} + \dfrac{\sqrt{23}}{6}i$, or $\dfrac{5}{6} \pm \dfrac{\sqrt{23}}{6}i$.

47. $x^2 - 8x + 5 = 0$

We use the quadratic formula. Here $a = 1$, $b = -8$, and $c = 5$.

$$x = \frac{-b \pm \sqrt{b^2 - 4ac}}{2a}$$

$$= \frac{-(-8) \pm \sqrt{(-8)^2 - 4 \cdot 1 \cdot 5}}{2 \cdot 1}$$

$$= \frac{8 \pm \sqrt{44}}{2} = \frac{8 \pm 2\sqrt{11}}{2}$$

$$= \frac{2(4 \pm \sqrt{11})}{2} = 4 \pm \sqrt{11}$$

The solutions are $4 - \sqrt{11}$ and $4 + \sqrt{11}$, or $4 \pm \sqrt{11}$.

49. $\quad 3x^2 + x = 5$

$\quad 3x^2 + x - 5 = 0$

We use the quadratic formula. We have $a = 3$, $b = 1$, and $c = -5$.

$$x = \frac{-b \pm \sqrt{b^2 - 4ac}}{2a}$$

$$= \frac{-1 \pm \sqrt{1^2 - 4 \cdot 3 \cdot (-5)}}{2 \cdot 3}$$

$$= \frac{-1 \pm \sqrt{61}}{6}$$

The solutions are $\dfrac{-1 - \sqrt{61}}{6}$ and $\dfrac{-1 + \sqrt{61}}{6}$, or $\dfrac{-1 \pm \sqrt{61}}{6}$.

51. $\quad 2x^2 + 1 = 5x$

$\quad 2x^2 - 5x + 1 = 0$

We use the quadratic formula. We have $a = 2$, $b = -5$, and $c = 1$.

$$x = \frac{-b \pm \sqrt{b^2 - 4ac}}{2a}$$

$$= \frac{-(-5) \pm \sqrt{(-5)^2 - 4 \cdot 2 \cdot 1}}{2 \cdot 2} = \frac{5 \pm \sqrt{17}}{4}$$

The solutions are $\dfrac{5 - \sqrt{17}}{4}$ and $\dfrac{5 + \sqrt{17}}{4}$, or $\dfrac{5 \pm \sqrt{17}}{4}$.

53. $\quad 5x^2 + 2x = -2$

$\quad 5x^2 + 2x + 2 = 0$

We use the quadratic formula. We have $a = 5$, $b = 2$, and $c = 2$.

$$x = \frac{-b \pm \sqrt{b^2 - 4ac}}{2a}$$

$$= \frac{-2 \pm \sqrt{2^2 - 4 \cdot 5 \cdot 2}}{2 \cdot 5}$$

$$= \frac{-2 \pm \sqrt{-36}}{10} = \frac{-2 \pm 6i}{10}$$

$$= \frac{2(-1 \pm 3i)}{2 \cdot 5} = \frac{-1 \pm 3i}{5}$$

$$= -\frac{1}{5} \pm \frac{3}{5}i$$

The solutions are $-\dfrac{1}{5} - \dfrac{3}{5}i$ and $-\dfrac{1}{5} + \dfrac{3}{5}i$, or $-\dfrac{1}{5} \pm \dfrac{3}{5}i$.

55. $\quad\quad\quad 4x^2 = 8x + 5$

$\quad 4x^2 - 8x - 5 = 0$

$\quad a = 4$, $b = -8$, $c = -5$

$\quad b^2 - 4ac = (-8)^2 - 4 \cdot 4(-5) = 144$

Since $b^2 - 4ac > 0$, there are two different real-number solutions.

57. $\quad x^2 + 3x + 4 = 0$

$\quad a = 1$, $b = 3$, $c = 4$

$\quad b^2 - 4ac = 3^2 - 4 \cdot 1 \cdot 4 = -7$

Since $b^2 - 4ac < 0$, there are two different imaginary-number solutions.

59. $\quad 5t^2 - 7t = 0$

$\quad a = 5$, $b = -7$, $c = 0$

$\quad b^2 - 4ac = (-7)^2 - 4 \cdot 5 \cdot 0 = 49$

Since $b^2 - 4ac > 0$, there are two different real-number solutions.

61. $\quad\quad x^2 + 6x + 5 = 0 \quad$ Setting $f(x) = 0$

$\quad (x + 5)(x + 1) = 0 \quad$ Factoring

$\quad x + 5 = 0 \quad or \quad x + 1 = 0$

$\quad\quad\quad x = -5 \quad or \quad\quad\quad x = -1$

The zeros of the function are -5 and -1.

63. $x^2 - 3x - 3 = 0$

$\quad a = 1$, $b = -3$, $c = -3$

$$x = \frac{-b \pm \sqrt{b^2 - 4ac}}{2a}$$

$$= \frac{-(-3) \pm \sqrt{(-3)^2 - 4 \cdot 1 \cdot (-3)}}{2 \cdot 1}$$

$$= \frac{3 \pm \sqrt{9 + 12}}{2}$$

$$= \frac{3 \pm \sqrt{21}}{2}$$

The zeros of the function are $\dfrac{3 - \sqrt{21}}{2}$ and $\dfrac{3 + \sqrt{21}}{2}$, or $\dfrac{3 \pm \sqrt{21}}{2}$.

65. $x^2 - 5x + 1 = 0$

$\quad a = 1$, $b = -5$, $c = 1$

$$x = \frac{-b \pm \sqrt{b^2 - 4ac}}{2a}$$

$$= \frac{-(-5) \pm \sqrt{(-5)^2 - 4 \cdot 1 \cdot 1}}{2 \cdot 1}$$

$$= \frac{5 \pm \sqrt{25 - 4}}{2}$$

$$= \frac{5 \pm \sqrt{21}}{2}$$

The zeros of the function are $\dfrac{5 - \sqrt{21}}{2}$ and $\dfrac{5 + \sqrt{21}}{2}$, or $\dfrac{5 \pm \sqrt{21}}{2}$.

67. $x^2 + 2x - 5 = 0$

$a = 1,\ b = 2,\ c = -5$

$x = \dfrac{-b \pm \sqrt{b^2 - 4ac}}{2a}$

$= \dfrac{-2 \pm \sqrt{2^2 - 4 \cdot 1 \cdot (-5)}}{2 \cdot 1}$

$= \dfrac{-2 \pm \sqrt{4 + 20}}{2} = \dfrac{-2 \pm \sqrt{24}}{2}$

$= \dfrac{-2 \pm 2\sqrt{6}}{2} = -1 \pm \sqrt{6}$

The zeros of the function are $-1 + \sqrt{6}$ and $-1 - \sqrt{6}$, or $-1 \pm \sqrt{6}$.

69. $2x^2 - x + 4 = 0$

$a = 2,\ b = -1,\ c = 4$

$x = \dfrac{-b \pm \sqrt{b^2 - 4ac}}{2a}$

$= \dfrac{-(-1) \pm \sqrt{(-1)^2 - 4 \cdot 2 \cdot 4}}{2 \cdot 2}$

$= \dfrac{1 \pm \sqrt{-31}}{4} = \dfrac{1 \pm \sqrt{31}i}{4}$

$= \dfrac{1}{4} \pm \dfrac{\sqrt{31}}{4}i$

The zeros of the function are $\dfrac{1}{4} - \dfrac{\sqrt{31}}{4}i$ and $\dfrac{1}{4} + \dfrac{\sqrt{31}}{4}i$, or $\dfrac{1}{4} \pm \dfrac{\sqrt{31}}{4}i$.

71. $3x^2 - x - 1 = 0$

$a = 3,\ b = -1,\ c = -1$

$x = \dfrac{-b \pm \sqrt{b^2 - 4ac}}{2a}$

$= \dfrac{-(-1) \pm \sqrt{(-1)^2 - 4 \cdot 3 \cdot (-1)}}{2 \cdot 3}$

$= \dfrac{1 \pm \sqrt{13}}{6}$

The zeros of the function are $\dfrac{1 - \sqrt{13}}{6}$ and $\dfrac{1 + \sqrt{13}}{6}$, or $\dfrac{1 \pm \sqrt{13}}{6}$.

73. $5x^2 - 2x - 1 = 0$

$a = 5,\ b = -2,\ c = -1$

$x = \dfrac{-b \pm \sqrt{b^2 - 4ac}}{2a}$

$= \dfrac{-(-2) \pm \sqrt{(-2)^2 - 4 \cdot 5 \cdot (-1)}}{2 \cdot 5}$

$= \dfrac{2 \pm \sqrt{24}}{10} = \dfrac{2 \pm 2\sqrt{6}}{10}$

$= \dfrac{2(1 \pm \sqrt{6})}{2 \cdot 5} = \dfrac{1 \pm \sqrt{6}}{5}$

The zeros of the function are $\dfrac{1 - \sqrt{6}}{5}$ and $\dfrac{1 + \sqrt{6}}{5}$, or $\dfrac{1 \pm \sqrt{6}}{5}$.

75. $4x^2 + 3x - 3 = 0$

$a = 4,\ b = 3,\ c = -3$

$x = \dfrac{-b \pm \sqrt{b^2 - 4ac}}{2a}$

$= \dfrac{-3 \pm \sqrt{3^2 - 4 \cdot 4 \cdot (-3)}}{2 \cdot 4}$

$= \dfrac{-3 \pm \sqrt{57}}{8}$

The zeros of the function are $\dfrac{-3 - \sqrt{57}}{8}$ and $\dfrac{-3 + \sqrt{57}}{8}$, or $\dfrac{-3 \pm \sqrt{57}}{8}$.

77. $x^4 - 3x^2 + 2 = 0$

Let $u = x^2$.

$u^2 - 3u + 2 = 0$ Substituting u for x^2

$(u - 1)(u - 2) = 0$

$u - 1 = 0\ \ or\ \ u - 2 = 0$

$u = 1\ \ or\ \ \ \ \ \ u = 2$

Now substitute x^2 for u and solve for x.

$x^2 = 1\ \ \ or\ \ x^2 = 2$

$x = \pm 1\ \ or\ \ x = \pm\sqrt{2}$

The solutions are $-1,\ 1,\ -\sqrt{2}$, and $\sqrt{2}$.

79. $\ \ \ \ \ \ \ \ x^4 + 3x^2 = 10$

$x^4 + 3x^2 - 10 = 0$

Let $u = x^2$.

$u^2 + 3u - 10 = 0$ Substituting u for x^2

$(u + 5)(u - 2) = 0$

$u + 5 = 0\ \ \ or\ \ u - 2 = 0$

$u = -5\ \ or\ \ \ \ \ \ u = 2$

Now substitute x^2 for u and solve for x.

$x^2 = -5\ \ \ \ or\ \ x^2 = 2$

$x = \pm\sqrt{5}i\ \ or\ \ x = \pm\sqrt{2}$

The solutions are $-\sqrt{5}i,\ \sqrt{5}i,\ -\sqrt{2}$, and $\sqrt{2}$.

81. $y^4 + 4y^2 - 5 = 0$

Let $u = y^2$.

$u^2 + 4u - 5 = 0$ Substituting u for y^2

$(u - 1)(u + 5) = 0$

$u - 1 = 0\ \ or\ \ u + 5 = 0$

$u = 1\ \ or\ \ \ \ \ \ u = -5$

Substitute y^2 for u and solve for y.

$y^2 = 1\ \ \ or\ \ y^2 = -5$

$y = \pm 1\ \ or\ \ y = \pm\sqrt{5}i$

The solutions are $-1,\ 1,\ -\sqrt{5}i$, and $\sqrt{5}i$.

83. $x - 3\sqrt{x} - 4 = 0$

Let $u = \sqrt{x}$.

$\quad u^2 - 3u - 4 = 0 \quad$ Substituting u for $\sqrt{x}$

$\quad (u+1)(u-4) = 0$

$\quad u + 1 = 0 \quad or \quad u - 4 = 0$

$\qquad u = -1 \quad or \qquad u = 4$

Now substitute $\sqrt{x}$ for u and solve for x.

$\quad \sqrt{x} = -1 \quad or \quad \sqrt{x} = 4$

$\quad$ No solution $\qquad x = 16$

Note that $\sqrt{x}$ must be nonnegative, so $\sqrt{x} = -1$ has no solution. The number 16 checks and is the solution. The solution is 16.

85. $m^{2/3} - 2m^{1/3} - 8 = 0$

Let $u = m^{1/3}$.

$\quad u^2 - 2u - 8 = 0 \quad$ Substituting u for $m^{1/3}$

$\quad (u+2)(u-4) = 0$

$\quad u + 2 = 0 \quad or \quad u - 4 = 0$

$\qquad u = -2 \quad or \qquad u = 4$

Now substitute $m^{1/3}$ for u and solve for m.

$\quad m^{1/3} = -2 \quad or \qquad m^{1/3} = 4$

$\quad (m^{1/3})^3 = (-2)^3 \quad or \quad (m^{1/3})^3 = 4^3 \quad$ Using the

$\qquad\qquad\qquad\qquad\qquad\qquad$ principle of powers

$\qquad m = -8 \quad or \qquad m = 64$

The solutions are -8 and 64.

87. $x^{1/2} - 3x^{1/4} + 2 = 0$

Let $u = x^{1/4}$.

$\quad u^2 - 3u + 2 = 0 \quad$ Substituting u for $x^{1/4}$

$\quad (u-1)(u-2) = 0$

$\quad u - 1 = 0 \quad or \quad u - 2 = 0$

$\qquad u = 1 \quad or \qquad u = 2$

Now substitute $x^{1/4}$ for u and solve for x.

$\quad x^{1/4} = 1 \quad or \qquad x^{1/4} = 2$

$\quad (x^{1/4})^4 = 1^4 \quad or \quad (x^{1/4})^4 = 2^4$

$\qquad x = 1 \quad or \qquad x = 16$

The solutions are 1 and 16.

89. $(2x-3)^2 - 5(2x-3) + 6 = 0$

Let $u = 2x - 3$.

$\quad u^2 - 5u + 6 = 0 \quad$ Substituting u for $2x - 3$

$\quad (u-2)(u-3) = 0$

$\quad u - 2 = 0 \quad or \quad u - 3 = 0$

$\qquad u = 2 \quad or \qquad u = 3$

Now substitute $2x - 3$ for u and solve for x.

$\quad 2x - 3 = 2 \quad or \quad 2x - 3 = 3$

$\qquad 2x = 5 \quad or \qquad 2x = 6$

$\qquad x = \dfrac{5}{2} \quad or \qquad x = 3$

The solutions are $\dfrac{5}{2}$ and 3.

91. $(2t^2 + t)^2 - 4(2t^2 + t) + 3 = 0$

Let $u = 2t^2 + t$.

$\quad u^2 - 4u + 3 = 0 \quad$ Substituting u for $2t^2 + t$

$\quad (u-1)(u-3) = 0$

$\quad u - 1 = 0 \quad or \quad u - 3 = 0$

$\qquad u = 1 \quad or \qquad u = 3$

Now substitute $2t^2 + t$ for u and solve for t.

$\quad 2t^2 + t = 1 \quad or \qquad 2t^2 + t = 3$

$\quad 2t^2 + t - 1 = 0 \quad or \qquad 2t^2 + t - 3 = 0$

$\quad (2t-1)(t+1) = 0 \quad or \quad (2t+3)(t-1) = 0$

$\quad 2t-1=0 \ or \ t+1=0 \ \ or \ 2t+3=0 \ \ or \ t-1 = 0$

$\quad t = \dfrac{1}{2} \ or \quad t = -1 \ or \quad t = -\dfrac{3}{2} or \quad t = 1$

The solutions are $\dfrac{1}{2}$, -1, $-\dfrac{3}{2}$ and 1.

93. **Familiarize and Translate**. We will use the formula $s = 16t^2$, substituting 2120 for s.

$$2120 = 16t^2$$

Carry out. We solve the equation.

$\qquad 2120 = 16t^2$

$\qquad 132.5 = t^2 \qquad$ Dividing by 16 on both sides

$\qquad 11.5 \approx t \qquad$ Taking the square root on both sides

Check. When $t = 11.5$, $s = 16(11.5)^2 = 2116 \approx 2120$. The answer checks.

State. It would take an object about 11.5 sec to reach the ground.

95. Substitute 9.7 for $w(x)$ and solve for x.

$\quad -0.01x^2 + 0.27x + 8.60 = 9.7$

$\quad -0.01x^2 + 0.27x - 1.1 = 0$

$\quad a = -0.01, \ b = 0.27, \ c = -1.1$

$\quad x = \dfrac{-b \pm \sqrt{b^2 - 4ac}}{2a}$

$\qquad = \dfrac{-0.27 \pm \sqrt{(0.27)^2 - 4(-0.01)(-1.1)}}{2(-0.01)}$

$\qquad = \dfrac{-0.27 \pm \sqrt{0.0289}}{-0.02}$

$\quad x \approx 5 \ or \ x \approx 22$

There were 9.7 million self-employed workers in the United States 5 years after 1980 and also 22 years after 1980, or in 1985 and in 2002.

97. **Familiarize**. Let w = the width of the rug. Then $w + 1 =$ the length.

Translate. We use the Pythagorean equation.

$$w^2 + (w+1)^2 = 5^2$$

Carry out. We solve the equation.

$$w^2 + (w+1)^2 = 5^2$$
$$w^2 + w^2 + 2w + 1 = 25$$
$$2w^2 + 2w + 1 = 25$$
$$2w^2 + 2w - 24 = 0$$
$$2(w+4)(w-3) = 0$$
$$w + 4 = 0 \quad or \quad w - 3 = 0$$
$$w = -4 \quad or \qquad w = 3$$

Since the width cannot be negative, we consider only 3. When $w = 3$, $w + 1 = 3 + 1 = 4$.

Check. The length, 4 ft, is 1 ft more than the width, 3 ft. The length of a diagonal of a rectangle with width 3 ft and length 4 ft is $\sqrt{3^2 + 4^2} = \sqrt{9 + 16} = \sqrt{25} = 5$. The answer checks.

State. The length is 4 ft, and the width is 3 ft.

99. Familiarize. Let $n =$ the smaller number. Then $n + 5 =$ the larger number.

Translate.

$\underbrace{\text{The product of the numbers}}$ is 36.

$\qquad\qquad\downarrow\qquad\qquad\qquad\downarrow\quad\downarrow$

$\qquad\quad n(n+5) \qquad\qquad = \quad 36$

Carry out.
$$n(n+5) = 36$$
$$n^2 + 5n = 36$$
$$n^2 + 5n - 36 = 0$$
$$(n+9)(n-4) = 0$$
$$n + 9 = 0 \quad or \quad n - 4 = 0$$
$$n = -9 \quad or \qquad n = 4$$

If $n = -9$, then $n + 5 = -9 + 5 = -4$. If $n = 4$, then $n + 5 = 4 + 5 = 9$.

Check. The number -4 is 5 more than -9 and $(-4)(-9) = 36$, so the pair -9 and -4 check. The number 9 is 5 more than 4 and $9 \cdot 4 = 36$, so the pair 4 and 9 also check.

State. The numbers are -9 and -4 or 4 and 9.

101. Familiarize. We add labels to the drawing in the text.

We let x represent the length of a side of the square in each corner. Then the length and width of the resulting base are represented by $20 - 2x$ and $10 - 2x$, respectively. Recall that for a rectangle, Area $=$ length $\times$ width.

Translate.

$\underbrace{\text{The area of the base}}$ is $\underbrace{96 \text{ cm}^2}$.

$\quad (20 - 2x)(10 - 2x) \quad = \qquad 96$

Carry out. We solve the equation.
$$200 - 60x + 4x^2 = 96$$
$$4x^2 - 60x + 104 = 0$$
$$x^2 - 15x + 26 = 0$$
$$(x - 13)(x - 2) = 0$$
$$x - 13 = 0 \quad or \quad x - 2 = 0$$
$$x = 13 \quad or \qquad x = 2$$

Check. When $x = 13$, both $20 - 2x$ and $10 - 2x$ are negative numbers, so we only consider $x = 2$. When $x = 2$, then $20 - 2x = 20 - 2 \cdot 2 = 16$ and $10 - 2x = 10 - 2 \cdot 2 = 6$, and the area of the base is $16 \cdot 6$, or 96 cm^2. The answer checks.

State. The length of the sides of the squares is 2 cm.

103. Familiarize. We have $P = 2l + 2w$, or $28 = 2l + 2w$. Solving for w, we have
$$28 = 2l + 2w$$
$$14 = l + w \quad \text{Dividing by 2}$$
$$14 - l = w.$$

Then we have $l =$ the length of the rug and $14 - l =$ the width, in feet. Recall that the area of a rectangle is the product of the length and the width.

Translate.

$\underbrace{\text{The area}}$ is $\underbrace{48 \text{ ft}^2}$.

$\quad\downarrow\qquad\quad\downarrow\qquad\downarrow$

$l(14 - l) \quad = \qquad 48$

Carry out. We solve the equation.
$$l(14 - l) = 48$$
$$14l - l^2 = 48$$
$$0 = l^2 - 14l + 48$$
$$0 = (l - 6)(l - 8)$$
$$l - 6 = 0 \quad or \quad l - 8 = 0$$
$$l = 6 \quad or \qquad l = 8$$

If $l = 6$, then $14 - l = 14 - 6 = 8$.

If $l = 8$, then $14 - l = 14 - 8 = 6$.

In either case, the dimensions are 8 ft by 6 ft. Since we usually consider the length to be greater than the width, we let 8 ft $=$ the length and 6 ft $=$ the width.

Check. The perimeter is $2 \cdot 8$ ft $+ 2 \cdot 6$ ft $= 16$ ft $+ 12$ ft $= 28$ ft. The answer checks.

State. The length of the rug is 8 ft, and the width is 6 ft.

105. $f(x) = 4 - 5x = -5x + 4$

The function can be written in the form $y = mx + b$, so it is a linear function.

107. $f(x) = 7x^2$

The function is in the form $f(x) = ax^2 + bx + c$, $a \neq 0$, so it is a quadratic function.

109. $f(x) = 1.2x - (3.6)^2$

The function is in the form $f(x) = mx + b$, so it is a linear function.

111. Graph $y = x^2 - 8x + 12$ and use the Zero feature twice.

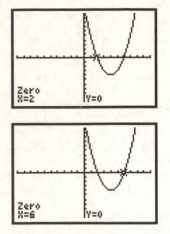

The solutions are 2 and 6.

113. Graph $y = 7x^2 - 43x + 6$ and use the Zero feature twice.

One solution is approximately 0.143 and the other is 6.

115. Graph $y_1 = 6x + 1$ and $y_2 = 4x^2$ and use the Intersect feature twice.

The solutions are approximately −0.151 and 1.651.

117. Graph $y = 2x^2 - 5x - 4$ and use the Zero feature twice.

The zeros are approximately −0.637 and 3.137.

119. Graph $y = 3x^2 + 2x - 4$ and use the Zero feature twice.

The zeros are approximately −1.535 and 0.869.

121. Graph $y = 5.02x^2 - 4.19x - 2.057$ and use the Zero feature twice.

The zeros are approximately −0.347 and 1.181.

123. Discussion and Writing

125. $1998 - 1980 = 18$, so we substitute 18 for x.
$a(18) = 9096(18) + 387,725 = 551,453$ associate's degrees

127. Test for symmetry with respect to the x-axis:

$$3x^2 + 4y^2 = 5 \quad \text{Original equation}$$
$$3x^2 + 4(-y)^2 = 5 \quad \text{Replacing } y \text{ by } -y$$
$$3x^2 + 4y^2 = 5 \quad \text{Simplifying}$$

The last equation is equivalent to the original equation, so the graph is symmetric with respect to the x-axis.

Test for symmetry with respect to the y-axis:

$$3x^2 + 4y^2 = 5 \quad \text{Original equation}$$
$$3(-x)^2 + 4y^2 = 5 \quad \text{Replacing } x \text{ by } -x$$
$$3x^2 + 4y^2 = 5 \quad \text{Simplifying}$$

The last equation is equivalent to the original equation, so the equation is symmetric with respect to the y-axis.

Test for symmetry with respect to the origin:

$$3x^2 + 4y^2 = 5 \quad \text{Original equation}$$
$$3(-x)^2 + 4(-y)^2 = 5 \quad \text{Replacing } x \text{ by } -x$$
$$\text{and } y \text{ by } -y$$
$$3x^2 + 4y^2 = 5 \quad \text{Simplifying}$$

The last equation is equivalent to the original equation, so the equation is symmetric with respect to the origin.

129. $f(x) = 2x^3 - x$
$f(-x) = 2(-x)^3 - (-x) = -2x^3 + x$
$-f(x) = -2x^3 + x$
$f(x) \neq f(-x)$ so f is not even
$f(-x) = -f(x)$, so f is odd.

131. a) $\quad kx^2 - 17x + 33 = 0$
$k(3)^2 - 17(3) + 33 = 0$ Substituting 3 for x
$9k - 51 + 33 = 0$
$9k = 18$
$k = 2$

b) $\quad 2x^2 - 17x + 33 = 0$ Substituting 2 for k
$(2x - 11)(x - 3) = 0$
$2x - 11 = 0 \quad \text{or} \quad x - 3 = 0$
$x = \dfrac{11}{2} \quad \text{or} \quad x = 3$

The other solution is $\dfrac{11}{2}$.

133. a) $\quad (1+i)^2 - k(1+i) + 2 = 0$ Substituting $1+i$ for x
$1 + 2i - 1 - k - ki + 2 = 0$
$2 + 2i = k + ki$
$2(1+i) = k(1+i)$
$2 = k$

b) $\quad x^2 - 2x + 2 = 0$ Substituting 2 for k
$$x = \frac{-(-2) \pm \sqrt{(-2)^2 - 4 \cdot 1 \cdot 2}}{2 \cdot 1}$$
$$= \frac{2 \pm \sqrt{-4}}{2}$$
$$= \frac{2 \pm 2i}{2} = 1 \pm i$$
The other solution is $1 - i$.

135. $(x-2)^3 = x^3 - 2$
$x^3 - 6x^2 + 12x - 8 = x^3 - 2$
$0 = 6x^2 - 12x + 6$
$0 = 6(x^2 - 2x + 1)$
$0 = 6(x-1)(x-1)$
$x - 1 = 0 \quad \text{or} \quad x - 1 = 0$
$x = 1 \quad \text{or} \quad x = 1$
The solution is 1.

137. $(6x^3 + 7x^2 - 3x)(x^2 - 7) = 0$
$x(6x^2 + 7x - 3)(x^2 - 7) = 0$
$x(3x - 1)(2x + 3)(x^2 - 7) = 0$
$x = 0$ or $3x - 1 = 0$ or $2x + 3 = 0$ or $x^2 - 7 = 0$
$x = 0$ or $x = \dfrac{1}{3}$ or $x = -\dfrac{3}{2}$ or $x = \sqrt{7}$ or $x = -\sqrt{7}$

The exact solutions are $-\sqrt{7}$, $-\dfrac{3}{2}$, 0, $\dfrac{1}{3}$, and $\sqrt{7}$.

139. $x^2 + x - \sqrt{2} = 0$
$$x = \frac{-b \pm \sqrt{b^2 - 4ac}}{2a}$$
$$= \frac{-1 \pm \sqrt{1^2 - 4 \cdot 1(-\sqrt{2})}}{2 \cdot 1} = \frac{-1 \pm \sqrt{1 + 4\sqrt{2}}}{2}$$
The solutions are $\dfrac{-1 \pm \sqrt{1 + 4\sqrt{2}}}{2}$.

141. $2t^2 + (t-4)^2 = 5t(t-4) + 24$
$2t^2 + t^2 - 8t + 16 = 5t^2 - 20t + 24$
$0 = 2t^2 - 12t + 8$
$0 = t^2 - 6t + 4$ Dividing by 2

Use the quadratic formula.
$$t = \frac{-b \pm \sqrt{b^2 - 4ac}}{2a}$$
$$= \frac{-(-6) \pm \sqrt{(-6)^2 - 4 \cdot 1 \cdot 4}}{2 \cdot 1}$$
$$= \frac{6 \pm \sqrt{20}}{2} = \frac{6 \pm 2\sqrt{5}}{2}$$
$$= \frac{2(3 \pm \sqrt{5})}{2} = 3 \pm \sqrt{5}$$
The solutions are $3 \pm \sqrt{5}$.

143. $\sqrt{x-3} - \sqrt[4]{x-3} = 2$

Substitute u for $\sqrt[4]{x-3}$.

$$u^2 - u - 2 = 0$$
$$(u-2)(u+1) = 0$$
$$u - 2 = 0 \quad or \quad u + 1 = 0$$
$$u = 2 \quad or \qquad u = -1$$

Substitute $\sqrt[4]{x-3}$ for u and solve for x.

$$\sqrt[4]{x-3} = 2 \quad or \quad \sqrt[4]{x-3} = 1$$
$$x - 3 = 16 \qquad \text{No solution}$$
$$x = 19$$

The value checks. The solution is 19.

145.
$$\left(y + \frac{2}{y}\right)^2 + 3y + \frac{6}{y} = 4$$
$$\left(y + \frac{2}{y}\right)^2 + 3\left(y + \frac{2}{y}\right) - 4 = 0$$

Substitute u for $y + \frac{2}{y}$.

$$u^2 + 3u - 4 = 0$$
$$(u+4)(u-1) = 0$$
$$u = -4 \quad or \quad u = 1$$

Substitute $y + \frac{2}{y}$ for u and solve for y.

$$y + \frac{2}{y} = -4 \quad or \qquad y + \frac{2}{y} = 1$$
$$y^2 + 2 = -4y \quad or \qquad y^2 + 2 = y$$
$$y^2 + 4y + 2 = 0 \quad or \quad y^2 - y + 2 = 0$$
$$y = \frac{-4 \pm \sqrt{4^2 - 4 \cdot 1 \cdot 2}}{2 \cdot 1} \quad or$$
$$y = \frac{-(-1) \pm \sqrt{(-1)^2 - 4 \cdot 1 \cdot 2}}{2 \cdot 1}$$
$$y = \frac{-4 \pm \sqrt{8}}{2} \qquad or \quad y = \frac{1 \pm \sqrt{-7}}{2}$$
$$y = \frac{-4 \pm 2\sqrt{2}}{2} \qquad or \quad y = \frac{1 \pm \sqrt{7}i}{2}$$
$$y = -2 \pm \sqrt{2} \qquad or \quad y = \frac{1}{2} \pm \frac{\sqrt{7}}{2}i$$

The solutions are $-2 \pm \sqrt{2}$ and $\frac{1}{2} \pm \frac{\sqrt{7}}{2}i$.

147. $\frac{1}{2}at + v_0 t + x_0 = 0$

Use the quadratic formula. Here $a = \frac{1}{2}a$, $b = v_0$, and $c = x_0$.

$$t = \frac{-v_0 \pm \sqrt{(v_0)^2 - 4 \cdot \frac{1}{2}a \cdot x_0}}{2 \cdot \frac{1}{2}a}$$
$$t = \frac{-v_0 \pm \sqrt{v_0^2 - 2ax_0}}{a}$$

Exercise Set 2.4

1. a) The minimum function value occurs at the vertex, so the vertex is $\left(-\frac{1}{2}, -\frac{9}{4}\right)$.

b) The axis of symmetry is a vertical line through the vertex. It is $x = -\frac{1}{2}$.

c) The minimum value of the function is $-\frac{9}{4}$.

3. $f(x) = x^2 - 8x + 12$ 16 completes the square for $x^2 - 8x$.

$$= x^2 - 8x + 16 - 16 + 12 \quad \text{Adding } 16 - 16$$
$$\text{on the right side}$$
$$= (x^2 - 8x + 16) - 16 + 12$$
$$= (x-4)^2 - 4 \qquad \text{Factoring and}$$
$$\text{simplifying}$$
$$= (x-4)^2 + (-4) \quad \text{Writing in the form}$$
$$f(x) = a(x-h)^2 + k$$

a) Vertex: $(4, -4)$

b) Axis of symmetry: $x = 4$

c) Minimum value: -4

d) We plot the vertex and find several points on either side of it. Then we plot these points and connect them with a smooth curve.

x	$f(x)$
4	-4
2	0
1	5
5	-3
6	0

$f(x) = x^2 - 8x + 12$

5. $f(x) = x^2 - 7x + 12$ $\frac{49}{4}$ completes the square for $x^2 - 7x$.

$$= x^2 - 7x + \frac{49}{4} - \frac{49}{4} + 12 \quad \text{Adding}$$
$$\frac{49}{4} - \frac{49}{4} \text{ on the right side}$$
$$= \left(x^2 - 7x + \frac{49}{4}\right) - \frac{49}{4} + 12$$
$$= \left(x - \frac{7}{2}\right)^2 - \frac{1}{4} \qquad \text{Factoring and}$$
$$\text{simplifying}$$
$$= \left(x - \frac{7}{2}\right)^2 + \left(-\frac{1}{4}\right) \quad \text{Writing in the}$$
$$\text{form } f(x) = a(x-h)^2 + k$$

a) Vertex: $\left(\frac{7}{2}, -\frac{1}{4}\right)$

b) Axis of symmetry: $x = \dfrac{7}{2}$

c) Minimum value: $-\dfrac{1}{4}$

d) We plot the vertex and find several points on either side of it. Then we plot these points and connect them with a smooth curve.

x	$f(x)$
$\dfrac{7}{2}$	$-\dfrac{1}{4}$
4	0
5	2
3	0
1	6

$f(x) = x^2 - 7x + 12$

7. $f(x) = x^2 + 4x + 5$ 4 completes the square for $x^2 + 4x$

$\quad = x^2 + 4x + 4 - 4 + 5$ Adding $4 - 4$ on the right side

$\quad = (x + 2)^2 + 1$ Factoring and simplifying

$\quad = [x - (-2)]^2 + 1$ Writing in the form $f(x) = a(x - h)^2 + k$

a) Vertex: $(-2, 1)$

b) Axis of symmetry: $x = -2$

c) Minimum value: 1

d) We plot the vertex and find several points on either side of it. Then we plot these points and connect them with a smooth curve.

x	$f(x)$
-2	1
-1	2
0	5
-3	2
-4	5

$f(x) = x^2 + 4x + 5$

9. $g(x) = \dfrac{x^2}{2} + 4x + 6$

$\quad = \dfrac{1}{2}(x^2 + 8x) + 6$ Factoring $\dfrac{1}{2}$ out of the first two terms

$\quad = \dfrac{1}{2}(x^2 + 8x + 16 - 16) + 6$ Adding $16 - 16$ inside the parentheses

$\quad = \dfrac{1}{2}(x^2 + 8x + 16) - \dfrac{1}{2} \cdot 16 + 6$ Removing -16 from within the parentheses

$\quad = \dfrac{1}{2}(x + 4)^2 - 2$ Factoring and simplifying

$\quad = \dfrac{1}{2}[x - (-4)]^2 + (-2)$

a) Vertex: $(-4, -2)$

b) Axis of symmetry: $x = -4$

c) Minimum value: -2

d) We plot the vertex and find several points on either side of it. Then we plot these points and connect them with a smooth curve.

x	$g(x)$
-4	-2
-2	0
0	6
-6	0
-8	6

$g(x) = \dfrac{x^2}{2} + 4x + 6$

11. $g(x) = 2x^2 + 6x + 8$

$\quad = 2(x^2 + 3x) + 8$ Factoring 2 out of the first two terms

$\quad = 2\left(x^2 + 3x + \dfrac{9}{4} - \dfrac{9}{4}\right) + 8$ Adding $\dfrac{9}{4} - \dfrac{9}{4}$ inside the parentheses

$\quad = 2\left(x^2 + 3x + \dfrac{9}{4}\right) - 2 \cdot \dfrac{9}{4} + 8$ Removing $-\dfrac{9}{4}$ from within the parentheses

$\quad = 2\left(x + \dfrac{3}{2}\right)^2 + \dfrac{7}{2}$ Factoring and simplifying

$\quad = 2\left[x - \left(-\dfrac{3}{2}\right)\right]^2 + \dfrac{7}{2}$

a) Vertex: $\left(-\dfrac{3}{2}, \dfrac{7}{2}\right)$

b) Axis of symmetry: $x = -\dfrac{3}{2}$

c) Minimum value: $\dfrac{7}{2}$

d) We plot the vertex and find several points on either side of it. Then we plot these points and connect them with a smooth curve.

x	$f(x)$
$-\dfrac{3}{2}$	$\dfrac{7}{2}$
-1	4
0	8
-2	4
-3	8

$g(x) = 2x^2 + 6x + 8$

13. $f(x) = -x^2 - 6x + 3$

$\qquad = -(x^2 + 6x) + 3$ 9 completes the square
$\qquad\qquad\qquad\qquad\qquad$ for $x^2 + 6x$.

$\qquad = -(x^2 + 6x + 9 - 9) + 3$

$\qquad = -(x + 3)^2 - (-9) + 3$ Removing -9 from
$\qquad\qquad\qquad\qquad\qquad\qquad$ the parentheses

$\qquad = -(x + 3)^2 + 9 + 3$

$\qquad = -[x - (-3)]^2 + 12$

a) Vertex: $(-3, 12)$

b) Axis of symmetry: $x = -3$

c) Maximum value: 12

d) We plot the vertex and find several points on either side of it. Then we plot these points and connect them with a smooth curve.

x	$f(x)$
-3	12
0	3
1	-4
-6	3
-7	-4

$f(x) = -x^2 - 6x + 3$

15. $g(x) = -2x^2 + 2x + 1$

$\qquad = -2(x^2 - x) + 1$ Factoring -2 out of the
$\qquad\qquad\qquad\qquad\qquad$ first two terms

$\qquad = -2\left(x^2 - x + \dfrac{1}{4} - \dfrac{1}{4}\right) + 1$ Adding $\dfrac{1}{4} - \dfrac{1}{4}$
$\qquad\qquad\qquad\qquad\qquad\qquad\qquad$ inside the parentheses

$\qquad = -2\left(x^2 - x + \dfrac{1}{4}\right) - 2\left(-\dfrac{1}{4}\right) + 1$

$\qquad$ Removing $-\dfrac{1}{4}$ from within the parentheses

$\qquad = -2\left(x - \dfrac{1}{2}\right)^2 + \dfrac{3}{2}$

a) Vertex: $\left(\dfrac{1}{2}, \dfrac{3}{2}\right)$

b) Axis of symmetry: $x = \dfrac{1}{2}$

c) Maximum value: $\dfrac{3}{2}$

d) We plot the vertex and find several points on either side of it. Then we plot these points and connect them with a smooth curve.

x	$f(x)$
$\dfrac{1}{2}$	$\dfrac{3}{2}$
1	1
2	-3
0	1
-1	-3

$g(x) = -2x^2 + 2x + 1$

17. The graph of $y = (x + 3)^2$ has vertex $(-3, 0)$ and opens up. It is graph (f).

19. The graph of $y = 2(x - 4)^2 - 1$ has vertex $(4, -1)$ and opens up. It is graph (b).

21. The graph of $y = -\dfrac{1}{2}(x + 3)^2 + 4$ has vertex $(-3, 4)$ and opens down. It is graph (h).

23. The graph of $y = -(x + 3)^2 + 4$ has vertex $(-3, 4)$ and opens down. It is graph (c).

25. $f(x) = x^2 - 6x + 5$

a) The x-coordinate of the vertex is
$$-\frac{b}{2a} = -\frac{-6}{2 \cdot 1} = 3.$$
Since $f(3) = 3^2 - 6 \cdot 3 + 5 = -4$, the vertex is $(3, -4)$.

b) Since $a = 1 > 0$, the graph opens up so the second coordinate of the vertex, -4, is the minimum value of the function.

c) The range is $[-4, \infty)$.

d) Since the graph opens up, function values decrease to the left of the vertex and increase to the right of the vertex. Thus, $f(x)$ is increasing on $(3, \infty)$ and decreasing on $(-\infty, 3)$.

27. $f(x) = 2x^2 + 4x - 16$

a) The x-coordinate of the vertex is
$$-\frac{b}{2a} = -\frac{4}{2 \cdot 2} = -1.$$
Since $f(-1) = 2(-1)^2 + 4(-1) - 16 = -18$, the vertex is $(-1, -18)$.

b) Since $a = 2 > 0$, the graph opens up so the second coordinate of the vertex, -18, is the minimum value of the function.

c) The range is $[-18, \infty)$.

d) Since the graph opens up, function values decrease to the left of the vertex and increase to the right of the vertex. Thus, $f(x)$ is increasing on $(-1, \infty)$ and decreasing on $(-\infty, -1)$.

29. $f(x) = -\dfrac{1}{2}x^2 + 5x - 8$

a) The x-coordinate of the vertex is
$$-\frac{b}{2a} = -\frac{5}{2\left(-\dfrac{1}{2}\right)} = 5.$$

Since $f(5) = -\frac{1}{2} \cdot 5^2 + 5 \cdot 5 - 8 = \frac{9}{2}$, the vertex is $\left(5, \frac{9}{2}\right)$.

b) Since $a = -\frac{1}{2} < 0$, the graph opens down so the second coordinate of the vertex, $\frac{9}{2}$, is the maximum value of the function.

c) The range is $\left(-\infty, \frac{9}{2}\right]$.

d) Since the graph opens down, function values increase to the left of the vertex and decrease to the right of the vertex. Thus, $f(x)$ is increasing on $(-\infty, 5)$ and decreasing on $(5, \infty)$.

31. $f(x) = 3x^2 + 6x + 5$

a) The x-coordinate of the vertex is
$$-\frac{b}{2a} = -\frac{6}{2 \cdot 3} = -1.$$
Since $f(-1) = 3(-1)^2 + 6(-1) + 5 = 2$, the vertex is $(-1, 2)$.

b) Since $a = 3 > 0$, the graph opens up so the second coordinate of the vertex, 2, is the minimum value of the function.

c) The range is $[2, \infty)$.

d) Since the graph opens up, function values decrease to the left of the vertex and increase to the right of the vertex. Thus, $f(x)$ is increasing on $(-1, \infty)$ and decreasing on $(-\infty, -1)$.

33. $g(x) = -4x^2 - 12x + 9$

a) The x-coordinate of the vertex is
$$-\frac{b}{2a} = -\frac{-12}{2(-4)} = -\frac{3}{2}.$$
Since $g\left(-\frac{3}{2}\right) = -4\left(-\frac{3}{2}\right)^2 - 12\left(-\frac{3}{2}\right) + 9 = 18$, the vertex is $\left(-\frac{3}{2}, 18\right)$.

b) Since $a = -4 < 0$, the graph opens down so the second coordinate of the vertex, 18, is the maximum value of the function.

c) The range is $(-\infty, 18]$.

d) Since the graph opens down, function values increase to the left of the vertex and decrease to the right of the vertex. Thus, $g(x)$ is increasing on $\left(-\infty, -\frac{3}{2}\right)$ and decreasing on $\left(-\frac{3}{2}, \infty\right)$.

35. Familiarize and Translate. The function $s(t) = -16t^2 + 20t + 6$ is given in the statement of the problem.

Carry out. The function $s(t)$ is quadratic and the coefficient of t^2 is negative, so $s(t)$ has a maximum value. It occurs at the vertex of the graph of the function. We find the first coordinate of the vertex. This is the time at which the ball reaches its maximum height.
$$t = -\frac{b}{2a} = -\frac{20}{2(-16)} = 0.625$$

The second coordinate of the vertex gives the maximum height.
$$s(0.625) = -16(0.625)^2 + 20(0.625) + 6 = 12.25$$

Check. Completing the square, we write the function in the form $s(t) = -16(t - 0.625)^2 + 12.25$. We see that the coordinates of the vertex are $(0.625, 12.25)$, so the answer checks.

State. The ball reaches its maximum height after 0.625 seconds. The maximum height is 12.25 ft.

37. Familiarize and Translate. The function $s(t) = -16t^2 + 120t + 80$ is given in the statement of the problem.

Carry out. The function $s(t)$ is quadratic and the coefficient of t^2 is negative, so $s(t)$ has a maximum value. It occurs at the vertex of the graph of the function. We find the first coordinate of the vertex. This is the time at which the rocket reaches its maximum height.
$$t = -\frac{b}{2a} = -\frac{120}{2(-16)} = 3.75$$

The second coordinate of the vertex gives the maximum height.
$$s(3.75) = -16(3.75)^2 + 120(3.75) + 80 = 305$$

Check. Completing the square, we write the function in the form $s(t) = -16(t - 3.75)^2 + 305$. We see that the coordinates of the vertex are $(3.75, 305)$, so the answer checks.

State. The rocket reaches its maximum height after 3.75 seconds. The maximum height is 305 ft.

39. Familiarize. Using the label in the text, we let $x =$ the height of the file. Then the length $= 10$ and the width $= 18 - 2x$.

Translate. Since the volume of a rectangular solid is length $\times$ width $\times$ height we have
$$V(x) = 10(18 - 2x)x, \text{ or } -20x^2 + 180x.$$

Carry out. Since $V(x)$ is a quadratic function with $a = -20 < 0$, the maximum function value occurs at the vertex of the graph of the function. The first coordinate of the vertex is
$$-\frac{b}{2a} = -\frac{180}{2(-20)} = 4.5.$$

Check. When $x = 4.5$, then $18 - 2x = 9$ and $V(x) = 10 \cdot 9(4.5)$, or 405. As a partial check, we can find $V(x)$ for a value of x less than 4.5 and for a value of x greater than 4.5. For instance, $V(4.4) = 404.8$ and $V(4.6) = 404.8$. Since both of these values are less than 405, our result appears to be correct.

State. The file should be 4.5 in. tall in order to maximize the volume.

41. Familiarize. Let $b =$ the length of the base of the triangle. Then the height $= 20 - b$.

Translate. Since the area of a triangle is $\frac{1}{2} \times$ base$\times$height, we have
$$A(b) = \frac{1}{2}b(20 - b), \text{ or } -\frac{1}{2}b^2 + 10b.$$

Carry out. Since $A(b)$ is a quadratic function with $a = -\dfrac{1}{2} < 0$, the maximum function value occurs at the vertex of the graph of the function. The first coordinate of the vertex is

$$-\frac{b}{2a} = -\frac{10}{2\left(-\dfrac{1}{2}\right)} = 10.$$

When $b = 10$, then $20 - b = 20 - 10 = 10$, and the area is $\dfrac{1}{2} \cdot 10 \cdot 10 = 50 \text{ cm}^2$.

Check. As a partial check, we can find $A(b)$ for a value of b less than 10 and for a value of b greater than 10. For instance, $V(9.9) = 49.995$ and $V(10.1) = 49.995$. Since both of these values are less than 50, our result appears to be correct.

State. The area is a maximum when the base and the height are both 10 cm.

43. $C(x) = 0.1x^2 - 0.7x + 2.425$

Since $C(x)$ is a quadratic function with $a = 0.1 > 0$, a minimum function value occurs at the vertex of the graph of $C(x)$. The first coordinate of the vertex is

$$-\frac{b}{2a} = -\frac{-0.7}{2(0.1)} = 3.5.$$

Thus, 3.5 hundred, or 350 bicycles should be built to minimize the average cost per bicycle.

45. $P(x) = R(x) - C(x)$

$P(x) = (50x - 0.5x^2) - (10x + 3)$

$P(x) = -0.5x^2 + 40x - 3$

Since $P(x)$ is a quadratic function with $a = -0.5 < 0$, a maximum function value occurs at the vertex of the graph of the function. The first coordinate of the vertex is

$$-\frac{b}{2a} = -\frac{40}{2(-0.5)} = 40.$$

$P(40) = -0.5(40)^2 + 40 \cdot 40 - 3 = 797$

Thus, the maximum profit is \$797. It occurs when 40 units are sold.

47. Familiarize. We let $s = $ the height of the elevator shaft, $t_1 = $ the time it takes the screwdriver to reach the bottom of the shaft, and $t_2 = $ the time it takes the sound to reach the top of the shaft.

Translate. We know that $t_1 + t_2 = 5$. Using the information in Example 4 we also know that

$$s = 16t_1^2, \quad or \quad t_1 = \frac{\sqrt{s}}{4} \text{ and}$$

$$s = 1100t_2, \quad or \quad t_2 = \frac{s}{1100}.$$

Then $\dfrac{\sqrt{s}}{4} + \dfrac{s}{1100} = 5.$

Carry out. We solve the last equation above.

$$\frac{\sqrt{s}}{4} + \frac{s}{1100} = 5$$

$$275\sqrt{s} + s = 5500 \quad \text{Multiplying by 1100}$$

$$s + 275\sqrt{s} - 5500 = 0$$

Let $u = \sqrt{s}$ and substitute.

$$u^2 + 275u - 5500 = 0$$

$$u = \frac{-b + \sqrt{b^2 - 4ac}}{2a} \quad \begin{array}{l} \text{We only want the} \\ \text{positive solution.} \end{array}$$

$$= \frac{-275 + \sqrt{275^2 - 4 \cdot 1(-5500)}}{2 \cdot 1}$$

$$= \frac{-275 + \sqrt{97,625}}{2} \approx 18.725$$

Since $u \approx 18.725$, we have $\sqrt{s} = 18.725$, so $s \approx 350.6$.

Check. If $s \approx 350.6$, then $t_1 = \dfrac{\sqrt{s}}{4} = \dfrac{\sqrt{350.6}}{4} \approx$ 4.68 and $t_2 = \dfrac{s}{1100} = \dfrac{350.6}{1100} \approx 0.32$, so $t_1 + t_2 = 4.68 + 0.32 = 5$.

The result checks.

State. The elevator shaft is about 350.6 ft tall.

49. Familiarize. Using the labels on the drawing in the text, we let $x = $ the width of each corral and $240 - 3x = $ the total length of the corrals.

Translate. Since the area of a rectangle is length $\times$ width, we have

$$A(x) = (240 - 3x)x = -3x^2 + 240x.$$

Carry out. Since $A(x)$ is a quadratic function with $a = -3 < 0$, the maximum function value occurs at the vertex of the graph of $A(x)$. The first coordinate of the vertex is

$$-\frac{b}{2a} = -\frac{240}{2(-3)} = 40.$$

$A(40) = -3(40)^2 + 240(40) = 4800$

Check. As a partial check we can find $A(x)$ for a value of x less than 40 and for a value of x greater than 40. For instance, $A(39.9) = 4799.97$ and $A(40.1) = 4799.97$. Since both of these values are less than 4800, our result appears to be correct.

State. The largest total area that can be enclosed is 4800 yd^2.

51. Left to the student

53. Left to the student

55. Discussion and Writing

57. Discussion and Writing

59. $f(x) = 2x^2 - x + 4$

$f(x + h) = 2(x + h)^2 - (x + h) + 4$

$\qquad = 2(x^2 + 2xh + h^2) - (x + h) + 4$

$\qquad = 2x^2 + 4xh + 2h^2 - x - h - 4$

$$\frac{f(x+h)-f(x)}{h}$$

$$=\frac{2x^2+4xh+2h^2-x-h-4-(2x^2-x+4)}{h}$$

$$=\frac{2x^2+4xh+2h^2-x-h-4-2x^2+x-4}{h}$$

$$=\frac{4xh+2h^2-h}{h}=\frac{h(4x+2h-1)}{h}$$

$$=4x+2h-1$$

61. The graph of $f(x)$ is stretched vertically and reflected across the x-axis.

$g(x)=-2f(x)$

63. $f(x)=-0.2x^2-3x+c$

The x-coordinate of the vertex of $f(x)$ is $-\dfrac{b}{2a}=$

$-\dfrac{-3}{2(-0.2)}=-7.5$. Now we find c such that

$f(-7.5)=$

-225.

$$-0.2(-7.5)^2-3(-7.5)+c=-225$$

$$-11.25+22.5+c=-225$$

$$c=-236.25$$

65.

$y=(|x|-5)^2-3$

Exercise Set 2.5

1. $\qquad \dfrac{1}{4}+\dfrac{1}{5}=\dfrac{1}{t}$, LCD is $20t$

$$20t\left(\frac{1}{4}+\frac{1}{5}\right)=20t\cdot\frac{1}{t}$$

$$20t\cdot\frac{1}{4}+20t\cdot\frac{1}{5}=20t\cdot\frac{1}{t}$$

$$5t+4t=20$$

$$9t=20$$

$$t=\frac{20}{9}$$

Check:

$$\frac{\dfrac{1}{4}+\dfrac{1}{5}=\dfrac{1}{t}}{\dfrac{1}{4}+\dfrac{1}{5} \ ? \ \dfrac{1}{\dfrac{20}{9}}}$$

$$\frac{5}{20}+\frac{4}{20} \ \bigg| \ 1\cdot\frac{9}{20}$$

$$\frac{9}{20} \ \bigg| \ \frac{9}{20} \qquad \text{TRUE}$$

The solution is $\dfrac{20}{9}$.

3. $\qquad \dfrac{x+2}{4}-\dfrac{x-1}{5}=15$, LCD is 20

$$20\left(\frac{x+2}{4}-\frac{x-1}{5}\right)=20\cdot 15$$

$$5(x+2)-4(x-1)=300$$

$$5x+10-4x+4=300$$

$$x+14=300$$

$$x=286$$

The solution is 286.

5. $\qquad \dfrac{1}{2}+\dfrac{2}{x}=\dfrac{1}{3}+\dfrac{3}{x}$, LCD is $6x$

$$6x\left(\frac{1}{2}+\frac{2}{x}\right)=6x\left(\frac{1}{3}+\frac{3}{x}\right)$$

$$3x+12=2x+18$$

$$3x-2x=18-12$$

$$x=6$$

Check:

$$\frac{\dfrac{1}{2}+\dfrac{2}{x}=\dfrac{1}{3}+\dfrac{3}{x}}{\dfrac{1}{2}+\dfrac{2}{6} \ ? \ \dfrac{1}{3}+\dfrac{3}{6}}$$

$$\frac{1}{2}+\frac{1}{3} \ \bigg| \ \frac{1}{3}+\frac{1}{2} \qquad \text{TRUE}$$

The solution is 6.

7. $\qquad \dfrac{5}{3x+2}=\dfrac{3}{2x}$, LCD is $2x(3x+2)$

$$2x(3x+2)\cdot\frac{5}{3x+2}=2x(3x+2)\cdot\frac{3}{2x}$$

$$2x\cdot 5=3(3x+2)$$

$$10x=9x+6$$

$$x=6$$

6 checks, so the solution is 6.

9. $\qquad x+\dfrac{6}{x}=5$, LCD is x

$$x\left(x+\frac{6}{x}\right)=x\cdot 5$$

$$x^2+6=5x$$

$$x^2-5x+6=0$$

$$(x-2)(x-3)=0$$

$x - 2 = 0 \ \ or \ \ x - 3 = 0$

$x = 2 \ \ or \ \ \ \ \ \ x = 3$

Both numbers check. The solutions are 2 and 3.

11.
$$\frac{6}{y+3} + \frac{2}{y} = \frac{5y-3}{y^2-9}$$

$$\frac{6}{y+3} + \frac{2}{y} = \frac{5y-3}{(y+3)(y-3)},$$

$$\text{LCD is } y(y+3)(y-3)$$

$$y(y+3)(y-3)\left(\frac{6}{y+3}+\frac{2}{y}\right) = y(y+3)(y-3)\cdot\frac{5y-3}{(y+3)(y-3)}$$

$$6y(y-3)+2(y+3)(y-3) = y(5y-3)$$

$$6y^2 - 18y + 2(y^2-9) = 5y^2 - 3y$$

$$6y^2 - 18y + 2y^2 - 18 = 5y^2 - 3y$$

$$8y^2 - 18y - 18 = 5y^2 - 3y$$

$$3y^2 - 15y - 18 = 0$$

$$y^2 - 5y - 6 = 0$$

$$(y-6)(y+1) = 0$$

$y - 6 = 0 \ \ or \ \ y + 1 = 0$

$y = 6 \ \ or \ \ \ \ \ \ y = -1$

Both numbers check. The solutions are 6 and -1.

13.
$$\frac{2x}{x-1} = \frac{5}{x-3}, \text{ LCD is } (x-1)(x-3)$$

$$(x-1)(x-3)\cdot\frac{2x}{x-1} = (x-1)(x-3)\cdot\frac{5}{x-3}$$

$$2x(x-3) = 5(x-1)$$

$$2x^2 - 6x = 5x - 5$$

$$2x^2 - 11x + 5 = 0$$

$$(2x-1)(x-5) = 0$$

$2x - 1 = 0 \ \ or \ \ x - 5 = 0$

$2x = 1 \ \ or \ \ \ \ \ \ x = 5$

$x = \dfrac{1}{2} \ \ or \ \ \ \ \ \ x = 5$

Both numbers check. The solutions are $\dfrac{1}{2}$ and 5.

15.
$$\frac{2}{x+5} + \frac{1}{x-5} = \frac{16}{x^2-25}$$

$$\frac{2}{x+5} + \frac{1}{x-5} = \frac{16}{(x+5)(x-5)},$$

$$\text{LCD is } (x+5)(x-5)$$

$$(x+5)(x-5)\left(\frac{2}{x+5}+\frac{1}{x-5}\right) = (x+5)(x-5)\cdot\frac{16}{(x+5)(x-5)}$$

$$2(x-5) + x + 5 = 16$$

$$2x - 10 + x + 5 = 16$$

$$3x - 5 = 16$$

$$3x = 21$$

$$x = 7$$

7 checks, so the solution is 7.

17.
$$\frac{3x}{x+2} + \frac{6}{x} = \frac{12}{x^2+2x}$$

$$\frac{3x}{x+2} + \frac{6}{x} = \frac{12}{x(x+2)}, \text{ LCD is } x(x+2)$$

$$x(x+2)\left(\frac{3x}{x+2}+\frac{6}{x}\right) = x(x+2)\cdot\frac{12}{x(x+2)}$$

$$3x\cdot x + 6(x+2) = 12$$

$$3x^2 + 6x + 12 = 12$$

$$3x^2 + 6x = 0$$

$$3x(x+2) = 0$$

$3x = 0 \ \ or \ \ x + 2 = 0$

$x = 0 \ \ or \ \ \ \ \ \ x = -2$

Neither 0 nor -2 checks, so the equation has no solution.

19.
$$\frac{1}{5x+20} - \frac{1}{x^2-16} = \frac{3}{x-4}$$

$$\frac{1}{5(x+4)} - \frac{1}{(x+4)(x-4)} = \frac{3}{x-4},$$

$$\text{LCD is } 5(x+4)(x-4)$$

$$5(x+4)(x-4)\left(\frac{1}{5(x+4)}-\frac{1}{(x+4)(x-4)}\right) = 5(x+4)(x-4)\cdot\frac{3}{x-4}$$

$$x - 4 - 5 = 15(x+4)$$

$$x - 9 = 15x + 60$$

$$-14x - 9 = 60$$

$$-14x = 69$$

$$x = -\frac{69}{14}$$

$-\dfrac{69}{14}$ checks, so the solution is $-\dfrac{69}{14}$.

21.
$$\frac{2}{5x+5} - \frac{3}{x^2-1} = \frac{4}{x-1}$$

$$\frac{2}{5(x+1)} - \frac{3}{(x+1)(x-1)} = \frac{4}{x-1},$$

$$\text{LCD is } 5(x+1)(x-1)$$

$$5(x+1)(x-1)\left(\frac{2}{5(x+1)}-\frac{3}{(x+1)(x-1)}\right) = 5(x+1)(x-1)\cdot\frac{4}{x-1}$$

$$2(x-1) - 5\cdot 3 = 20(x+1)$$

$$2x - 2 - 15 = 20x + 20$$

$$2x - 17 = 20x + 20$$

$$-18x - 17 = 20$$

$$-18x = 37$$

$$x = -\frac{37}{18}$$

$-\dfrac{37}{18}$ checks, so the solution is $-\dfrac{37}{18}$.

23.
$$\frac{8}{x^2 - 2x + 4} = \frac{x}{x+2} + \frac{24}{x^3 + 8},$$
$$\text{LCD is } (x+2)(x^2 - 2x + 4)$$

$$(x+2)(x^2-2x+4)\cdot \frac{8}{x^2-2x+4} =$$
$$(x+2)(x^2-2x+4)\left(\frac{x}{x+2} + \frac{24}{(x+2)(x^2-2x+4)}\right)$$
$$8(x+2) = x(x^2-2x+4)+24$$
$$8x + 16 = x^3 - 2x^2 + 4x + 24$$
$$0 = x^3 - 2x^2 - 4x + 8$$
$$0 = x^2(x-2) - 4(x-2)$$
$$0 = (x-2)(x^2-4)$$
$$0 = (x-2)(x+2)(x-2)$$

$$x - 2 = 0 \ \text{ or } \ x + 2 = 0 \ \text{ or } \ x - 2 = 0$$
$$x = 2 \ \text{ or } \quad x = -2 \ \text{ or } \quad x = 2$$

Only 2 checks. The solution is 2.

25.
$$\frac{x}{x-4} - \frac{4}{x+4} = \frac{32}{x^2 - 16}$$
$$\frac{x}{x-4} - \frac{4}{x+4} = \frac{32}{(x+4)(x-4)},$$
$$\text{LCD is } (x+4)(x-4)$$

$$(x+4)(x-4)\left(\frac{x}{x-4} - \frac{4}{x+4}\right) = (x+4)(x-4)\cdot \frac{32}{(x+4)(x-4)}$$
$$x(x+4) - 4(x-4) = 32$$
$$x^2 + 4x - 4x + 16 = 32$$
$$x^2 + 16 = 32$$
$$x^2 = 16$$
$$x = \pm 4$$

Neither 4 nor -4 checks, so the equation has no solution.

27.
$$\frac{1}{x-6} - \frac{1}{x} = \frac{6}{x^2 - 6x}$$
$$\frac{1}{x-6} - \frac{1}{x} = \frac{6}{x(x-6)}, \text{ LCD is } x(x-6)$$
$$x(x-6)\left(\frac{1}{x-6} - \frac{1}{x}\right) = x(x-6)\cdot \frac{6}{x(x-6)}$$
$$x - (x-6) = 6$$
$$x - x + 6 = 6$$
$$6 = 6$$

We get an equation that is true for all real numbers. Note, however, that when $x = 6$ or $x = 0$, division by 0 occurs in the original equation. Thus, the solution set is $\{x | x \text{ is a real number } and \ x \neq 6 \ and \ x \neq 0\}$, or $(-\infty, 0) \cup (0, 6) \cup (6, \infty)$.

29.
$$\sqrt{3x - 4} = 1$$
$$(\sqrt{3x-4})^2 = 1^2$$
$$3x - 4 = 1$$
$$3x = 5$$
$$x = \frac{5}{3}$$

Check:
$$\frac{\sqrt{3x-4} = 1}{\sqrt{3\cdot \frac{5}{3} - 4} \ ? \ 1}$$
$$\sqrt{5-4}$$
$$\sqrt{1}$$
$$1 \ \bigg| \ 1 \quad \text{TRUE}$$

The solution is $\frac{5}{3}$.

31.
$$\sqrt{2x - 5} = 2$$
$$(\sqrt{2x-5})^2 = 2^2$$
$$2x - 5 = 4$$
$$2x = 9$$
$$x = \frac{9}{2}$$

Check:
$$\frac{\sqrt{2x-5} = 2}{\sqrt{2\cdot \frac{9}{2} - 5} \ ? \ 2}$$
$$\sqrt{9-5}$$
$$\sqrt{4}$$
$$2 \ \bigg| \ 2 \quad \text{TRUE}$$

The solution is $\frac{9}{2}$.

33.
$$\sqrt{7 - x} = 2$$
$$(\sqrt{7-x})^2 = 2^2$$
$$7 - x = 4$$
$$-x = -3$$
$$x = 3$$

Check:
$$\frac{\sqrt{7-x} = 2}{\sqrt{7-3} \ ? \ 2}$$
$$\sqrt{4}$$
$$2 \ \bigg| \ 2 \quad \text{TRUE}$$

The solution is 3.

35.
$$\sqrt{1 - 2x} = 3$$
$$(\sqrt{1-2x})^2 = 3^2$$
$$1 - 2x = 9$$
$$-2x = 8$$
$$x = -4$$

Check:
$$\frac{\sqrt{1-2x} = 3}{\sqrt{1-2(-4)} \ ? \ 3}$$
$$\sqrt{1+8}$$
$$\sqrt{9}$$
$$3 \ \bigg| \ 3 \quad \text{TRUE}$$

The solution is -4.

37.
$$\sqrt[3]{5x - 2} = -3$$
$$(\sqrt[3]{5x - 2})^3 = (-3)^3$$
$$5x - 2 = -27$$
$$5x = -25$$
$$x = -5$$

Check:

$$\begin{array}{c|c} \sqrt[3]{5x - 2} = -3 \\ \hline \sqrt[3]{5(-5) - 2} \;?\; -3 \\ \sqrt[3]{-25 - 2} \\ \sqrt[3]{-27} \\ -3 & -3 \quad \text{TRUE} \end{array}$$

The solution is -5.

39.
$$\sqrt[4]{x^2 - 1} = 1$$
$$(\sqrt[4]{x^2 - 1})^4 = 1^4$$
$$x^2 - 1 = 1$$
$$x^2 = 2$$
$$x = \pm\sqrt{2}$$

Check:

$$\begin{array}{c|c} \sqrt[4]{x^2 - 1} = 1 \\ \hline \sqrt[4]{(\pm\sqrt{2})^2 - 1} \;?\; 1 \\ \sqrt[4]{2 - 1} \\ \sqrt[4]{1} \\ 1 & 1 \quad \text{TRUE} \end{array}$$

The solutions are $\pm\sqrt{2}$.

41.
$$\sqrt{y - 1} + 4 = 0$$
$$\sqrt{y - 1} = -4$$

The principal square root is never negative. Thus, there is no solution.

If we do not observe the above fact, we can continue and reach the same answer.

$$(\sqrt{y - 1})^2 = (-4)^2$$
$$y - 1 = 16$$
$$y = 17$$

Check:

$$\begin{array}{c|c} \sqrt{y - 1} + 4 = 0 \\ \hline \sqrt{17 - 1} + 4 \;?\; 0 \\ \sqrt{16} + 4 \\ 4 + 4 \\ 8 & 0 \quad \text{FALSE} \end{array}$$

Since 17 does not check, there is no solution.

43.
$$\sqrt{b + 3} - 2 = 1$$
$$\sqrt{b + 3} = 3$$
$$(\sqrt{b + 3})^2 = 3^2$$
$$b + 3 = 9$$
$$b = 6$$

Check:

$$\begin{array}{c|c} \sqrt{b + 3} - 2 = 1 \\ \hline \sqrt{6 + 3} - 2 \;?\; 1 \\ \sqrt{9} - 2 \\ 3 - 2 \\ 1 & 1 \quad \text{TRUE} \end{array}$$

The solution is 6.

45.
$$\sqrt{z + 2} + 3 = 4$$
$$\sqrt{z + 2} = 1$$
$$(\sqrt{z + 2})^2 = 1^2$$
$$z + 2 = 1$$
$$z = -1$$

Check:

$$\begin{array}{c|c} \sqrt{z + 2} + 3 = 4 \\ \hline \sqrt{-1 + 2} + 3 \;?\; 4 \\ \sqrt{1} + 3 \\ 1 + 3 \\ 4 & 4 \quad \text{TRUE} \end{array}$$

The solution is -1.

47.
$$\sqrt{2x + 1} - 3 = 3$$
$$\sqrt{2x + 1} = 6$$
$$(\sqrt{2x + 1})^2 = 6^2$$
$$2x + 1 = 36$$
$$2x = 35$$
$$x = \frac{35}{2}$$

Check:

$$\begin{array}{c|c} \sqrt{2x + 1} - 3 = 3 \\ \hline \sqrt{2 \cdot \frac{35}{2} + 1} - 3 \;?\; 3 \\ \sqrt{35 + 1} - 3 \\ \sqrt{36} - 3 \\ 6 - 3 \\ 3 & 3 \quad \text{TRUE} \end{array}$$

The solution is $\dfrac{35}{2}$.

49.
$$\sqrt{2 - x} - 4 = 6$$
$$\sqrt{2 - x} = 10$$
$$(\sqrt{2 - x})^2 = 10^2$$
$$2 - x = 100$$
$$-x = 98$$
$$x = -98$$

Check:

$$\frac{\sqrt{2-x}-4=6}{\begin{array}{c|c}\sqrt{2-(-98)}-4 \ ? \ 6 \\ \sqrt{100}-4 \\ 10-4 \\ 6 & 6 \end{array}} \quad \text{TRUE}$$

The solution is -98.

51. $\sqrt[3]{6x+9}+8=5$

$\sqrt[3]{6x+9}=-3$

$(\sqrt[3]{6x+9})^3=(-3)^3$

$6x+9=-27$

$6x=-36$

$x=-6$

Check:

$$\frac{\sqrt[3]{6x+9}+8=5}{\begin{array}{c|c}\sqrt[3]{6(-6)+9}+8 \ ? \ 5 \\ \sqrt[3]{-27}+8 \\ -3+8 \\ 5 & 5 \end{array}} \quad \text{TRUE}$$

The solution is -6.

53. $\sqrt{x+4}+2=x$

$\sqrt{x+4}=x-2$

$(\sqrt{x+4})^2=(x-2)^2$

$x+4=x^2-4x+4$

$0=x^2-5x$

$0=x(x-5)$

$x=0 \ \ or \ \ x-5=0$

$x=0 \ \ or \ \ \ \ \ \ x=5$

Check:

For 0:

$$\frac{\sqrt{x+4}+2=x}{\begin{array}{c|c}\sqrt{0+4}+2 \ ? \ 0 \\ 2+2 \\ 4 & 0 \end{array}} \quad \text{FALSE}$$

For 5:

$$\frac{\sqrt{x+4}+2=x}{\begin{array}{c|c}\sqrt{5+4}+2 \ ? \ 5 \\ \sqrt{9}+2 \\ 3+2 \\ 5 & 5 \end{array}} \quad \text{TRUE}$$

The number 5 checks but 0 does not. The solution is 5.

55. $\sqrt{x-3}+5=x$

$\sqrt{x-3}=x-5$

$(\sqrt{x-3})^2=(x-5)^2$

$x-3=x^2-10x+25$

$0=x^2-11x+28$

$0=(x-4)(x-7)$

$x-4=0 \ \ or \ \ x-7=0$

$x=4 \ \ or \ \ \ \ \ \ x=7$

Check:

For 4:

$$\frac{\sqrt{x-3}+5=x}{\begin{array}{c|c}\sqrt{4-3}+5 \ ? \ 4 \\ \sqrt{1}+5 \\ 1+5 \\ 6 & 4 \end{array}} \quad \text{FALSE}$$

For 7:

$$\frac{\sqrt{x-3}+5=x}{\begin{array}{c|c}\sqrt{7-3}+5 \ ? \ 7 \\ \sqrt{4}+5 \\ 2+5 \\ 7 & 7 \end{array}} \quad \text{TRUE}$$

The number 7 checks but 4 does not. The solution is 7.

57. $\sqrt{x+7}=x+1$

$(\sqrt{x+7})^2=(x+1)^2$

$x+7=x^2+2x+1$

$0=x^2+x-6$

$0=(x+3)(x-2)$

$x+3=0 \ \ \ or \ \ x-2=0$

$x=-3 \ \ or \ \ \ \ \ \ x=2$

Check:

For -3:

$$\frac{\sqrt{x+7}=x+1}{\begin{array}{c|c}\sqrt{-3+7} \ ? \ -3+1 \\ \sqrt{4} & -2 \\ 2 & -2 \end{array}} \quad \text{FALSE}$$

For 2:

$$\frac{\sqrt{x+7}=x+1}{\begin{array}{c|c}\sqrt{2+7} \ ? \ 2+1 \\ \sqrt{9} & 3 \\ 3 & 3 \end{array}} \quad \text{TRUE}$$

The number 2 checks but -3 does not. The solution is 2.

59. $\sqrt{3x+3}=x+1$

$(\sqrt{3x+3})^2=(x+1)^2$

$3x+3=x^2+2x+1$

$0=x^2-x-2$

$0=(x-2)(x+1)$

$$x - 2 = 0 \quad or \quad x + 1 = 0$$
$$x = 2 \quad or \quad x = -1$$

Check:

For 2:

$$\sqrt{3x + 3} = x + 1$$

$$\sqrt{3 \cdot 2 + 3} \ ? \ 2 + 1$$
$$\sqrt{9} \ \bigg| \ 3$$
$$3 \ \bigg| \ 3 \quad \text{TRUE}$$

For -1:

$$\sqrt{3x + 3} = x + 1$$

$$\sqrt{3(-1) + 3} \ ? \ -1 + 1$$
$$\sqrt{0} \ \bigg| \ 0$$
$$0 \ \bigg| \ 0 \quad \text{TRUE}$$

Both numbers check. The solutions are 2 and -1.

61.
$$\sqrt{5x + 1} = x - 1$$
$$(\sqrt{5x + 1})^2 = (x - 1)^2$$
$$5x + 1 = x^2 - 2x + 1$$
$$0 = x^2 - 7x$$
$$0 = x(x - 7)$$
$$x = 0 \quad or \quad x - 7 = 0$$
$$x = 0 \quad or \quad x = 7$$

Check:

For 0:

$$\sqrt{5x + 1} = x - 1$$

$$\sqrt{5 \cdot 0 + 1} \ ? \ 0 - 1$$
$$\sqrt{1} \ \bigg| \ -1$$
$$1 \ \bigg| \ -1 \quad \text{FALSE}$$

For 7:

$$\sqrt{5x + 1} = x - 1$$

$$\sqrt{5 \cdot 7 + 1} \ ? \ 7 - 1$$
$$\sqrt{36} \ \bigg| \ 6$$
$$6 \ \bigg| \ 6 \quad \text{TRUE}$$

The number 7 checks but 0 does not. The solution is 7.

63.
$$\sqrt{x - 3} + \sqrt{x + 2} = 5$$
$$\sqrt{x + 2} = 5 - \sqrt{x - 3}$$
$$(\sqrt{x + 2})^2 = (5 - \sqrt{x - 3})^2$$
$$x + 2 = 25 - 10\sqrt{x - 3} + (x - 3)$$
$$x + 2 = 22 - 10\sqrt{x - 3} + x$$
$$10\sqrt{x - 3} = 20$$
$$\sqrt{x - 3} = 2$$
$$(\sqrt{x - 3})^2 = 2^2$$
$$x - 3 = 4$$
$$x = 7$$

Check:

$$\sqrt{x - 3} + \sqrt{x + 2} = 5$$

$$\sqrt{7 - 3} + \sqrt{7 + 2} \ ? \ 5$$
$$\sqrt{4} + \sqrt{9} \ \bigg|$$
$$2 + 3 \ \bigg|$$
$$5 \ \bigg| \ 5 \quad \text{TRUE}$$

The solution is 7.

65.
$$\sqrt{3x - 5} + \sqrt{2x + 3} + 1 = 0$$
$$\sqrt{3x - 5} + \sqrt{2x + 3} = -1$$

The principal square root is never negative. Thus the sum of two principal square roots cannot equal -1. There is no solution.

67.
$$\sqrt{x} - \sqrt{3x - 3} = 1$$
$$\sqrt{x} = \sqrt{3x - 3} + 1$$
$$(\sqrt{x})^2 = (\sqrt{3x - 3} + 1)^2$$
$$x = (3x - 3) + 2\sqrt{3x - 3} + 1$$
$$2 - 2x = 2\sqrt{3x - 3}$$
$$1 - x = \sqrt{3x - 3}$$
$$(1 - x)^2 = (\sqrt{3x - 3})^2$$
$$1 - 2x + x^2 = 3x - 3$$
$$x^2 - 5x + 4 = 0$$
$$(x - 4)(x - 1) = 0$$
$$x = 4 \quad or \quad x = 1$$

The number 4 does not check, but 1 does. The solution is 1.

69.
$$\sqrt{2y - 5} - \sqrt{y - 3} = 1$$
$$\sqrt{2y - 5} = \sqrt{y - 3} + 1$$
$$(\sqrt{2y - 5})^2 = (\sqrt{y - 3} + 1)^2$$
$$2y - 5 = (y - 3) + 2\sqrt{y - 3} + 1$$
$$y - 3 = 2\sqrt{y - 3}$$
$$(y - 3)^2 = (2\sqrt{y - 3})^2$$
$$y^2 - 6y + 9 = 4(y - 3)$$
$$y^2 - 6y + 9 = 4y - 12$$
$$y^2 - 10y + 21 = 0$$
$$(y - 7)(y - 3) = 0$$
$$y = 7 \quad or \quad y = 3$$

Both numbers check. The solutions are 7 and 3.

71.
$$\sqrt{y + 4} - \sqrt{y - 1} = 1$$
$$\sqrt{y + 4} = \sqrt{y - 1} + 1$$
$$(\sqrt{y + 4})^2 = (\sqrt{y - 1} + 1)^2$$
$$y + 4 = y - 1 + 2\sqrt{y - 1} + 1$$
$$4 = 2\sqrt{y - 1}$$
$$2 = \sqrt{y - 1} \qquad \text{Dividing by 2}$$
$$2^2 = (\sqrt{y - 1})^2$$
$$4 = y - 1$$
$$5 = y$$

The answer checks. The solution is 5.

73. $\sqrt{x+5} + \sqrt{x+2} = 3$

$\sqrt{x+5} = 3 - \sqrt{x+2}$

$(\sqrt{x+5})^2 = (3 - \sqrt{x+2})^2$

$x + 5 = 9 - 6\sqrt{x+2} + x + 2$

$-6 = -6\sqrt{x+2}$

$1 = \sqrt{x+2}$ Dividing by -6

$1^2 = (\sqrt{x+2})^2$

$1 = x + 2$

$-1 = x$

The answer checks. The solution is -1.

75. $x^{1/3} = -2$

$(x^{1/3})^3 = (-2)^3$ $(x^{1/3} = \sqrt[3]{x})$

$x = -8$

The value checks. The solution is -8.

77. $t^{1/4} = 3$

$(t^{1/4})^4 = 3^4$ $(t^{1/4} = \sqrt[4]{t})$

$t = 81$

The value checks. The solution is 81.

79. $|x| = 7$

The solutions are those numbers whose distance from 0 on a number line is 7. They are -7 and 7. That is,

$x = -7$ or $x = 7$.

The solutions are -7 and 7.

81. $|x| = -10.7$

The absolute value of a number is nonnegative. Thus, the equation has no solution.

83. $|x - 1| = 4$

$x - 1 = -4$ or $x - 1 = 4$

$x = -3$ or $x = 5$

The solutions are -3 and 5.

85. $|3x| = 1$

$3x = -1$ or $3x = 1$

$x = -\dfrac{1}{3}$ or $x = \dfrac{1}{3}$

The solutions are $-\dfrac{1}{3}$ and $\dfrac{1}{3}$.

87. $|x| = 0$

The distance of 0 from 0 on a number line is 0. That is,

$x = 0$.

The solution is 0.

89. $|3x + 2| = 1$

$3x + 2 = -1$ or $3x + 2 = 1$

$3x = -3$ or $3x = -1$

$x = -1$ or $x = -\dfrac{1}{3}$

The solutions are -1 and $-\dfrac{1}{3}$.

91. $\left|\dfrac{1}{2}x - 5\right| = 17$

$\dfrac{1}{2}x - 5 = -17$ or $\dfrac{1}{2}x - 5 = 17$

$\dfrac{1}{2}x = -12$ or $\dfrac{1}{2}x = 22$

$x = -24$ or $x = 44$

The solutions are -24 and 44.

93. $|x - 1| + 3 = 6$

$|x - 1| = 3$

$x - 1 = -3$ or $x - 1 = 3$

$x = -2$ or $x = 4$

The solutions are -2 and 4.

95. $|x + 3| - 2 = 8$

$|x + 3| = 10$

$x + 3 = -10$ or $x + 3 = 10$

$x = -13$ or $x = 7$

The solutions are -13 and 7.

97. $|3x + 1| - 4 = -1$

$|3x + 1| = 3$

$3x + 1 = -3$ or $3x + 1 = 3$

$3x = -4$ or $3x = 2$

$x = -\dfrac{4}{3}$ or $x = \dfrac{2}{3}$

The solutions are $-\dfrac{4}{3}$ and $\dfrac{2}{3}$.

99. $|4x - 3| + 1 = 7$

$|4x - 3| = 6$

$4x - 3 = -6$ or $4x - 3 = 6$

$4x = -3$ or $4x = 9$

$x = -\dfrac{3}{4}$ or $x = \dfrac{9}{4}$

The solutions are $-\dfrac{3}{4}$ and $\dfrac{9}{4}$.

101. $12 - |x + 6| = 5$

$-|x + 6| = -7$

$|x + 6| = 7$ Multiplying by -1

$x + 6 = -7$ or $x + 6 = 7$

$x = -13$ or $x = 1$

The solutions are -13 and 1.

103. $\dfrac{P_1 V_1}{T_1} = \dfrac{P_2 V_2}{T_2}$

$P_1 V_1 T_2 = P_2 V_2 T_1$ Multiplying by $T_1 T_2$ on both sides

$\dfrac{P_1 V_1 T_2}{P_2 V_2} = T_1$ Dividing by $P_2 V_2$ on both sides

105.
$$\frac{1}{R} = \frac{1}{R_1} + \frac{1}{R_2}$$

$$RR_1R_2 \cdot \frac{1}{R} = RR_1R_2\left(\frac{1}{R_1} + \frac{1}{R_2}\right)$$
Multiplying by RR_1R_2 on both sides

$$R_1R_2 = RR_2 + RR_1$$

$R_1R_2 - RR_2 = RR_1$ Subtracting RR_2 on both sides

$R_2(R_1 - R) = RR_1$ Factoring

$$R_2 = \frac{RR_1}{R_1 - R}$$ Dividing by $R_1 - R$ on both sides

107.
$$\frac{1}{F} = \frac{1}{m} + \frac{1}{p}$$

$$Fmp \cdot \frac{1}{F} = Fmp\left(\frac{1}{m} + \frac{1}{p}\right)$$ Multiplying by Fmp on both sides

$$mp = Fp + Fm$$

$mp - Fp = Fm$ Subtracting Fp on both sides

$p(m - F) = Fm$ Factoring

$$p = \frac{Fm}{m - F}$$ Dividing by $m - F$ on both sides

109. Left to the student

111. Discussion and Writing

113. $15 - 2x = 0$ Setting $f(x) = 0$

$$15 = 2x$$

$$\frac{15}{2} = x, \text{ or}$$

$$7.5 = x$$

The zero of the function is $\frac{15}{2}$, or 7.5.

115. Familiarize. Let p = the number of prescriptions for sleeping pills filled in 2000, in millions. Then the number of prescriptions filled in 2005 was $p + 60\% \cdot p$, or $p + 0.6p$, or $1.6p$.

Translate.

$$\underbrace{\text{Number of prescriptions filled in 2005}}_{\downarrow} \quad \underbrace{\text{was}}_{\downarrow} \quad \underbrace{\text{42 million.}}_{\downarrow}$$
$$1.6p \qquad\qquad = \qquad 42$$

Carry out. We solve the equation.

$$1.6p = 42$$

$$p = 26.25$$

Check. 60% of 26.25 is 0.6(26.25), or 15.75, and $26.25 + 15.75 = 42$. The answer checks.

State. 26.25 million prescriptions for sleeping pills were filled in 2000.

117.
$$(x - 3)^{2/3} = 2$$
$$[(x - 3)^{2/3}]^3 = 2^3$$
$$(x - 3)^2 = 8$$
$$x^2 - 6x + 9 = 8$$
$$x^2 - 6x + 1 = 0$$
$$a = 1, b = -6, c = 1$$
$$x = \frac{-b \pm \sqrt{b^2 - 4ac}}{2a}$$
$$= \frac{-(-6) \pm \sqrt{(-6)^2 - 4 \cdot 1 \cdot 1}}{2 \cdot 1}$$
$$= \frac{6 \pm \sqrt{32}}{2} = \frac{6 \pm 4\sqrt{2}}{2}$$
$$= \frac{2(3 \pm 2\sqrt{2})}{2} = 3 \pm 2\sqrt{2}$$

Both values check. The solutions are $3 \pm 2\sqrt{2}$.

119.
$$\sqrt{x + 5} + 1 = \frac{6}{\sqrt{x + 5}}, \text{ LCD is } \sqrt{x + 5}$$
$$x + 5 + \sqrt{x + 5} = 6 \qquad \text{Multiplying by } \sqrt{x + 5}$$
$$\sqrt{x + 5} = 1 - x$$
$$x + 5 = 1 - 2x + x^2$$
$$0 = x^2 - 3x - 4$$
$$0 = (x - 4)(x + 1)$$

$x = 4 \ or \ x = -1$

Only -1 checks. The solution set is -1.

Exercise Set 2.6

1. $x + 6 < 5x - 6$

$6 + 6 < 5x - x$ Subtracting x and adding 6 on both sides

$12 < 4x$

$\frac{12}{4} < x$ Dividing by 4 on both sides

$3 < x$

This inequality could also be solved as follows:

$x + 6 < 5x - 6$

$x - 5x < -6 - 6$ Subtracting $5x$ and 6 on both sides

$-4x < -12$

$x > \frac{-12}{-4}$ Dividing by -4 on both sides and reversing the inequality symbol

$x > 3$

The solution set is $\{x | x > 3\}$, or $(3, \infty)$. The graph is shown below.

3. $3x - 3 + 2x \geq 1 - 7x - 9$

$5x - 3 \geq -7x - 8$ Collecting like terms

$5x + 7x \geq -8 + 3$ Adding $7x$ and 3
on both sides

$12x \geq -5$

$x \geq -\dfrac{5}{12}$ Dividing by 12 on both sides

The solution set is $\left\{x \middle| x \geq -\dfrac{5}{12}\right\}$, or $\left[-\dfrac{5}{12}, \infty\right)$. The graph is shown below.

5. $14 - 5y \leq 8y - 8$

$14 + 8 \leq 8y + 5y$

$22 \leq 13y$

$\dfrac{22}{13} \leq y$

This inequality could also be solved as follows:

$14 - 5y \leq 8y - 8$

$-5y - 8y \leq -8 - 14$

$-13y \leq -22$

$y \geq \dfrac{22}{13}$ Dividing by -13 on both sides and reversing the inequality symbol

The solution set is $\left\{y \middle| y \geq \dfrac{22}{13}\right\}$, or $\left[\dfrac{22}{13}, \infty\right)$. The graph is shown below.

7. $-\dfrac{3}{4}x \geq -\dfrac{5}{8} + \dfrac{2}{3}x$

$\dfrac{5}{8} \geq \dfrac{3}{4}x + \dfrac{2}{3}x$

$\dfrac{5}{8} \geq \dfrac{9}{12}x + \dfrac{8}{12}x$

$\dfrac{5}{8} \geq \dfrac{17}{12}x$

$\dfrac{12}{17} \cdot \dfrac{5}{8} \geq \dfrac{12}{17} \cdot \dfrac{17}{12}x$

$\dfrac{15}{34} \geq x$

The solution set is $\left\{x \middle| x \leq \dfrac{15}{34}\right\}$, or $\left(-\infty, \dfrac{15}{34}\right]$. The graph is shown below.

9. $4x(x - 2) < 2(2x - 1)(x - 3)$

$4x(x - 2) < 2(2x^2 - 7x + 3)$

$4x^2 - 8x < 4x^2 - 14x + 6$

$-8x < -14x + 6$

$-8x + 14x < 6$

$6x < 6$

$x < \dfrac{6}{6}$

$x < 1$

The solution set is $\{x | x < 1\}$, or $(-\infty, 1)$. The graph is shown below.

11. $-2 \leq x + 1 < 4$

$-3 \leq x < 3$ Subtracting 1

The solution set is $[-3, 3)$. The graph is shown below.

13. $5 \leq x - 3 \leq 7$

$8 \leq x \leq 10$ Adding 3

The solution set is $[8, 10]$. The graph is shown below.

15. $-3 \leq x + 4 \leq 3$

$-7 \leq x \leq -1$ Subtracting 4

The solution set is $[-7, -1]$. The graph is shown below.

17. $-2 < 2x + 1 < 5$

$-3 < 2x < 4$ Adding -1

$-\dfrac{3}{2} < x < 2$ Multiplying by $\dfrac{1}{2}$

The solution set is $\left(-\dfrac{3}{2}, 2\right)$. The graph is shown below.

19. $-4 \leq 6 - 2x < 4$

$-10 \leq -2x < -2$ Adding -6

$5 \geq x > 1$ Multiplying by $-\dfrac{1}{2}$

or $1 < x \leq 5$

The solution set is $(1, 5]$. The graph is shown below.

21. $-5 < \frac{1}{2}(3x + 1) < 7$

$\quad$ $-10 < 3x + 1 < 14$ $\qquad$ Multiplying by 2

$\quad$ $-11 < 3x < 13$ $\qquad$ Adding -1

$\quad$ $-\frac{11}{3} < x < \frac{13}{3}$ $\qquad$ Multiplying by $\frac{1}{3}$

The solution set is $\left(-\frac{11}{3}, \frac{13}{3}\right)$. The graph is shown below.

23. $3x \leq -6 \;or\; x - 1 > 0$

$\quad$ $x \leq -2 \;or\;\quad x > 1$

The solution set is $(-\infty, -2] \cup (1, \infty)$. The graph is shown below.

25. $2x + 3 \leq -4 \;\; or \;\; 2x + 3 \geq 4$

$\quad$ $2x \leq -7 \;or\;\quad 2x \geq 1$

$\quad$ $x \leq -\frac{7}{2} \;or\;\quad x \geq \frac{1}{2}$

The solution set is $\left(-\infty, -\frac{7}{2}\right] \cup \left[\frac{1}{2}, \infty\right)$. The graph is shown below.

27. $2x - 20 < -0.8 \;or\; 2x - 20 > 0.8$

$\quad$ $2x < 19.2 \;or\;\quad 2x > 20.8$

$\quad$ $x < 9.6 \;\;or\;\quad x > 10.4$

The solution set is $(-\infty, 9.6) \cup (10.4, \infty)$. The graph is shown below.

29. $x + 14 \leq -\frac{1}{4} \;\; or \; x + 14 \geq \frac{1}{4}$

$\quad$ $x \leq -\frac{57}{4} \;or\;\quad x \geq -\frac{55}{4}$

The solution set is $\left(-\infty, -\frac{57}{4}\right] \cup \left[-\frac{55}{4}, \infty\right)$. The graph is shown below.

31. $|x| < 7$

To solve we look for all numbers x whose distance from 0 is less than 7. These are the numbers between -7 and 7. That is, $-7 < x < 7$. The solution set and its graph are as follows:

$(-7, 7)$

33. $|x| \geq 4.5$

To solve we look for all numbers x whose distance from 0 is greater than or equal to 4.5. That is, $x \leq -4.5$ or $x \geq 4.5$. The solution set and its graph are as follows.

$\{x | x \leq -4.5 \text{ or } x \geq 4.5\}$, or $(-\infty, -4.5] \cup [4.5, \infty)$

35. $|x + 8| < 9$

$\quad$ $-9 < x + 8 < 9$

$\quad$ $-17 < x < 1$ $\qquad$ Subtracting 8

The solution set is $(-17, 1)$. The graph is shown below.

37. $|x + 8| \geq 9$

$\quad$ $x + 8 \leq -9 \;\; or \; x + 8 \geq 9$

$\quad$ $x \leq -17 \;or\;\quad x \geq 1$ $\quad$ Subtracting 8

The solution set is $(-\infty, -17] \cup [1, \infty)$. The graph is shown below.

39. $\left|x - \frac{1}{4}\right| < \frac{1}{2}$

$\quad$ $-\frac{1}{2} < x - \frac{1}{4} < \frac{1}{2}$

$\quad$ $-\frac{1}{4} < x < \frac{3}{4}$ $\qquad$ Adding $\frac{1}{4}$

The solution set is $\left(-\frac{1}{4}, \frac{3}{4}\right)$. The graph is shown below.

41. $|3x| < 1$

$\quad$ $-1 < 3x < 1$

$\quad$ $-\frac{1}{3} < x < \frac{1}{3}$ $\qquad$ Dividing by 3

The solution set is $\left(-\frac{1}{3}, \frac{1}{3}\right)$. The graph is shown below.

43. $|2x + 3| \leq 9$

$\quad$ $-9 \leq 2x + 3 \leq 9$

$\quad$ $-12 \leq 2x \leq 6$ $\qquad$ Subtracting 3

$\quad$ $-6 \leq x \leq 3$ $\qquad$ Dividing by 2

The solution set is $[-6, 3]$. The graph is shown below.

45. $|x - 5| > 0.1$

$\quad x - 5 < -0.1 \quad or \quad x - 5 > 0.1$

$\qquad x < 4.9 \quad or \qquad x > 5.1 \qquad$ Adding 5

The solution set is $(-\infty, 4.9) \cup (5.1, \infty)$. The graph is shown below.

47. $|6 - 4x| \le 8$

$\quad -8 \le 6 - 4x \le 8$

$\quad -14 \le -4x \le 2 \quad$ Subtracting 6

$\dfrac{14}{4} \ge x \ge -\dfrac{2}{4} \quad$ Dividing by -4 and reversing the inequality symbols

$\dfrac{7}{2} \ge x \ge -\dfrac{1}{2} \quad$ Simplifying

The solution set is $\left[-\dfrac{1}{2}, \dfrac{7}{2}\right]$. The graph is shown below.

49. $\left|x + \dfrac{2}{3}\right| \le \dfrac{5}{3}$

$\quad -\dfrac{5}{3} \le x + \dfrac{2}{3} \le \dfrac{5}{3}$

$\quad -\dfrac{7}{3} \le x \le 1 \qquad$ Subtracting $\dfrac{2}{3}$

The solution set is $\left[-\dfrac{7}{3}, 1\right]$. The graph is shown below.

51. $\left|\dfrac{2x + 1}{3}\right| > 5$

$\quad \dfrac{2x + 1}{3} < -5 \quad or \quad \dfrac{2x + 1}{3} > 5$

$\quad 2x + 1 < -15 \ or \ 2x + 1 > 15 \ $ Multiplying by 3

$\quad 2x < -16 \ or \qquad 2x > 14 \ $ Subtracting 1

$\qquad x < -8 \ \ or \qquad\quad x > 7 \ $ Dividing by 2

The solution set is $\{x | x < -8 \ or \ x > 7\}$, or $(-\infty, -8) \cup (7, \infty)$. The graph is shown below.

53. $|2x - 4| < -5$

Since $|2x - 4| \ge 0$ for all x, there is no x such that $|2x - 4|$ would be less than -5. There is no solution.

55. *Familiarize and Translate.* Spending is given by the equation $y = 12.7x + 15.2$. We want to know when the spending will be more than \$66 billion, so we have

$\quad 12.7x + 15.2 > 66.$

Carry out. We solve the inequality.

$\quad 12.7x + 15.2 > 66$

$\qquad 12.7x > 50.8$

$\qquad\quad x > 4$

Check. When $x = 4$, the spending is $12.7(4) + 15.2 = 66$. As a partial check, we could try a value of x less than 4 and one greater than 4. When $x = 3.9$, we have $y = 12.7(3.9) + 15.2 = 64.73 < 66$; when $x = 4.1$, we have $y = 12.7(4.1) + 15.2 = 67.27 > 66$. Since $y = 66$ when $x = 4$ and $y > 66$ when $x = 4.1 > 4$, the answer is probably correct.

State. The spending will be more than \$66 billion more than 4 yr after 2002.

57. *Familiarize.* Let $t =$ the number of hours worked. Then Acme Movers charge $100 + 30t$ and Hank's Movers charge $55t$.

Translate.

$\underbrace{\text{Hank's charge}}_{55t} \ \underbrace{\text{is less than}}_{<} \ \underbrace{\text{Acme's charge.}}_{100 + 30t}$

Carry out. We solve the inequality.

$\quad 55t < 100 + 30t$

$\quad 25t < 100$

$\quad\ \ t < 4$

Check. When $t = 4$, Hank's Movers charge $55 \cdot 4$, or \$220 and Acme Movers charge $100 + 30 \cdot 4 = 100 + 120 = \$220$, so the charges are the same. As a partial check, we find the charges for a value of $t < 4$. When $t = 3.5$, Hank's Movers charge $55(3.5) = \$192.50$ and Acme Movers charge $100 + 30(3.5) = 100 + 105 = \$205$. Since Hank's charge is less than Acme's, the answer is probably correct.

State. For times less than 4 hr it costs less to hire Hank's Movers.

59. *Familiarize.* Let $x =$ the amount invested at 4%. Then $7500 - x =$ the amount invested at 5%. Using the simple-interest formula, $I = Prt$, we see that in one year the 4% investment earns $0.04x$ and the 5% investment earns $0.05(7500 - x)$.

Translate.

$\underbrace{\text{Interest at 4\%}}_{0.04x} \ \text{plus} \ \underbrace{\text{interest at 5\%}}_{0.05(7500 - x)} \ \underbrace{\text{is at least}}_{\ge} \ \underbrace{\$325.}_{325}$

Carry out. We solve the inequality.

$\quad 0.04x + 0.05(7500 - x) \ge 325$

$\quad 0.04x + 375 - 0.05x \ge 325$

$\qquad\quad -0.01x + 375 \ge 325$

$\qquad\qquad\quad -0.01x \ge -50$

$\qquad\qquad\qquad\quad x \le 5000$

Check. When \$5000 is invested at 4%, then \$7500−\$5000, or \$2500, is invested at 5%. In one year the 4% investment earns $0.04(\$5000)$, or \$200, in simple interest and the 5% investment earns $0.05(\$2500)$, or \$125, so the total interest is \$200 + \$125, or \$325. As a partial check, we determine the total interest when an amount greater than

$5000 is invested at 4%. Suppose $5001 is invested at 4%. Then $2499 is invested at 5%, and the total interest is $0.04(\$5001) + 0.05(\$2499)$, or $324.99. Since this amount is less than $325, the answer is probably correct.

State. The most that can be invested at 4% is $5000.

61. Familiarize. Let $c =$ the number of checks written per month. Then the No Frills plan costs $0.35c$ per month and the Simple Checking plan costs $5 + 0.10c$ per month.

Translate.

Simple Checking cost	is less than	No Frills cost.
$\downarrow$	$\downarrow$	$\downarrow$
$5 + 0.10c$	$<$	$0.35c$

Carry out. We solve the inequality.

$$5 + 0.10c < 0.35c$$
$$5 < 0.25c$$
$$20 < c$$

Check. When 20 checks are written the No Frills plan costs $0.35(20)$, or $7 per month and the Simple Checking plan costs $5 + 0.10(20)$, or $7, so the costs are the same. As a partial check, we compare the cost for some number of checks greater than 20. When 21 checks are written, the No Frills plan costs $0.35(21)$, or $7.35 and the Simple Checking plan costs $5+0.10(21)$, or $7.10. Since the Simple Checking plan costs less than the No Frills plan, the answer is probably correct.

State. The Simple Checking plan costs less when more than 20 checks are written per month.

63. Familiarize. Let $s =$ the monthly sales. Then the amount of sales in excess of $8000 is $s - 8000$.

Translate.

Income from plan B	is greater than	income from plan A.
$\downarrow$	$\downarrow$	$\downarrow$
$1200 + 0.15(s - 8000)$	$>$	$900 + 0.1s$

Carry out. We solve the inequality.

$$1200 + 0.15(s - 8000) > 900 + 0.1s$$
$$1200 + 0.15s - 1200 > 900 + 0.1s$$
$$0.15s > 900 + 0.1s$$
$$0.05s > 900$$
$$s > 18,000$$

Check. For sales of $18,000 the income from plan A is $900+0.1(\$18,000)$, or $2700, and the income from plan B is $1200 + 0.15(18,000 - 8000)$, or $2700 so the incomes are the same. As a partial check we can compare the incomes for an amount of sales greater than $18,000. For sales of $18,001, for example, the income from plan A is $900 + 0.1(\$18,001)$, or $2700.10, and the income from plan B is $1200 + 0.15(\$18,001 - \$8000)$, or $2700.15. Since plan B is better than plan A in this case, the answer is probably correct.

State. Plan B is better than plan A for monthly sales greater than $18,000.

65. Left to the student

67. Discussion and Writing

69. y-intercept

71. relation

73. horizontal line

75. decreasing

77. $2x \le 5 - 7x < 7 + x$

$$2x \le 5 - 7x \quad and \quad 5 - 7x < 7 + x$$
$$9x \le 5 \quad\quad and \quad\quad -8x < 2$$
$$x \le \frac{5}{9} \quad\quad and \quad\quad x > -\frac{1}{4}$$

The solution set is $\left(-\frac{1}{4}, \frac{5}{9} \right]$.

79. $|3x - 1| > 5x - 2$

$$3x - 1 < -(5x - 2) \quad or \quad 3x - 1 > 5x - 2$$
$$3x - 1 < -5x + 2 \quad or \quad\quad 1 > 2x$$
$$8x < 3 \quad\quad\quad or \quad\quad \frac{1}{2} > x$$
$$x < \frac{3}{8} \quad\quad\quad or \quad\quad \frac{1}{2} > x$$

The solution set is $\left(-\infty, \frac{3}{8} \right) \cup \left(-\infty, \frac{1}{2} \right)$. This is equivalent to $\left(-\infty, \frac{1}{2} \right)$.

81. $|p - 4| + |p + 4| < 8$

If $p < -4$, then $|p - 4| = -(p - 4)$ and $|p + 4| = -(p + 4)$.

Solve: $-(p - 4) + [-(p + 4)] < 8$
$$-p + 4 - p - 4 < 8$$
$$-2p < 8$$
$$p > -4$$

Since this is false for all values of p in the interval $(-\infty, -4)$ there is no solution in this interval.

If $p \ge -4$, then $|p + 4| = p + 4$.

Solve: $|p - 4| + p + 4 < 8$
$$|p - 4| < 4 - p$$
$$p - 4 > -(4 - p) \quad and \quad p - 4 < 4 - p$$
$$p - 4 > p - 4 \quad\quad and \quad\quad 2p < 8$$
$$-4 > -4 \quad\quad\quad and \quad\quad p < 4$$

Since $-4 > -4$ is false for all values of p, there is no solution in the interval $[-4, \infty)$.

Thus, $|p - 4| + |p + 4| < 8$ has no solution.

83. $|x - 3| + |2x + 5| > 6$

Divide the set of real numbers into three intervals:

$$\left(-\infty, -\frac{5}{2} \right), \left[-\frac{5}{2}, 3 \right), \text{ and } [3, \infty).$$

Find the solution set of $|x - 3| + |2x + 5| > 6$ in each interval. Then find the union of the three solution sets.

If $x < -\frac{5}{2}$, then $|x - 3| = -(x - 3)$ and $|2x + 5| = -(2x + 5)$.

Solve: $x < -\dfrac{5}{2}$ and $-(x-3) + [-(2x+5)] > 6$

$\qquad x < -\dfrac{5}{2}$ and $\qquad -x+3-2x-5 > 6$

$\qquad x < -\dfrac{5}{2}$ and $\qquad -3x > 8$

$\qquad x < -\dfrac{5}{2}$ and $\qquad x < -\dfrac{8}{3}$

The solution set in this interval is $\left(-\infty, -\dfrac{8}{3}\right)$.

If $-\dfrac{5}{2} \le x < 3$, then $|x-3| = -(x-3)$ and $|2x+5| = 2x+5$.

Solve: $-\dfrac{5}{2} \le x < 3$ and $-(x-3) + 2x + 5 > 6$

$\qquad -\dfrac{5}{2} \le x < 3$ and $\quad -x+3+2x+5 > 6$

$\qquad -\dfrac{5}{2} \le x < 3$ and $\qquad x > -2$

The solution set in this interval is $(-2, 3)$.

If $x \ge 3$, then $|x-3| = x-3$ and $|2x+5| = 2x+5$.

Solve: $x \ge 3$ and $x-3+2x+5 > 6$

$\qquad x \ge 3$ and $\qquad 3x > 4$

$\qquad x \ge 3$ and $\qquad x > \dfrac{4}{3}$

The solution set in this interval is $[3, \infty)$.

The union of the above solution sets is

$\left(-\infty, -\dfrac{8}{3}\right) \cup (-2, \infty)$. This is the solution set of $|x-3| + |2x+5| > 6$.

Chapter 2 Review Exercises

1. The statement is false. We find the zeros of a function $f(x)$ by finding the values of x for which $f(x) = 0$.

3. The statement is true. See page 205 in the text.

5. The statement is false. For example, $3^2 = (-3)^2$, but $3 \ne -3$.

7. $\quad 4y - 5 = 1$

$\qquad 4y = 6$

$\qquad y = \dfrac{3}{2}$

The solution is $\dfrac{3}{2}$.

9. $\quad 5(3x+1) = 2(x-4)$

$\qquad 15x + 5 = 2x - 8$

$\qquad 13x = -13$

$\qquad x = -1$

The solution is -1.

11. $(2y+5)(3y-1) = 0$

$\quad 2y + 5 = 0 \quad$ or $\quad 3y - 1 = 0$

$\qquad 2y = -5 \quad$ or $\qquad 3y = 1$

$\qquad y = -\dfrac{5}{2} \quad$ or $\qquad y = \dfrac{1}{3}$

The solutions are $-\dfrac{5}{2}$ and $\dfrac{1}{3}$.

13. $\qquad 3x^2 + 2x = 8$

$\qquad 3x^2 + 2x - 8 = 0$

$\qquad (x+2)(3x-4) = 0$

$\quad x + 2 = 0 \quad$ or $\quad 3x - 4 = 0$

$\qquad x = -2 \quad$ or $\qquad 3x = 4$

$\qquad x = -2 \quad$ or $\qquad x = \dfrac{4}{3}$

The solutions are -2 and $\dfrac{4}{3}$.

15. $\quad x^2 - 10 = 0$

$\qquad x^2 = 10$

$\quad x = -\sqrt{10} \quad$ or $\quad x = \sqrt{10}$

The solutions are $-\sqrt{10}$ and $\sqrt{10}$.

17. $\quad x - 4 = 0$

$\qquad x = 4$

The zero of the function is 4.

19. $\quad 8 - 2x = 0$

$\qquad -2x = -8$

$\qquad x = 4$

The zero of the function is 4.

21. $\qquad x^2 + 2x - 15 = 0$

$\qquad (x+5)(x-3) = 0$

$\quad x + 5 = 0 \quad$ or $\quad x - 3 = 0$

$\qquad x = -5 \quad$ or $\qquad x = 3$

The zeros of the function are -5 and 3.

23. $3x^2 + 2x - 3 = 0$

$a = 3$, $b = 2$, $c = -3$

$x = \dfrac{-b \pm \sqrt{b^2 - 4ac}}{2a}$

$x = \dfrac{-2 \pm \sqrt{2^2 - 4 \cdot 3 \cdot (-3)}}{2 \cdot 3}$

$ = \dfrac{-2 \pm \sqrt{40}}{6} = \dfrac{-2 \pm 2\sqrt{10}}{6} = \dfrac{-1 \pm \sqrt{10}}{3}$

The zeros of the function are $\dfrac{-1 \pm \sqrt{10}}{3}$.

25.
$$\frac{3}{8x+1} + \frac{8}{2x+5} = 1$$
$$\text{LCD is } (8x+1)(2x+5)$$
$$(8x+1)(2x+5)\left(\frac{3}{8x+1} + \frac{8}{2x+5}\right) = (8x+1)(2x+5)\cdot 1$$
$$3(2x+5) + 8(8x+1) = (8x+1)(2x+5)$$
$$6x + 15 + 64x + 8 = 16x^2 + 42x + 5$$
$$70x + 23 = 16x^2 + 42x + 5$$
$$0 = 16x^2 - 28x - 18$$
$$0 = 2(8x^2 - 14x - 9)$$
$$0 = 2(2x+1)(4x-9)$$

$$2x + 1 = 0 \quad or \quad 4x - 9 = 0$$
$$2x = -1 \quad or \quad 4x = 9$$
$$x = -\frac{1}{2} \quad or \quad x = \frac{9}{4}$$

Both numbers check. The solutions are $-\frac{1}{2}$ and $\frac{9}{4}$.

27. $\sqrt{x-1} - \sqrt{x-4} = 1$
$$\sqrt{x-1} = \sqrt{x-4} + 1$$
$$(\sqrt{x-1})^2 = (\sqrt{x-4} + 1)^2$$
$$x - 1 = x - 4 + 2\sqrt{x-4} + 1$$
$$x - 1 = x - 3 + 2\sqrt{x-4}$$
$$2 = 2\sqrt{x-4}$$
$$1 = \sqrt{x-4} \qquad \text{Dividing by 2}$$
$$1^2 = (\sqrt{x-4})^2$$
$$1 = x - 4$$
$$5 = x$$

This number checks. The solution is 5.

29. $|2y + 7| = 9$
$$2y + 7 = -9 \quad or \quad 2y + 7 = 9$$
$$2y = -16 \quad or \quad 2y = 2$$
$$y = -8 \quad or \quad y = 1$$
The solutions are -8 and 1.

31. $-2 < 5x - 4 \le 6$
$$2 < 5x \le 10 \qquad \text{Adding 4}$$
$$\frac{2}{5} < x \le 2 \qquad \text{Dividing by 5}$$

The solution set is $\left(\frac{2}{5}, 2\right]$.

33. $3x + 7 \le 2 \quad or \quad 2x + 3 \ge 5$
$$3x \le -5 \quad or \quad 2x \ge 2$$
$$x \le -\frac{5}{3} \quad or \quad x \ge 1$$

The solution set is $\left(-\infty, -\frac{5}{3}\right] \cup [1, \infty)$.

35. $|x + 4| \ge 2$
$$x + 4 \le -2 \quad or \quad x + 4 \ge 2$$
$$x \le -6 \quad or \quad x \ge -2$$
The solution is $(-\infty, -6] \cup [-2, \infty)$.

37.
$$M = n + 0.3s$$
$$M - n = 0.3s$$
$$\frac{M-n}{0.3} = s$$

39.
$$\frac{1}{a} + \frac{1}{b} = \frac{1}{t}, \text{ LCD is } abt$$
$$abt\left(\frac{1}{a} + \frac{1}{b}\right) = abt \cdot \frac{1}{t}$$
$$bt + at = ab$$
$$t(b + a) = ab$$
$$t = \frac{ab}{a+b}$$

41. $\sqrt{-12} \cdot \sqrt{-20} = \sqrt{-1} \cdot \sqrt{4} \cdot \sqrt{3} \cdot \sqrt{-1} \cdot \sqrt{4} \cdot \sqrt{5}$
$$= 2i\sqrt{3} \cdot 2i\sqrt{5}$$
$$= 4i^2\sqrt{3 \cdot 5}$$
$$= -4\sqrt{15}$$

43. $(6 + 2i)(-4 - 3i) = -24 - 18i - 8i - 6i^2$
$$= -24 - 26i + 6$$
$$= -18 - 26i$$

45. $(3 - 5i) - (2 - i) = (3 - 2) + [-5i - (-i)]$
$$= 1 - 4i$$

47. $i^{23} = (i^2)^{11} \cdot i = (-1)^{11} \cdot i = -1 \cdot i = -i$

49.
$$x^2 - 3x = 18$$
$$x^2 - 3x + \frac{9}{4} = 18 + \frac{9}{4} \quad \left(\frac{1}{2}(-3) = -\frac{3}{2} \text{ and } \left(-\frac{3}{2}\right)^2 = \frac{9}{4}\right)$$
$$\left(x - \frac{3}{2}\right)^2 = \frac{81}{4}$$
$$x - \frac{3}{2} = \pm\frac{9}{2}$$
$$x = \frac{3}{2} \pm \frac{9}{2}$$
$$x = \frac{3}{2} - \frac{9}{2} \quad or \quad x = \frac{3}{2} + \frac{9}{2}$$
$$x = -3 \quad or \quad x = 6$$

The solutions are -3 and 6.

51.
$$3x^2 + 10x = 8$$
$$3x^2 + 10x - 8 = 0$$
$$(x + 4)(3x - 2) = 0$$

$x + 4 = 0$ or $3x - 2 = 0$

$x = -4$ or $3x = 2$

$x = -4$ or $x = \dfrac{2}{3}$

The solutions are -4 and $\dfrac{2}{3}$.

53.
$$x^2 = 10 + 3x$$
$$x^2 - 3x - 10 = 0$$
$$(x + 2)(x - 5) = 0$$

$x + 2 = 0$ or $x - 5 = 0$

$x = -2$ or $x = 5$

The solutions are -2 and 5.

55. $y^4 - 3y^2 + 1 = 0$

Let $u = y^2$.

$u^2 - 3u + 1 = 0$

$u = \dfrac{-(-3) \pm \sqrt{(-3)^2 - 4 \cdot 1 \cdot 1}}{2 \cdot 1} = \dfrac{3 \pm \sqrt{5}}{2}$

Substitute y^2 for u and solve for y.

$y^2 = \dfrac{3 \pm \sqrt{5}}{2}$

$y = \pm\sqrt{\dfrac{3 \pm \sqrt{5}}{2}}$

The solutions are $\pm\sqrt{\dfrac{3 \pm \sqrt{5}}{2}}$.

57. $(p - 3)(3p + 2)(p + 2) = 0$

$p - 3 = 0$ or $3p + 2 = 0$ or $p + 2 = 0$

$p = 3$ or $3p = -2$ or $p = -2$

$p = 3$ or $p = -\dfrac{2}{3}$ or $p = -2$

The solutions are -2, $-\dfrac{2}{3}$ and 3.

59. $f(x) = -4x^2 + 3x - 1$

$= -4\left(x^2 - \dfrac{3}{4}x\right) - 1$

$= -4\left(x^2 - \dfrac{3}{4}x + \dfrac{9}{64} - \dfrac{9}{64}\right) - 1$

$= -4\left(x^2 - \dfrac{3}{4}x + \dfrac{9}{64}\right) - 4\left(-\dfrac{9}{64}\right) - 1$

$= -4\left(x^2 - \dfrac{3}{4}x + \dfrac{9}{64}\right) + \dfrac{9}{16} - 1$

$= -4\left(x - \dfrac{3}{8}\right)^2 - \dfrac{7}{16}$

a) Vertex: $\left(\dfrac{3}{8}, -\dfrac{7}{16}\right)$

b) Axis of symmetry: $x = \dfrac{3}{8}$

c) Maximum value: $-\dfrac{7}{16}$

d) Range: $\left(-\infty, -\dfrac{7}{16}\right]$

e) Since the graph opens down, function values increase to the left of the vertex and decrease to the right of the vertex. Thus, $f(x)$ is increasing on $\left(-\infty, \dfrac{3}{8}\right)$ and decreasing on $\left(\dfrac{3}{8}, \infty\right)$.

f)

$f(x) = -4x^2 + 3x - 1$

61. The graph of $y = (x - 2)^2$ has vertex $(2, 0)$ and opens up. It is graph (d).

63. The graph of $y = -2(x + 3)^2 + 4$ has vertex $(-3, 4)$ and opens down. It is graph (b).

65. *Familiarize.* Using the labels in the textbook, the legs of the right triangle are represented by x and $x + 10$.

Translate. We use the Pythagorean theorem.

$$x^2 + (x + 10)^2 = 50^2$$

Carry out. We solve the equation.

$$x^2 + (x + 10)^2 = 50^2$$
$$x^2 + x^2 + 20x + 100 = 2500$$
$$2x^2 + 20x - 2400 = 0$$
$$2(x^2 + 10x - 1200) = 0$$
$$2(x + 40)(x - 30) = 0$$

$x + 40 = 0$ or $x - 30 = 0$

$x = -40$ or $x = 30$

Check. Since the length cannot be negative, we need to check only 30. If $x = 30$, then $x + 10 = 30 + 10 = 40$. Since $30^2 + 40^2 = 900 + 1600 = 2500 = 50^2$, the answer checks.

State. The lengths of the legs are 30 ft and 40 ft.

67. *Familiarize.* Let r = the speed of the second train, in km/h. After 1 hr the first train has traveled 60 km, and the second train has traveled r km, and they are 100 km apart. We make a drawing.

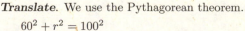

Translate. We use the Pythagorean theorem.

$$60^2 + r^2 = 100^2$$

Carry out. We solve the equation.

$$60^2 + r^2 = 100^2$$
$$3600 + r^2 = 10,000$$
$$r^2 = 6400$$
$$r = \pm 80$$

Check. Since the speed cannot be negative, we need to check only 80. We see that $60^2 + 80^2 = 3600 + 6400 = 10,000 = 100^2$, so the answer checks.

State. The second train is traveling 80 km/h.

69. *Familiarize*. Let $l =$ the length of the toy corral, in ft. Then the width is $\dfrac{24 - 2l}{2}$, or $12 - l$. The height of the corral is 2 ft.

Translate. We use the formula for the volume of a rectangular solid, $V = lwh$.

$$V(l) = l(12 - l)(2)$$
$$= 24l - 2l^2$$
$$= -2l^2 + 24l$$

Carry out. Since $V(l)$ is a quadratic function with $a = -2 < 0$, the maximum function value occurs at the vertex of the graph of the function. The first coordinate of the vertex is

$$-\frac{b}{2a} = -\frac{24}{2(-2)} = 6$$

When $l = 6$, then $12 - l = 12 - 6 = 6$.

Check. The volume of a corral with length 6 ft, width 6 ft, and height 2 ft is $6 \cdot 6 \cdot 2$, or 72 ft³. As a partial check, we can find $V(l)$ for a value of l less than 6 and for a value of l greater than 6. For instance, $V(5.9) = 71.98$ and $V(6.1) = 71.98$. Since both of these values are less than 72, our result appears to be correct.

State. The dimensions of the corral should be 6 ft by 6 ft.

71. *Familiarize and Translate*. The number of faculty, in thousands, is given by the equation $y = 6x + 121$. We want to know when this number will exceed 325 thousand, so we have

$$6x + 121 > 325.$$

Carry out. We solve the inequality.

$$6x + 121 > 325$$
$$6x > 204$$
$$x > 34$$

Check. When $x = 34$. the number of faculty is $6 \cdot 34 + 121 = 325$. As a partial check, we could try a value of x less than 34 and one greater than 34. When $x = 33.9$ we have $y = 6(33.9) + 121 = 324.4 < 325$; when $x = 34.1$ we have $y = 6(34.1) + 121 = 325.6 > 325$. Since $y = 325$ when $x = 34$ and $y > 325$ when $x > 34$, the answer is probably correct.

State. There will be more than 325 thousand faculty members more than 34 years after 1970, or in years after 2004.

73. $2x^2 - 5x + 1 = 0$

$$a = 2, \ b = -5, \ c = 1$$

$$x = \frac{-b \pm \sqrt{b^2 - 4ac}}{2a}$$

$$= \frac{-(-5) \pm \sqrt{(-5)^2 - 4 \cdot 2 \cdot 1}}{2 \cdot 2} = \frac{5 \pm \sqrt{25 - 8}}{4}$$

$$= \frac{5 \pm \sqrt{17}}{4}$$

Answer B is correct.

75. Left to the student

77. Left to the student

79. If an equation contains no fractions, using the addition principle before using the multiplication principle eliminates the need to add or subtract fractions.

81.
$$\sqrt{\sqrt{\sqrt{x}}} = 2$$
$$\left(\sqrt{\sqrt{\sqrt{x}}} \right)^2 = 2^2$$
$$\sqrt{\sqrt{x}} = 4$$
$$\left(\sqrt{\sqrt{x}} \right)^2 = 4^2$$
$$\sqrt{x} = 16$$
$$(\sqrt{x})^2 = 16^2$$
$$x = 256$$

The answer checks. The solution is 256.

83.
$$(t - 4)^{4/5} = 3$$
$$\left[(t - 4)^{4/5} \right]^5 = 3^5$$
$$(t - 4)^4 = 243$$
$$t - 4 = \pm \sqrt[4]{243}$$
$$t = 4 \pm \sqrt[4]{243}$$

The exact solutions are $4 + \sqrt[4]{243}$ and $4 - \sqrt[4]{243}$. The approximate solutions are 7.948 and 0.052.

85.
$$(2y - 2)^2 + y - 1 = 5$$
$$4y^2 - 8y + 4 + y - 1 = 5$$
$$4y^2 - 7y + 3 = 5$$
$$4y^2 - 7y - 2 = 0$$
$$(4y + 1)(y - 2) = 0$$

$$4y + 1 = 0 \quad or \quad y - 2 = 0$$
$$4y = -1 \quad or \qquad y = 2$$
$$y = -\frac{1}{4} \quad or \qquad y = 2$$

The solutions are $-\dfrac{1}{4}$ and 2.

87. *Familiarize*. When principal P is deposited in an account at interest rate r, compounded annually, the amount A to which it grows in t years is given by $A = P(1 + r)^t$. In 2 years the \$3500 deposit had grown to $3500(1 + r)^2$. In one year the \$4000 deposit had grown to $4000(1 + r)$.

Translate. The amount in the account at the end of 2 years was $8518.35, so we have

$$3500(1+r)^2 + 4000(1+r) = 8518.35.$$

Carry out. We solve the equation. Let $u = 1 + r$.

$$3500u^2 + 4000u = 8518.35$$

$$3500u^2 + 4000u - 8518.35 = 0$$

Using the quadratic formula, we find that $u = 1.09$ or $u \approx -2.23$. Substitute $1 + r$ for u and solve for r.

$$1 + r = 1.09 \quad or \quad 1 + r = -2.23$$
$$r = 0.09 \quad or \quad r = -3.23$$

Check. Since the interest rate cannot be negative, we need to check only 0.09. At 9%, the $3500 deposit would grow to $3500(1 + 0.09)^2$, or $4158.35. The $4000 deposit would grow to $4000(1 + 0.09)$, or $4360. Since $4158.35 + $4360 = $8518.35, the answer checks.

State. The interest rate was 9%.

Chapter 2 Test

1. $6x + 7 = 1$
$$6x = -6$$
$$x = -1$$
The solution is -1.

2. $3y - 4 = 5y + 6$
$$-4 = 2y + 6$$
$$-10 = 2y$$
$$-5 = y$$
The solution is -5.

3. $2(4x + 1) = 8 - 3(x - 5)$
$$8x + 2 = 8 - 3x + 15$$
$$8x + 2 = 23 - 3x$$
$$11x + 2 = 23$$
$$11x = 21$$
$$x = \frac{21}{11}$$
The solution is $\frac{21}{11}$.

4. $(2x - 1)(x + 5) = 0$
$$2x - 1 = 0 \quad or \quad x + 5 = 0$$
$$2x = 1 \quad or \qquad x = -5$$
$$x = \frac{1}{2} \quad or \qquad x = -5$$
The solutions are $\frac{1}{2}$ and -5.

5. $6x^2 - 36 = 0$
$$6x^2 = 36$$
$$x^2 = 6$$
$$x = -\sqrt{6} \quad or \quad x = \sqrt{6}$$
The solutions are $-\sqrt{6}$ and $\sqrt{6}$.

6. $x^2 + 4 = 0$
$$x^2 = -4$$
$$x = \pm\sqrt{-4}$$
$$x = -2i \quad or \quad x = 2i$$
The solutions are $-2i$ and $2i$.

7. $x^2 - 2x - 3 = 0$
$$(x + 1)(x - 3) = 0$$
$$x + 1 = 0 \quad or \quad x - 3 = 0$$
$$x = -1 \quad or \qquad x = 3$$
The solutions are -1 and 3.

8. $x^2 - 5x + 3 = 0$
$a = 1, b = -5, c = 3$
$$x = \frac{-b \pm \sqrt{b^2 - 4ac}}{2a}$$
$$x = \frac{-(-5) \pm \sqrt{(-5)^2 - 4 \cdot 1 \cdot 3}}{2 \cdot 1}$$
$$= \frac{5 \pm \sqrt{13}}{2}$$
The solutions are $\frac{5 + \sqrt{13}}{2}$ and $\frac{5 - \sqrt{13}}{2}$.

9. $2t^2 - 3t + 4 = 0$
$a = 2, b = -3, c = 4$
$$x = \frac{-b \pm \sqrt{b^2 - 4ac}}{2a}$$
$$x = \frac{-(-3) \pm \sqrt{(-3)^2 - 4 \cdot 2 \cdot 4}}{2 \cdot 2}$$
$$= \frac{3 \pm \sqrt{-23}}{4} = \frac{3 \pm i\sqrt{23}}{4}$$
$$= \frac{3}{4} \pm \frac{\sqrt{23}}{4}i$$
The solutions are $\frac{3}{4} + \frac{\sqrt{23}}{4}i$ and $\frac{3}{4} - \frac{\sqrt{23}}{4}i$.

10. $x + 5\sqrt{x} - 36 = 0$
Let $u = \sqrt{x}$.
$$u^2 + 5u - 36 = 0$$
$$(u + 9)(u - 4) = 0$$
$$u + 9 = 0 \quad or \quad u - 4 = 0$$
$$u = -9 \quad or \qquad u = 4$$
Substitute $\sqrt{x}$ for u and solve for x.
$$\sqrt{x} = -9 \quad or \quad \sqrt{x} = 4$$
No solution $\qquad x = 16$
The number 16 checks. It is the solution.

11.
$$\frac{3}{3x+4} + \frac{2}{x-1} = 2, \text{ LCD is } (3x+4)(x-1)$$

$$(3x+4)(x-1)\left(\frac{3}{3x+4}+\frac{2}{x-1}\right) = (3x+4)(x-1)(2)$$

$$3(x-1)+2(3x+4) = 2(3x^2+x-4)$$

$$3x-3+6x+8 = 6x^2+2x-8$$

$$9x+5 = 6x^2+2x-8$$

$$0 = 6x^2-7x-13$$

$$0 = (x+1)(6x-13)$$

$$x+1=0 \quad or \quad 6x-13=0$$

$$x=-1 \quad or \quad 6x=13$$

$$x=-1 \quad or \quad x=\frac{13}{6}$$

Both numbers check. The solutions are -1 and $\frac{13}{6}$.

12.
$$\sqrt{x+4}-2 = 1$$

$$\sqrt{x+4} = 3$$

$$(\sqrt{x+4})^2 = 3^2$$

$$x+4 = 9$$

$$x = 5$$

This number checks. The solution is 5.

13.
$$\sqrt{x+4}-\sqrt{x-4} = 2$$

$$\sqrt{x+4} = \sqrt{x-4}+2$$

$$(\sqrt{x+4})^2 = (\sqrt{x-4}+2)^2$$

$$x+4 = x-4+4\sqrt{x-4}+4$$

$$4 = 4\sqrt{x-4}$$

$$1 = \sqrt{x-4}$$

$$1^2 = (\sqrt{x-4})^2$$

$$1 = x-4$$

$$5 = x$$

This number checks. The solution is 5.

14. $|4y-3| = 5$

$$4y-3=-5 \quad or \quad 4y-3=5$$

$$4y=-2 \quad or \quad 4y=8$$

$$y=-\frac{1}{2} \quad or \quad y=2$$

The solutions are $-\frac{1}{2}$ and 2.

15.
$$-7 < 2x+3 < 9$$

$$-10 < 2x < 6 \qquad \text{Subtracting 3}$$

$$-5 < x < 3 \qquad \text{Dividing by 2}$$

The solution set is $(-5, 3)$.

16. $2x-1 \le 3 \quad or \quad 5x+6 \ge 26$

$$2x \le 4 \quad or \qquad 5x \ge 20$$

$$x \le 2 \quad or \qquad x \ge 4$$

The solution set is $(-\infty, 2] \cup [4, \infty)$.

17. $|x+3| \le 4$

$$-4 \le x+3 \le 4$$

$$-7 \le x \le 1$$

The solution set is $[-7, 1]$.

18. $|x+5| > 2$

$$x+5 < -2 \quad or \quad x+5 > 2$$

$$x < -7 \quad or \qquad x > -3$$

The solution set is $(-\infty, -7) \cup (-3, \infty)$.

19.
$$V = \frac{2}{3}\pi r^2 h$$

$$\frac{3V}{2} = \pi r^2 h \qquad \text{Multiplying by } \frac{3}{2}$$

$$\frac{3V}{2\pi r^2} = h \qquad \text{Dividing by } \pi r^2$$

20.
$$R = \sqrt{3np}$$

$$R^2 = (\sqrt{3np})^2$$

$$R^2 = 3np$$

$$\frac{R^2}{3p} = n$$

21.
$$x^2+4x = 1$$

$$x^2+4x+4 = 1+4 \qquad \left(\frac{1}{2}(4)=2 \text{ and } 2^2=4\right)$$

$$(x+2)^2 = 5$$

$$x+2 = \pm\sqrt{5}$$

$$x = -2 \pm \sqrt{5}$$

The solutions are $-2+\sqrt{5}$ and $-2-\sqrt{5}$.

22. *Familiarize*. Let $l = $ the length, in meters. Then $\frac{3}{4}l = $ the width. Recall that the formula for the perimeter P of a rectangle with length l and width w is $P = 2l+2w$.

***Translate*.**

$$\underbrace{\text{The perimeter}}_{\downarrow} \quad \text{is} \quad \underbrace{\text{210 m.}}_{\downarrow}$$

$$2l + 2\cdot\frac{3}{4}l \quad = \quad 210$$

Carry out. We solve the equation.

$$2l + 2 \cdot \frac{3}{4}l = 210$$

$$2l + \frac{3}{2}l = 210$$

$$\frac{7}{2}l = 210$$

$$l = 60$$

If $l = 60$, then $\frac{3}{4}l = \frac{3}{4} \cdot 60 = 45$.

Check. The width, 45 m, is three-fourths of the length, 60 m. Also, $2 \cdot 60$ m $+ 2 \cdot 45$ m $= 210$ m, so the answer checks.

State. The length is 60 m and the width is 45 m.

23. *Familiarize*. Let $c =$ the speed of the current, in km/h. The boat travels downstream at a speed of $12 + c$ and upstream at a speed of $12 - c$. Using the formula $d = rt$ in the form $t = \frac{d}{r}$, we see that the travel time downstream is $\frac{45}{12+c}$ and the time upstream is $\frac{45}{12-c}$.

Translate.

$$\underbrace{\text{Total travel time}}_{} \;\; \text{is} \;\; \underbrace{\text{8 hr.}}_{}$$

$$\frac{45}{12+c} + \frac{45}{12-c} \;\; = \;\; 8$$

Carry out. We solve the equation. First we multiply both sides by the LCD, $(12+c)(12-c)$.

$$\frac{45}{12+c} + \frac{45}{12-c} = 8$$

$$(12+c)(12-c)\left(\frac{45}{12+c} + \frac{45}{12-c}\right) = (12+c)(12-c)(8)$$

$$45(12-c) + 45(12+c) = 8(144 - c^2)$$

$$540 - 45c + 540 + 45c = 1152 - 8c^2$$

$$1080 = 1152 - 8c^2$$

$$0 = 72 - 8c^2$$

$$0 = 8(9 - c^2)$$

$$0 = 8(3+c)(3-c)$$

$$3 + c = 0 \quad \text{or} \quad 3 - c = 0$$
$$c = -3 \quad \text{or} \quad 3 = c$$

Check. Since the speed of the current cannot be negative, we need to check only 3. If the speed of the current is 3 km/h, then the boat's speed downstream is $12 + 3$, or 15 km/h, and the speed upstream is $12 - 3$, or 9 km/h. At 15 km/h, it takes the boat $\frac{45}{15}$, or 3 hr, to travel 45 km downstream. At 9 km/h, it takes the boat $\frac{45}{9}$, or 5 hr, to travel 45 km upstream. The total travel time is 3 hr + 5 hr, or 8 hr, so the answer checks.

State. The speed of the current is 3 km/h.

24. *Familiarize*. Let $p =$ the wholesale price of the juice.

Translate. We express 25¢ as $0.25.

Wholesale price	plus	50% of wholesale price	plus	$0.25	is	$2.95.
↓	↓	↓	↓	↓	↓	↓
p	$+$	$0.5p$	$+$	0.25	$=$	2.95

Carry out. We solve the equation.

$$p + 0.5p + 0.25 = 2.95$$

$$1.5p + 0.25 = 2.95$$

$$1.5p = 2.7$$

$$p = 1.8$$

Check. 50% of $1.80 is $0.90 and $1.80 + $0.90 + $0.25 = $2.95, so the answer checks.

State. The wholesale price of a bottle of juice is $1.80.

25. $\sqrt{-43} = \sqrt{-1} \cdot \sqrt{43} = i\sqrt{43}$, or $\sqrt{43}i$

26. $-\sqrt{-25} = -\sqrt{-1} \cdot \sqrt{25} = -5i$

27. $(5 - 2i) - (2 + 3i) = (5 - 2) + (-2i - 3i)$
$$= 3 - 5i$$

28. $(3 + 4i)(2 - i) = 6 - 3i + 8i - 4i^2$
$$= 6 + 5i + 4 \quad (i^2 = -1)$$
$$= 10 + 5i$$

29.
$$\frac{1-i}{6+2i} = \frac{1-i}{6+2i} \cdot \frac{6-2i}{6-2i}$$

$$= \frac{6 - 2i - 6i + 2i^2}{36 - 4i^2}$$

$$= \frac{6 - 8i - 2}{36 + 4}$$

$$= \frac{4 - 8i}{40}$$

$$= \frac{4}{40} - \frac{8}{40}i$$

$$= \frac{1}{10} - \frac{1}{5}i$$

30. $i^{33} = (i^2)^{16} \cdot i = (-1)^{16} \cdot i = 1 \cdot i = i$

31. $3x + 9 = 0$
$$3x = -9$$
$$x = -3$$

The zero of the function is -3.

32. $4x^2 - 11x - 3 = 0$
$$(4x + 1)(x - 3) = 0$$

$$4x + 1 = 0 \quad \text{or} \quad x - 3 = 0$$
$$4x = -1 \quad \text{or} \quad x = 3$$
$$x = -\frac{1}{4} \quad \text{or} \quad x = 3$$

The zeros of the functions are $-\frac{1}{4}$ and 3.

33. $2x^2 - x - 7 = 0$

$a = 2,\ b = -1,\ c = -7$

$x = \dfrac{-b \pm \sqrt{b^2 - 4ac}}{2a}$

$x = \dfrac{-(-1) \pm \sqrt{(-1)^2 - 4 \cdot 2 \cdot (-7)}}{2 \cdot 2}$

$ = \dfrac{1 \pm \sqrt{57}}{4}$

The solutions are $\dfrac{1 + \sqrt{57}}{4}$ and $\dfrac{1 - \sqrt{57}}{4}$.

34. $f(x) = -x^2 + 2x + 8$

$ = -(x^2 - 2x) + 8$

$ = -(x^2 - 2x + 1 - 1) + 8$

$ = -(x^2 - 2x + 1) - (-1) + 8$

$ = -(x^2 - 2x + 1) + 1 + 8$

$ = -(x - 1)^2 + 9$

a) Vertex: $(1, 9)$

b) Axis of symmetry: $x = 1$

c) Maximum value: 9

d) Range: $(-\infty, 9]$

e) The graph opens down, so function values increase to the left of the vertex and decrease to the right of the vertex. Thus, $f(x)$ is increasing on $(-\infty, 1)$ and decreasing on $(1, \infty)$.

f)

$f(x) = -x^2 + 2x + 8$

35. *Familiarize.* We make a drawing, letting $w =$ the width of the rectangle, in ft. This leaves $80 - w - w$, or $80 - 2w$ ft of fencing for the length.

Translating. The area of a rectangle is given by length times width.

$A(w) = (80 - 2w)w$

$ = 80w - 2w^2,\ \text{or}\ -2w^2 + 80w$

Carry out. This is a quadratic function with $a < 0$, so it has a maximum value that occurs at the vertex of the graph of the function. The first coordinate of the vertex is

$w = -\dfrac{b}{2a} = -\dfrac{80}{2(-2)} = 20.$

If $w = 20$, then $80 - 2w = 80 - 2 \cdot 20 = 40$.

Check. The area of a rectangle with length 40 ft and width 20 ft is $40 \cdot 20$, or 800 ft². As a partial check, we can find $A(w)$ for a value of w less than 20 and for a value of w greater than 20. For instance, $A(19.9) = 799.98$ and $A(20.1) = 799.98$. Since both of these values are less than 800, the result appears to be correct.

State. The dimensions for which the area is a maximum are 20 ft by 40 ft.

36. *Familiarize.* Let $t =$ the number of hours a move requires. Then Morgan Movers charges $90 + 25t$ to make a move and McKinley Movers charges $40t$.

Translate.

Morgan Movers' charge	is less than	McKinley Movers' charge.
↓	↓	↓
$90 + 25t$	$<$	$40t$

Carry out. We solve the inequality.

$90 + 25t < 40t$

$90 < 15t$

$6 < t$

Check. For $t = 6$, Morgan Movers charge $90 + 25 \cdot 6$, or \$240, and McKinley Movers charge $40 \cdot 6$, or \$240, so the charge is the same for 6 hours. As a partial check, we can find the charges for a value of t greater than 6. For instance, for 6.5 hr Morgan Movers charge $90 + 25(6.5)$, or \$252.50, and McKinley Movers charge $40(6.5)$, or \$260. Since Morgan Movers cost less for a value of t greater than 6, the answer is probably correct.

State. It costs less to hire Morgan Movers when a move takes more than 6 hr.

37. The maximum value occurs at the vertex. The first coordinate of the vertex is $-\dfrac{b}{2a} = -\dfrac{(-4)}{2a} = \dfrac{2}{a}$ and $f\left(\dfrac{2}{a}\right) = 12$. Then we have:

$a\left(\dfrac{2}{a}\right)^2 - 4\left(\dfrac{2}{a}\right) + 3 = 12$

$a \cdot \dfrac{4}{a^2} - \dfrac{8}{a} + 3 = 12$

$\dfrac{4}{a} - \dfrac{8}{a} + 3 = 12$

$-\dfrac{4}{a} + 3 = 12$

$-\dfrac{4}{a} = 9$

$-4 = 9a$

$-\dfrac{4}{9} = a$

Chapter 3

Polynomial and Rational Functions

Exercise Set 3.1

1. $g(x) = \dfrac{1}{2}x^3 - 10x + 8$

The degree of the polynomial is 3, so the polynomial is cubic. The leading term is $\dfrac{1}{2}x^3$ and the leading coefficient is $\dfrac{1}{2}$.

3. $h(x) = 0.9x - 0.13$

The degree of the polynomial is 1, so the polynomial is linear. The leading term is $0.9x$ and the leading coefficient is 0.9.

5. $g(x) = 305x^4 + 4021$

The degree of the polynomial is 4, so the polynomial is quartic. The leading term is $305x^4$ and the leading coefficient is 305.

7. $h(x) = -5x^2 + 7x^3 + x^4 = x^4 + 7x^3 - 5x^2$

The degree of the polynomial is 4, so the polynomial is quartic. The leading term is x^4 and the leading coefficient is 1 ($x^4 = 1 \cdot x^4$).

9. $g(x) = 4x^3 - \dfrac{1}{2}x^2 + 8$

The degree of the polynomial is 3, so the polynomial is cubic. The leading term is $4x^3$ and the leading coefficient is 4.

11. $f(x) = -3x^3 - x + 4$

The leading term is $-3x^3$. The degree, 3, is odd and the leading coefficient, -3, is negative. Thus the end behavior of the graph is like that of (d).

13. $f(x) = -x^6 + \dfrac{3}{4}x^4$

The leading term is $-x^6$. The degree, 6, is even and leading coefficient, -1, is negative. Thus the end behavior of the graph is like that of (b).

15. $f(x) = -3.5x^4 + x^6 + 0.1x^7 = 0.1x^7 + x^6 - 3.5x^4$

The leading term is $0.1x^7$. The degree, 7, is odd and the leading coefficient, 0.1, is positive. Thus the end behavior of the graph is like that of (c).

17. $f(x) = 10 + \dfrac{1}{10}x^4 - \dfrac{2}{5}x^3 = \dfrac{1}{10}x^4 - \dfrac{2}{5}x^3 + 10$

The leading term is $\dfrac{1}{10}x^4$. The degree, 4, is even and the leading coefficient, $\dfrac{1}{10}$, is positive. Thus the end behavior of the graph is like that of (a).

19. $f(x) = -x^6 + 2x^5 - 7x^2$

The leading term is $-x^6$. The degree, 6, is even and the leading coefficient, -1, is negative. Thus, (c) is the correct graph.

21. $f(x) = x^5 + \dfrac{1}{10}x - 3$

The leading term is x^5. The degree, 5, is odd and the leading coefficient, 1, is positive. Thus, (d) is the correct graph.

23. $f(x) = x^3 - 9x^2 + 14x + 24$

$f(4) = 4^3 - 9 \cdot 4^2 + 14 \cdot 4 + 24 = 0$

Since $f(4) = 0$, 4 is a zero of $f(x)$.

$f(5) = 5^3 - 9 \cdot 5^2 + 14 \cdot 5 + 24 = -6$

Since $f(5) \neq 0$, 5 is not a zero of $f(x)$.

$f(-2) = (-2)^3 - 9(-2)^2 + 14(-2) + 24 = -48$

Since $f(-2) \neq 0$, -2 is not a zero of $f(x)$.

25. $g(x) = x^4 - 6x^3 + 8x^2 + 6x - 9$

$g(2) = 2^4 - 6 \cdot 2^3 + 8 \cdot 2^2 + 6 \cdot 2 - 9 = 3$

Since $g(2) \neq 0$, 2 is not a zero of $g(x)$.

$g(3) = 3^4 - 6 \cdot 3^3 + 8 \cdot 3^2 + 6 \cdot 3 - 9 = 0$

Since $g(3) = 0$, 3 is a zero of $g(x)$.

$g(-1) = (-1)^4 - 6(-1)^3 + 8(-1)^2 + 6(-1) - 9 = 0$

Since $g(-1) = 0$, -1 is a zero of $g(x)$.

27. $f(x) = (x+3)^2(x-1) = (x+3)(x+3)(x-1)$

To solve $f(x) = 0$ we use the principle of zero products, solving $x + 3 = 0$ and $x - 1 = 0$. The zeros of $f(x)$ are -3 and 1.

The factor $x + 3$ occurs twice. Thus the zero -3 has a multiplicity of two.

The factor $x - 1$ occurs only one time. Thus the zero 1 has a multiplicity of one.

29. $f(x) = -2(x-4)(x-4)(x-4)(x+6) = -2(x-4)^3(x+6)$

To solve $f(x) = 0$ we use the principle of zero products, solving $x - 4 = 0$ and $x + 6 = 0$. The zeros of $f(x)$ are 4 and -6.

The factor $x - 4$ occurs three times. Thus the zero 4 has a multiplicity of 3.

The factor $x + 6$ occurs only one time. Thus the zero -6 has a multiplicity of 1.

31. $f(x) = (x^2 + 9)^3 = [(x+3i)(x-3i)]^3 = (x+3i)^3(x-3i)^3$

To solve $f(x) = 0$ we use the principle of zero products, solving $x + 3i = 0$ and $x - 3i = 0$. The zeros of $f(x)$ are $-3i$ and $3i$.

The factors $x + 3i$ and $x - 3i$ each occur three times so each zero has a multiplicity of 3.

33. $f(x) = x^3(x-1)^2(x+4)$

To solve $f(x) = 0$ we use the principle of zero products, solving $x = 0$, $x - 1 = 0$, and $x + 4 = 0$. The zeros of $f(x)$ are 0, 1, and -4.

The factor x occurs three times. Thus the zero 0 has a multiplicity of three.

The factor $x - 1$ occurs twice. Thus the zero 1 has a multiplicity of two.

The factor $x + 4$ occurs only one time. Thus the zero -4 has a multiplicity of one.

35. $f(x) = -8(x-3)^2(x+4)^3x^4$

To solve $f(x) = 0$ we use the principle of zero products, solving $x - 3 = 0$, $x + 4 = 0$, and $x = 0$. The zeros of $f(x)$ are 3, -4, and 0.

The factor $x - 3$ occurs twice. Thus the zero 3 has a multiplicity of 2.

The factor $x + 4$ occurs three times. Thus the zero -4 has a multiplicity of 3.

The factor x occurs four times. Thus the zero 0 has a multiplicity of 4.

37. $f(x) = x^4 - 4x^2 + 3$

We factor as follows:
$$f(x) = (x^2 - 3)(x^2 - 1)$$
$$= (x - \sqrt{3})(x + \sqrt{3})(x - 1)(x + 1)$$

The zeros of the function are $\sqrt{3}$, $-\sqrt{3}$, 1, and -1. Each has a multiplicity of 1.

39. $f(x) = x^3 + 3x^2 - x - 3$

We factor by grouping:
$$f(x) = x^2(x+3) - (x+3)$$
$$= (x^2 - 1)(x + 3)$$
$$= (x - 1)(x + 1)(x + 3)$$

The zeros of the function are 1, -1, and -3. Each has a multiplicity of 1.

41. $f(x) = 2x^3 - x^2 - 8x + 4$
$$= x^2(2x - 1) - 4(2x - 1)$$
$$= (2x - 1)(x^2 - 4)$$
$$= (2x - 1)(x + 2)(x - 2)$$

The zeros of the function are $\frac{1}{2}$, -2, and 2. Each has a multiplicity of 1.

43. Graphing the function, we see that the graph touches the x-axis at $(3, 0)$ but does not cross it, so the statement is false.

45. Graphing the function, we see that the statement is true.

47. In 1960, $x = 60$:

$f(60) = 0.0013924589(60)^4 - 0.2121392316(60)^3 + 3.414011377(60)^2 + 310.7243489(60) + 16,209.42591 \approx 19,368$

There were about 19,368 thousand, or 19,368,000, milk cows in 1960.

In 1980, $x = 80$:

$f(80) = 0.0013924589(80)^4 - 0.2121392316(80)^3 + 3.414011377(80)^2 + 310.7243489(80) + 16,209.42591 \approx 11,337$

There were about 11,337 thousand, or 11,337,000, milk cows in 1980.

49. In 2002, $x = 1$:

$f(1) = 2.375(1)^2 - 7.395(1) + 28.205 \approx 23.2$

There were about 23.2 million real Christmas trees in 2002.

In 2004, $x = 3$:

$f(3) = 2.375(3)^2 - 7.395(3) + 28.205 \approx 27.4$

There were about 27.4 million real Christmas trees in 2004.

51. First find the number of games played.
$$N(x) = x^2 - x$$
$$N(9) = 9^2 - 9 = 72$$

Now multiply the number of games by the cost per game to find the total cost.
$$72 \cdot 45 = 3240$$

It will cost $3240 to play the entire schedule.

53. In 1996, $x = 1$:

$f(1) = -0.2607808858(1)^4 + 5.281857032(1)^3 - 30.77185315(1)^2 + 60.89063714(1) + 663.8111888 \approx \$699$

In 1999, $x = 4$:

$f(4) = -0.2607808858(4)^4 + 5.281857032(4)^3 - 30.77185315(4)^2 + 60.89063714(4) + 663.8111888 \approx \$686$

In 2004, $x = 9$:

$f(9) = -0.2607808858(9)^4 + 5.281857032(9)^3 - 30.77185315(9)^2 + 60.89063714(9) + 663.8111888 \approx \$859$

55. $A = P(1 + i)^t$

a) $4368.10 = 4000(1 + i)^2$ Substituting

$\dfrac{4368.10}{4000} = (1 + i)^2$

$\pm 1.045 = 1 + i$ Taking the square root on both sides

$-1 \pm 1.045 = i$

$-1 - 1.045 = i$ or $-1 + 1.045 = i$

$-2.045 = i$ or $0.045 = i$

Only the positive result has meaning in this application. The interest rate is 0.045, or 4.5%.

b) $13,310 = 10,000(1 + i)^3$ Substituting

$$\frac{13,310}{10,000} = (1 + i)^3$$

$1.10 \approx 1 + i$ Taking the cube root on both sides

$0.10 \approx i$

The interest rate is 0.1, or 10%.

57. Graph $y = x^3 - 3x - 1$ and use the Zero feature three times. The real-number zeros are about -1.532, -0.347, and 1.879.

59. Graph $y = x^4 - 2x^2$ and use the Zero feature three times. The real-number zeros are about -1.414, 0, and 1.414.

61. Graph $y = x^3 - x$ and use the Zero feature three times. The real-number zeros are -1, 0, and 1.

63. Graph $y = x^8 + 8x^7 - 28x^6 - 56x^5 + 70x^4 + 56x^3 - 28x^2 - 8x + 1$ and use the Zero feature eight times. The real-number zeros are about -10.153, -1.871, -0.821, -0.303, 0.098, 0.535, 1.219, and 3.297.

65. $g(x) = x^3 - 1.2x + 1$

Graph the function and use the Zero, Maximum, and Minimum features.

Relative maximum: 1.506 when $x \approx -0.632$

Relative minimum: 0.494 when $x \approx 0.632$

Range: $(-\infty, \infty)$

67. $f(x) = x^6 - 3.8$

Graph the function and use the Zero and Minimum features.

There is no relative maximum.

Relative minimum: -3.8 when $x = 0$

Range: $[-3.8, \infty)$

69. $f(x) = x^2 + 10x - x^5$

Graph the function and use the Zero, Maximum, and Minimum features.

Relative maximum: 11.012 when $x \approx 1.258$

Relative minimum: -8.183 when $x \approx -1.116$

Range: $(-\infty, \infty)$

71. The sales drop and then rise. This suggests that a quadratic function that opens up might fit the data. Thus, we choose (b).

73. The sales rise and then drop. This suggests that a quadratic function that opens down might fit the data. Thus, we choose (c).

75. The sales fall steadily. Thus, we choose (a), a linear function.

77. a) For each function, x represents the number of years after 1900.

Cubic: $y = 0.000053215724x^3 - 0.0059922487x^2 + 0.0288132843x + 14.25265259$

Quartic: $y = -0.0000007674506x^4 + 0.0002128619x^3 - 0.0163703506x^2 + 0.2450880151x + 13.59031965$

b) Evaluate each function for $x = 2010 - 1900 = 110$.

Cubic: $y \approx 15.7\%$

Quartic: $y \approx 13.4\%$

Given the trend shown in the table, many would say that the quartic function gives the more realistic estimate. Answers will vary.

79. a) For each function, x represents the number of years after 1996.

Cubic: $y = -1.590909091x^3 + 27.21678322x^2 - 113.3951049x + 421.3426573$

Quartic: $y = -0.2966200466x^4 + 4.341491841x^3 - 9.860722611x^2 - 39.24009324x + 399.986014$

b) Evaluate each function for $x = 2008 - 1996 = 12$.

Cubic: $y \approx 231$ thousand

Quartic: $y \approx -139$ thousand

Since the quartic function gives a negative value, the cubic function provides a more realistic estimate.

81. Discussion and Writing

83. $d = \sqrt{(x_2 - x_1)^2 + (y_2 - y_1)^2}$

$= \sqrt{[-1 - (-5)]^2 + (0 - 3)^2}$

$= \sqrt{4^2 + (-3)^2} = \sqrt{16 + 9}$

$= \sqrt{25} = 5$

85. $(x - 3)^2 + (y + 5)^2 = 49$

$(x - 3)^2 + [y - (-5)]^2 = 7^2$

Center: $(3, -5)$; radius: 7

87. $2y - 3 \geq 1 - y + 5$

$2y - 3 \geq 6 - y$ Collecting like terms

$3y - 3 \geq 6$ Adding y

$3y \geq 9$ Adding 3

$y \geq 3$ Dividing by 3

The solution set is $\{y | y \geq 3\}$, or $[3, \infty)$.

89. $|x + 6| \geq 7$

$x + 6 \leq -7$ or $x + 6 \geq 7$

$x \leq -13$ or $x \geq 1$

The solution set is $\{x | x \leq -13 \text{ or } x \geq 1\}$, or $(-\infty, -13] \cup [1, \infty)$.

91. *Familiarize*. We will use the compound interest formula. The $2000 deposit is invested for two years and grows to an amount A_1 given by $A_1 = 2000(1 + i)^2$. The $1200 deposit is invested for one year and grows to an amount A_2 given by $A_2 = 1200(1 + i)$.

Translate. There is a total of $3573.80 in both accounts at the end of the second year, so we have

$A_1 + A_2 = 3573.80$, or

$2000(1 + i)^2 + 1200(1 + i) = 3573.80$.

Carry out. We solve the equation.

$$2000(1+i)^2 + 1200(1+i) - 3573.8 = 0$$

Substitute u for $1+i$.

$$2000u^2 + 1200u - 3573.8 = 0$$

Using the quadratic formula we find that $u = 1.07$ or $u = -3.34$. Only the positive value has meaning in this application. Then since $u = 1.07$, we have $1+i = 1.07$, or $i = 0.07$.

Check. At an interest rate of 0.07, or 7%, in two years $2000 would grow to $2000(1 + 0.07)^2$, or $2289.80. In one year $1200 would grow to $1200(1 + 0.07)$, or $1284. Now $2289.80 + $1284 = $3573.80, so the result checks.

State. The interest rate is 7%.

Exercise Set 3.2

1. $f(x) = x^5 - x^2 + 6$

   a) This function has degree 5, so its graph can have at most 5 real zeros.

   b) This function has degree 5, so its graph can have at most 5 x-intercepts.

   c) This function has degree 5, so its graph can have at most $5 - 1$, or 4, turning points.

3. $f(x) = x^{10} - 2x^5 + 4x - 2$

   a) This function has degree 10, so its graph can have at most 10 real zeros.

   b) This function has degree 10, so its graph can have at most 10 x-intercepts.

   c) This function has degree 10, so its graph can have at most $10 - 1$, or 9, turning points.

5. $f(x) = -x - x^3 = -x^3 - x$

   a) This function has degree 3, so its graph can have at most 3 real zeros.

   b) This function has degree 3, so its graph can have at most 3 x-intercepts.

   c) This function has degree 3, so its graph can have at most $3 - 1$, or 2, turning points.

7. $f(x) = \frac{1}{4}x^2 - 5$

The leading term is $\frac{1}{4}x^2$. The sign of the leading coefficient, $\frac{1}{4}$, is positive and the degree, 2, is even, so we would choose either graph (b) or graph (d). Note also that $f(0) = -5$, so the y-intercept is $(0, -5)$. Thus, graph (d) is the graph of this function.

9. $f(x) = x^5 - x^4 + x^2 + 4$

The leading term is x^5. The sign of the leading coefficient, 1, is positive and the degree, 5, is odd. Thus, graph (f) is the graph of this function.

11. $f(x) = x^4 - 2x^3 + 12x^2 + x - 20$

The leading term is x^4. The sign of the leading coefficient, 1, is positive and the degree, 4, is even, so we would choose either graph (b) or graph (d). Note also that $f(0) = -20$, so the y-intercept is $(0, -20)$. Thus, graph (b) is the graph of this function.

13. $f(x) = -x^3 - 2x^2$

   1. The leading term is $-x^3$. The degree, 3, is odd and the leading coefficient, -1, is negative so as $x \to \infty$, $f(x) \to -\infty$ and as $x \to -\infty$, $f(x) \to \infty$.

   2. We solve $f(x) = 0$.
$$-x^3 - 2x^2 = 0$$
$$-x^2(x + 2) = 0$$
$$-x^2 = 0 \quad or \quad x + 2 = 0$$
$$x^2 = 0 \quad or \qquad x = -2$$
$$x = 0 \quad or \qquad x = -2$$

   The zeros of the function are 0 and -2, so the x-intercepts of the graph are $(0, 0)$ and $(-2, 0)$.

   3. The zeros divide the x-axis into 3 intervals, $(-\infty, -2)$, $(-2, 0)$, and $(0, \infty)$. We choose a value for x from each interval and find $f(x)$. This tells us the sign of $f(x)$ for all values of x in that interval.

   In $(-\infty, -2)$, test -3:
$$f(-3) = -(-3)^3 - 2(-3)^2 = 9 > 0$$

   In $(-2, 0)$, test -1:
$$f(-1) = -(-1)^3 - 2(-1)^2 = -1 < 0$$

   In $(0, \infty)$, test 1:
$$f(1) = -1^3 - 2 \cdot 1^2 = -3 < 0$$

   Thus the graph lies above the x-axis on $(-\infty, -2)$ and below the x-axis on $(-2, 0)$ and $(0, \infty)$. We also know the points $(-3, 9)$, $(-1, -1)$, and $(1, -3)$ are on the graph.

   4. From Step 2 we see that the y-intercept is $(0, 0)$.

   5. We find additional points on the graph and then draw the graph.

x	$f(x)$
-2.5	3.125
-1.5	-1.125
1.5	-7.875

$f(x) = -x^3 - 2x^2$

   6. Checking the graph as described on page 272 in the text, we see that it appears to be correct.

15. $h(x) = x^5 - 4x^3$

   1. The leading term is x^5. The degree, 5, is odd and the leading coefficient, 1, is positive so as $x \to \infty$, $h(x) \to \infty$ and as $x \to -\infty$, $h(x) \to -\infty$.

2. We solve $h(x) = 0$.

$$x^5 - 4x^3 = 0$$
$$x^3(x^2 - 4) = 0$$
$$x^3(x + 2)(x - 2) = 0$$

$$x^3 = 0 \ \ or \ \ x + 2 = 0 \ \ \ or \ \ x - 2 = 0$$

$$x = 0 \ \ or \ \ \ \ \ \ x = -2 \ or \ \ \ \ \ \ x = 2$$

The zeros of the function are 0, -2, and 2 so the x-intercepts of the graph are $(0,0)$, $(-2,0)$, and $(2,0)$.

3. The zeros divide the x-axis into 4 intervals, $(-\infty, -2)$, $(-2, 0)$, $(0, 2)$, and $(2, \infty)$. We choose a value for x from each interval and find $h(x)$. This tells us the sign of $h(x)$ for all values of x in that interval.

In $(-\infty, -2)$, test -3:

$$h(-3) = (-3)^5 - 4(-3)^3 = -135 < 0$$

In $(-2, 0)$, test -1:

$$h(-1) = (-1)^5 - 4(-1)^3 = 3 > 0$$

In $(0, 2)$, test 1:

$$h(1) = 1^5 - 4 \cdot 1^3 = -3 < 0$$

In $(2, \infty)$, test 3:

$$h(3) = 3^5 - 4 \cdot 3^3 = 135 > 0$$

Thus the graph lies below the x-axis on $(-\infty, -2)$ and on $(0, 2)$. It lies above the x-axis on $(-2, 0)$ and on $(2, \infty)$. We also know the points $(-3, -135)$, $(-1, 3)$, $(1, -3)$, and $(3, 135)$ are on the graph.

4. From Step 2 we see that the y-intercept is $(0, 0)$.

5. We find additional points on the graph and then draw the graph.

x	$h(x)$
-2.5	-35.2
-1.5	5.9
1.5	-5.9
2.5	35.2

$h(x) = x^5 - 4x^3$

6. Checking the graph as described on page 272 in the text, we see that it appears to be correct.

17. $h(x) = x(x - 4)(x + 1)(x - 2)$

1. The leading term is $x \cdot x \cdot x \cdot x$, or x^4. The degree, 4, is even and the leading coefficient, 1, is positive so as $x \to \infty$, $h(x) \to \infty$ and as $x \to -\infty$, $h(x) \to \infty$.

2. We see that the zeros of the function are 0, 4, -1, and 2 so the x-intercepts of the graph are $(0,0)$, $(4,0)$, $(-1,0)$, and $(2,0)$.

3. The zeros divide the x-axis into 5 intervals, $(-\infty, -1)$, $(-1, 0)$, $(0, 2)$, $(2, 4)$, and $(4, \infty)$. We choose a value for x from each interval and find $h(x)$. This tells us the sign of $h(x)$ for all values of x in that interval.

In $(-\infty, -1)$, test -2:

$$h(-2) = -2(-2 - 4)(-2 + 1)(-2 - 2) = 48 > 0$$

In $(-1, 0)$, test -0.5:

$$h(-0.5) = (-0.5)(-0.5 - 4)(-0.5 + 1)(-0.5 - 2) = -2.8125 < 0$$

In $(0, 2)$, test 1:

$$h(1) = 1(1 - 4)(1 + 1)(1 - 2) = 6 > 0$$

In $(2, 4)$, test 3:

$$h(3) = 3(3 - 4)(3 + 1)(3 - 2) = -12 < 0$$

In $(4, \infty)$, test 5:

$$h(5) = 5(5 - 4)(5 + 1)(5 - 2) = 90 > 0$$

Thus the graph lies above the x-axis on $(-\infty, -1)$, $(0, 2)$, and $(4, \infty)$. It lies below the x-axis on $(-1, 0)$ and on $(2, 4)$. We also know the points $(-2, 48)$, $(-0.5, -2.8125)$, $(1, 6)$, $(3, -12)$, and $(5, 90)$ are on the graph.

4. From Step 2 we see that the y-intercept is $(0, 0)$.

5. We find additional points on the graph and then draw the graph.

x	$h(x)$
-1.5	14.4
1.5	4.7
2.5	-6.6
4.5	30.9

$h(x) = x(x - 4)(x + 1)(x - 2)$

6. Checking the graph as described on page 272 in the text, we see that it appears to be correct.

19. $g(x) = -x^4 - 2x^3$

1. The leading term is $-x^4$. The degree, 4, is even and the leading coefficient, -1, is negative so as $x \to \infty$, $g(x) \to -\infty$ and as $x \to -\infty$, $g(x) \to -\infty$.

2. We solve $g(x) = 0$.

$$-x^4 - 2x^3 = 0$$
$$-x^3(x + 2) = 0$$
$$-x^3 = 0 \ \ or \ \ x + 2 = 0$$
$$x = 0 \ \ or \ \ \ \ \ \ x = -2$$

The zeros of the function are 0 and -2, so the x-intercepts of the graph are $(0, 0)$ and $(-2, 0)$.

3. The zeros divide the x-axis into 3 intervals, $(-\infty, -2)$, $(-2, 0)$, and $(0, \infty)$. We choose a value for x from each interval and find $g(x)$. This tells us the sign of $g(x)$ for all values of x in that interval.

In $(-\infty, -2)$, test -3:

$$g(-3) = -(-3)^4 - 2(-3)^3 = -27 < 0$$

In $(-2, 0)$, test -1:

$$g(-1) = -(-1)^4 - 2(-1)^3 = 1 > 0$$

In $(0, \infty)$, test 1:

$$g(1) = -1^4 - 2 \cdot 1^3 = -3 < 0$$

Thus the graph lies below the x-axis on $(-\infty, -2)$ and $(0, \infty)$ and above the x-axis on $(-2, 0)$. We also know the points $(-3, -27)$, $(-1, 1)$, and $(1, -3)$ are on the graph.

4. From Step 2 we see that the y-intercept is $(0, 0)$.

5. We find additional points on the graph and then draw the graph.

x	$g(x)$
-2.5	-7.8
-1.5	1.7
0.5	-0.3
2	-32

$g(x) = -x^4 - 2x^3$

6. Checking the graph as described on page 272 in the text, we see that it appears to be correct.

21. $f(x) = -\dfrac{1}{2}(x - 2)(x + 1)^2(x - 1)$

1. The leading term is $-\dfrac{1}{2} \cdot x \cdot x \cdot x \cdot x$, or $-\dfrac{1}{2}x^4$. The degree, 4, is even and the leading coefficient, $-\dfrac{1}{2}$, is negative so as $x \to \infty$, $f(x) \to -\infty$ and as $x \to -\infty$, $f(x) \to -\infty$.

2. We see that the zeros of the function are 2, -1, and 1 so the x-intercepts of the graph are $(2, 0)$, $(-1, 0)$, and $(1, 0)$.

3. The zeros divide the x-axis into 4 intervals, $(-\infty, -1)$, $(-1, 1)$, $(1, 2)$, and $(2, \infty)$. We choose a value for x from each interval and find $f(x)$. This tells us the sign of $f(x)$ for all values of x in that interval.

   In $(-\infty, -1)$, test -2:
   $$f(-2) = -\frac{1}{2}(-2 - 2)(-2 + 1)^2(-2 - 1) = -6 < 0$$

   In $(-1, 1)$, test 0:
   $$f(0) = -\frac{1}{2}(0 - 2)(0 + 1)^2(0 - 1) = -1 < 0$$

   In $(1, 2)$, test 1.5:
   $$g(1.5) = -\frac{1}{2}(1.5 - 2)(1.5 + 1)^2(1.5 - 1) = 0.78125 > 0$$

   In $(2, \infty)$, test 3:
   $$f(3) = -\frac{1}{2}(3 - 2)(3 + 1)^2(3 - 1) = -16 < 0$$

   Thus the graph lies below the x-axis on $(-\infty, -1)$, $(-1, 1)$, and $(2, \infty)$ and above the x-axis on $(1, 2)$. We also know that the points $(-2, -6)$, $(0, -1)$, $(1.5, 0.78125)$, and $(3, -16)$ are on the graph.

4. From Step 3 we see that $f(0) = -1$, so the y-intercept is $(0, -1)$.

5. We find additional points on the graph and then draw the graph.

x	$f(x)$
-3	-40
-1.5	-1.1
-0.5	-0.5
0.5	-0.8

$f(x) = -\dfrac{1}{2}(x - 2)(x + 1)^2(x - 1)$

6. Checking the graph as described on page 272 in the text, we see that it appears to be correct.

23. $g(x) = -x(x - 1)^2(x + 4)^2$

1. The leading term is $-x \cdot x \cdot x \cdot x \cdot x$, or $-x^5$. The degree, 5, is odd and the leading coefficient, -1, is negative so as $x \to \infty$, $g(x) \to -\infty$ and as $x \to -\infty$, $g(x) \to \infty$.

2. We see that the zeros of the function are 0, 1, and -4 so the x-intercepts are $(0, 0)$, $(1, 0)$, and $(-4, 0)$.

3. The zeros divide the x-axis into 4 intervals, $(-\infty, -4)$, $(-4, 0)$, $(0, 1)$, and $(1, \infty)$. We choose a value of x from each interval and find $g(x)$. This tells us the sign of $g(x)$ for all values of x in that interval.

   In $(-\infty, -4)$, test -5:
   $$g(-5) = -(-5)(-5 - 1)^2(-5 + 4)^2 = 180 > 0$$

   In $(-4, 0)$, test -1:
   $$g(-1) = -(-1)(-1 - 1)^2(-1 + 4)^2 = 36 > 0$$

   In $(0, 1)$, test 0.5:
   $$g(0.5) = -0.5(0.5 - 1)^2(0.5 + 4)^2 = -2.53125 < 0$$

   In $(1, \infty)$, test 2:
   $$g(2) = -2(2 - 1)^2(2 + 4)^2 = -72 < 0$$

   Thus the graph lies above the x-axis on $(-\infty, -4)$ and $(-4, 0)$ and below the x-axis on $(0, 1)$ and $(1, \infty)$. We also know that the points $(-5, 180)$, $(-1, 36)$, $(0.5, -2.53125)$, and $(2, -72)$ are on the graph.

4. From Step 2 we know that the y-intercept is $(0, 0)$.

5. We find additional points on the graph and then draw the graph.

x	$g(x)$
-3	48
-2	72
-0.5	13.8
1.5	-11.3

$g(x) = -x(x - 1)^2(x + 4)^2$

6. Checking the graph as described on page 272 in the text, we see that it appears to be correct.

25. $f(x) = (x-2)^2(x+1)^4$

1. The leading term is $x \cdot x \cdot x \cdot x \cdot x \cdot x$, or x^6. The degree, 6, is even and the leading coefficient, 1, is positive so as $x \to \infty$, $f(x) \to \infty$ and as $x \to -\infty$, $f(x) \to \infty$.

2. We see that the zeros of the function are 2 and -1 so the x-intercepts of the graph are $(2,0)$ and $(-1,0)$.

3. The zeros divide the x-axis into 3 intervals, $(-\infty, -1)$, $(-1, 2)$, and $(2, \infty)$. We choose a value for x from each interval and find $f(x)$. This tells us the sign of $f(x)$ for all values of x in that interval.
   In $(-\infty, -1)$, test -2:
   $$f(-2) = (-2-2)^2(-2+1)^4 = 16 > 0$$
   In $(-1, 2)$, test 0:
   $$f(0) = (0-2)^2(0+1)^4 = 4 > 0$$
   In $(2, \infty)$, test 3:
   $$f(3) = (3-2)^2(3+1)^4 = 256 > 0$$
   Thus the graph lies above the x-axis on all 3 intervals. We also know the points $(-2, 16)$, $(0, 4)$, and $(3, 256)$ are on the graph.

4. From Step 3 we know that $f(0) = 4$ so the y-intercept is $(0, 4)$.

5. We find additional points on the graph and then draw the graph.

x	$f(x)$
-1.5	0.8
-0.5	0.4
1	16
1.5	9.8

$f(x) = (x-2)^2(x+1)^4$

6. Checking the graph as described on page 272 in the text, we see that it appears to be correct.

27. $g(x) = -(x-1)^4$

1. The leading term is $-1 \cdot x \cdot x \cdot x \cdot x$, or $-x^4$. The degree, 4, is even and the leading coefficient, -1, is negative so as $x \to \infty$, $g(x) \to -\infty$ and as $x \to -\infty$, $g(x) \to -\infty$.

2. We see that the zero of the function is 1, so the x-intercept is $(1, 0)$.

3. The zero divides the x-axis into 2 intervals, $(-\infty, 1)$ and $(1, \infty)$. We choose a value for x from each interval and find $g(x)$. This tells us the sign of $g(x)$ for all values of x in that interval.
   In $(-\infty, 1)$, test 0:
   $$g(0) = -(0-1)^4 = -1 < 0$$
   In $(1, \infty)$, test 2:
   $$g(2) = -(2-1)^4 = -1 < 0$$

Thus the graph lies below the x-axis on both intervals. We also know the points $(0, -1)$ and $(2, -1)$ are on the graph.

4. From Step 3 we know that $g(0) = -1$ so the y-intercept is $(0, -1)$.

5. We find additional points on the graph and then draw the graph.

x	$g(x)$
-1	-16
-0.5	-5.1
1.5	0.1
3	-16

$g(x) = -(x-1)^4$

6. Checking the graph as described on page 272 in the text, we see that it appears to be correct.

29. $h(x) = x^3 + 3x^2 - x - 3$

1. The leading term is x^3. The degree, 3, is odd and the leading coefficient, 1, is positive so as $x \to \infty$, $h(x) \to \infty$ and as $x \to -\infty$, $h(x) \to -\infty$.

2. We solve $h(x) = 0$.
   $$x^3 + 3x^2 - x - 3 = 0$$
   $$x^2(x+3) - (x+3) = 0$$
   $$(x+3)(x^2 - 1) = 0$$
   $$(x+3)(x+1)(x-1) = 0$$
   $$x+3 = 0 \quad or \quad x+1 = 0 \quad or \quad x-1 = 0$$
   $$x = -3 \quad or \qquad x = -1 \quad or \qquad x = 1$$
   The zeros of the function are -3, -1, and 1 so the x-intercepts of the graph are $(-3, 0)$, $(-1, 0)$, and $(1, 0)$.

3. The zeros divide the x-axis into 4 intervals, $(-\infty, -3)$, $(-3, -1)$, $(-1, 1)$, and $(1, \infty)$. We choose a value for x from each interval and find $h(x)$. This tells us the sign of $h(x)$ for all values of x in that interval.
   In $(-\infty, -3)$, test -4:
   $$h(-4) = (-4)^3 + 3(-4)^2 - (-4) - 3 = -15 < 0$$
   In $(-3, -1)$, test -2:
   $$h(-2) = (-2)^3 + 3(-2)^2 - (-2) - 3 = 3 > 0$$
   In $(-1, 1)$, test 0:
   $$h(0) = 0^3 + 3 \cdot 0^2 - 0 - 3 = -3 < 0$$
   In $(1, \infty)$, test 2:
   $$h(2) = 2^3 + 3 \cdot 2^2 - 2 - 3 = 15 > 0$$
   Thus the graph lies below the x-axis on $(-\infty, -3)$ and on $(-1, 1)$ and above the x-axis on $(-3, -1)$ and on $(1, \infty)$. We also know the points $(-4, -15)$, $(-2, 3)$, $(0, -3)$, and $(2, 15)$ are on the graph.

4. From Step 3 we know that $h(0) = -3$ so the y-intercept is $(0, -3)$.

5. We find additional points on the graph and then draw the graph.

x	$h(x)$
-4.5	-28.9
-2.5	2.6
0.5	-2.6
2.5	28.9

$$h(x) = x^3 + 3x^2 - x - 3$$

6. Checking the graph as described on page 272 in the text, we see that it appears to be correct.

31. $f(x) = 6x^3 - 8x^2 - 54x + 72$

1. The leading term is $6x^3$. The degree, 3, is odd and the leading coefficient, 6, is positive so as $x \to \infty$, $f(x) \to \infty$ and as $x \to -\infty$, $f(x) \to -\infty$.

2. We solve $f(x) = 0$.

$$6x^3 - 8x^2 - 54x + 72 = 0$$
$$2(3x^3 - 4x^2 - 27x + 36) = 0$$
$$2[x^2(3x - 4) - 9(3x - 4)] = 0$$
$$2(3x - 4)(x^2 - 9) = 0$$
$$2(3x - 4)(x + 3)(x - 3) = 0$$

$$3x - 4 = 0 \quad or \quad x + 3 = 0 \quad or \quad x - 3 = 0$$
$$x = \frac{4}{3} \quad or \quad x = -3 \quad or \quad x = 3$$

The zeros of the function are $\frac{4}{3}$, -3, and 3, so the x-intercepts of the graph are $\left(\frac{4}{3}, 0\right)$, $(-3, 0)$, and $(3, 0)$.

3. The zeros divide the x-axis into 4 intervals, $(-\infty, -3)$, $\left(-3, \frac{4}{3}\right)$, $\left(\frac{4}{3}, 3\right)$, and $(3, \infty)$. We choose a value for x from each interval and find $f(x)$. This tells us the sign of $f(x)$ for all values of x in that interval.

In $(-\infty, -3)$, test -4:
$$f(-4) = 6(-4)^3 - 8(-4)^2 - 54(-4) + 72 = -224 < 0$$

In $\left(-3, \frac{4}{3}\right)$, test 0:
$$f(0) = 6 \cdot 0^3 - 8 \cdot 0^2 - 54 \cdot 0 + 72 = 72 > 0$$

In $\left(\frac{4}{3}, 3\right)$, test 2:
$$f(2) = 6 \cdot 2^3 - 8 \cdot 2^2 - 54 \cdot 2 + 72 = -20 < 0$$

In $(3, \infty)$, test 4:
$$f(4) = 6 \cdot 4^3 - 8 \cdot 4^2 - 54 \cdot 4 + 72 = 112 > 0$$

Thus the graph lies below the x-axis on $(-\infty, -3)$ and on $\left(\frac{4}{3}, 3\right)$ and above the x-axis on $\left(-3, \frac{4}{3}\right)$ and on $(3, \infty)$. We also know the points $(-4, -224)$, $(0, 72)$, $(2, -20)$, and $(4, 112)$ are on the graph.

4. From Step 3 we know that $f(0) = 72$ so the y-intercept is $(0, 72)$.

5. We find additional points on the graph and then draw the graph.

x	$f(x)$
-1	112
1	16
3.5	42.25

$$f(x) = 6x^3 - 8x^2 - 54x + 72$$

6. Checking the graph as described on page 272 in the text, we see that it appears to be correct.

33. $f(-5) = (-5)^3 + 3(-5)^2 - 9(-5) - 13 = -18$
$f(-4) = (-4)^3 + 3(-4)^2 - 9(-4) - 13 = 7$

By the intermediate value theorem, since $f(-5)$ and $f(-4)$ have opposite signs then $f(x)$ has a zero between -5 and -4.

35. $f(-3) = 3(-3)^2 - 2(-3) - 11 = 22$
$f(-2) = 3(-2)^2 - 2(-2) - 11 = 5$

Since both $f(-3)$ and $f(-2)$ are positive, we cannot use the intermediate value theorem to determine if there is a zero between -3 and -2.

37. $f(2) = 2^4 - 2 \cdot 2^2 - 6 = 2$
$f(3) = 3^4 - 2 \cdot 3^2 - 6 = 57$

Since both $f(2)$ and $f(3)$ are positive, we cannot use the intermediate value theorem to determine if there is a zero between 2 and 3.

39. $f(4) = 4^3 - 5 \cdot 4^2 + 4 = -12$
$f(5) = 5^3 - 5 \cdot 5^2 + 4 = 4$

By the intermediate value theorem, since $f(4)$ and $f(5)$ have opposite signs then $f(x)$ has a zero between 4 and 5.

41. Discussion and Writing

43. The graph of $y = x$, or $y = x + 0$, has y-intercept $(0, 0)$, so (d) is the correct answer.

45. The graph of $y - 2x = 6$, or $y = 2x + 6$, has y-intercept $(0, 6)$, so (e) is the correct answer.

47. The graph of $y = 1 - x$, or $y = -x + 1$, has y-intercept $(0, 1)$, so (b) is the correct answer.

49.
$$2x - \frac{1}{2} = 4 - 3x$$
$$5x - \frac{1}{2} = 4 \qquad \text{Adding } 3x$$
$$5x = \frac{9}{2} \qquad \text{Adding } \frac{1}{2}$$
$$x = \frac{1}{5} \cdot \frac{9}{2} \qquad \text{Multiplying by } \frac{1}{5}$$
$$x = \frac{9}{10}$$

The solution is $\frac{9}{10}$.

51. $6x^2 - 23x - 55 = 0$

$(3x + 5)(2x - 11) = 0$

$3x + 5 = 0 \quad or \quad 2x - 11 = 0$

$3x = -5 \quad or \quad 2x = 11$

$x = -\dfrac{5}{3} \quad or \quad x = \dfrac{11}{2}$

The solutions are $-\dfrac{5}{3}$ and $\dfrac{11}{2}$.

Exercise Set 3.3

1. a)
$$\begin{array}{r} x^3 - 7x^2 + 8x + 16 \\ x+1 \overline{\smash{\big)}\ x^4 - 6x^3 + x^2 + 24x - 20} \\ \underline{x^4 + x^3} \\ -7x^3 + x^2 \\ \underline{-7x^3 - 7x^2} \\ 8x^2 + 24x \\ \underline{8x^2 + 8x} \\ 16x - 20 \\ \underline{16x + 16} \\ -4 \end{array}$$

Since the remainder is not 0, $x + 1$ is not a factor of $f(x)$.

b)
$$\begin{array}{r} x^3 - 4x^2 - 7x + 10 \\ x-2 \overline{\smash{\big)}\ x^4 - 6x^3 + x^2 + 24x - 20} \\ \underline{x^4 - 2x^3} \\ -4x^3 + x^2 \\ \underline{-4x^3 + 8x^2} \\ -7x^2 + 24x \\ \underline{-7x^2 + 14x} \\ 10x - 20 \\ \underline{10x - 20} \\ 0 \end{array}$$

Since the remainder is 0, $x - 2$ is a factor of $f(x)$.

c)
$$\begin{array}{r} x^3 - 11x^2 + 56x - 256 \\ x+5 \overline{\smash{\big)}\ x^4 - 6x^3 + x^2 + 24x - 20} \\ \underline{x^4 + 5x^3} \\ -11x^3 + x^2 \\ \underline{-11x^3 - 55x^2} \\ 56x^2 + 24x \\ \underline{56x^2 + 280x} \\ -256x - 20 \\ \underline{-256x - 1280} \\ 1260 \end{array}$$

Since the remainder is not 0, $x + 5$ is not a factor of $f(x)$.

3. a)
$$\begin{array}{r} x^2 + 2x - 3 \\ x-4 \overline{\smash{\big)}\ x^3 - 2x^2 - 11x + 12} \\ \underline{x^3 - 4x^2} \\ 2x^2 - 11x \\ \underline{2x^2 - 8x} \\ -3x + 12 \\ \underline{-3x + 12} \\ 0 \end{array}$$

Since the remainder is 0, $x - 4$ is a factor of $g(x)$.

b)
$$\begin{array}{r} x^2 + x - 8 \\ x-3 \overline{\smash{\big)}\ x^3 - 2x^2 - 11x + 12} \\ \underline{x^3 - 3x^2} \\ x^2 - 11x \\ \underline{x^2 - 3x} \\ -8x + 12 \\ \underline{-8x + 24} \\ -12 \end{array}$$

Since the remainder is not 0, $x - 3$ is not a factor of $g(x)$.

c)
$$\begin{array}{r} x^2 - x - 12 \\ x-1 \overline{\smash{\big)}\ x^3 - 2x^2 - 11x + 12} \\ \underline{x^3 - x^2} \\ -x^2 - 11x \\ \underline{-x^2 + x} \\ -12x + 12 \\ \underline{-12x + 12} \\ 0 \end{array}$$

Since the remainder is 0, $x - 1$ is a factor of $g(x)$.

5.
$$\begin{array}{r} x^2 - 2x + 4 \\ x+2 \overline{\smash{\big)}\ x^3 + 0x^2 + 0x - 8} \\ \underline{x^3 + 2x^2} \\ -2x^2 + 0x \\ \underline{-2x^2 - 4x} \\ 4x - 8 \\ \underline{4x + 8} \\ -16 \end{array}$$

$x^3 - 8 = (x + 2)(x^2 - 2x + 4) - 16$

7.
$$\begin{array}{r} x^2 - 3x + 2 \\ x+9 \overline{\smash{\big)}\ x^3 + 6x^2 - 25x + 18} \\ \underline{x^3 + 9x^2} \\ -3x^2 - 25x \\ \underline{-3x^2 - 27x} \\ 2x + 18 \\ \underline{2x + 18} \\ 0 \end{array}$$

$x^3 + 6x^2 - 25x + 18 = (x + 9)(x^2 - 3x + 2) + 0$

9.
$$\begin{array}{r} x^3 - 2x^2 + 2x - 4 \\ x+2 \overline{\smash{\big)}\ x^4 + 0x^3 - 2x^2 + 0x + 3} \\ \underline{x^4 + 2x^3} \\ -2x^3 - 2x^2 \\ \underline{-2x^3 - 4x^2} \\ 2x^2 + 0x \\ \underline{2x^2 + 4x} \\ -4x + 3 \\ \underline{-4x - 8} \\ 11 \end{array}$$

$x^4 - 2x^2 + 3 = (x + 2)(x^3 - 2x^2 + 2x - 4) + 11$

11. $(2x^4 + 7x^3 + x - 12) \div (x + 3)$

$= (2x^4 + 7x^3 + 0x^2 + x - 12) \div [x - (-3)]$

$$\begin{array}{r|rrrrr} -3 & 2 & 7 & 0 & 1 & -12 \\ & & -6 & -3 & 9 & -30 \\ \hline & 2 & 1 & -3 & 10 & -42 \end{array}$$

The quotient is $2x^3 + x^2 - 3x + 10$. The remainder is -42.

13. $(x^3 - 2x^2 - 8) \div (x + 2)$

$= (x^3 - 2x^2 + 0x - 8) \div [x - (-2)]$

$$\begin{array}{r|rrrr} -2 & 1 & -2 & 0 & -8 \\ & & -2 & 8 & -16 \\ \hline & 1 & -4 & 8 & |-24 \end{array}$$

The quotient is $x^2 - 4x + 8$. The remainder is -24.

15. $(3x^3 - x^2 + 4x - 10) \div (x + 1)$

$= (3x^3 - x^2 + 4x - 10) \div [x - (-1)]$

$$\begin{array}{r|rrrr} -1 & 3 & -1 & 4 & -10 \\ & & -3 & 4 & -8 \\ \hline & 3 & -4 & 8 & |-18 \end{array}$$

The quotient is $3x^2 - 4x + 8$. The remainder is -18.

17. $(x^5 + x^3 - x) \div (x - 3)$

$= (x^5 + 0x^4 + x^3 + 0x^2 - x + 0) \div (x - 3)$

$$\begin{array}{r|rrrrrr} 3 & 1 & 0 & 1 & 0 & -1 & 0 \\ & & 3 & 9 & 30 & 90 & 267 \\ \hline & 1 & 3 & 10 & 30 & 89 & |267 \end{array}$$

The quotient is $x^4 + 3x^3 + 10x^2 + 30x + 89$.

The remainder is 267.

19. $(x^4 - 1) \div (x - 1)$

$= (x^4 + 0x^3 + 0x^2 + 0x - 1) \div (x - 1)$

$$\begin{array}{r|rrrrr} 1 & 1 & 0 & 0 & 0 & -1 \\ & & 1 & 1 & 1 & 1 \\ \hline & 1 & 1 & 1 & 1 & |\;0 \end{array}$$

The quotient is $x^3 + x^2 + x + 1$. The remainder is 0.

21. $(2x^4 + 3x^2 - 1) \div \left(x - \dfrac{1}{2}\right)$

$(2x^4 + 0x^3 + 3x^2 + 0x - 1) \div \left(x - \dfrac{1}{2}\right)$

$$\begin{array}{r|rrrrr} \frac{1}{2} & 2 & 0 & 3 & 0 & -1 \\ & & 1 & \frac{1}{2} & \frac{7}{4} & \frac{7}{8} \\ \hline & 2 & 1 & \frac{7}{2} & \frac{7}{4} & |-\frac{1}{8} \end{array}$$

The quotient is $2x^3 + x^2 + \dfrac{7}{2}x + \dfrac{7}{4}$. The remainder is $-\dfrac{1}{8}$.

23. $f(x) = x^3 - 6x^2 + 11x - 6$

Find $f(1)$.

$$\begin{array}{r|rrrr} 1 & 1 & -6 & 11 & -6 \\ & & 1 & -5 & 6 \\ \hline & 1 & -5 & 6 & |\;0 \end{array}$$

$f(1) = 0$

Find $f(-2)$.

$$\begin{array}{r|rrrr} -2 & 1 & -6 & 11 & -6 \\ & & -2 & 16 & -54 \\ \hline & 1 & -8 & 27 & |-60 \end{array}$$

$f(-2) = -60$

Find $f(3)$.

$$\begin{array}{r|rrrr} 3 & 1 & -6 & 11 & -6 \\ & & 3 & -9 & 6 \\ \hline & 1 & -3 & 2 & |\;0 \end{array}$$

$f(3) = 0$

25. $f(x) = x^4 - 3x^3 + 2x + 8$

Find $f(-1)$.

$$\begin{array}{r|rrrrr} -1 & 1 & -3 & 0 & 2 & 8 \\ & & -1 & 4 & -4 & 2 \\ \hline & 1 & -4 & 4 & -2 & |10 \end{array}$$

$f(-1) = 10$

Find $f(4)$.

$$\begin{array}{r|rrrrr} 4 & 1 & -3 & 0 & 2 & 8 \\ & & 4 & 4 & 16 & 72 \\ \hline & 1 & 1 & 4 & 18 & |80 \end{array}$$

$f(4) = 80$

Find $f(-5)$.

$$\begin{array}{r|rrrrr} -5 & 1 & -3 & 0 & 2 & 8 \\ & & -5 & 40 & -200 & 990 \\ \hline & 1 & -8 & 40 & -198 & |998 \end{array}$$

$f(-5) = 998$

27. $f(x) = 2x^5 - 3x^4 + 2x^3 - x + 8$

Find $f(20)$.

$$\begin{array}{r|rrrrrr} 20 & 2 & -3 & 2 & 0 & -1 & 8 \\ & & 40 & 740 & 14,840 & 296,800 & 5,935,980 \\ \hline & 2 & 37 & 742 & 14,840 & 296,799 & |5,935,988 \end{array}$$

$f(20) = 5,935,988$

Find $f(-3)$.

$$\begin{array}{r|rrrrrr} -3 & 2 & -3 & 2 & 0 & -1 & 8 \\ & & -6 & 27 & -87 & 261 & -780 \\ \hline & 2 & -9 & 29 & -87 & 260 & |-772 \end{array}$$

$f(-3) = -772$

29. $f(x) = x^4 - 16$

Find $f(2)$.

$$\begin{array}{r|rrrrr} 2 & 1 & 0 & 0 & 0 & -16 \\ & & 2 & 4 & 8 & 16 \\ \hline & 1 & 2 & 4 & 8 & |\;0 \end{array}$$

$f(2) = 0$

Find $f(-2)$.

$$\begin{array}{r|rrrrr} -2 & 1 & 0 & 0 & 0 & -16 \\ & & -2 & 4 & -8 & 16 \\ \hline & 1 & -2 & 4 & -8 & |\;0 \end{array}$$

$f(-2) = 0$

Find $f(3)$.

$$\begin{array}{r|rrrrr} 3 & 1 & 0 & 0 & 0 & -16 \\ & & 3 & 9 & 27 & 81 \\ \hline & 1 & 3 & 9 & 27 & |\;65 \end{array}$$

$f(3) = 65$

Find $f(1 - \sqrt{2})$.

$$\begin{array}{r|rrrrr} 1-\sqrt{2} & 1 & 0 & 0 & 0 & -16 \\ & & 1-\sqrt{2} & 3-2\sqrt{2} & 7-5\sqrt{2} & 17-12\sqrt{2} \\ \hline & 1 & 1-\sqrt{2} & 3-2\sqrt{2} & 7-5\sqrt{2} & |1-12\sqrt{2} \end{array}$$

$f(1 - \sqrt{2}) = 1 - 12\sqrt{2}$

31. $f(x) = 3x^3 + 5x^2 - 6x + 18$

If -3 is a zero of $f(x)$, then $f(-3) = 0$. Find $f(-3)$ using synthetic division.

$$
\begin{array}{r|rrrr}
-3 & 3 & 5 & -6 & 18 \\
   &   & -9 & 12 & -18 \\
\hline
   & 3 & -4 & 6 & 0
\end{array}
$$

Since $f(-3) = 0$, -3 is a zero of $f(x)$.

If 2 is a zero of $f(x)$, then $f(2) = 0$. Find $f(2)$ using synthetic division.

$$
\begin{array}{r|rrrr}
2 & 3 & 5 & -6 & 18 \\
  &   & 6 & 22 & 32 \\
\hline
  & 3 & 11 & 16 & 50
\end{array}
$$

Since $f(2) \neq 0$, 2 is not a zero of $f(x)$.

33. $h(x) = x^4 + 4x^3 + 2x^2 - 4x - 3$

If -3 is a zero of $h(x)$, then $h(-3) = 0$. Find $h(-3)$ using synthetic division.

$$
\begin{array}{r|rrrrr}
-3 & 1 & 4 & 2 & -4 & -3 \\
   &   & -3 & -3 & 3 & 3 \\
\hline
   & 1 & 1 & -1 & -1 & 0
\end{array}
$$

Since $h(-3) = 0$, -3 is a zero of $h(x)$.

If 1 is a zero of $h(x)$, then $h(1) = 0$. Find $h(1)$ using synthetic division.

$$
\begin{array}{r|rrrrr}
1 & 1 & 4 & 2 & -4 & -3 \\
  &   & 1 & 5 & 7 & 3 \\
\hline
  & 1 & 5 & 7 & 3 & 0
\end{array}
$$

Since $h(1) = 0$, 1 is a zero of $h(x)$.

35. $g(x) = x^3 - 4x^2 + 4x - 16$

If i is a zero of $g(x)$, then $g(i) = 0$. Find $g(i)$ using synthetic division. Keep in mind that $i^2 = -1$.

$$
\begin{array}{r|rrrr}
i & 1 & -4 & 4 & -16 \\
  &   & i & -4i-1 & 3i+4 \\
\hline
  & 1 & -4+i & 3-4i & -12+3i
\end{array}
$$

Since $g(i) \neq 0$, i is not a zero of $g(x)$.

If $-2i$ is a zero of $g(x)$, then $g(-2i) = 0$. Find $g(-2i)$ using synthetic division. Keep in mind that $i^2 = -1$.

$$
\begin{array}{r|rrrr}
-2i & 1 & -4 & 4 & -16 \\
    &   & -2i & 8i-4 & 16 \\
\hline
    & 1 & -4-2i & 8i & 0
\end{array}
$$

Since $g(-2i) = 0$, $-2i$ is a zero of $g(x)$.

37. $f(x) = x^3 - \dfrac{7}{2}x^2 + x - \dfrac{3}{2}$

If -3 is a zero of $f(x)$, then $f(-3) = 0$. Find $f(-3)$ using synthetic division.

$$
\begin{array}{r|rrrr}
-3 & 1 & -\frac{7}{2} & 1 & -\frac{3}{2} \\
   &   & -3 & \frac{39}{2} & -\frac{123}{2} \\
\hline
   & 1 & -\frac{13}{2} & \frac{41}{2} & -63
\end{array}
$$

Since $f(-3) \neq 0$, -3 is not a zero of $f(x)$.

If $\dfrac{1}{2}$ is a zero of $f(x)$, then $f\left(\dfrac{1}{2}\right) = 0$.

Find $f\left(\dfrac{1}{2}\right)$ using synthetic division.

$$
\begin{array}{r|rrrr}
\frac{1}{2} & 1 & -\frac{7}{2} & 1 & -\frac{3}{2} \\
            &   & \frac{1}{2} & -\frac{3}{2} & -\frac{1}{4} \\
\hline
            & 1 & -3 & -\frac{1}{2} & -\frac{7}{4}
\end{array}
$$

Since $f\left(\dfrac{1}{2}\right) \neq 0$, $\dfrac{1}{2}$ is not a zero of $f(x)$.

39. $f(x) = x^3 + 4x^2 + x - 6$

Try $x - 1$. Use synthetic division to see whether $f(1) = 0$.

$$
\begin{array}{r|rrrr}
1 & 1 & 4 & 1 & -6 \\
  &   & 1 & 5 & 6 \\
\hline
  & 1 & 5 & 6 & 0
\end{array}
$$

Since $f(1) = 0$, $x - 1$ is a factor of $f(x)$. Thus $f(x) = (x-1)(x^2 + 5x + 6)$.

Factoring the trinomial we get

$f(x) = (x - 1)(x + 2)(x + 3)$.

To solve the equation $f(x) = 0$, use the principle of zero products.

$(x - 1)(x + 2)(x + 3) = 0$

$x - 1 = 0$ or $x + 2 = 0$ or $x + 3 = 0$

$x = 1$ or $x = -2$ or $x = -3$

The solutions are 1, -2, and -3.

41. $f(x) = x^3 - 6x^2 + 3x + 10$

Try $x - 1$. Use synthetic division to see whether $f(1) = 0$.

$$
\begin{array}{r|rrrr}
1 & 1 & -6 & 3 & 10 \\
  &   & 1 & -5 & -2 \\
\hline
  & 1 & -5 & -2 & 8
\end{array}
$$

Since $f(1) \neq 0$, $x - 1$ is not a factor of $P(x)$.

Try $x + 1$. Use synthetic division to see whether $f(-1) = 0$.

$$
\begin{array}{r|rrrr}
-1 & 1 & -6 & 3 & 10 \\
   &   & -1 & 7 & -10 \\
\hline
   & 1 & -7 & 10 & 0
\end{array}
$$

Since $f(-1) = 0$, $x + 1$ is a factor of $f(x)$.

Thus $f(x) = (x + 1)(x^2 - 7x + 10)$.

Factoring the trinomial we get

$f(x) = (x + 1)(x - 2)(x - 5)$.

To solve the equation $f(x) = 0$, use the principle of zero products.

$(x + 1)(x - 2)(x - 5) = 0$

$x + 1 = 0$ or $x - 2 = 0$ or $x - 5 = 0$

$x = -1$ or $x = 2$ or $x = 5$

The solutions are -1, 2, and 5.

43. $f(x) = x^3 - x^2 - 14x + 24$

Try $x + 1$, $x - 1$, and $x + 2$. Using synthetic division we find that $f(-1) \neq 0$, $f(1) \neq 0$ and $f(-2) \neq 0$. Thus $x + 1$, $x - 1$, and $x + 2$, are not factors of $f(x)$.

Try $x - 2$. Use synthetic division to see whether $f(2) = 0$.

$$
\begin{array}{r|rrrr}
2 & 1 & -1 & -14 & 24 \\
  &   & 2 & 2 & -24 \\
\hline
  & 1 & 1 & -12 & 0
\end{array}
$$

Since $f(2) = 0$, $x - 2$ is a factor of $f(x)$. Thus $f(x) = (x - 2)(x^2 + x - 12)$.

Factoring the trinomial we get

$$f(x) = (x - 2)(x + 4)(x - 3)$$

To solve the equation $f(x) = 0$, use the principle of zero products.

$$(x - 2)(x + 4)(x - 3) = 0$$

$x - 2 = 0 \;\; or \;\; x + 4 = 0 \;\; or \;\; x - 3 = 0$

$\quad x = 2 \;\; or \qquad x = -4 \;\; or \qquad x = 3$

The solutions are 2, -4, and 3.

45. $f(x) = x^4 - 7x^3 + 9x^2 + 27x - 54$

Try $x + 1$ and $x - 1$. Using synthetic division we find that $f(-1) \neq 0$ and $f(1) \neq 0$. Thus $x + 1$ and $x - 1$ are not factors of $f(x)$. Try $x + 2$. Use synthetic division to see whether $f(-2) = 0$.

$$
\begin{array}{r|rrrrr}
-2 & 1 & -7 & 9 & 27 & -54 \\
   &   & -2 & 18 & -54 & 54 \\
\hline
   & 1 & -9 & 27 & -27 & 0
\end{array}
$$

Since $f(-2) = 0$, $x + 2$ is a factor of $f(x)$. Thus $f(x) = (x + 2)(x^3 - 9x^2 + 27x - 27)$.

We continue to use synthetic division to factor $g(x) = x^3 - 9x^2 + 27x - 27$. Trying $x + 2$ again and $x - 2$ we find that $g(-2) \neq 0$ and $g(2) \neq 0$. Thus $x + 2$ and $x - 2$ are not factors of $g(x)$. Try $x - 3$.

$$
\begin{array}{r|rrrr}
3 & 1 & -9 & 27 & -27 \\
  &   & 3 & -18 & 27 \\
\hline
  & 1 & -6 & 9 & 0
\end{array}
$$

Since $g(3) = 0$, $x - 3$ is a factor of $x^3 - 9x^2 + 27x - 27$. Thus $f(x) = (x + 2)(x - 3)(x^2 - 6x + 9)$.

Factoring the trinomial we get

$f(x) = (x + 2)(x - 3)(x - 3)^2$, or $f(x) = (x + 2)(x - 3)^3$.

To solve the equation $f(x) = 0$, use the principle of zero products.

$$(x + 2)(x - 3)(x - 3)(x - 3) = 0$$

$x + 2 = 0 \;\; or \;\; x - 3 = 0 \; or \; x - 3 = 0 \; or \; x - 3 = 0$

$\quad x = -2 \; or \qquad x = 3 \; or \qquad x = 3 \; or \qquad x = 3$

The solutions are -2 and 3.

47. $f(x) = x^4 - x^3 - 19x^2 + 49x - 30$

Try $x - 1$. Use synthetic division to see whether $f(1) = 0$.

$$
\begin{array}{r|rrrrr}
1 & 1 & -1 & -19 & 49 & -30 \\
  &   & 1 & 0 & -19 & 30 \\
\hline
  & 1 & 0 & -19 & 30 & 0
\end{array}
$$

Since $f(1) = 0$, $x - 1$ is a factor of $f(x)$. Thus $f(x) = (x - 1)(x^3 - 19x + 30)$.

We continue to use synthetic division to factor $g(x) = x^3 - 19x + 30$. Trying $x - 1$, $x + 1$, and $x + 2$ we find that $g(1) \neq 0$, $g(-1) \neq 0$, and $g(-2) \neq 0$. Thus $x - 1$, $x + 1$, and $x + 2$ are not factors of $x^3 - 19x + 30$. Try $x - 2$.

$$
\begin{array}{r|rrrr}
2 & 1 & 0 & -19 & 30 \\
  &   & 2 & 4 & -30 \\
\hline
  & 1 & 2 & -15 & 0
\end{array}
$$

Since $g(2) = 0$, $x - 2$ is a factor of $x^3 - 19x + 30$.

Thus $f(x) = (x - 1)(x - 2)(x^2 + 2x - 15)$.

Factoring the trinomial we get

$$f(x) = (x - 1)(x - 2)(x - 3)(x + 5).$$

To solve the equation $f(x) = 0$, use the principle of zero products.

$$(x - 1)(x - 2)(x - 3)(x + 5) = 0$$

$x - 1 = 0 \; or \; x - 2 = 0 \; or \; x - 3 = 0 \; or \; x + 5 = 0$

$\quad x = 1 \; or \qquad x = 2 \; or \qquad x = 3 \; or \qquad x = -5$

The solutions are 1, 2, 3, and -5.

49. $f(x) = x^4 - x^3 - 7x^2 + x + 6$

1. The leading term is x^4. The degree, 4, is even and the leading coefficient, 1, is positive so as $x \to \infty$, $f(x) \to \infty$ and as $x \to -\infty$, $f(x) \to \infty$.

2. Find the zeros of the function. We first use synthetic division to determine if $f(1) = 0$.

$$
\begin{array}{r|rrrrr}
1 & 1 & -1 & -7 & 1 & 6 \\
  &   & 1 & 0 & -7 & -6 \\
\hline
  & 1 & 0 & -7 & -6 & 0
\end{array}
$$

1 is a zero of the function and we have $f(x) = (x - 1)(x^3 - 7x - 6)$.

Synthetic division shows that -1 is a zero of $g(x) = x^3 - 7x - 6$.

$$
\begin{array}{r|rrrr}
-1 & 1 & 0 & -7 & -6 \\
   &   & -1 & 1 & 6 \\
\hline
   & 1 & -1 & -6 & 0
\end{array}
$$

Then we have $f(x) = (x - 1)(x + 1)(x^2 - x - 6)$.

To find the other zeros we solve the following equation:

$$x^2 - x - 6 = 0$$

$$(x - 3)(x + 2) = 0$$

$$x - 3 = 0 \;\; or \;\; x + 2 = 0$$

$$x = 3 \;\; or \qquad x = -2$$

The zeros of the function are 1, -1, 3, and -2 so the x-intercepts of the graph are $(1, 0)$, $(-1, 0)$, $(3, 0)$, and $(-2, 0)$.

3. The zeros divide the x-axis into five intervals, $(-\infty, -2)$, $(-2, -1)$, $(-1, 1)$, $(1, 3)$, and $(3, \infty)$. We choose a value for x from each interval and find $f(x)$. This tells us the sign of $f(x)$ for all values of x in the interval.

In $(-\infty, -2)$, test -3:

$$f(-3) = (-3)^4 - (-3)^3 - 7(-3)^2 + (-3) + 6 = 48 > 0$$

In $(-2, -1)$, test -1.5:

$$f(-1.5) = (-1.5)^4 - (-1.5)^3 - 7(-1.5)^2 + (-1.5) + 6 = -2.8125 < 0$$

In $(-1, 1)$, test 0:

$$f(0) = 0^4 - 0^3 - 7 \cdot 0^2 + 0 + 6 = 6 > 0$$

In $(1, 3)$, test 2:

$$f(2) = 2^4 - 2^3 - 7 \cdot 2^2 + 2 + 6 = -12 < 0$$

In $(3, \infty)$, test 4:

$f(4) = 4^4 - 4^3 - 7 \cdot 4^2 + 4 + 6 = 90 > 0$

Thus the graph lies above the x-axis on $(-\infty, -2)$, on $(-1, 1)$, and on $(3, \infty)$. It lies below the x-axis on $(-2, -1)$ and on $(1, 3)$. We also know the points $(-3, 48)$, $(-1.5, -2.8125)$, $(0, 6)$, $(2, -12)$, and $(4, 90)$ are on the graph.

4. From Step 3 we see that $f(0) = 6$ so the y-intercept is $(0, 6)$.

5. We find additional points on the graph and draw the graph.

x	$f(x)$
-2.5	14.3
-0.5	3.9
0.5	4.7
2.5	-11.8

$f(x) = x^4 - x^3 - 7x^2 + x + 6$

6. Checking the graph as described on page 272 in the text, we see that it appears to be correct.

51. $f(x) = x^3 - 7x + 6$

1. The leading term is x^3. The degree, 3, is odd and the leading coefficient, 1, is positive so as $x \to \infty$, $f(x) \to \infty$ and as $x \to -\infty$, $f(x) \to -\infty$.

2. Find the zeros of the function. We first use synthetic division to determine if $f(1) = 0$.

```
1 | 1   0  -7   6
  |     1   1  -6
  -----------------
    1   1  -6 | 0
```

1 is a zero of the function and we have $f(x) = (x - 1)(x^2 + x - 6)$. To find the other zeros we solve the following equation.

$$x^2 + x - 6 = 0$$
$$(x + 3)(x - 2) = 0$$
$$x + 3 = 0 \quad or \quad x - 2 = 0$$
$$x = -3 \quad or \qquad x = 2$$

The zeros of the function are 1, -3, and 2 so the x-intercepts of the graph are $(1, 0)$, $(-3, 0)$, and $(2, 0)$.

3. The zeros divide the x-axis into four intervals, $(-\infty, -3)$, $(-3, 1)$, $(1, 2)$, and $(2, \infty)$. We choose a value for x from each interval and find $f(x)$. This tells us the sign of $f(x)$ for all values of x in the interval.

In $(-\infty, -3)$, test -4:

$f(-4) = (-4)^3 - 7(-4) + 6 = -30 < 0$

In $(-3, 1)$, test 0:

$f(0) = 0^3 - 7 \cdot 0 + 6 = 6 > 0$

In $(1, 2)$, test 1.5:

$f(1.5) = (1.5)^3 - 7(1.5) + 6 = -1.125 < 0$

In $(2, \infty)$, test 3:

$f(3) = 3^3 - 7 \cdot 3 + 6 = 12 > 0$

Thus the graph lies below the x-axis on $(-\infty, -3)$ and on $(1, 2)$. It lies above the x-axis on $(-3, 1)$ and on $(2, \infty)$. We also know the points $(-4, -30)$, $(0, 6)$, $(1.5, -1.125)$, and $(3, 12)$ are on the graph.

4. From Step 3 we see that $f(0) = 6$ so the y-intercept is $(0, 6)$.

5. We find additional points on the graph and draw the graph.

x	$f(x)$
-3.5	-12.4
-2	12
2.5	4.1
4	42

$f(x) = x^3 - 7x + 6$

6. Checking the graph as described on page 272 in the text, we see that it appears to be correct.

53. $f(x) = -x^3 + 3x^2 + 6x - 8$

1. The leading term is $-x^3$. The degree, 3, is odd and the leading coefficient, -1, is negative so as $x \to \infty$, $f(x) \to -\infty$ and as $x \to -\infty$, $f(x) \to \infty$.

2. Find the zeros of the function. We first use synthetic division to determine if $f(1) = 0$.

```
1 | -1   3   6  -8
  |     -1   2   8
  -----------------
   -1   2   8 | 0
```

1 is a zero of the function and we have $f(x) = (x - 1)(-x^2 + 2x + 8)$. To find the other zeros we solve the following equation.

$$-x^2 + 2x + 8 = 0$$
$$x^2 - 2x - 8 = 0$$
$$(x - 4)(x + 2) = 0$$
$$x - 4 = 0 \quad or \quad x + 2 = 0$$
$$x = 4 \quad or \qquad x = -2$$

The zeros of the function are 1, 4, and -2 so the x-intercepts of the graph are $(1, 0)$, $(4, 0)$, and $(-2, 0)$.

3. The zeros divide the x-axis into four intervals, $(-\infty, -2)$, $(-2, 1)$, $(1, 4)$, and $(4, \infty)$. We choose a value for x from each interval and find $f(x)$. This tells us the sign of $f(x)$ for all values of x in the interval.

In $(-\infty, -2)$, test -3:

$f(-3) = (-3)^3 + 3(-3)^2 + 6(-3) - 8 = 28 > 0$

In $(-2, 1)$, test 0:

$f(0) = -0^3 + 3 \cdot 0^2 + 6 \cdot 0 - 8 = -8 < 0$

In $(1, 4)$, test 2:

$f(2) = -2^3 + 3 \cdot 2^2 + 6 \cdot 2 - 8 = 8 > 0$

In $(4, \infty)$, test 5:

$f(5) = -5^3 + 3 \cdot 5^2 + 6 \cdot 5 - 8 = -28 < 0$

Thus the graph lies above the x-axis on $(-\infty, -2)$ and on $(1, 4)$. It lies below the x-axis on $(-2, 1)$ and on $(4, \infty)$. We also know the points $(-3, 28)$, $(0, -8)$, $(2, 8)$, and $(5, -28)$ are on the graph.

4. From Step 3 we see that $f(0) = -8$ so the y-intercept is $(0, -8)$.

5. We find additional points on the graph and draw the graph.

x	$f(x)$
-2.5	11.4
-1	-10
3	10
4.5	-11.4

$f(x) = -x^3 + 3x^2 + 6x - 8$

6. Checking the graph as described on page 272 in the text, we see that it appears to be correct.

55. Left to the student

57. Discussion and Writing

59.
$$2x^2 + 12 = 5x$$
$$2x^2 - 5x + 12 = 0$$
$$a = 2, \ b = -5, \ c = 12$$
$$x = \frac{-b \pm \sqrt{b^2 - 4ac}}{2a}$$
$$x = \frac{-(-5) \pm \sqrt{(-5)^2 - 4 \cdot 2 \cdot 12}}{2 \cdot 2}$$
$$= \frac{5 \pm \sqrt{-71}}{4}$$
$$= \frac{5 \pm i\sqrt{71}}{4} = \frac{5}{4} \pm \frac{\sqrt{71}}{4}i$$

The solutions are $\frac{5}{4} + \frac{\sqrt{71}}{4}i$ and $\frac{5}{4} - \frac{\sqrt{71}}{4}i$, or $\frac{5}{4} \pm \frac{\sqrt{71}}{4}i$.

61. We substitute -14 for $g(x)$ and solve for x.
$$-14 = x^2 + 5x - 14$$
$$0 = x^2 + 5x$$
$$0 = x(x + 5)$$
$$x = 0 \ \text{ or } \ x + 5 = 0$$
$$x = 0 \ \text{ or } \ \quad\quad x = -5$$

When the output is -14, the input is 0 or -5.

63. We substitute -20 for $g(x)$ and solve for x.
$$-20 = x^2 + 5x - 14$$
$$0 = x^2 + 5x + 6$$
$$0 = (x + 3)(x + 2)$$
$$x + 3 = 0 \ \text{ or } \ x + 2 = 0$$
$$x = -3 \ \text{ or } \quad\quad x = -2$$

When the output is -20, the input is -3 or -2.

65. Let b and h represent the length of the base and the height of the triangle, respectively.

$b + h = 30$, so $b = 30 - h$.
$$A = \frac{1}{2}bh = \frac{1}{2}(30 - h)h = -\frac{1}{2}h^2 + 15h$$

Find the value of h for which A is a maximum:
$$h = \frac{-15}{2(-1/2)} = 15$$

When $h = 15$, $b = 30 - 15 = 15$.

The area is a maximum when the base and the height are each 15 in.

67. a) -4, -3, 2, and 5 are zeros of the function, so $x + 4$, $x + 3$, $x - 2$, and $x - 5$ are factors.

b) We first write the product of the factors:
$$P(x) = (x + 4)(x + 3)(x - 2)(x - 5)$$

Note that $P(0) = 4 \cdot 3(-2)(-5) > 0$ and the graph shows a positive y-intercept, so this function is a correct one.

c) Yes; two examples are $f(x) = c \cdot P(x)$ for any non-zero constant c and $g(x) = (x - a)P(x)$.

d) No; only the function in part (b) has the given graph.

69. Divide $x^3 - kx^2 + 3x + 7k$ by $x + 2$.

$$
\begin{array}{r|rrrr}
-2 & 1 & -k & 3 & 7k \\
   &   & -2 & 2k+4 & -4k-14 \\
\hline
   & 1 & -k-2 & 2k+7 & 3k-14 \\
\end{array}
$$

Thus $P(-2) = 3k - 14$.

We know that if $x + 2$ is a factor of $f(x)$, then $f(-2) = 0$. We solve $0 = 3k - 14$ for k.
$$0 = 3k - 14$$
$$\frac{14}{3} = k$$

71. $\dfrac{2x^2}{x^2 - 1} + \dfrac{4}{x + 3} = \dfrac{32}{3x^2 - x - 3}$,
$$\text{LCM is } (x + 1)(x - 1)(x + 3)$$
$$(x + 1)(x - 1)(x + 3)\left[\frac{2x^2}{(x + 1)(x - 1)} + \frac{4}{x + 3}\right] =$$
$$(x + 1)(x - 1)(x + 3) \cdot \frac{32}{(x + 1)(x - 1)(x + 3)}$$
$$2x^2(x + 3) + 4(x + 1)(x - 1) = 32$$
$$2x^3 + 6x^2 + 4x^2 - 4 = 32$$
$$2x^3 + 10x^2 - 36 = 0$$
$$x^3 + 5x^2 - 18 = 0$$

Using synthetic division and several trials, we find that -3 is a factor of $f(x) = x^3 + 5x^2 - 18$:

$$
\begin{array}{r|rrrr}
-3 & 1 & 5 & 0 & -18 \\
   &   & -3 & -6 & 18 \\
\hline
   & 1 & 2 & -6 & 0 \\
\end{array}
$$

Then we have:
$$(x + 3)(x^2 + 2x - 6) = 0$$

$$x + 3 = 0 \quad \text{or} \quad x^2 + 2x - 6 = 0$$
$$x = -3 \quad \text{or} \qquad x = -1 \pm \sqrt{7}$$

Only $-1 \pm \sqrt{7}$ check.

73. Answers may vary. One possibility is $P(x) = x^{15} - x^{14}$.

75.
$$\begin{array}{r|rrr} i & 1 & -3 & 7 \quad (i^2 = -1) \\ & & i & -3i - 1 \\ \hline & 1 & -3 + i & 6 - 3i \end{array}$$

The answer is $x - 3 + i$, R $6 - 3i$.

Exercise Set 3.4

1. Find a polynomial function of degree 3 with -2, 3, and 5 as zeros.

Such a function has factors $x + 2$, $x - 3$, and $x - 5$, so we have $f(x) = a_n(x + 2)(x - 3)(x - 5)$.

The number a_n can be any nonzero number. The simplest polynomial will be obtained if we let it be 1. Multiplying the factors, we obtain
$$\begin{aligned} f(x) &= (x + 2)(x - 3)(x - 5) \\ &= (x^2 - x - 6)(x - 5) \\ &= x^3 - 6x^2 - x + 30. \end{aligned}$$

3. Find a polynomial function of degree 3 with -3, $2i$, and $-2i$ as zeros.

Such a function has factors $x + 3$, $x - 2i$, and $x + 2i$, so we have $f(x) = a_n(x + 3)(x - 2i)(x + 2i)$.

The number a_n can be any nonzero number. The simplest polynomial will be obtained if we let it be 1. Multiplying the factors, we obtain
$$\begin{aligned} f(x) &= (x + 3)(x - 2i)(x + 2i) \\ &= (x + 3)(x^2 + 4) \\ &= x^3 + 3x^2 + 4x + 12. \end{aligned}$$

5. Find a polynomial function of degree 3 with $\sqrt{2}$, $-\sqrt{2}$, and 3 as zeros.

Such a function has factors $x - \sqrt{2}$, $x + \sqrt{2}$, and $x - 3$, so we have $f(x) = a_n(x - \sqrt{2})(x + \sqrt{2})(x - 3)$.

The number a_n can be any nonzero number. The simplest polynomial will be obtained if we let it be 1. Multiplying the factors, we obtain
$$\begin{aligned} f(x) &= (x - \sqrt{2})(x + \sqrt{2})(x - 3) \\ &= (x^2 - 2)(x - 3) \\ &= x^3 - 3x^2 - 2x + 6. \end{aligned}$$

7. Find a polynomial function of degree 3 with $1 - \sqrt{3}$, $1 + \sqrt{3}$, and -2 as zeros.

Such a function has factors $x - (1 - \sqrt{3})$, $x - (1 + \sqrt{3})$, and $x + 2$, so we have
$$f(x) = a_n[x - (1 - \sqrt{3})][x - (1 + \sqrt{3})](x + 2).$$

The number a_n can be any nonzero number. The simplest polynomial will be obtained if we let it be 1. Multiplying the factors, we obtain

$$\begin{aligned} f(x) &= [x - (1 - \sqrt{3})][x - (1 + \sqrt{3})](x + 2) \\ &= [(x - 1) + \sqrt{3}][(x - 1) - \sqrt{3}](x + 2) \\ &= [(x - 1)^2 - (\sqrt{3})^2](x + 2) \\ &= (x^2 - 2x + 1 - 3)(x + 2) \\ &= (x^2 - 2x - 2)(x + 2) \\ &= x^3 - 2x^2 - 2x + 2x^2 - 4x - 4 \\ &= x^3 - 6x - 4. \end{aligned}$$

9. Find a polynomial function of degree 3 with $1 + 6i$, $1 - 6i$, and -4 as zeros.

Such a function has factors $x - (1 + 6i)$, $x - (1 - 6i)$, and $x + 4$, so we have
$$f(x) = a_n[x - (1 + 6i)][x - (1 - 6i)](x + 4).$$

The number a_n can be any nonzero number. The simplest polynomial will be obtained if we let it be 1. Multiplying the factors, we obtain
$$\begin{aligned} f(x) &= [x - (1 + 6i)][x - (1 - 6i)](x + 4) \\ &= [(x - 1) - 6i][(x - 1) + 6i](x + 4) \\ &= [(x - 1)^2 - (6i)^2](x + 4) \\ &= (x^2 - 2x + 1 + 36)(x + 4) \\ &= (x^2 - 2x + 37)(x + 4) \\ &= x^3 - 2x^2 + 37x + 4x^2 - 8x + 148 \\ &= x^3 + 2x^2 + 29x + 148. \end{aligned}$$

11. Find a polynomial function of degree 3 with $-\dfrac{1}{3}$, 0, and 2 as zeros.

Such a function has factors $x + \dfrac{1}{3}$, $x - 0$ (or x), and $x - 2$ so we have
$$f(x) = a_n\left(x + \frac{1}{3}\right)(x)(x - 2).$$

The number a_n can be any nonzero number. The simplest polynomial will be obtained if we let it be 1. Multiplying the factors, we obtain
$$\begin{aligned} f(x) &= \left(x + \frac{1}{3}\right)(x)(x - 2) \\ &= \left(x^2 + \frac{1}{3}x\right)(x - 2) \\ &= x^3 - \frac{5}{3}x^2 - \frac{2}{3}x. \end{aligned}$$

13. A polynomial function of degree 5 has at most 5 zeros. Since 5 zeros are given, these are all of the zeros of the desired function. We proceed as in Exercises 1-11, letting $a_n = 1$.
$$\begin{aligned} f(x) &= (x + 1)^3(x - 0)(x - 1) \\ &= (x^3 + 3x^2 + 3x + 1)(x^2 - x) \\ &= x^5 + 2x^4 - 2x^2 - x \end{aligned}$$

15. A polynomial function of degree 4 has at most 4 zeros. Since 4 zeros are given, these are all of the zeros of the desired function. We proceed as in Exercises 1-11, letting $a_n = 1$.

$$f(x) = (x+1)^3(x-0)$$
$$= (x^3 + 3x^2 + 3x + 1)(x)$$
$$= x^4 + 3x^3 + 3x^2 + x$$

17. A polynomial function of degree 4 can have at most 4 zeros. Since $f(x)$ has rational coefficients, in addition to the three zeros given, the other zero is the conjugate of $\sqrt{3}$, or $-\sqrt{3}$.

19. A polynomial function of degree 4 can have at most 4 zeros. Since $f(x)$ has rational coefficients, the other zeros are the conjugates of the given zeros. They are i and $2 + \sqrt{5}$.

21. A polynomial function of degree 4 can have at most 4 zeros. Since $f(x)$ has rational coefficients, in addition to the three zeros given, the other zero is the conjugate of $3i$, or $-3i$.

23. A polynomial function of degree 4 can have at most 4 zeros. Since $f(x)$ has rational coefficients, the other zeros are the conjugates of the given zeros. They are $-4+3i$ and $2+\sqrt{3}$.

25. A polynomial function $f(x)$ of degree 5 has at most 5 zeros. Since $f(x)$ has rational coefficients, in addition to the 3 given zeros, the other zeros are the conjugates of $\sqrt{5}$ and $-4i$, or $-\sqrt{5}$ and $4i$.

27. A polynomial function $f(x)$ of degree 5 has at most 5 zeros. Since $f(x)$ has rational coefficients, the other zero is the conjugate of $2 - i$, or $2 + i$.

29. A polynomial function $f(x)$ of degree 5 has at most 5 zeros. Since $f(x)$ has rational coefficients, in addition to the 3 given zeros, the other zeros are the conjugates of $-3 + 4i$ and $4 - \sqrt{5}$, or $-3 - 4i$ and $4 + \sqrt{5}$.

31. A polynomial function $f(x)$ of degree 5 has at most 5 zeros. Since $f(x)$ has rational coefficients, the other zero is the conjugate of $4 - i$, or $4 + i$.

33. Find a polynomial function of lowest degree with rational coefficients that has $1 + i$ and 2 as some of its zeros. $1 - i$ is also a zero.

Thus the polynomial function is
$$f(x) = a_n(x-2)[x-(1+i)][x-(1-i)].$$
If we let $a_n = 1$, we obtain
$$f(x) = (x-2)[(x-1)-i][(x-1)+i]$$
$$= (x-2)[(x-1)^2 - i^2]$$
$$= (x-2)(x^2 - 2x + 1 + 1)$$
$$= (x-2)(x^2 - 2x + 2)$$
$$= x^3 - 4x^2 + 6x - 4.$$

35. Find a polynomial function of lowest degree with rational coefficients that has $4i$ as one of its zeros. $-4i$ is also a zero.

Thus the polynomial function is
$$f(x) = a_n(x - 4i)(x + 4i).$$
If we let $a_n = 1$, we obtain
$$f(x) = (x - 4i)(x + 4i) = x^2 + 16.$$

37. Find a polynomial function of lowest degree with rational coefficients that has $-4i$ and 5 as some of its zeros.

$4i$ is also a zero.

Thus the polynomial function is
$$f(x) = a_n(x-5)(x+4i)(x-4i).$$
If we let $a_n = 1$, we obtain
$$f(x) = (x-5)[x^2 - (4i)^2]$$
$$= (x-5)(x^2 + 16)$$
$$= x^3 - 5x^2 + 16x - 80$$

39. Find a polynomial function of lowest degree with rational coefficients that has $1 - i$ and $-\sqrt{5}$ as some of its zeros. $1 + i$ and $\sqrt{5}$ are also zeros.

Thus the polynomial function is
$$f(x) = a_n[x-(1-i)][x-(1+i)](x+\sqrt{5})(x-\sqrt{5}).$$
If we let $a_n = 1$, we obtain
$$f(x) = [x-(1-i)][x-(1+i)](x+\sqrt{5})(x-\sqrt{5})$$
$$= [(x-1)+i][(x-1)-i](x+\sqrt{5})(x-\sqrt{5})$$
$$= (x^2 - 2x + 1 + 1)(x^2 - 5)$$
$$= (x^2 - 2x + 2)(x^2 - 5)$$
$$= x^4 - 2x^3 + 2x^2 - 5x^2 + 10x - 10$$
$$= x^4 - 2x^3 - 3x^2 + 10x - 10$$

41. Find a polynomial function of lowest degree with rational coefficients that has $\sqrt{5}$ and $-3i$ as some of its zeros.

$-\sqrt{5}$ and $3i$ are also zeros.

Thus the polynomial function is
$$f(x) = a_n(x - \sqrt{5})(x + \sqrt{5})(x + 3i)(x - 3i).$$
If we let $a_n = 1$, we obtain
$$f(x) = (x^2 - 5)(x^2 + 9)$$
$$= x^4 + 4x^2 - 45$$

43. $f(x) = x^3 + 5x^2 - 2x - 10$

Since -5 is a zero of $f(x)$, we have $f(x) = (x+5) \cdot Q(x)$. We use synthetic division to find $Q(x)$.

$$\begin{array}{r|rrrr} -5 & 1 & 5 & -2 & -10 \\ & & -5 & 0 & 10 \\ \hline & 1 & 0 & -2 & 0 \end{array}$$

Then $f(x) = (x+5)(x^2 - 2)$. To find the other zeros we solve $x^2 - 2 = 0$.
$$x^2 - 2 = 0$$
$$x^2 = 2$$
$$x = \pm\sqrt{2}$$

The other zeros are $-\sqrt{2}$ and $\sqrt{2}$.

45. If $-i$ is a zero of $f(x) = x^4 - 5x^3 + 7x^2 - 5x + 6$, i is also a zero. Thus $x+i$ and $x-i$ are factors of the polynomial. Since $(x+i)(x-i) = x^2 + 1$, we know that $f(x) = (x^2 + 1) \cdot Q(x)$. Divide $x^4 - 5x^3 + 7x^2 - 5x + 6$ by $x^2 + 1$.

$$\begin{array}{r} x^2 - 5x + 6 \\ x^2 + 1 \overline{\smash)\, x^4 - 5x^3 + 7x^2 - 5x + 6} \\ \underline{x^4 + x^2} \\ -5x^3 + 6x^2 - 5x \\ \underline{-5x^3 - 5x} \\ 6x^2 + 6 \\ \underline{6x^2 + 6} \\ 0 \end{array}$$

Thus

$$x^4 - 5x^3 + 7x^2 - 5x + 6 = (x+i)(x-i)(x^2 - 5x + 6)$$
$$= (x+i)(x-i)(x-2)(x-3)$$

Using the principle of zero products we find the other zeros to be i, 2, and 3.

47. $x^3 - 6x^2 + 13x - 20 = 0$

If 4 is a zero, then $x - 4$ is a factor. Use synthetic division to find another factor.

$$\begin{array}{r|rrr} 4 & 1 & -6 & 13 & -20 \\ & & 4 & -8 & 20 \\ \hline & 1 & -2 & 5 & 0 \end{array}$$

$$(x-4)(x^2 - 2x + 5) = 0$$

$x - 4 = 0 \quad or \quad x^2 - 2x + 5 = 0 \qquad$ Principle of zero products

$$x = 4 \quad or \qquad\qquad x = \frac{2 \pm \sqrt{4-20}}{2}$$

$\qquad\qquad\qquad\qquad\qquad\qquad$ Quadratic formula

$$x = 4 \quad or \qquad\qquad x = \frac{2 \pm 4i}{2} = 1 \pm 2i$$

The other zeros are $1 + 2i$ and $1 - 2i$.

49. $f(x) = x^5 - 3x^2 + 1$

According to the rational zeros theorem, any rational zero of f must be of the form p/q, where p is a factor of the constant term, 1, and q is a factor of the coefficient of x^5, 1.

$$\frac{\text{Possibilities for } p}{\text{Possibilities for } q} : \frac{\pm 1}{\pm 1}$$

Possibilities for p/q: 1, −1

51. $f(x) = 2x^4 - 3x^3 - x + 8$

According to the rational zeros theorem, any rational zero of f must be of the form p/q, where p is a factor of the constant term, 8, and q is a factor of the coefficient of x^4, 2.

$$\frac{\text{Possibilities for } p}{\text{Possibilities for } q} : \frac{\pm 1, \pm 2, \pm 4, \pm 8}{\pm 1, \pm 2}$$

Possibilities for p/q: 1, −1, 2, −2, 4, −4, 8, −8, $\frac{1}{2}$, $-\frac{1}{2}$

53. $f(x) = 15x^6 + 47x^2 + 2$

According to the rational zeros theorem, any rational zero of f must be of the form p/q, where p is a factor of 2 and q is a factor of 15.

$$\frac{\text{Possibilities for } p}{\text{Possibilities for } q} : \frac{\pm 1, \pm 2}{\pm 1, \pm 3, \pm 5, \pm 15}$$

Possibilities for p/q: 1, −1, 2, −2, $\frac{1}{3}$, $-\frac{1}{3}$, $\frac{2}{3}$, $-\frac{2}{3}$, $\frac{1}{5}$,

$-\frac{1}{5}$, $\frac{2}{5}$, $-\frac{2}{5}$, $\frac{1}{15}$, $-\frac{1}{15}$, $\frac{2}{15}$, $-\frac{2}{15}$

55. $f(x) = x^3 + 3x^2 - 2x - 6$

a) $\dfrac{\text{Possibilities for } p}{\text{Possibilities for } q} : \dfrac{\pm 1, \pm 2, \pm 3, \pm 6}{\pm 1}$

Possibilities for p/q: 1, −1, 2, −2, 3, −3, 6, −6

From the graph of $y = x^3 + 3x^2 - 2x - 6$, we see that, of the possibilities above, only −3 might be a zero. We use synthetic division to determine whether −3 is indeed a zero.

$$\begin{array}{r|rrr} -3 & 1 & 3 & -2 & -6 \\ & & -3 & 0 & 6 \\ \hline & 1 & 0 & -2 & 0 \end{array}$$

Then we have $f(x) = (x+3)(x^2 - 2)$.

We find the other zeros:

$$x^2 - 2 = 0$$
$$x^2 = 2$$
$$x = \pm\sqrt{2}.$$

There is only one rational zero, −3. The other zeros are $\pm\sqrt{2}$. (Note that we could have used factoring by grouping to find this result.)

b) $f(x) = (x+3)(x - \sqrt{2})(x + \sqrt{2})$

57. $f(x) = x^3 - 3x + 2$

a) $\dfrac{\text{Possibilities for } p}{\text{Possibilities for } q} : \dfrac{\pm 1, \pm 2}{\pm 1}$

Possibilities for p/q: 1, −1, 2, −2

From the graph of $y = x^3 - 3x + 2$, we see that, of the possibilities above, −2 and 1 might be a zeros. We use synthetic division to determine whether −2 is a zero.

$$\begin{array}{r|rrr} -2 & 1 & 0 & -3 & 2 \\ & & -2 & 4 & -2 \\ \hline & 1 & -2 & 1 & 0 \end{array}$$

Then we have $f(x) = (x + 2)(x^2 - 2x + 1) = (x + 2)(x - 1)^2$.

Now $(x - 1)^2 = 0$ for $x = 1$. Thus, the rational zeros are −2 and 1. (The zero 1 has a multiplicity of 2.) These are the only zeros.

b) $f(x) = (x + 2)(x - 1)^2$

59. $f(x) = x^3 - 5x^2 + 11x + 17$

a) $\dfrac{\text{Possibilities for } p}{\text{Possibilities for } q} : \dfrac{\pm 1, \pm 17}{\pm 1}$

Possibilities for p/q: 1, −1, 17, −17

From the graph of $y = x^3 - 5x^2 + 11x + 17$, we see that, of the possibilities above, we see that only −1 might be a zero. We use synthetic division to determine whether −1 is indeed a zero.

$$\begin{array}{r|rrr} -1 & 1 & -5 & 11 & 17 \\ & & -1 & 6 & -17 \\ \hline & 1 & -6 & 17 & 0 \end{array}$$

Then we have $f(x) = (x + 1)(x^2 - 6x + 17)$. We use the quadratic formula to find the other zeros.

$$x^2 - 6x + 17 = 0$$

$$x = \frac{-(-6) \pm \sqrt{(-6)^2 - 4 \cdot 1 \cdot 17}}{2 \cdot 1}$$

$$= \frac{6 \pm \sqrt{-32}}{2} = \frac{6 \pm 4\sqrt{2}i}{2}$$

$$= 3 \pm 2\sqrt{2}i$$

The only rational zero is -1. The other zeros are $3 \pm 2\sqrt{2}i$.

b) $f(x) = (x+1)[x - (3 + 2\sqrt{2}i)][x - (3 - 2\sqrt{2}i)]$

$\qquad = (x+1)(x - 3 - 2\sqrt{2}i)(x - 3 + 2\sqrt{2}i)$

61. $f(x) = 5x^4 - 4x^3 + 19x^2 - 16x - 4$

a) $\dfrac{\text{Possibilities for } p}{\text{Possibilities for } q} : \dfrac{\pm 1, \pm 2, \pm 4}{\pm 1, \pm 5}$

Possibilities for p/q: $1, -1, 2, -2, 4, -4, \dfrac{1}{5}, -\dfrac{1}{5}$

$\qquad\qquad\qquad \dfrac{2}{5}, -\dfrac{2}{5}, \dfrac{4}{5}, -\dfrac{4}{5}$

From the graph of $y = 5x^4 - 4x^3 + 19x^2 - 16x - 4$, we see that, of the possibilities above, only $-\dfrac{2}{5}, -\dfrac{1}{5}$ and 1 might be zeros. We use synthetic division to determine whether 1 is a zero.

$$\begin{array}{r|rrrrr} 1 & 5 & -4 & 19 & -16 & -4 \\ & & 5 & 1 & 20 & 4 \\ \hline & 5 & 1 & 20 & 4 & 0 \end{array}$$

Then we have

$$f(x) = (x-1)(5x^3 + x^2 + 20x + 4)$$

$$= (x-1)[x^2(5x+1) + 4(5x+1)]$$

$$= (x-1)(5x+1)(x^2+4).$$

We find the other zeros:

$$5x + 1 = 0 \quad or \quad x^2 + 4 = 0$$

$$5x = -1 \quad or \quad x^2 = -4$$

$$x = -\frac{1}{5} \quad or \quad x = \pm 2i$$

The rational zeros are $-\dfrac{1}{5}$ and 1. The other zeros are $\pm 2i$.

b) From part (a) we see that

$$f(x) = (5x+1)(x-1)(x+2i)(x-2i).$$

63. $f(x) = x^4 - 3x^3 - 20x^2 - 24x - 8$

a) $\dfrac{\text{Possibilities for } p}{\text{Possibilities for } q} : \dfrac{\pm 1, \pm 2, \pm 4, \pm 8}{\pm 1}$

Possibilities for p/q: $1, -1, 2, -2, 4, -4, 8, -8$

From the graph of $y = x^4 - 3x^3 - 20x^2 - 24x - 8$, we see that, of the possibilities above, only -2 and -1 might be zeros. We use synthetic division to determine if -2 is a zero.

$$\begin{array}{r|rrrrr} -2 & 1 & -3 & -20 & -24 & -8 \\ & & -2 & 10 & 20 & 8 \\ \hline & 1 & -5 & -10 & -4 & 0 \end{array}$$

We see that -2 is a zero. Now we determine whether -1 is a zero.

$$\begin{array}{r|rrrr} -1 & 1 & -5 & -10 & -4 \\ & & -1 & 6 & 4 \\ \hline & 1 & -6 & -4 & 0 \end{array}$$

Then we have $f(x) = (x+2)(x+1)(x^2 - 6x - 4)$.

Use the quadratic formula to find the other zeros.

$$x^2 - 6x - 4 = 0$$

$$x = \frac{-(-6) \pm \sqrt{(-6)^2 - 4 \cdot 1 \cdot (-4)}}{2 \cdot 1}$$

$$= \frac{6 \pm \sqrt{52}}{2} = \frac{6 \pm 2\sqrt{13}}{2}$$

$$= 3 \pm \sqrt{13}$$

The rational zeros are -2 and -1. The other zeros are $3 \pm \sqrt{13}$.

b) $f(x) = (x+2)(x+1)[x - (3 + \sqrt{13})][x - (3 - \sqrt{13})]$

$\qquad = (x+2)(x+1)(x - 3 - \sqrt{13})(x - 3 + \sqrt{13})$

65. $f(x) = x^3 - 4x^2 + 2x + 4$

a) $\dfrac{\text{Possibilities for } p}{\text{Possibilities for } q} : \dfrac{\pm 1, \pm 2, \pm 4}{\pm 1}$

Possibilities for p/q: $1, -1, 2, -2, 4, -4$

From the graph of $y = x^3 - 4x^2 + 2x + 4$, we see that, of the possibilities above, only -1, 1, and 2 might be zeros. Synthetic division shows that neither -1 nor 1 is a zero. Try 2.

$$\begin{array}{r|rrrr} 2 & 1 & -4 & 2 & 4 \\ & & 2 & -4 & -4 \\ \hline & 1 & -2 & -2 & 0 \end{array}$$

Then we have $f(x) = (x-2)(x^2 - 2x - 2)$. Use the quadratic formula to find the other zeros.

$$x^2 - 2x - 2 = 0$$

$$x = \frac{-(-2) \pm \sqrt{(-2)^2 - 4 \cdot 1 \cdot (-2)}}{2 \cdot 1}$$

$$= \frac{2 \pm \sqrt{12}}{2} = \frac{2 \pm 2\sqrt{3}}{2}$$

$$= 1 \pm \sqrt{3}$$

The only rational zero is 2. The other zeros are $1 \pm \sqrt{3}$.

b) $f(x) = (x-2)[x - (1 + \sqrt{3})][x - (1 - \sqrt{3})]$

$\qquad = (x-2)(x - 1 - \sqrt{3})(x - 1 + \sqrt{3})$

67. $f(x) = x^3 + 8$

a) $\dfrac{\text{Possibilities for } p}{\text{Possibilities for } q} : \dfrac{\pm 1, \pm 2, \pm 4, \pm 8}{\pm 1}$

Possibilities for p/q: $1, -1, 2, -2, 4, -4, 8, -8$

From the graph of $y = x^3 + 8$, we see that, of the possibilities above, only -2 might be a zero. We use synthetic division to see if it is.

$$\begin{array}{r|rrrr} -2 & 1 & 0 & 0 & 8 \\ & & -2 & 4 & -8 \\ \hline & 1 & -2 & 4 & 0 \end{array}$$

We have $f(x) = (x+2)(x^2 - 2x + 4)$. Use the quadratic formula to find the other zeros.

$$x^2 - 2x + 4 = 0$$

$$x = \frac{-(-2) \pm \sqrt{(-2)^2 - 4 \cdot 1 \cdot 4}}{2 \cdot 1}$$

$$= \frac{2 \pm \sqrt{-12}}{2} = \frac{2 \pm 2\sqrt{3}i}{2}$$

$$= 1 \pm \sqrt{3}i$$

The only rational zero is -2. The other zeros are $1 \pm \sqrt{3}i$.

b) $f(x) = (x+2)[x - (1 + \sqrt{3}i)][x - (1 - \sqrt{3}i)]$
$= (x+2)(x - 1 - \sqrt{3}i)(x - 1 + \sqrt{3}i)$

69. $f(x) = \frac{1}{3}x^3 - \frac{1}{2}x^2 - \frac{1}{6}x + \frac{1}{6}$

$$= \frac{1}{6}(2x^3 - 3x^2 - x + 1)$$

a) The second form of the equation is equivalent to the first and has the advantage of having integer coefficients. Thus, we can use the rational zeros theorem for $g(x) = 2x^3 - 3x^2 - x + 1$. The zeros of $g(x)$ are the same as the zeros of $f(x)$. We find the zeros of $g(x)$.

$$\frac{\text{Possibilities for } p}{\text{Possibilities for } q} : \frac{\pm 1}{\pm 1, \pm 2}$$

Possibilities for p/q: $1, -1, \frac{1}{2}, -\frac{1}{2}$

From the graph of $y = 2x^3 - 3x^2 - x + 1$, we see that, of the possibilities above, only $-\frac{1}{2}$ and $\frac{1}{2}$ might be zeros. Synthetic division shows that $-\frac{1}{2}$ is not a zero. Try $\frac{1}{2}$.

```
1/2 | 2  -3  -1   1
    |     1  -1  -1
    ---------------
      2  -2  -2 | 0
```

We have $g(x) = \left(x - \frac{1}{2}\right)(2x^2 - 2x - 2) =$

$\left(x - \frac{1}{2}\right)(2)(x^2 - x - 1)$. Use the quadratic formula to find the other zeros.

$$x^2 - x - 1 = 0$$

$$x = \frac{-(-1) \pm \sqrt{(-1)^2 - 4 \cdot 1 \cdot (-1)}}{2 \cdot 1}$$

$$= \frac{1 \pm \sqrt{5}}{2}$$

The only rational zero is $\frac{1}{2}$. The other zeros are $\frac{1 \pm \sqrt{5}}{2}$.

b) $f(x) = \frac{1}{6}g(x)$

$$= \frac{1}{6}\left(x - \frac{1}{2}\right)(2)\left[x - \frac{1+\sqrt{5}}{2}\right]\left[x - \frac{1-\sqrt{5}}{2}\right]$$

$$= \frac{1}{3}\left(x - \frac{1}{2}\right)\left(x - \frac{1+\sqrt{5}}{2}\right)\left(x - \frac{1-\sqrt{5}}{2}\right)$$

71. $f(x) = x^4 + 2x^3 - 5x^2 - 4x + 6$

According to the rational zeros theorem, the possible rational zeros are ± 1, ± 2, ± 3, and ± 6. Synthetic division shows that only 1 and -3 are zeros.

73. $f(x) = x^3 - x^2 - 4x + 3$

According to the rational zeros theorem, the possible rational zeros are ± 1 and ± 3. Synthetic division shows that none of these is a zero. Thus, there are no rational zeros.

75. $f(x) = x^4 + 2x^3 + 2x^2 - 4x - 8$

According to the rational zeros theorem, the possible rational zeros are ± 1, ± 2, ± 4, and ± 8. Synthetic division shows that none of the possibilities is a zero. Thus, there are no rational zeros.

77. $f(x) = x^5 - 5x^4 + 5x^3 + 15x^2 - 36x + 20$

According to the rational zeros theorem, the possible rational zeros are ± 1, ± 2, ± 4, ± 5, ± 10, and ± 20. Synthetic division shows that only -2, 1 and 2 are zeros.

79. $f(x) = 3x^5 - 2x^2 + x - 1$

The number of variations in sign in $f(x)$ is 3. Then the number of positive real zeros is either 3 or less than 3 by 2, 4, 6, and so on. Thus, the number of positive real zeros is 3 or 1.

$$f(-x) = 3(-x)^5 - 2(-x)^2 + (-x) - 1$$

$$= -3x^5 - 2x^2 - x - 1$$

There are no variations in sign in $f(-x)$, so there are 0 negative real zeros.

81. $h(x) = 6x^7 + 2x^2 + 5x + 4$

There are no variations in sign in $h(x)$, so there are 0 positive real zeros.

$$h(-x) = 6(-x)^7 + 2(-x)^2 + 5(-x) + 4$$

$$= -6x^7 + 2x^2 - 5x + 4$$

The number of variations in sign in $h(-x)$ is 3. Thus, there are 3 or 1 negative real zeros.

83. $F(p) = 3p^{18} + 2p^4 - 5p^2 + p + 3$

There are 2 variations in sign in $F(p)$, so there are 2 or 0 positive real zeros.

$$F(-p) = 3(-p)^{18} + 2(-p)^4 - 5(-p)^2 + (-p) + 3$$

$$= 3p^{18} + 2p^4 - 5p^2 - p + 3$$

There are 2 variations in sign in $F(-p)$, so there are 2 or 0 negative real zeros.

85. $C(x) = 7x^6 + 3x^4 - x - 10$

There is 1 variation in sign in $C(x)$, so there is 1 positive real zero.

$$C(-x) = 7(-x)^6 + 3(-x)^4 - (-x) - 10$$
$$= 7x^6 + 3x^4 + x - 10$$

There is 1 variation in sign in $C(-x)$, so there is 1 negative real zero.

87. $h(t) = -4t^5 - t^3 + 2t^2 + 1$

There is 1 variation in sign in $h(t)$, so there is 1 positive real zero.

$$h(-t) = -4(-t)^5 - (-t)^3 + 2(-t)^2 + 1$$
$$= 4t^5 + t^3 + 2t^2 + 1$$

There are no variations in sign in $h(-t)$, so there are 0 negative real zeros.

89. $f(y) = y^4 + 13y^3 - y + 5$

There are 2 variations in sign in $f(y)$, so there are 2 or 0 positive real zeros.

$$f(-y) = (-y)^4 + 13(-y)^3 - (-y) + 5$$
$$= y^4 - 13y^3 + y + 5$$

There are 2 variations in sign in $f(-y)$, so there are 2 or 0 negative real zeros.

91. $r(x) = x^4 - 6x^2 + 20x - 24$

There are 3 variations in sign in $r(x)$, so there are 3 or 1 positive real zeros.

$$r(-x) = (-x)^4 - 6(-x)^2 + 20(-x) - 24$$
$$= x^4 - 6x^2 - 20x - 24$$

There is 1 variation in sign in $r(-x)$, so there is 1 negative real zero.

93. $R(x) = 3x^5 - 5x^3 - 4x$

There is 1 variation in sign in $R(x)$, so there is 1 positive real zero.

$$R(-x) = 3(-x)^5 - 5(-x)^3 - 4(-x)$$
$$= -3x^5 + 5x^3 + 4x$$

There is 1 variation in sign in $R(-x)$, so there is 1 negative real zero.

95. $f(x) = 4x^3 + x^2 - 8x - 2$

1. The leading term is $4x^3$. The degree, 3, is odd and the leading coefficient, 4, is positive so as $x \to \infty$, $f(x) \to \infty$ and $x \to -\infty$, $f(x) \to -\infty$.

2. We find the rational zeros p/q of $f(x)$.

$$\frac{\text{Possibilities for } p}{\text{Possibilities for } q} : \frac{\pm 1, \pm 2}{\pm 1, \pm 2, \pm 4}$$

Possibilities for p/q: $1, -1, 2, -2, \dfrac{1}{2}, -\dfrac{1}{2}, \dfrac{1}{4}, -\dfrac{1}{4}$

Synthetic division shows that $-\dfrac{1}{4}$ is a zero.

$$
\begin{array}{r|rrrr}
-\frac{1}{4} & 4 & 1 & -8 & -2 \\
 & & -1 & 0 & 2 \\
\hline
 & 4 & 0 & -8 & 0 \\
\end{array}
$$

We have $f(x) = \left(x + \dfrac{1}{4}\right)(4x^2 - 8) =$

$4\left(x + \dfrac{1}{4}\right)(x^2 - 2)$. Solving $x^2 - 2 = 0$ we get

$x = \pm\sqrt{2}$. Thus the zeros of the function are $-\dfrac{1}{4}$, $-\sqrt{2}$, and $\sqrt{2}$ so the x-intercepts of the graph are $\left(-\dfrac{1}{4}, 0\right)$, $(-\sqrt{2}, 0)$, and $(\sqrt{2}, 0)$.

3. The zeros divide the x-axis into 4 intervals, $(-\infty, -\sqrt{2})$, $\left(-\sqrt{2}, -\dfrac{1}{4}\right)$, $\left(-\dfrac{1}{4}, \sqrt{2}\right)$, and $(\sqrt{2}, \infty)$. We choose a value for x from each interval and find $f(x)$. This tells us the sign of $f(x)$ for all values of x in that interval.

In $(-\infty, -\sqrt{2})$, test -2:
$$f(-2) = 4(-2)^3 + (-2)^2 - 8(-2) - 2 = -14 < 0$$

In $\left(-\sqrt{2}, -\dfrac{1}{4}\right)$, test -1:
$$f(-1) = 4(-1)^3 + (-1)^2 - 8(-1) - 2 = 3 > 0$$

In $\left(-\dfrac{1}{4}, \sqrt{2}\right)$, test 0:
$$f(0) = 4 \cdot 0^3 + 0^2 - 8 \cdot 0 - 2 = -2 < 0$$

In $(\sqrt{2}, \infty)$, test 2:
$$f(2) = 4 \cdot 2^3 + 2^2 - 8 \cdot 2 - 2 = 18 > 0$$

Thus the graph lies below the x-axis on $(-\infty, -\sqrt{2})$ and on $\left(-\dfrac{1}{4}, \sqrt{2}\right)$. It lies above the x-axis on $\left(-\sqrt{2}, -\dfrac{1}{4}\right)$ and on $(\sqrt{2}, \infty)$. We also know the points $(-2, -14)$, $(-1, 3)$, $(0, -2)$, and $(2, 18)$ are on the graph.

4. From Step 3 we see that $f(0) = -2$ so the y-intercept is $(0, -2)$.

5. We find additional points on the graph and then draw the graph.

x	$f(x)$
-1.5	-1.25
-0.5	1.75
1	-5
1.5	1.75

$f(x) = 4x^3 + x^2 - 8x - 2$

6. Checking the graph as described on page 272 in the text, we see that it appears to be correct.

97. $f(x) = 2x^4 - 3x^3 - 2x^2 + 3x$

1. The leading term is $2x^4$. The degree, 4, is even and the leading coefficient, 2, is positive so as $x \to \infty$, $f(x) \to \infty$ and as $x \to -\infty$, $f(x) \to \infty$.

2. We find the rational zeros p/q of $f(x)$. First note that $f(x) = x(2x^3 - 3x^2 - 2x + 3)$, so 0 is a zero. Now consider $g(x) = 2x^3 - 3x^2 - 2x + 3$.

$$\frac{\text{Possibilities for } p}{\text{Possibilities for } q} : \frac{\pm 1, \pm 3}{\pm 1, \pm 2}$$

Possibilities for p/q: $1, -1, 3, -3, \frac{1}{2}, -\frac{1}{2}, \frac{3}{2}, -\frac{3}{2}$

We try 1.

$$
\begin{array}{r|rrrr}
1 & 2 & -3 & -2 & 3 \\
  &   & 2 & -1 & -3 \\
\hline
  & 2 & -1 & -3 & 0
\end{array}
$$

Then $f(x) = x(x-1)(2x^2 - x - 3)$. Using the principle of zero products to solve $2x^2 - x - 3 = 0$, we get $x = \frac{3}{2}$ or $x = -1$.

Thus the zeros of the function are $0, 1, \frac{3}{2}$, and -1 so the x-intercepts of the graph are $(0,0)$, $(1,0)$, $\left(\frac{3}{2}, 0\right)$, and $(-1, 0)$.

3. The zeros divide the x-axis into 5 intervals, $(-\infty, -1)$, $(-1, 0)$, $(0, 1)$, $\left(1, \frac{3}{2}\right)$, and $\left(\frac{3}{2}, \infty\right)$. We choose a value for x from each interval and find $f(x)$. This tells us the sign of $f(x)$ for all values of x in that interval.

In $(-\infty, -1)$, test -2:
$f(-2) = 2(-2)^4 - 3(-2)^3 - 2(-2)^2 + 3(-2) = 42 > 0$

In $(-1, 0)$, test -0.5:
$f(-0.5) = 2(-0.5)^4 - 3(-0.5)^3 - 2(-0.5)^2 + 3(-0.5) = -1.5 < 0$

In $(0, 1)$, test 0.5:
$f(0.5) = 2(0.5)^4 - 3(0.5)^3 - 2(0.5)^2 + 3(0.5) = 0.75 > 0$

In $\left(1, \frac{3}{2}\right)$, test 1.25:
$f(1.25) = 2(1.25)^4 - 3(1.25)^3 - 2(1.25)^2 + 3(1.25) = -0.3515625 < 0$

In $\left(\frac{3}{2}, \infty\right)$, test 2:
$f(2) = 2 \cdot 2^4 - 3 \cdot 2^3 - 2 \cdot 2^2 + 3 \cdot 2 = 6 > 0$

Thus the graph lies above the x-axis on $(-\infty, -1)$, on $(0, 1)$, and on $\left(\frac{3}{2}, \infty\right)$. It lies below the x-axis on $(-1, 0)$ and on $\left(1, \frac{3}{2}\right)$. We also know the points $(-2, 42)$, $(-0.5, -1.5)$, $(0.5, 0.75)$, $(1.25, -0.3515625)$, and $(2, 6)$ are on the graph.

4. From Step 2 we know that $f(0) = 0$ so the y-intercept is $(0, 0)$.

5. We find additional points on the graph and then draw the graph.

x	$f(x)$
-1.5	11.25
2.5	26.25
3	72

$$f(x) = 2x^4 - 3x^3 - 2x^2 + 3x$$

6. Checking the graph as described on page 272 in the text, we see that it appears to be correct.

99. Left to the student

101. Discussion and Writing

103. $f(x) = x^2 - 8x + 10$

a) $-\dfrac{b}{2a} = -\dfrac{-8}{2 \cdot 1} = -(-4) = 4$

$f(4) = 4^2 - 8 \cdot 4 + 10 = -6$

The vertex is $(4, -6)$.

b) The axis of symmetry is $x = 4$.

c) Since the coefficient of x^2 is positive, there is a minimum function value. It is the second coordinate of the vertex, -6. It occurs when $x = 4$.

105.
$$-\frac{4}{5}x + 8 = 0$$

$$-\frac{4}{5}x = -8 \qquad \text{Subtracting 8}$$

$$-\frac{5}{4}\left(-\frac{4}{5}x\right) = -\frac{5}{4}(-8) \quad \text{Multiplying by } -\frac{5}{4}$$

$$x = 10$$

The zero is 10.

107. $g(x) = -x^3 - 2x^2$

The degree is 3, so the function is cubic.

Leading term: $-x^3$; leading coefficient: -1

Since the degree is odd and the leading coefficient is negative, as $x \to \infty$, $g(x) \to -\infty$ and as $x \to -\infty$, $g(x) \to \infty$.

109. $f(x) = -\dfrac{4}{9}$

The degree is 0, so this is a constant function.

Leading term: $-\dfrac{4}{9}$; leading coefficient: $-\dfrac{4}{9}$; for all x, $f(x) = -\dfrac{4}{9}$

111. $g(x) = x^4 - 2x^3 + x^2 - x + 2$

The degree is 4, so the function is quartic.

Leading term: x^4; leading coefficient: 1

Since the degree is even and the leading coefficient is positive, as $x \to \infty$, $g(x) \to \infty$ and as $x \to -\infty$, $g(x) \to \infty$.

113.
$$(x^4 - y^4) \div (x - y)$$
$$= (x^4 + 0x^3 + 0x^2 + 0x - y^4) \div (x - y)$$

$$\begin{array}{r|rrrrr} y & 1 & 0 & 0 & 0 & -y^4 \\ & & y & y^2 & y^3 & y^4 \\ \hline & 1 & y & y^2 & y^3 & 0 \end{array}$$

The quotient is $x^3 + x^2y + xy^2 + y^3$. The remainder is 0.

115. $f(x) = 2x^3 - 5x^2 - 4x + 3$

a) $2x^3 - 5x^2 - 4x + 3 = 0$

$\dfrac{\text{Possibilities for } p}{\text{Possibilities for } q} : \dfrac{\pm 1, \pm 3}{\pm 1, \pm 2}$

Possibilities for p/q: $1, -1, 3, -3, \dfrac{1}{2}, -\dfrac{1}{2}, \dfrac{3}{2}, -\dfrac{3}{2}$

The first possibility that is a solution of $f(x) = 0$ is -1:

$$\begin{array}{r|rrr} -1 & 2 & -5 & -4 & 3 \\ & & -2 & 7 & -3 \\ \hline & 2 & -7 & 3 & 0 \end{array}$$

Thus, -1 is a solution.

Then we have:

$(x + 1)(2x^2 - 7x + 3) = 0$

$(x + 1)(2x - 1)(x - 3) = 0$

The other solutions are $\dfrac{1}{2}$ and 3.

b) The graph of $y = f(x - 1)$ is the graph of $y = f(x)$ shifted 1 unit right. Thus, we add 1 to each solution of $f(x) = 0$ to find the solutions of $f(x - 1) = 0$. The solutions are $-1 + 1$, or 0; $\dfrac{1}{2} + 1$, or $\dfrac{3}{2}$; and $3 + 1$, or 4.

c) The graph of $y = f(x + 2)$ is the graph of $y = f(x)$ shifted 2 units left. Thus, we subtract 2 from each solution of $f(x) = 0$ to find the solutions of $f(x + 2) = 0$. The solutions are $-1 - 2$, or -3; $\dfrac{1}{2} - 2$, or $-\dfrac{3}{2}$; and $3 - 2$, or 1.

d) The graph of $y = f(2x)$ is a horizontal shrinking of the graph of $y = f(x)$ by a factor of 2. We divide each solution of $f(x) = 0$ by 2 to find the solutions of $f(2x) = 0$. The solutions are $\dfrac{-1}{2}$ or $-\dfrac{1}{2}$; $\dfrac{1/2}{2}$, or $\dfrac{1}{4}$; and $\dfrac{3}{2}$.

117. $P(x) = 2x^5 - 33x^4 - 84x^3 + 2203x^2 - 3348x - 10,080$

a) $2x^5 - 33x^4 - 84x^3 + 2203x^2 - 3348x - 10,080 = 0$

Trying some of the many possibilities for p/q, we find that 4 is a zero.

$$\begin{array}{r|rrrrrr} 4 & 2 & -33 & -84 & 2203 & -3348 & -10,080 \\ & & 8 & -100 & -736 & 5868 & 10,080 \\ \hline & 2 & -25 & -184 & 1467 & 2520 & 0 \end{array}$$

Then we have:

$(x - 4)(2x^4 - 25x^3 - 184x^2 + 1467x + 2520) = 0$

We now use the fourth degree polynomial above to find another zero. Synthetic division shows that 4 is not a double zero, but 7 is a zero.

$$\begin{array}{r|rrrrr} 7 & 2 & -25 & -184 & 1467 & 2520 \\ & & 14 & -77 & -1827 & -2520 \\ \hline & 2 & -11 & -261 & -360 & 0 \end{array}$$

Now we have:

$(x - 4)(x - 7)(2x^3 - 11x^2 - 261x - 360) = 0$

Use the third degree polynomial above to find a third zero. Synthetic division shows that 7 is not a double zero, but 15 is a zero.

$$\begin{array}{r|rrrr} 15 & 2 & -11 & -261 & -360 \\ & & 30 & 285 & 360 \\ \hline & 2 & 19 & 24 & 0 \end{array}$$

We have:

$$\begin{aligned} P(x) &= (x - 4)(x - 7)(x - 15)(2x^2 + 19x + 24) \\ &= (x - 4)(x - 7)(x - 15)(2x + 3)(x + 8) \end{aligned}$$

The rational zeros are 4, 7, 15, $-\dfrac{3}{2}$, and -8.

Exercise Set 3.5

1. Graph (d) is the graph of $f(x) = \dfrac{8}{x^2 - 4}$.

$x^2 - 4 = 0$ when $x = \pm 2$, so $x = -2$ and $x = 2$ are vertical asymptotes.

The x-axis, $y = 0$, is the horizontal asymptote because the degree of the numerator is less than the degree of the denominator.

There is no oblique asymptote.

3. Graph (e) is the graph of $f(x) = \dfrac{8x}{x^2 - 4}$.

As in Exercise 1, $x = -2$ and $x = 2$ are vertical asymptotes.

The x-axis, $y = 0$, is the horizontal asymptote because the degree of the numerator is less than the degree of the denominator.

There is no oblique asymptote.

5. Graph (c) is the graph of $f(x) = \dfrac{8x^3}{x^2 - 4}$.

As in Exercise 1, $x = -2$ and $x = 2$ are vertical asymptotes.

The degree of the numerator is greater than the degree of the denominator, so there is no horizontal asymptote but there is an oblique asymptote. To find it we first divide to find an equivalent expression.

$$\begin{array}{r} 8x \\ x^2 - 4 \overline{\smash{)}8x^3 } \\ \underline{8x^3 - 32x} \\ 32x \end{array}$$

$\dfrac{8x^3}{x^2 - 4} = 8x + \dfrac{32x}{x^2 - 4}$

Now we multiply by 1, using $(1/x^2)/(1/x^2)$.

$$\frac{32x}{x^2-4}\cdot\frac{\frac{1}{x^2}}{\frac{1}{x^2}}=\frac{\frac{32}{x}}{1-\frac{4}{x^2}}$$

As $|x|$ becomes very large, each expression with x in the denominator tends toward zero.

Then, as $|x|\to\infty$, we have

$$\frac{\frac{32}{x}}{1-\frac{4}{x^2}}\to\frac{0}{1-0},\text{ or }0.$$

Thus, as $|x|$ becomes very large, the graph of $f(x)$ gets very close to the graph of $y=8x$, so $y=8x$ is the oblique asymptote.

7. $g(x)=\dfrac{1}{x^2}$

The zero of the denominator is 0, so the vertical asymptote is $x=0$.

9. $h(x)=\dfrac{x+7}{2-x}$

$2-x=0$ when $x=2$, so the vertical asymptote is $x=2$.

11. $f(x)=\dfrac{3-x}{(x-4)(x+6)}$

The zeros of the denominator are 4 and -6, so the vertical asymptotes are $x=4$ and $x=-6$.

13. $g(x)=\dfrac{x^2}{2x^2-x-3}=\dfrac{x^2}{(2x-3)(x+1)}$

The zeros of the denominator are $\dfrac{3}{2}$ and -1, so the vertical asymptotes are $x=\dfrac{3}{2}$ and $x=-1$.

15. $f(x)=\dfrac{3x^2+5}{4x^2-3}$

The numerator and denominator have the same degree and the ratio of the leading coefficients is $\dfrac{3}{4}$, so $y=\dfrac{3}{4}$ is the horizontal asymptote.

17. $h(x)=\dfrac{x^2-4}{2x^4+3}$

The degree of the numerator is less than the degree of the denominator, so $y=0$ is the horizontal asymptote.

19. $g(x)=\dfrac{x^3-2x^2+x-1}{x^2-16}$

The degree of the numerator is greater than the degree of the denominator, so there is no horizontal asymptote.

21. $g(x)=\dfrac{x^2+4x-1}{x+3}$

$$\begin{array}{r}x+1\\x+3\overline{\smash{\big)}\,x^2+4x-1}\\\underline{x^2+3x}\\x-1\\\underline{x+3}\\-4\end{array}$$

Then $g(x)=x+1+\dfrac{-4}{x+3}$. The oblique asymptote is $y=x+1$.

23. $h(x)=\dfrac{x^4-2}{x^3+1}$

$$\begin{array}{r}x\\x^3+1\overline{\smash{\big)}\,x^4+0x^3+0x^2+0x}\\\underline{x^4+x}\\-x\end{array}$$

Then $h(x)=x+\dfrac{-x}{x^3+1}$. The oblique asymptote is $y=x$.

25. $f(x)=\dfrac{x^3-x^2+x-4}{x^2+2x-1}$

$$\begin{array}{r}x-3\\x^2+2x-1\overline{\smash{\big)}\,x^3-x^2+x-4}\\\underline{x^3+2x^2-x}\\-3x^2+2x-4\\\underline{-3x^2-6x+3}\\8x-7\end{array}$$

Then $f(x)=x-3+\dfrac{8x-7}{x^2+2x-1}$. The oblique asymptote is $y=x-3$.

27. $f(x)=\dfrac{1}{x}$

1. 0 is the zero of the denominator, so the domain excludes 0. It is $(-\infty,0)\cup(0,\infty)$. The line $x=0$, or the y-axis, is the vertical asymptote.

2. Because the degree of the numerator is less than the degree of the denominator, the x-axis, or $y=0$, is the horizontal asymptote. There are no oblique asymptotes.

3. The numerator has no zeros, so there is no x-intercept.

4. Since 0 is not in the domain of the function, there is no y-intercept.

5. Find other function values to determine the shape of the graph and then draw it.

29. $h(x)=-\dfrac{4}{x^2}$

1. 0 is the zero of the denominator, so the domain excludes 0. It is $(-\infty,0)\cup(0,\infty)$. The line $x=0$, or the y-axis, is the vertical asymptote.

2. Because the degree of the numerator is less than the degree of the denominator, the x-axis, or $y=0$, is the horizontal asymptote. There is no oblique asymptote.

3. The numerator has no zeros, so there is no x-intercept.

4. Since 0 is not in the domain of the function, there is no y-intercept.

5. Find other function values to determine the shape of the graph and then draw it.

$$y = 0 \quad \quad x = 0$$

$$h(x) = -\frac{4}{x^2}$$

31. $g(x) = \dfrac{x^2 - 4x + 3}{x + 1}$

1. The denominator, $x + 1$, is 0 when $x = -1$, so the domain excludes -1. It is $(-\infty, -1) \cup (-1, \infty)$. The line $x = -1$ is the vertical asymptote.

2. The degree of the numerator is 1 greater than the degree of the denominator, so we divide to find the oblique asymptote.

$$
\begin{array}{r}
x - 5 \\
x + 1 \overline{\smash{\big)}\ x^2 - 4x + 3} \\
\underline{x^2 + x} \\
-5x + 3 \\
\underline{-5x - 5} \\
8
\end{array}
$$

The oblique asymptote is $y = x - 5$. There is no horizontal asymptote.

3. $x^2 - 4x + 3 = (x - 1)(x - 3)$, so the zeros of the numerator are 1 and 3. Thus the x-intercepts are $(1, 0)$ and $(3, 0)$.

4. $g(0) = \dfrac{0^2 - 4 \cdot 0 + 3}{0 + 1} = 3$, so the y-intercept is $(0, 3)$.

5. Find other function values to determine the shape of the graph and then draw it.

$$x = -1 \quad \quad y = x - 5$$

$$g(x) = \frac{x^2 - 4x + 3}{x + 1}$$

33. $f(x) = \dfrac{1}{x + 3}$

1. -3 is the zero of the denominator, so the domain excludes -3. It is $(-\infty, -3) \cup (-3, \infty)$. The line $x = -3$ is the vertical asymptote.

2. Because the degree of the numerator is less than the degree of the denominator, the x-axis, or $y = 0$, is the horizontal asymptote. There is no oblique asymptote.

3. The numerator has no zeros, so there is no x-intercept.

4. $f(0) = \dfrac{1}{0 + 3} = \dfrac{1}{3}$, so $\left(0, \dfrac{1}{3}\right)$ is the y-intercept.

5. Find other function values to determine the shape of the graph and then draw it.

$$x = -3 \quad \quad f(x) = \frac{1}{x + 3}$$

$$y = 0$$

35. $f(x) = \dfrac{-2}{x - 5}$

1. 5 is the zero of the denominator, so the domain excludes 5. It is $(-\infty, 5) \cup (5, \infty)$. The line $x = 5$ is the vertical asymptote.

2. Because the degree of the numerator is less than the degree of the denominator, the x-axis, or $y = 0$, is the horizontal asymptote. There is no oblique asymptote.

3. The numerator has no zeros, so there is no x-intercept.

4. $f(0) = \dfrac{-2}{0 - 5} = \dfrac{2}{5}$, so $\left(0, \dfrac{2}{5}\right)$ is the y-intercept.

5. Find other function values to determine the shape of the graph and then draw it.

$$f(x) = \frac{-2}{x - 5}$$

$$y = 0$$

$$x = 5$$

37. $f(x) = \dfrac{2x + 1}{x}$

1. 0 is the zero of the denominator, so the domain excludes 0. It is $(-\infty, 0) \cup (0, \infty)$. The line $x = 0$, or the y-axis, is the vertical asymptote.

2. The numerator and denominator have the same degree, so the horizontal asymptote is determined by the ratio of the leading coefficients, $2/1$, or 2. Thus, $y = 2$ is the horizontal asymptote. There is no oblique asymptote.

3. The zero of the numerator is the solution of $2x + 1 = 0$, or $-\dfrac{1}{2}$. The x-intercept is $\left(-\dfrac{1}{2}, 0\right)$.

4. Since 0 is not in the domain of the function, there is no y-intercept.

5. Find other function values to determine the shape of the graph and then draw it.

39. $f(x) = \dfrac{1}{(x-2)^2}$

1. 2 is the zero of the denominator, so the domain excludes 2. It is $(-\infty, 2) \cup (2, \infty)$. The line $x = 2$ is the vertical asymptote.

2. Because the degree of the numerator is less than the degree of the denominator, the x-axis, or $y = 0$, is the horizontal asymptote. There is no oblique asymptote.

3. The numerator has no zeros, so there is no x-intercept.

4. $f(0) = \dfrac{1}{(0-2)^2} = \dfrac{1}{4}$, so $\left(0, \dfrac{1}{4}\right)$ is the y-intercept.

5. Find other function values to determine the shape of the graph and then draw it.

41. $f(x) = -\dfrac{1}{x^2}$

1. 0 is the zero of the denominator, so the domain excludes 0. It is $(-\infty, 0) \cup (0, \infty)$. The line $x = 0$, or the y-axis, is the vertical asymptote.

2. Because the degree of the numerator is less than the degree of the denominator, the x-axis, or $y = 0$, is the horizontal asymptote. There is no oblique asymptote.

3. The numerator has no zeros, so there is no x-intercept.

4. Since 0 is not in the domain of the function, there is no y-intercept.

5. Find other function values to determine the shape of the graph and then draw it.

43. $f(x) = \dfrac{1}{x^2 + 3}$

1. The denominator has no real-number zeros, so the domain is $(-\infty, \infty)$ and there is no vertical asymptote.

2. Because the degree of the numerator is less than the degree of the denominator, the x-axis, or $y = 0$, is the horizontal asymptote. There is no oblique asymptote.

3. The numerator has no zeros, so there is no x-intercept.

4. $f(0) = \dfrac{1}{0^2 + 3} = \dfrac{1}{3}$, so $\left(0, \dfrac{1}{3}\right)$ is the y-intercept.

5. Find other function values to determine the shape of the graph and then draw it.

45. $f(x) = \dfrac{x^2 - 4}{x - 2} = \dfrac{(x+2)(x-2)}{x-2} = x + 2, \; x \neq 2$

The graph is the same as the graph of $f(x) = x + 2$ except at $x = 2$, where there is a hole. The zero of $f(x) = x + 2$ is -2, so the x-intercept is $(-2, 0)$; $f(0) = 2$, so the y-intercept is $(0, 2)$.

47. $f(x) = \dfrac{x - 1}{x + 2}$

1. -2 is the zero of the denominator, so the domain excludes -2. It is $(-\infty, -2) \cup (-2, \infty)$. The line $x = -2$ is the vertical asymptote.

2. The numerator and denominator have the same degree, so the horizontal asymptote is determined by the ratio of the leading coefficients, $1/1$, or 1. Thus, $y = 1$ is the horizontal asymptote. There is no oblique asymptote.

3. The zero of the numerator is 1, so the x-intercept is $(1, 0)$.

4. $f(0) = \dfrac{0-1}{0+2} = -\dfrac{1}{2}$, so $\left(0, -\dfrac{1}{2}\right)$ is the y-intercept.

5. Find other function values to determine the shape of the graph and then draw it.

49. $f(x) = \dfrac{x+3}{2x^2 - 5x - 3}$

1. The zeros of the denominator are the solutions of $2x^2 - 5x - 3 = 0$. Since $2x^2 - 5x - 3 = (2x+1)(x-3)$, the zeros are $-\dfrac{1}{2}$ and 3. Thus, the domain is $\left(-\infty, -\dfrac{1}{2}\right) \cup \left(-\dfrac{1}{2}, 3\right) \cup (3, \infty)$ and the lines $x = -\dfrac{1}{2}$ and $x = 3$ are vertical asymptotes.

2. Because the degree of the numerator is less than the degree of the denominator, the x-axis, or $y = 0$, is the horizontal asymptote. There is no oblique asymptote.

3. -3 is the zero of the numerator, so $(-3, 0)$ is the x-intercept.

4. $f(0) = \dfrac{0+3}{2 \cdot 0^2 - 5 \cdot 0 - 3} = -1$, so $(0, -1)$ is the y-intercept.

5. Find other function values to determine the shape of the graph and then draw it.

51. $f(x) = \dfrac{x^2 - 9}{x + 1}$

1. -1 is the zero of the denominator, so the domain excludes -1. It is $(-\infty, -1) \cup (-1, \infty)$. The line $x = -1$ is the vertical asymptote.

2. Because the degree of the numerator is one greater than the degree of the denominator, there is an oblique asymptote. Using division, we find that $\dfrac{x^2 - 9}{x + 1} = x - 1 + \dfrac{-8}{x + 1}$. As $|x|$ becomes very large, the graph of $f(x)$ gets close to the graph of $y = x - 1$. Thus, the line $y = x - 1$ is the oblique asymptote.

3. Since $x^2 - 9 = (x + 3)(x - 3)$, the zeros of the numerator are -3 and 3. Thus, the x-intercepts are $(-3, 0)$ and $(3, 0)$.

4. $f(0) = \dfrac{0^2 - 9}{0 + 1} = -9$, so $(0, -9)$ is the y-intercept.

5. Find other function values to determine the shape of the graph and then draw it.

53. $f(x) = \dfrac{x^2 + x - 2}{2x^2 + 1}$

1. The denominator has no real-number zeros, so the domain is $(-\infty, \infty)$ and there is no vertical asymptote.

2. The numerator and denominator have the same degree, so the horizontal asymptote is determined by the ratio of the leading coefficients, $1/2$. Thus, $y = 1/2$ is the horizontal asymptote. There is no oblique asymptote.

3. Since $x^2 + x - 2 = (x + 2)(x - 1)$, the zeros of the numerator are -2 and 1. Thus, the x-intercepts are $(-2, 0)$ and $(1, 0)$.

4. $f(0) = \dfrac{0^2 + 0 - 2}{2 \cdot 0^2 + 1} = -2$, so $(0, -2)$ is the y-intercept.

5. Find other function values to determine the shape of the graph and then draw it.

55. $g(x) = \dfrac{3x^2 - x - 2}{x - 1} = \dfrac{(3x + 2)(x - 1)}{x - 1} = 3x + 2,$ $x \neq 1$

The graph is the same as the graph of $g(x) = 3x + 2$ except at $x = 1$, where there is a hole.

The zero of $g(x) = 3x + 2$ is $-\dfrac{2}{3}$, so the x-intercept is $\left(-\dfrac{2}{3}, 0\right)$; $g(0) = 2$, so the y-intercept is $(0, 2)$.

$$g(x) = \frac{3x^2 - x - 2}{x - 1}$$

57. $f(x) = \dfrac{x - 1}{x^2 - 2x - 3}$

1. The zeros of the denominator are the solutions of $x^2 - 2x - 3 = 0$. Since $x^2 - 2x - 3 = (x + 1)(x - 3)$, the zeros are -1 and 3. Thus, the domain is $(-\infty, -1) \cup (-1, 3) \cup (3, \infty)$ and the lines $x = -1$ and $x = 3$ are the vertical asymptotes.

2. Because the degree of the numerator is less than the degree of the denominator, the x-axis, or $y = 0$, is the horizontal asymptote. There is no oblique asymptote.

3. 1 is the zero of the numerator, so $(1, 0)$ is the x-intercept.

4. $f(0) = \dfrac{0 - 1}{0^2 - 2 \cdot 0 - 3} = \dfrac{1}{3}$, so $\left(0, \dfrac{1}{3}\right)$ is the y-intercept.

5. Find other function values to determine the shape of the graph and then draw it.

$$f(x) = \frac{x - 1}{x^2 - 2x - 3}$$

59. $f(x) = \dfrac{x - 3}{(x + 1)^3}$

1. -1 is the zero of the denominator, so the domain excludes -1. It is $(-\infty, -1) \cup (-1, \infty)$. The line $x = -1$ is the vertical asymptote.

2. Because the degree of the numerator is less than the degree of the denominator, the x-axis, or $y = 0$, is the horizontal asymptote. There is no oblique asymptote.

3. 3 is the zero of the numerator, so $(3, 0)$ is the x-intercept.

4. $f(0) = \dfrac{0 - 3}{(0 + 1)^3} = -3$, so $(0, -3)$ is the y-intercept.

5. Find other function values to determine the shape of the graph and then draw it.

$$f(x) = \frac{x - 3}{(x + 1)^3}$$

61. $f(x) = \dfrac{x^3 + 1}{x}$

1. 0 is the zero of the denominator, so the domain excludes 0. It is $(-\infty, 0) \cup (0, \infty)$. The line $x = 0$, or the y-axis, is the vertical asymptote.

2. Because the degree of the numerator is more than one greater than the degree of the denominator, there is no horizontal or oblique asymptote.

3. The real-number zero of the numerator is -1, so the x-intercept is $(-1, 0)$.

4. Since 0 is not in the domain of the function, there is no y-intercept.

5. Find other function values to determine the shape of the graph and then draw it.

$$f(x) = \frac{x^3 + 1}{x}$$

63. $f(x) = \dfrac{x^3 + 2x^2 - 15x}{x^2 - 5x - 14}$

1. The zeros of the denominator are the solutions of $x^2 - 5x - 14 = 0$. Since $x^2 - 5x - 14 = (x + 2)(x - 7)$, the zeros are -2 and 7. Thus, the domain is $(-\infty, -2) \cup (-2, 7) \cup (7, \infty)$ and the lines $x = -2$ and $x = 7$ are the vertical asymptotes.

2. Because the degree of the numerator is one greater than the degree of the denominator, there is an oblique asymptote. Using division, we find that $\dfrac{x^3 + 2x^2 - 15x}{x^2 - 5x - 14} = x + 7 + \dfrac{34x + 98}{x^2 - 5x - 14}$. As $|x|$ becomes very large, the graph of $f(x)$ gets close to the graph of $y = x + 7$. Thus, the line $y = x + 7$ is the oblique asymptote.

3. The zeros of the numerator are the solutions of $x^3 + 2x^2 - 15x = 0$. Since $x^3 + 2x^2 - 15x = x(x + 5)(x - 3)$, the zeros are 0, -5, and 3. Thus, the x-intercepts are $(-5, 0)$, $(0, 0)$, and $(3, 0)$.

4. From part (3) we see that $(0, 0)$ is the y-intercept.

5. Find other function values to determine the shape of the graph and then draw it.

$$f(x) = \frac{x^3 + 2x^2 - 15x}{x^2 - 5x - 14}$$

65. $f(x) = \dfrac{5x^4}{x^4 + 1}$

1. The denominator has no real-number zeros, so the domain is $(-\infty, \infty)$ and there is no vertical asymptote.

2. The numerator and denominator have the same degree, so the horizontal asymptote is determined by the ratio of the leading coefficients, 5/1, or 5. Thus, $y = 5$ is the horizontal asymptote. There is no oblique asymptote.

3. The zero of the numerator is 0, so $(0, 0)$ is the x-intercept.

4. From part (3) we see that $(0, 0)$ is the y-intercept.

5. Find other function values to determine the shape of the graph and then draw it.

$$f(x) = \frac{5x^4}{x^4 + 1}$$

67. $f(x) = \dfrac{x^2}{x^2 - x - 2}$

1. The zeros of the denominator are the solutions of $x^2 - x - 2 = 0$. Since $x^2 - x - 2 = (x+1)(x-2)$, the zeros are -1 and 2. Thus, the domain is $(-\infty, -1) \cup (-1, 2) \cup (2, \infty)$ and the lines $x = -1$ and $x = 2$ are the vertical asymptotes.

2. The numerator and denominator have the same degree, so the horizontal asymptote is determined by the ratio of the leading coefficients, 1/1, or 1. Thus, $y = 1$ is the horizontal asymptote. There is no oblique asymptote.

3. The zero of the numerator is 0, so the x-intercept is $(0, 0)$.

4. From part (3) we see that $(0, 0)$ is the y-intercept.

5. Find other function values to determine the shape of the graph and then draw it.

$$f(x) = \frac{x^2}{x^2 - x - 2}$$

69. Answers may vary. The numbers -4 and 5 must be zeros of the denominator. A function that satisfies these conditions is
$$f(x) = \frac{1}{(x+4)(x-5)}, \text{ or } f(x) = \frac{1}{x^2 - x - 20}.$$

71. Answers may vary. The numbers -4 and 5 must be zeros of the denominator and -2 must be a zero of the numerator. In addition, the numerator and denominator must have the same degree and the ratio of their leading coefficients must be 3/2. A function that satisfies these conditions is
$$f(x) = \frac{3x(x+2)}{2(x+4)(x-5)}, \text{ or } f(x) = \frac{3x^2 + 6x}{2x^2 - 2x - 40}.$$
Another function that satisfies these conditions is
$$g(x) = \frac{3(x+2)^2}{2(x+4)(x-5)}, \text{ or } g(x) = \frac{3x^2 + 12x + 12}{2x^2 - 2x - 40}.$$

73. a) The horizontal asymptote of $N(t)$ is the ratio of the leading coefficients of the numerator and denominator, 0.8/5, or 0.16. Thus, $N(t) \to 0.16$ as $t \to \infty$.

b) The medication never completely disappears from the body; a trace amount remains.

75. a) $P(0) = 0$; $P(1) = 45.455$ thousand, or $45,455$;

$P(3) = 55.556$ thousand, or $55,556$;

$P(8) = 29.197$ thousand, or $29,197$

b) The degree of the numerator is less than the degree of the denominator, so the x-axis is the horizontal asymptote. Thus, $P(t) \to 0$ as $t \to \infty$.

c) Eventually, no one will live in Lordsburg.

77. Using the Maximum feature on a graphing calculator, we find that the maximum population is about 58.926 thousand, or 58,926. This occurs when $t \approx 2.12$ months.

79. Discussion and Writing

81. slope

83. point-slope equation

85. $f(-x) = -f(x)$

87. midpoint formula

89. $f(x) = \dfrac{x^5 + 2x^3 + 4x^2}{x^2 + 2} = x^3 + 4 + \dfrac{-8}{x^2 + 2}$

As $|x| \to \infty$, $\dfrac{-8}{x^2 + 2} \to 0$ and the value of $f(x) \to x^3 + 4$. Thus, the nonlinear asymptote is $y = x^3 + 4$.

91.

$$f(x) = \frac{2x^3 + x^2 - 8x - 4}{x^3 + x^2 - 9x - 9}$$

93. $f(x) = \sqrt{\dfrac{72}{x^2 - 4x - 21}}$

The radicand must be nonnegative and the denominator must be nonzero. Thus, the values of x for which $x^2 - 4x - 21 > 0$ comprise the domain. By inspecting the graph of $y = x^2 - 4x - 21$ we see that the domain is $\{x | x < -3 \text{ or } x > 7\}$, or $(-\infty, -3) \cup (7, \infty)$.

Exercise Set 3.6

1.
$$x^2 + 2x - 15 = 0$$
$$(x + 5)(x - 3) = 0$$
$$x + 5 = 0 \quad or \quad x - 3 = 0$$
$$x = -5 \quad or \quad x = 3$$

The solution set is $\{-5, 3\}$.

3. Solve $x^2 + 2x - 15 \le 0$.

From Exercise 2 we know the solution set of $x^2 + 2x - 15 < 0$ is $(-5, 3)$. The solution set of $x^2 + 2x - 15 \le 0$ includes the endpoints of this interval. Thus the solution set is $[-5, 3]$.

5. Solve $x^2 + 2x - 15 \ge 0$.

From Exercise 4 we know the solution set of $x^2 + 2x - 15 > 0$ is $(-\infty, -5) \cup (3, \infty)$. The solution set of $x^2 + 2x - 15 \ge 0$ includes the endpoints -5 and 3. Thus the solution set is $(-\infty, -5] \cup [3, \infty)$.

7. Solve $x^5 - 9x^3 < 0$.

From Exercise 6 we know the solutions of the related equation are -3, 0, and 3. These numbers divide the x-axis into the intervals $(-\infty, -3)$, $(-3, 0)$, $(0, 3)$, and $(3, \infty)$. We test a value in each interval.

$(-\infty, -3)$: $g(-4) = -448 < 0$

$(-3, 0)$: $g(-1) = 8 > 0$

$(0, 3)$: $g(1) = -8 < 0$

$(3, \infty)$: $g(4) = 448 > 0$

Function values are negative on $(-\infty, -3)$ and on $(0, 3)$. The solution set is $(-\infty, -3) \cup (0, 3)$.

9. Solve $x^5 - 9x^3 > 0$.

From our work in Exercise 7 we see that the solution set is $(-3, 0) \cup (3, \infty)$.

11. First we find an equivalent inequality with 0 on one side.
$$x^3 + 6x^2 < x + 30$$
$$x^3 + 6x^2 - x - 30 < 0$$

From the graph we see that the x-intercepts of the related function occur at $x = -5$, $x = -3$, and $x = 2$. They divide the x-axis into the intervals $(-\infty, -5)$, $(-5, -3)$, $(-3, 2)$, and $(2, \infty)$. From the graph we see that the function has negative values only on $(-\infty, -5)$ and $(-3, 2)$. Thus, the solution set is $(-\infty, -5) \cup (-3, 2)$.

13. By observing the graph or the denominator of the function, we see that the function is not defined for $x = -2$ or $x = 2$. We also see that 0 is a zero of the function. These numbers divide the x-axis into the intervals $(-\infty, -2)$, $(-2, 0)$, $(0, 2)$, and $(2, \infty)$. From the graph we see that the function has positive values only on $(-2, 0)$ and $(2, \infty)$. Since the inequality symbol is $\ge$, 0 must be included in the solution set. It is $(-2, 0] \cup (2, \infty)$.

15. $(x - 1)(x + 4) < 0$

The related equation is $(x - 1)(x + 4) = 0$. Using the principle of zero products, we find that the solutions of the related equation are 1 and -4. These numbers divide the x-axis into the intervals $(-\infty, -4)$, $(-4, 1)$, and $(1, \infty)$. We let $f(x) = (x - 1)(x + 4)$ and test a value in each interval.

$(-\infty, -4)$: $f(-5) = 6 > 0$

$(-4, 1)$: $f(0) = -4 < 0$

$(1, \infty)$: $f(2) = 6 > 0$

Function values are negative only in the interval $(-4, 1)$. The solution set is $(-4, 1)$.

17. $(x - 4)(x + 2) \ge 0$

The related equation is $(x - 4)(x + 2) = 0$. Using the principle of zero products, we find that the solutions of the related equation are 4 and -2. These numbers divide the x-axis into the intervals $(-\infty, -2)$, $(-2, 4)$, and $(4, \infty)$. We let $f(x) = (x - 4)(x + 2)$ and test a value in each interval.

$(-\infty, -2)$: $f(-3) = 7 > 0$

$(-2, 4)$: $f(0) = -8 < 0$

$(4, \infty)$: $f(5) = 7 > 0$

Function values are positive on $(-\infty, -2)$ and $(4, \infty)$. Since the inequality symbol is $\ge$, the endpoints of the intervals must be included in the solution set. It is $(-\infty, -2] \cup [4, \infty)$.

19.
$$x^2 + x - 2 > 0 \quad \text{Polynomial inequality}$$
$$x^2 + x - 2 = 0 \quad \text{Related equation}$$
$$(x + 2)(x - 1) = 0 \quad \text{Factoring}$$

Using the principle of zero products, we find that the solutions of the related equation are -2 and 1. These numbers divide the x-axis into the intervals $(-\infty, -2)$, $(-2, 1)$, and $(1, \infty)$. We let $f(x) = x^2 + x - 2$ and test a value in each interval.

$(-\infty, -2)$: $f(-3) = 4 > 0$

$(-2, 1)$: $f(0) = -2 < 0$

$(1, \infty)$: $f(2) = 4 > 0$

Function values are positive on $(-\infty, -2)$ and $(1, \infty)$. The solution set is $(-\infty, -2) \cup (1, \infty)$.

21.
$$x^2 > 25 \quad \text{Polynomial inequality}$$
$$x^2 - 25 > 0 \quad \text{Equivalent inequality with } 0 \text{ on one side}$$
$$x^2 - 25 = 0 \quad \text{Related equation}$$
$$(x + 5)(x - 5) = 0 \quad \text{Factoring}$$

Using the principle of zero products, we find that the solutions of the related equation are -5 and 5. These numbers divide the x-axis into the intervals $(-\infty, -5)$, $(-5, 5)$, and $(5, \infty)$. We let $f(x) = x^2 - 25$ and test a value in each interval.

$(-\infty, -5)$: $f(-6) = 11 > 0$

$(-5, 5)$: $f(0) = -25 < 0$

$(5, \infty)$: $f(6) = 11 > 0$

Function values are positive on $(-\infty, -5)$ and $(5, \infty)$. The solution set is $(-\infty, -5) \cup (5, \infty)$.

23.
$$4 - x^2 \leq 0 \quad \text{Polynomial inequality}$$
$$4 - x^2 = 0 \quad \text{Related equation}$$
$$(2 + x)(2 - x) = 0 \quad \text{Factoring}$$

Using the principle of zero products, we find that the solutions of the related equation are -2 and 2. These numbers divide the x-axis into the intervals $(-\infty, -2)$, $(-2, 2)$, and $(2, \infty)$. We let $f(x) = 4 - x^2$ and test a value in each interval.

$(-\infty, -2)$: $f(-3) = -5 < 0$

$(-2, 2)$: $f(0) = 4 > 0$

$(2, \infty)$: $f(3) = -5 < 0$

Function values are negative on $(-\infty, -2)$ and $(2, \infty)$. Since the inequality symbol is $\leq$, the endpoints of the intervals must be included in the solution set. It is $(-\infty, -2] \cup [2, \infty)$.

25.
$$6x - 9 - x^2 < 0 \quad \text{Polynomial inequality}$$
$$6x - 9 - x^2 = 0 \quad \text{Related equation}$$
$$-(x^2 - 6x + 9) = 0 \quad \text{Factoring out } -1 \text{ and rearranging}$$
$$-(x - 3)(x - 3) = 0 \quad \text{Factoring}$$

Using the principle of zero products, we find that the solution of the related equation is 3. This number divides the x-axis into the intervals $(-\infty, 3)$ and $(3, \infty)$. We let $f(x) = 6x - 9 - x^2$ and test a value in each interval.

$(-\infty, 3)$: $f(-4) = -49 < 0$

$(3, \infty)$: $f(4) = -1 < 0$

Function values are negative on both intervals. The solution set is $(-\infty, 3) \cup (3, \infty)$.

27.
$$x^2 + 12 < 4x \quad \text{Polynomial inequality}$$
$$x^2 - 4x + 12 < 0 \quad \text{Equivalent inequality with } 0 \text{ on one side}$$
$$x^2 - 4x + 12 = 0 \quad \text{Related equation}$$

Using the quadratic formula, we find that the related equation has no real-number solutions. The graph lies entirely above the x-axis, so the inequality has no solution. We could determine this algebraically by letting $f(x) = x^2 - 4x + 12$ and testing any real number (since there are no real-number solutions of $f(x) = 0$ to divide the x-axis into intervals). For example, $f(0) = 12 > 0$, so we see algebraically that the inequality has no solution. The solution set is $\emptyset$.

29.
$$4x^3 - 7x^2 \leq 15x \quad \text{Polynomial inequality}$$
$$4x^3 - 7x^2 - 15x \leq 0 \quad \text{Equivalent inequality with } 0 \text{ on one side}$$
$$4x^3 - 7x^2 - 15x = 0 \quad \text{Related equation}$$
$$x(4x + 5)(x - 3) = 0 \quad \text{Factoring}$$

Using the principle of zero products, we find that the solutions of the related equation are 0, $-\dfrac{5}{4}$, and 3. These numbers divide the x-axis into the intervals $\left(-\infty, -\dfrac{5}{4}\right)$, $\left(-\dfrac{5}{4}, 0\right)$, $(0, 3)$, and $(3, \infty)$. We let $f(x) = 4x^3 - 7x^2 - 15x$ and test a value in each interval.

$\left(-\infty, -\dfrac{5}{4}\right)$: $f(-2) = -30 < 0$

$\left(-\dfrac{5}{4}, 0\right)$: $f(-1) = 4 > 0$

$(0, 3)$ $f(1) = -18 < 0$

$(3, \infty)$: $f(4) = 84 > 0$

Function values are negative on $\left(-\infty, -\dfrac{5}{4}\right)$ and $(0, 3)$. Since the inequality symbol is $\leq$, the endpoints of the intervals must be included in the solution set. It is $\left(-\infty, -\dfrac{5}{4}\right] \cup [0, 3]$.

31.
$$x^3 + 3x^2 - x - 3 \geq 0 \quad \text{Polynomial inequality}$$
$$x^3 + 3x^2 - x - 3 = 0 \quad \text{Related equation}$$
$$x^2(x + 3) - (x + 3) = 0 \quad \text{Factoring}$$
$$(x^2 - 1)(x + 3) = 0$$
$$(x + 1)(x - 1)(x + 3) = 0$$

Using the principle of zero products, we find that the solutions of the related equation are -1, 1, and -3. These numbers divide the x-axis into the intervals $(-\infty, -3)$, $(-3, -1)$, $(-1, 1)$, and $(1, \infty)$. We let $f(x) = x^3 + 3x^2 - x - 3$ and test a value in each interval.

$(-\infty, -3)$: $f(-4) = -15 < 0$

$(-3, -1)$: $f(-2) = 3 > 0$

$(-1, 1)$: $f(0) = -3 < 0$

$(1, \infty)$: $f(2) = 15 > 0$

Function values are positive on $(-3, -1)$ and $(1, \infty)$. Since the inequality symbol is $\geq$, the endpoints of the intervals must be included in the solution set. It is $[-3, -1] \cup [1, \infty)$.

33.
$$x^3 - 2x^2 < 5x - 6 \quad \text{Polynomial inequality}$$

$$x^3 - 2x^2 - 5x + 6 < 0 \quad \begin{array}{l}\text{Equivalent inequality with 0 on one side}\end{array}$$

$$x^3 - 2x^2 - 5x + 6 = 0 \quad \text{Related equation}$$

Using the techniques of Section 3.3, we find that the solutions of the related equation are -2, 1, and 3. They divide the x-axis into the intervals $(-\infty, -2)$, $(-2, 1)$, $(1, 3)$, and $(3, \infty)$. Let $f(x) = x^3 - 2x^2 - 5x + 6$ and test a value in each interval.

$(-\infty, -2)$: $f(-3) = -24 < 0$

$(-2, 1)$: $f(0) = 6 > 0$

$(1, 3)$: $f(2) = -4 < 0$

$(3, \infty)$: $f(4) = 18 > 0$

Function values are negative on $(-\infty, -2)$ and $(1, 3)$. The solution set is $(-\infty, -2) \cup (1, 3)$.

35.
$$x^5 + x^2 \geq 2x^3 + 2 \quad \text{Polynomial inequality}$$

$$x^5 - 2x^3 + x^2 - 2 \geq 0 \quad \begin{array}{l}\text{Related inequality with 0 on one side}\end{array}$$

$$x^5 - 2x^3 + x^2 - 2 = 0 \quad \begin{array}{l}\text{Related equation}\end{array}$$

$$x^3(x^2 - 2) + x^2 - 2 = 0 \quad \text{Factoring}$$

$$(x^3 + 1)(x^2 - 2) = 0$$

Using the principle of zero products, we find that the real-number solutions of the related equation are -1, $-\sqrt{2}$, and $\sqrt{2}$. These numbers divide the x-axis into the intervals $(-\infty, -\sqrt{2})$, $(-\sqrt{2}, -1)$, $(-1, \sqrt{2})$, and $(\sqrt{2}, \infty)$. We let $f(x) = x^5 - 2x^3 + x^2 - 2$ and test a value in each interval.

$(-\infty, -\sqrt{2})$: $f(-2) = -14 < 0$

$(-\sqrt{2}, -1)$: $f(-1.3) \approx 0.37107 > 0$

$(-1, \sqrt{2})$: $f(0) = -2 < 0$

$(\sqrt{2}, \infty)$: $f(2) = 18 > 0$

Function values are positive on $(-\sqrt{2}, -1)$ and $(\sqrt{2}, \infty)$. Since the inequality symbol is $\geq$, the endpoints of the intervals must be included in the solution set. It is $[-\sqrt{2}, -1] \cup [\sqrt{2}, \infty)$.

37.
$$2x^3 + 6 \leq 5x^2 + x \quad \text{Polynomial inequality}$$

$$2x^3 - 5x^2 - x + 6 \leq 0 \quad \begin{array}{l}\text{Equivalent inequality with 0 on one side}\end{array}$$

$$2x^3 - 5x^2 - x + 6 = 0 \quad \text{Related equation}$$

Using the techniques of Section 3.3, we find that the solutions of the related equation are -1, $\dfrac{3}{2}$, and 2. We can also use the graph of $y = 2x^3 - 5x^2 - x + 6$ to find these solutions. They divide the x-axis into the intervals $(-\infty, -1)$, $\left(-1, \dfrac{3}{2}\right)$, $\left(\dfrac{3}{2}, 2\right)$, and $(2, \infty)$. Let $f(x) = 2x^3 - 5x^2 - x + 6$ and test a value in each interval.

$(-\infty, -1)$: $f(-2) = -28 < 0$

$\left(-1, \dfrac{3}{2}\right)$: $f(0) = 6 > 0$

$\left(\dfrac{3}{2}, 2\right)$: $f(1.6) = -0.208 < 0$

$(2, \infty)$: $f(3) = 12 > 0$

Function values are negative in $(-\infty, -1)$ and $\left(\dfrac{3}{2}, 2\right)$. Since the inequality symbol is $\leq$, the endpoints of the intervals must be included in the solution set. The solution set is $\left(-\infty, -1\right] \cup \left[\dfrac{3}{2}, 2\right]$.

39.
$$x^3 + 5x^2 - 25x \leq 125 \quad \text{Polynomial inequality}$$

$$x^3 + 5x^2 - 25x - 125 \leq 0 \quad \begin{array}{l}\text{Equivalent inequality with 0 on one side}\end{array}$$

$$x^3 + 5x^2 - 25x - 125 = 0 \quad \text{Related equation}$$

$$x^2(x + 5) - 25(x + 5) = 0 \quad \text{Factoring}$$

$$(x^2 - 25)(x + 5) = 0$$

$$(x + 5)(x - 5)(x + 5) = 0$$

Using the principle of zero products, we find that the solutions of the related equation are -5 and 5. These numbers divide the x-axis into the intervals $(-\infty, -5)$, $(-5, 5)$, and $(5, \infty)$. We let $f(x) = x^3 + 5x^2 - 25x - 125$ and test a value in each interval.

$(-\infty, -5)$: $f(-6) = -11 < 0$

$(-5, 5)$: $f(0) = -125 < 0$

$(5, \infty)$: $f(6) = 121 > 0$

Function values are negative on $(-\infty, -5)$ and $(-5, 5)$. Since the inequality symbol is $\leq$, the endpoints of the intervals must be included in the solution set. It is $(-\infty, -5] \cup [-5, 5]$ or $(-\infty, 5]$.

41.
$$\dfrac{1}{x + 4} > 0 \quad \text{Rational inequality}$$

$$\dfrac{1}{x + 4} = 0 \quad \text{Related equation}$$

The denominator of $f(x) = \dfrac{1}{x + 4}$ is 0 when $x = -4$, so the function is not defined for $x = -4$. The related equation has no solution. Thus, the only critical value is -4. It divides the x-axis into the intervals $(-\infty, -4)$ and $(-4, \infty)$. We test a value in each interval.

$(-\infty, -4)$: $f(-5) = -1 < 0$

$(-4, \infty)$: $f(0) = \dfrac{1}{4} > 0$

Function values are positive on $(-4, \infty)$. This can also be determined from the graph of $y = \dfrac{1}{x + 4}$. The solution set is $(-4, \infty)$.

43. $\dfrac{-4}{2x+5} < 0$ Rational inequality

$\dfrac{-4}{2x+5} = 0$ Related equation

The denominator of $f(x) = \dfrac{-4}{2x+5}$ is 0 when $x = -\dfrac{5}{2}$, so the function is not defined for $x = -\dfrac{5}{2}$. The related equation has no solution. Thus, the only critical value is $-\dfrac{5}{2}$. It divides the x-axis into the intervals $\left(-\infty, -\dfrac{5}{2}\right)$ and $\left(-\dfrac{5}{2}, \infty\right)$. We test a value in each interval.

$\left(-\infty, -\dfrac{5}{2}\right)$: $f(-3) = 4 > 0$

$\left(-\dfrac{5}{2}, \infty\right)$: $f(0) = -\dfrac{4}{5} < 0$

Function values are negative on $\left(-\dfrac{5}{2}, \infty\right)$. The solution set is $\left(-\dfrac{5}{2}, \infty\right)$.

45. $\dfrac{x-4}{x+3} - \dfrac{x+2}{x-1} \le 0$

The denominator of $f(x) = \dfrac{x-4}{x+3} - \dfrac{x+2}{x-1}$ is 0 when $x = -3$ or $x = 1$, so the function is not defined for these values of x. We solve the related equation $f(x) = 0$.

$$\dfrac{x-4}{x+3} - \dfrac{x+2}{x-1} = 0$$

$$(x+3)(x-1)\left(\dfrac{x-4}{x+3} - \dfrac{x+2}{x-1}\right) = (x+3)(x-1)\cdot 0$$

$$(x-1)(x-4) - (x+3)(x+2) = 0$$

$$x^2 - 5x + 4 - (x^2 + 5x + 6) = 0$$

$$-10x - 2 = 0$$

$$-10x = 2$$

$$x = -\dfrac{1}{5}$$

The critical values are -3, $-\dfrac{1}{5}$, and 1. They divide the x-axis into the intervals $(-\infty, -3)$, $\left(3, -\dfrac{1}{5}\right)$, $\left(-\dfrac{1}{5}, 1\right)$, and $(1, \infty)$. We test a value in each interval.

$(-\infty, -3)$: $f(-4) = 7.6 > 0$

$\left(-3, -\dfrac{1}{5}\right)$: $f(-1) = -2 < 0$

$\left(-\dfrac{1}{5}, 1\right)$: $f(0) = \dfrac{2}{3} > 0$

$(1, \infty)$: $f(2) = -4.4 < 0$

Function values are negative on $\left(-3, -\dfrac{1}{5}\right)$ and $(1, \infty)$. Note that since the inequality symbol is $\le$ and $f\left(-\dfrac{1}{5}\right) = 0$, then $-\dfrac{1}{5}$ must be included in the solution set. Note also that since neither -3 nor 1 is in the domain

of $f(x)$, they are not included in the solution set. It is $\left(-3, -\dfrac{1}{5}\right] \cup (1, \infty)$.

47. $\dfrac{2x-1}{x+3} \ge \dfrac{x+1}{3x+1}$ Rational inequality

$\dfrac{2x-1}{x+3} - \dfrac{x+1}{3x+1} \ge 0$ Equivalent inequality with 0 on one side

The denominator of $f(x) = \dfrac{2x-1}{x+3} - \dfrac{x+1}{3x+1}$ is 0 when $x = -3$ or $x = -\dfrac{1}{3}$, so the function is not defined for these values of x. We solve the related equation $f(x) = 0$.

$$\dfrac{2x-1}{x+3} - \dfrac{x+1}{3x+1} = 0$$

$$(x+3)(3x+1)\left(\dfrac{2x-1}{x+3} - \dfrac{x+1}{3x+1}\right) = (x+3)(3x+1)\cdot 0$$

$$(3x+1)(2x-1) - (x+3)(x+1) = 0$$

$$6x^2 - x - 1 - (x^2 + 4x + 3) = 0$$

$$5x^2 - 5x - 4 = 0$$

Using the quadratic formula we find that $x = \dfrac{5 \pm \sqrt{105}}{10}$. Then the critical values are -3, $\dfrac{5 - \sqrt{105}}{10}$, $-\dfrac{1}{3}$, and $\dfrac{5 + \sqrt{105}}{10}$. They divide the x-axis into the intervals $(-\infty, -3)$, $\left(-3, \dfrac{5 - \sqrt{105}}{10}\right)$, $\left(\dfrac{5 - \sqrt{105}}{10}, -\dfrac{1}{3}\right)$, $\left(-\dfrac{1}{3}, \dfrac{5 + \sqrt{105}}{10}\right)$, and $\left(\dfrac{5 + \sqrt{105}}{10}, \infty\right)$. We test a value in each interval.

$(-\infty, -3)$: $f(-4) \approx 8.7273 > 0$

$\left(-3, \dfrac{5 - \sqrt{105}}{10}\right)$: $f(-2) = -5.2 < 0$

$\left(\dfrac{5 - \sqrt{105}}{10}, -\dfrac{1}{3}\right)$: $f(-0.4) \approx 2.3077 > 0$

$\left(-\dfrac{1}{3}, \dfrac{5 + \sqrt{105}}{10}\right)$: $f(0) \approx -1.333 < 0$

$\left(\dfrac{5 + \sqrt{105}}{10}, \infty\right)$: $f(2) \approx 0.1714 > 0$

Function values are positive on $(-\infty, -3)$, $\left(\dfrac{5 - \sqrt{105}}{10}, -\dfrac{1}{3}\right)$ and $\left(\dfrac{5 + \sqrt{105}}{10}, \infty\right)$. Note that since the inequality symbol is $\ge$ and $f\left(\dfrac{5 \pm \sqrt{105}}{10}\right) = 0$, then $\dfrac{5 \pm \sqrt{105}}{10}$ must be included in the solution set. Note also that since neither -3 nor $-\dfrac{1}{3}$ is in the domain of $f(x)$, they are not included in the solution set. It is

$(-\infty, -3) \cup \left[\dfrac{5 - \sqrt{105}}{10}, -\dfrac{1}{3}\right) \cup \left[\dfrac{5 + \sqrt{105}}{10}, \infty\right)$.

49. $\dfrac{x+1}{x-2} \geq 3$ Rational inequality

$\dfrac{x+1}{x-2} - 3 \geq 0$ Equivalent inequality with 0 on one side

The denominator of $f(x) = \dfrac{x+1}{x-2} - 3$ is 0 when $x = 2$, so the function is not defined for this value of x. We solve the related equation $f(x) = 0$.

$$\frac{x+1}{x-2} - 3 = 0$$

$$(x-2)\left(\frac{x+1}{x-2} - 3\right) = (x-2)\cdot 0$$

$$x + 1 - 3(x-2) = 0$$

$$x + 1 - 3x + 6 = 0$$

$$-2x + 7 = 0$$

$$-2x = -7$$

$$x = \frac{7}{2}$$

The critical values are 2 and $\dfrac{7}{2}$. They divide the x-axis into the intervals $(-\infty, 2)$, $\left(2, \dfrac{7}{2}\right)$, and $\left(\dfrac{7}{2}, \infty\right)$. We test a value in each interval.

$(-\infty, 2)$: $f(0) = -3.5 < 0$

$\left(2, \dfrac{7}{2}\right)$: $f(3) = 1 > 0$

$\left(\dfrac{7}{2}, \infty\right)$: $f(4) = -0.5 < 0$

Function values are positive on $\left(2, \dfrac{7}{2}\right)$. Note that since the inequality symbol is $\geq$ and $f\left(\dfrac{7}{2}\right) = 0$, then $\dfrac{7}{2}$ must be included in the solution set. Note also that since 2 is not in the domain of $f(x)$, it is not included in the solution set. It is $\left(2, \dfrac{7}{2}\right]$.

51. $x - 2 > \dfrac{1}{x}$ Rational inequality

$x - 2 - \dfrac{1}{x} > 0$ Equivalent inequality with 0 on one side

The denominator of $f(x) = x - 2 - \dfrac{1}{x}$ is 0 when $x = 0$, so the function is not defined for this value of x. We solve the related equation $f(x) = 0$.

$$x - 2 - \frac{1}{x} = 0$$

$$x\left(x - 2 - \frac{1}{x}\right) = x \cdot 0$$

$$x^2 - 2x - x \cdot \frac{1}{x} = 0$$

$$x^2 - 2x - 1 = 0$$

Using the quadratic formula we find that $x = 1 \pm \sqrt{2}$. The critical values are $1 - \sqrt{2}$, 0, and $1 + \sqrt{2}$. They divide the x-axis into the intervals $(-\infty, 1 - \sqrt{2})$, $(1 - \sqrt{2}, 0)$,

$(0, 1 + \sqrt{2})$, and $(1 + \sqrt{2}, \infty)$. We test a value in each interval.

$(-\infty, 1 - \sqrt{2})$: $f(-1) = -2 < 0$

$(1 - \sqrt{2}, 0)$: $f(-0.1) = 7.9 > 0$

$(0, 1 + \sqrt{2})$: $f(1) = -2 < 0$

$(1 + \sqrt{2}, \infty)$: $f(3) = \dfrac{2}{3} > 0$

Function values are positive on $(1 - \sqrt{2}, 0)$ and $(1 + \sqrt{2}, \infty)$. The solution set is $(1 - \sqrt{2}, 0) \cup (1 + \sqrt{2}, \infty)$.

53. $\dfrac{2}{x^2 - 4x + 3} \leq \dfrac{5}{x^2 - 9}$

$\dfrac{2}{x^2 - 4x + 3} - \dfrac{5}{x^2 - 9} \leq 0$

$\dfrac{2}{(x-1)(x-3)} - \dfrac{5}{(x+3)(x-3)} \leq 0$

The denominator of $f(x) = \dfrac{2}{(x-1)(x-3)} - \dfrac{5}{(x+3)(x-3)}$ is 0 when $x = 1$, 3, or -3, so the function is not defined for these values of x. We solve the related equation $f(x) = 0$.

$$\frac{2}{(x-1)(x-3)} - \frac{5}{(x+3)(x-3)} = 0$$

$$(x-1)(x-3)(x+3)\left(\frac{2}{(x-1)(x-3)} - \frac{5}{(x+3)(x-3)}\right)$$
$$= (x-1)(x-3)(x+3)\cdot 0$$

$$2(x+3) - 5(x-1) = 0$$

$$2x + 6 - 5x + 5 = 0$$

$$-3x + 11 = 0$$

$$-3x = -11$$

$$x = \frac{11}{3}$$

The critical values are -3, 1, 3, and $\dfrac{11}{3}$. They divide the x-axis into the intervals $(-\infty, -3)$, $(-3, 1)$, $(1, 3)$, $\left(3, \dfrac{11}{3}\right)$, and $\left(\dfrac{11}{3}, \infty\right)$. We test a value in each interval.

$(-\infty, -3)$: $f(-4) \approx -0.6571 < 0$

$(-3, 1)$: $f(0) \approx 1.2222 > 0$

$(1, 3)$: $f(2) = -1 < 0$

$\left(3, \dfrac{11}{3}\right)$: $f(3.5) \approx 0.6154 > 0$

$\left(\dfrac{11}{3}, \infty\right)$: $f(4) \approx -0.0476 < 0$

Function values are negative on $(-\infty, -3)$, $(1, 3)$, and $\left(\dfrac{11}{3}, \infty\right)$. Note that since the inequality symbol is $\leq$ and $f\left(\dfrac{11}{3}\right) = 0$, then $\dfrac{11}{3}$ must be included in the solution set. Note also that since -3, 1, and 3 are not in the domain of $f(x)$, they are not included in the solution set. It is $(-\infty, -3) \cup (1, 3) \cup \left[\dfrac{11}{3}, \infty\right)$.

55.
$$\frac{3}{x^2+1} \ge \frac{6}{5x^2+2}$$

$$\frac{3}{x^2+1} - \frac{6}{5x^2+2} \ge 0$$

The denominator of $f(x) = \frac{3}{x^2+1} - \frac{6}{5x^2+2}$ has no real-number zeros. We solve the related equation $f(x) = 0$.

$$\frac{3}{x^2+1} - \frac{6}{5x^2+2} = 0$$

$$(x^2+1)(5x^2+2)\left(\frac{3}{x^2+1} - \frac{6}{5x^2+2}\right) =$$
$$(x^2+1)(5x^2+2)\cdot 0$$

$$3(5x^2+2) - 6(x^2+1) = 0$$
$$15x^2 + 6 - 6x^2 - 6 = 0$$
$$9x^2 = 0$$
$$x^2 = 0$$
$$x = 0$$

The only critical value is 0. It divides the x-axis into the intervals $(-\infty, 0)$ and $(0, \infty)$. We test a value in each interval.

$(-\infty, 0)$: $f(-1) \approx 0.64286 > 0$

$(0, \infty)$: $f(1) \approx 0.64286 > 0$

Function values are positive on both intervals. Note that since the inequality symbol is $\ge$ and $f(0) = 0$, then 0 must be included in the solution set. It is $(-\infty, 0] \cup [0, \infty)$, or $(-\infty, \infty)$.

57.
$$\frac{5}{x^2+3x} < \frac{3}{2x+1}$$

$$\frac{5}{x^2+3x} - \frac{3}{2x+1} < 0$$

$$\frac{5}{x(x+3)} - \frac{3}{2x+1} < 0$$

The denominator of $f(x) = \frac{5}{x(x+3)} - \frac{3}{2x+1}$ is 0 when $x = 0, -3$, or $-\frac{1}{2}$, so the function is not defined for these values of x. We solve the related equation $f(x) = 0$.

$$\frac{5}{x(x+3)} - \frac{3}{2x+1} = 0$$

$$x(x+3)(2x+1)\left(\frac{5}{x(x+3)} - \frac{3}{2x+1}\right) =$$
$$x(x+3)(2x+1)\cdot 0$$

$$5(2x+1) - 3x(x+3) = 0$$
$$10x + 5 - 3x^2 - 9x = 0$$
$$-3x^2 + x + 5 = 0$$

Using the quadratic formula we find that $x = \frac{1 \pm \sqrt{61}}{6}$. The critical values are -3, $\frac{1-\sqrt{61}}{6}$, $-\frac{1}{2}$, 0, and $\frac{1+\sqrt{61}}{6}$. They divide the x-axis into the intervals $(-\infty, -3)$, $\left(-3, \frac{1-\sqrt{61}}{6}\right)$,

$\left(\frac{1-\sqrt{61}}{6}, -\frac{1}{2}\right)$, $\left(-\frac{1}{2}, 0\right)$, $\left(0, \frac{1+\sqrt{61}}{6}\right)$, and $\left(\frac{1+\sqrt{61}}{6}, \infty\right)$.

We test a value in each interval.

$(-\infty, -3)$: $f(-4) \approx 1.6786 > 0$

$\left(-3, \frac{1-\sqrt{61}}{6}\right)$: $f(-2) = -1.5 < 0$

$\left(\frac{1-\sqrt{61}}{6}, -\frac{1}{2}\right)$: $f(-1) = 0.5 > 0$

$\left(-\frac{1}{2}, 0\right)$: $f(-0.1) \approx -20.99 < 0$

$\left(0, \frac{1+\sqrt{61}}{6}\right)$: $f(1) = 0.25 > 0$

$\left(\frac{1+\sqrt{61}}{6}, \infty\right)$: $f(2) = -0.1 < 0$

Function values are negative on $\left(-3, \frac{1-\sqrt{61}}{6}\right)$, $\left(-\frac{1}{2}, 0\right)$ and $\left(\frac{1+\sqrt{61}}{6}, \infty\right)$. The solution set is $\left(-3, \frac{1-\sqrt{61}}{6}\right) \cup \left(-\frac{1}{2}, 0\right) \cup \left(\frac{1+\sqrt{61}}{6}, \infty\right)$.

59.
$$\frac{5x}{7x-2} > \frac{x}{x+1}$$

$$\frac{5x}{7x-2} - \frac{x}{x+1} > 0$$

The denominator of $f(x) = \frac{5x}{7x-2} - \frac{x}{x+1}$ is 0 when $x = \frac{2}{7}$ or $x = -1$, so the function is not defined for these values of x. We solve the related equation $f(x) = 0$.

$$\frac{5x}{7x-2} - \frac{x}{x+1} = 0$$

$$(7x-2)(x+1)\left(\frac{5x}{7x-2} - \frac{x}{x+1}\right) = (7x-2)(x+1)\cdot 0$$

$$5x(x+1) - x(7x-2) = 0$$
$$5x^2 + 5x - 7x^2 + 2x = 0$$
$$-2x^2 + 7x = 0$$
$$-x(2x-7) = 0$$

$$x = 0 \quad or \quad x = \frac{7}{2}$$

The critical values are $-1, 0, \frac{2}{7}$, and $\frac{7}{2}$. They divide the x-axis into the intervals $(-\infty, -1)$, $(-1, 0)$, $\left(0, \frac{2}{7}\right)$, $\left(\frac{2}{7}, \frac{7}{2}\right)$, and $\left(\frac{7}{2}, \infty\right)$. We test a value in each interval.

$(-\infty, -1)$: $f(-2) = -1.375 < 0$

$(-1, 0)$: $f(-0.5) \approx 1.4545 > 0$

$\left(0, \frac{2}{7}\right)$: $f(0.1) \approx -0.4755 < 0$

$\left(\frac{2}{7}, \frac{7}{2}\right)$: $f(1) = 0.5 > 0$

$\left(\dfrac{7}{2}, \infty\right)$: $f(4) \approx -0.0308 < 0$

Function values are positive on $(-1, 0)$ and $\left(\dfrac{2}{7}, \dfrac{7}{2}\right)$. The solution set is $(-1, 0) \cup \left(\dfrac{2}{7}, \dfrac{7}{2}\right)$.

61.

$$\dfrac{\dfrac{x}{x^2+4x-5} + \dfrac{3}{x^2-25}}{\dfrac{2x}{x^2-6x+5}} \leq$$

$$\dfrac{x}{x^2+4x-5} + \dfrac{3}{x^2-25} - \dfrac{2x}{x^2-6x+5} \leq 0$$

$$\dfrac{x}{(x+5)(x-1)} + \dfrac{3}{(x+5)(x-5)} - \dfrac{2x}{(x-5)(x-1)} \leq 0$$

The denominator of

$$f(x) = \dfrac{x}{(x+5)(x-1)} + \dfrac{3}{(x+5)(x-5)} - \dfrac{2x}{(x-5)(x-1)}$$

is 0 when $x = -5$, 1, or 5, so the function is not defined for these values of x. We solve the related equation $f(x) = 0$.

$$\dfrac{x}{(x+5)(x-1)} + \dfrac{3}{(x+5)(x-5)} - \dfrac{2x}{(x-5)(x-1)} = 0$$

$$x(x-5) + 3(x-1) - 2x(x+5) = 0$$

Multiplying by $(x+5)(x-1)(x-5)$

$$x^2 - 5x + 3x - 3 - 2x^2 - 10x = 0$$

$$-x^2 - 12x - 3 = 0$$

$$x^2 + 12x + 3 = 0$$

Using the quadratic formula, we find that $x = -6 \pm \sqrt{33}$. The critical values are $-6 - \sqrt{33}$, -5, $-6 + \sqrt{33}$, 1, and 5. They divide the x-axis into the intervals $(-\infty, -6 - \sqrt{33})$, $(-6 - \sqrt{33}, -5)$, $(-5, -6 + \sqrt{33})$, $(-6 + \sqrt{33}, 1)$, $(1, 5)$, and $(5, \infty)$. We test a value in each interval.

$(-\infty, -6 - \sqrt{33})$: $f(-12) \approx 0.00194 > 0$

$(-6 - \sqrt{33}, -5)$: $f(-6) \approx -0.4286 < 0$

$(-5, -6 + \sqrt{33})$: $f(-1) \approx 0.16667 > 0$

$(-6 + \sqrt{33}, 1)$: $f(0) = -0.12 < 0$

$(1, 5)$: $f(2) \approx 1.4762 > 0$

$(5, \infty)$: $f(6) \approx -2.018 < 0$

Function values are negative on $(-6 - \sqrt{33}, -5)$, $(-6 + \sqrt{33}, 1)$, and $(5, \infty)$. Note that since the inequality symbol is $\leq$ and $f(-6 \pm \sqrt{33}) = 0$, then $-6 - \sqrt{33}$ and $-6 + \sqrt{33}$ must be included in the solution set. Note also that since -5, 1, and 5 are not in the domain of $f(x)$, they are not included in the solution set. It is $[-6 - \sqrt{33}, -5) \cup [-6 + \sqrt{33}, 1) \cup (5, \infty)$.

63. We write and solve a rational inequality.

$$\dfrac{4t}{t^2+1} + 98.6 > 100$$

$$\dfrac{4t}{t^2+1} - 1.4 > 0$$

The denominator of $f(t) = \dfrac{4t}{t^2+1} - 1.4$ has no real-number zeros. We solve the related equation $f(t) = 0$.

$$\dfrac{4t}{t^2+1} - 1.4 = 0$$

$$4t - 1.4(t^2+1) = 0 \quad \text{Multiplying by } t^2+1$$

$$4t - 1.4t^2 - 1.4 = 0$$

Using the quadratic formula, we find that $t = \dfrac{4 \pm \sqrt{8.16}}{2.8}$; that is, $t \approx 0.408$ or $t \approx 2.449$. These numbers divide the t-axis into the intervals $(-\infty, 0.408)$, $(0.408, 2.449)$, and $(2.449, \infty)$. We test a value in each interval.

$(-\infty, 0.408)$: $f(0) = -1.4 < 0$

$(0.408, 2.449)$: $f(1) = 0.6 > 0$

$(2.449, \infty)$: $f(3) = -0.2 < 0$

Function values are positive on $(0.408, 2.449)$. The solution set is $(0.408, 2.449)$.

65. a) We write and solve a polynomial inequality.

$$-3x^2 + 630x - 6000 > 0 \quad (x \geq 0)$$

We first solve the related equation.

$$-3x^2 + 630x - 6000 = 0$$

$$x^2 - 210x + 2000 = 0 \quad \text{Dividing by } -3$$

$$(x - 10)(x - 200) = 0 \quad \text{Factoring}$$

Using the principle of zero products or by observing the graph of $y = -3x^2 + 630 - 6000$, we see that the solutions of the related equation are 10 and 200. These numbers divide the x-axis into the intervals $(-\infty, 10)$, $(10, 200)$, and $(200, \infty)$. Since we are restricting our discussion to nonnegative values of x, we consider the intervals $[0, 10)$, $(10, 200)$, and $(200, \infty)$.

We let $f(x) = -3x^2 + 630x - 6000$ and test a value in each interval.

$[0, 10)$: $f(0) = -6000 < 0$

$(10, 200)$: $f(11) = 567 > 0$

$(200, \infty)$: $f(201) = -573 < 0$

Function values are positive only on $(10, 200)$. The solution set is $\{x | 10 < x < 200\}$, or $(10, 200)$.

b) From part (a), we see that function values are negative on $[0, 10)$ and $(200, \infty)$. Thus, the solution set is $\{x | 0 < x < 10 \text{ or } x > 200\}$, or $(0, 10) \cup (200, \infty)$.

67. We write an inequality.

$$27 \leq \dfrac{n(n-3)}{2} \leq 230$$

$$54 \leq n(n-3) \leq 460 \quad \text{Multiplying by 2}$$

$$54 \leq n^2 - 3n \leq 460$$

We write this as two inequalities.

$$54 \leq n^2 - 3n \quad and \quad n^2 - 3n \leq 460$$

Solve each inequality.

$$n^2 - 3n \geq 54$$

$$n^2 - 3n - 54 \geq 0$$

$$n^2 - 3n - 54 = 0 \quad \text{Related equation}$$

$$(n + 6)(n - 9) = 0$$

$n = -6 \ \text{ or } \ n = 9$

Since only positive values of n have meaning in this application, we consider the intervals $(0,9)$ and $(9,\infty)$. Let $f(n) = n^2 - 3n - 54$ and test a value in each interval.

$(0,9)$: $f(1) = -56 < 0$

$(9,\infty)$: $f(10) = 16 > 0$

Function values are positive on $(9, \infty)$. Since the inequality symbol is $\geq$, 9 must also be included in the solution set for this portion of the inequality. It is $\{n|n \geq 9\}$.

Now solve the second inequality.

$$n^2 - 3n \leq 460$$
$$n^2 - 3n - 460 \leq 0$$
$$n^2 - 3n - 460 = 0 \quad \text{Related equation}$$
$$(n + 20)(n - 23) = 0$$
$$n = -20 \ \text{ or } \ n = 23$$

We consider only positive values of n as above. Thus, we consider the intervals $(0, 23)$ and $(23, \infty)$. Let $f(n) = n^2 - 3n - 460$ and test a value in each interval.

$(0, 23)$: $f(1) = -462 < 0$

$(23, \infty)$: $f(24) = 44 > 0$

Function values are negative on $(0, 23)$. Since the inequality symbol is $\leq$, 23 must also be included in the solution set for this portion of the inequality. It is $\{n|0 < n \leq 23\}$.

The solution set of the original inequality is $\{n|n \geq 9 \ \text{ and } \ 0 < n \leq 23\}$, or $\{n|9 \leq n \leq 23\}$.

69. Left to the student

71. Discussion and Writing

73.
$$(x - h)^2 + (y - k)^2 = r^2$$
$$[x - (-2)]^2 + (y - 4)^2 = 3^2$$
$$(x + 2)^2 + (y - 4)^2 = 9$$

75. $h(x) = -2x^2 + 3x - 8$

a) $-\dfrac{b}{2a} = -\dfrac{3}{2(-2)} = \dfrac{3}{4}$

$h\left(\dfrac{3}{4}\right) = -2\left(\dfrac{3}{4}\right)^2 + 3 \cdot \dfrac{3}{4} - 8 = -\dfrac{55}{8}$

The vertex is $\left(\dfrac{3}{4}, -\dfrac{55}{8}\right)$.

b) The coefficient of x^2 is negative, so there is a maximum value. It is the second coordinate of the vertex, $-\dfrac{55}{8}$. It occurs at $x = \dfrac{3}{4}$.

c) The range is $\left(-\infty, -\dfrac{55}{8}\right]$.

77.
$$x^4 + 3x^2 > 4x - 15$$
$$x^4 + 3x^2 - 4x + 15 > 0$$

The graph of $y = x^4 + 3x^2 - 4x + 15$ lies entirely above the x-axis. Thus, the solution set is the set of all real numbers, or $(-\infty, \infty)$.

79. $|x^2 - 5| = |5 - x^2| = 5 - x^2$ when $5 - x^2 \geq 0$. Thus we solve $5 - x^2 \geq 0$.

$$5 - x^2 \geq 0$$
$$5 - x^2 = 0 \quad \text{Related equation}$$
$$5 = x^2$$
$$\pm\sqrt{5} = x$$

Let $f(x) = 5 - x^2$ and test a value in each of the intervals determined by the solutions of the related equation.

$(-\infty, -\sqrt{5})$: $f(-3) = -4 < 0$

$(-\sqrt{5}, \sqrt{5})$: $f(0) = 5 > 0$

$(\sqrt{5}, \infty)$: $f(3) = -4 < 0$

Function values are positive on $(-\sqrt{5}, \sqrt{5})$. Since the inequality symbol is $\geq$, the endpoints of the interval must be included in the solution set. It is $\left[-\sqrt{5}, \sqrt{5}\right]$.

81.
$$2|x|^2 - |x| + 2 \leq 5$$
$$2|x|^2 - |x| - 3 \leq 0$$
$$2|x|^2 - |x| - 3 = 0 \quad \text{Related equation}$$
$$(2|x| - 3)(|x| + 1) = 0 \quad \text{Factoring}$$
$$2|x| - 3 = 0 \ \text{ or } \ |x| + 1 = 0$$
$$|x| = \dfrac{3}{2} \ \text{ or } \quad |x| = -1$$

The solution of the first equation is $x = -\dfrac{3}{2}$ or $x = \dfrac{3}{2}$. The second equation has no solution. Let $f(x) = 2|x|^2 - |x| - 3$ and test a value in each interval determined by the solutions of the related equation.

$\left(-\infty, -\dfrac{3}{2}\right)$: $f(-2) = 3 > 0$

$\left(-\dfrac{3}{2}, \dfrac{3}{2}\right)$: $f(0) = -3 < 0$

$\left(\dfrac{3}{2}, \infty\right)$: $f(2) = 3 > 0$

Function values are negative on $\left(-\dfrac{3}{2}, \dfrac{3}{2}\right)$. Since the inequality symbol is $\leq$, the endpoints of the interval must also be included in the solution set. It is $\left[-\dfrac{3}{2}, \dfrac{3}{2}\right]$.

83.
$$\left|1 + \dfrac{1}{x}\right| < 3$$
$$-3 < 1 + \dfrac{1}{x} < 3$$
$$-3 < 1 + \dfrac{1}{x} \ \text{ and } \ 1 + \dfrac{1}{x} < 3$$

First solve $-3 < 1 + \dfrac{1}{x}$.

$$0 < 4 + \dfrac{1}{x}, \ \text{ or } \ \dfrac{1}{x} + 4 > 0$$

The denominator of $f(x) = \dfrac{1}{x} + 4$ is 0 when $x = 0$, so the function is not defined for this value of x. Now solve the related equation.

$\dfrac{1}{x} + 4 = 0$

$1 + 4x = 0$ Multiplying by x

$x = -\dfrac{1}{4}$

The critical values are $-\dfrac{1}{4}$ and 0. Test a value in each of the intervals determined by them.

$\left(\infty, -\dfrac{1}{4}\right)$: $f(-1) = 3 > 0$

$\left(-\dfrac{1}{4}, 0\right)$: $f(-0.1) = -6 < 0$

$(0, \infty)$: $f(1) = 5 > 0$

The solution set for this portion of the inequality is $\left(-\infty, -\dfrac{1}{4}\right) \cup (0, \infty)$.

Next solve $1 + \dfrac{1}{x} < 3$, or $\dfrac{1}{x} - 2 < 0$. The denominator of $f(x) = \dfrac{1}{x} - 2$ is 0 when $x = 0$, so the function is not defined for this value of x. Now solve the related equation.

$\dfrac{1}{x} - 2 = 0$

$1 - 2x = 0$ Multiplying by x

$x = \dfrac{1}{2}$

The critical values are 0 and $\dfrac{1}{2}$. Test a value in each of the intervals determined by them.

$(-\infty, 0)$: $f(-1) = -3 < 0$

$\left(0, \dfrac{1}{2}\right)$: $f(0.1) = 8 > 0$

$\left(\dfrac{1}{2}, \infty\right)$: $f(1) = -1 < 0$

The solution set for this portion of the inequality is $(-\infty, 0) \cup \left(\dfrac{1}{2}, \infty\right)$.

The solution set of the original inequality is

$\left(\left(-\infty, -\dfrac{1}{4}\right) \cup (0, \infty)\right) and \left((-\infty, 0) \cup \left(\dfrac{1}{2}, \infty\right)\right)$,

or $\left(-\infty, -\dfrac{1}{4}\right) \cup \left(\dfrac{1}{2}, \infty\right)$.

85. $|x^2 + 3x - 1| < 3$

$-3 < x^2 + 3x - 1 < 3$

$-3 < x^2 + 3x - 1$ and $x^2 + 3x - 1 < 3$

First solve $-3 < x^2 + 3x - 1$, or $x^2 + 3x - 1 > -3$.

$x^2 + 3x - 1 > -3$

$x^2 + 3x + 2 > 0$

$x^2 + 3x + 2 = 0$ Related equation

$(x + 2)(x + 1) = 0$

$x = -2$ or $x = -1$

Let $f(x) = x^2 + 3x + 2$ and test a value in each of the intervals determined by the solution of the related equation.

$(-\infty, -2)$: $f(-3) = 2 > 0$

$(-2, -1)$: $f(-1.5) = -0.25 < 0$

$(-1, \infty)$: $f(0) = 2 > 0$

The solution set for this portion of the inequality is $(-\infty, -2) \cup (-1, \infty)$.

Next solve $x^2 + 3x - 1 < 3$.

$x^2 + 3x - 1 < 3$

$x^2 + 3x - 4 < 0$

$x^2 + 3x - 4 = 0$ Related equation

$(x + 4)(x - 1) = 0$

$x = -4$ or $x = 1$

Let $f(x) = x^2 + 3x - 4$ and test a value in each of the intervals determined by the solution of the related equation.

$(-\infty, -4)$: $f(-5) = 6 > 0$

$(-4, 1)$: $f(0) = -4 < 0$

$(1, \infty)$: $f(2) = 6 > 0$

The solution set for this portion of the inequality is $(-4, 1)$.

The solution set of the original inequality is $\left((-\infty, -2) \cup (-1, \infty)\right) and (-4, 1)$, or $(-4, -2) \cup (-1, 1)$.

87. First find a quadratic equation with solutions -4 and 3.

$(x + 4)(x - 3) = 0$

$x^2 + x - 12 = 0$

Test a point in each of the three intervals determined by -4 and 3.

$(-\infty, -4)$: $(-5 + 4)(-5 - 3) = 8 > 0$

$(-4, 3)$: $(0 + 4)(0 - 3) = -12 < 0$

$(3, \infty)$: $(4 + 4)(4 - 3) = 8 > 0$

Then a quadratic inequality for which the solution set is $(-4, 3)$ is $x^2 + x - 12 < 0$. Answers may vary.

Exercise Set 3.7

1. $y = kx$

$54 = k \cdot 12$

$\dfrac{54}{12} = k$, or $k = \dfrac{9}{2}$

The variation constant is $\dfrac{9}{2}$, or 4.5. The equation of variation is $y = \dfrac{9}{2}x$, or $y = 4.5x$.

3. $y = \dfrac{k}{x}$

$3 = \dfrac{k}{12}$

$36 = k$

The variation constant is 36. The equation of variation is $y = \dfrac{36}{x}$.

5. $y = kx$

$1 = k \cdot \dfrac{1}{4}$

$4 = k$

The variation constant is 4. The equation of variation is $y = 4x$.

7. $y = \dfrac{k}{x}$

$32 = \dfrac{k}{\frac{1}{8}}$

$\dfrac{1}{8} \cdot 32 = k$

$4 = k$

The variation constant is 4. The equation of variation is $y = \dfrac{4}{x}$.

9. $y = kx$

$\dfrac{3}{4} = k \cdot 2$

$\dfrac{1}{2} \cdot \dfrac{3}{4} = k$

$\dfrac{3}{8} = k$

The variation constant is $\dfrac{3}{8}$. The equation of variation is $y = \dfrac{3}{8}x$.

11. $y = \dfrac{k}{x}$

$1.8 = \dfrac{k}{0.3}$

$0.54 = k$

The variation constant is 0.54. The equation of variation is $y = \dfrac{0.54}{x}$.

13. $T = \dfrac{k}{P}$ T varies inversely as P.

$5 = \dfrac{k}{7}$ Substituting

$35 = k$ Variation constant

$T = \dfrac{35}{P}$ Equation of variation

$T = \dfrac{35}{10}$ Substituting

$T = 3.5$

It will take 10 bricklayers 3.5 hr to complete the job.

15. Let $F =$ the number of grams of fat and $w =$ the weight.

$F = kw$ F varies directly as w.

$60 = k \cdot 120$ Substituting

$\dfrac{60}{120} = k$, or   Solving for k

$\dfrac{1}{2} = k$ Variation constant

$F = \dfrac{1}{2}w$ Equation of variation

$F = \dfrac{1}{2} \cdot 180$ Substituting

$F = 90$

The maximum daily fat intake for a person weighing 180 lb is 90 g.

17. $W = \dfrac{k}{L}$ W varies inversely as L.

$1200 = \dfrac{k}{8}$ Substituting

$9600 = k$ Variation constant

$W = \dfrac{9600}{L}$ Equation of variation

$W = \dfrac{9600}{14}$ Substituting

$W = 685\dfrac{5}{7}$

A 14-m beam can support $685\dfrac{5}{7}$ kg, or about 686 kg.

19. $M = kE$ M varies directly as E.

$38 = k \cdot 95$ Substituting

$\dfrac{2}{5} = k$ Variation constant

$M = \dfrac{2}{5}E$ Equation of variation

$M = \dfrac{2}{5} \cdot 100$ Substituting

$M = 40$

A 100-lb person would weigh 40 lb on Mars.

21. $d = km$ d varies directly as m.

$40 = k \cdot 3$ Substituting

$\dfrac{40}{3} = k$ Variation constant

$d = \dfrac{40}{3}m$ Equation of variation

$d = \dfrac{40}{3} \cdot 5 = \dfrac{200}{3}$ Substituting

$d = 66\dfrac{2}{3}$

A 5-kg mass will stretch the spring $66\dfrac{2}{3}$ cm.

23. $P = \dfrac{k}{W}$ P varies inversely as W.

$330 = \dfrac{k}{3.2}$ Substituting

$1056 = k$ Variation constant

$$P = \frac{1056}{W} \quad \text{Equation of variation}$$

$$550 = \frac{1056}{W} \quad \text{Substituting}$$

$$550W = 1056 \quad \text{Multiplying by } W$$

$$W = \frac{1056}{550} \quad \text{Dividing by 550}$$

$$W = 1.92 \quad \text{Simplifying}$$

A tone with a pitch of 550 vibrations per second has a wavelength of 1.92 ft.

25.
$$y = \frac{k}{x^2}$$

$$0.15 = \frac{k}{(0.1)^2} \quad \text{Substituting}$$

$$0.15 = \frac{k}{0.01}$$

$$0.15(0.01) = k$$

$$0.0015 = k$$

The equation of variation is $y = \frac{0.0015}{x^2}$.

$$y = \frac{54}{x^2}$$

27.
$$y = kx^2$$

$$0.15 = k(0.1)^2 \quad \text{Substituting}$$

$$0.15 = 0.01k$$

$$\frac{0.15}{0.01} = k$$

$$15 = k$$

The equation of variation is $y = 15x^2$.

29. $y = kxz$

$$56 = k \cdot 7 \cdot 8 \quad \text{Substituting}$$

$$56 = 56k$$

$$1 = k$$

The equation of variation is $y = xz$.

31. $y = kxz^2$

$$105 = k \cdot 14 \cdot 5^2 \quad \text{Substituting}$$

$$105 = 350k$$

$$\frac{105}{350} = k$$

$$\frac{3}{10} = k$$

The equation of variation is $y = \frac{3}{10}xz^2$.

33.
$$y = k\frac{xz}{wp}$$

$$\frac{3}{28} = k\frac{3 \cdot 10}{7 \cdot 8} \quad \text{Substituting}$$

$$\frac{3}{28} = k \cdot \frac{30}{56}$$

$$\frac{3}{28} \cdot \frac{56}{30} = k$$

$$\frac{1}{5} = k$$

The equation of variation is $y = \frac{xz}{5wp}$.

35.
$$I = \frac{k}{d^2}$$

$$90 = \frac{k}{5^2} \quad \text{Substituting}$$

$$90 = \frac{k}{25}$$

$$2250 = k$$

The equation of variation is $I = \frac{2250}{d^2}$.

Substitute 40 for I and find d.
$$40 = \frac{2250}{d^2}$$

$$40d^2 = 2250$$

$$d^2 = 56.25$$

$$d = 7.5$$

The distance from 5 m to 7.5 m is $7.5 - 5$, or 2.5 m, so it is 2.5 m further to a point where the intensity is 40 W/m^2.

37.
$$d = kr^2$$

$$200 = k \cdot 60^2 \quad \text{Substituting}$$

$$200 = 3600k$$

$$\frac{200}{3600} = k$$

$$\frac{1}{18} = k$$

The equation of variation is $d = \frac{1}{18}r^2$.

Substitute 72 for d and find r.
$$72 = \frac{1}{18}r^2$$

$$1296 = r^2$$

$$36 = r$$

A car can travel 36 mph and still stop in 72 ft.

39. $E = \frac{kR}{I}$

We first find k.
$$5.69 = \frac{k \cdot 59}{93.3} \quad \text{Substituting}$$

$$5.69\left(\frac{93.3}{59}\right) = k \quad \text{Multiplying by } \frac{93.3}{59}$$

$$9 \approx k$$

The equation of variation is $E = \frac{9R}{I}$.

Substitute 5.69 for E and 200 for I and solve for R.

$$5.69 = \frac{9R}{200}$$

$$5.69\left(\frac{200}{9}\right) = R \qquad \text{Multiplying by } \frac{200}{9}$$

$$126 \approx R$$

Curt Schilling would have given up about 126 earned runs if he had pitched 200 innings.

41. Discussion and Writing

43.

45. Test for symmetry with respect to the x-axis.

$$y^2 = x \quad \text{Original equation}$$
$$(-y)^2 = x \quad \text{Replacing } y \text{ by } -y$$
$$y^2 = x \quad \text{Simplifying}$$

The last equation is equivalent to the original equation, so the graph is symmetric with respect to the x-axis.

Test for symmetry with respect to the y-axis:

$$y^2 = x \quad \text{Original equation}$$
$$y^2 = -x \quad \text{Replacing } x \text{ by } -x$$

The last equation is not equivalent to the original equation, so the graph is not symmetric with respect to the y-axis.

Test for symmetry with respect to the origin:

$$y^2 = x \quad \text{Original equation}$$
$$(-y)^2 = -x \quad \text{Replacing } x \text{ by } -x \text{ and}$$
$$\qquad\qquad\qquad y \text{ by } -y$$
$$y^2 = -x \quad \text{Simplifying}$$

The last equation is not equivalent to the original equation, so the graph is not symmetric with respect to the origin.

47. Let V represent the volume and p represent the price of a jar of peanut butter.

$$V = kp \qquad V \text{ varies directly as } p.$$
$$\pi\left(\frac{3}{2}\right)^2(5) = k(1.8) \qquad \text{Substituting}$$
$$6.25\pi = k \qquad \text{Variation constant}$$
$$V = 6.25\pi p \qquad \text{Equation of variation}$$
$$\pi(1.625)^2(5.5) = 6.25\pi p \qquad \text{Substituting}$$
$$2.32 \approx p$$

If cost is directly proportional to volume, the larger jar should cost \$2.32.

Now let W represent the weight and p represent the price of a jar of peanut butter.

$$W = kp$$
$$18 = k(1.8) \qquad \text{Substituting}$$
$$10 = k \qquad \text{Variation constant}$$
$$W = 10p \qquad \text{Equation of variation}$$
$$28 = 10p \qquad \text{Substituting}$$
$$2.8 = p$$

If cost is directly proportional to weight, the larger jar should cost \$2.80.

49. We are told $A = kd^2$, and we know $A = \pi r^2$ so we have:

$$kd^2 = \pi r^2$$
$$kd^2 = \pi\left(\frac{d}{2}\right)^2 \qquad r = \frac{d}{2}$$
$$kd^2 = \frac{\pi d^2}{4}$$
$$k = \frac{\pi}{4} \qquad \text{Variation constant}$$

Chapter 3 Review Exercises

1. $f(-b) = (-b+a)(-b+b)(-b-c) = (-b+a)\cdot 0 \cdot(-b-c) = 0$, so the statement is true.

3. In addition to the given possibilities, 9 and -9 are also possible rational zeros. The statement is false.

5. The domain of the function is the set of all real numbers except -2 and 3, or $\{x | x \neq -2 \text{ and } x \neq 3\}$. The statement is false.

7. $h(x) = -25$

The degree of the polynomial is 0, so the polynomial is constant. The leading term is -25 and the leading coefficient is -25.

9. $f(x) = \frac{1}{3}x^3 - 2x + 3$

The degree of the polynomial is 3, so the polynomial is cubic. The leading term is $\frac{1}{3}x^3$ and the leading coefficient is $\frac{1}{3}$.

11. $f(x) = x^5 + 2x^3 - x^2 + 5x + 4$

The leading term is x^5. The degree, 5, is odd and the leading coefficient, 1, is positive. As $x \to \infty$, $f(x) \to \infty$, and as $x \to -\infty$, $f(x) \to -\infty$.

13. $f(x) = x^4 - 26x^2 + 25$
$$= (x^2 - 1)(x^2 - 25)$$
$$= (x + 1)(x - 1)(x + 5)(x - 5)$$

$\pm 1, \pm 5$; each has multiplicity 1

15. $A = P(1+r)^t$

a)
$$6760 = 6250(1+r)^2$$
$$\frac{6760}{6250} = (1+r)^2$$
$$\pm 1.04 = 1+r$$
$$-1 \pm 1.04 = r$$
$$-2.04 = r \quad or \quad 0.04 = r$$

Only 0.04 has meaning in this application. The interest rate is 0.04 or 4%.

b)
$$1,215,506.25 = 1,000,000(1+r)^4$$
$$\frac{1,215,506.25}{1,000,000} = (1+r)^4$$
$$\pm 1.05 = 1+r$$
$$-1 \pm 1.05 = r$$
$$-2.05 = r \quad or \quad 0.05 = r$$

Only 0.05 has meaning in this application. The interest rate is 0.05 or 5%.

17. $g(x) = (x-1)^3(x+2)^2$

1. The leading term is $x \cdot x \cdot x \cdot x \cdot x$, or x^5. The degree, 5, is odd and the leading coefficient, 1, is positive so as $x \to \infty$, $g(x) \to \infty$ and as $x \to -\infty$, $g(x) \to -\infty$.

2. We see that the zeros of the function are 1 and -2, so the x-intercepts of the graph are $(1,0)$ and $(-2,0)$.

3. The zeros divide the x-axis into 3 intervals, $(-\infty, -2)$, $(-2, 1)$, and $(1, \infty)$. We choose a value for x from each interval and find $g(x)$. This tells us the sign of $g(x)$ for all values of x in that interval.

   In $(-\infty, -2)$, test -3:
   $$g(-3) = (-3-1)^3(-3+2)^2 = -64 < 0$$

   In $(-2, 1)$, test 0:
   $$g(0) = (0-1)^3(0+2)^2 = -4 < 0$$

   In $(1, \infty)$, test 2:
   $$g(2) = (2-1)^3(2+2)^2 = 16 > 0$$

   Thus the graph lies below the x-axis on $(-\infty, -2)$ and on $(-2, 1)$ and above the x-axis on $(1, \infty)$. We also know that the points $(-3, -64)$, $(0, -4)$, and $(2, 16)$ are on the graph.

4. From Step 3 we know that $g(0) = -4$, so the y-intercept is $(0, -4)$.

5. We find additional points on the graph and then draw the graph.

x	$g(x)$
-2.5	-10.7
-1	-8
-0.5	-7.6
0.5	-0.8

$g(x) = (x-1)^3(x+2)^2$

6. Checking the graph as described on page 272 in the text, we see that it appears to be correct.

19. $f(x) = x^4 - 5x^3 + 6x^2 + 4x - 8$

1. The leading term is x^4. The degree, 4, is even and the leading coefficient, 1, is positive so as $x \to \infty$, $f(x) \to \infty$ and as $x \to -\infty$, $f(x) \to -\infty$.

2. We solve $f(x) = 0$, or $x^4 - 5x^3 + 6x^2 + 4x - 8 = 0$. The possible rational zeros are ± 1, ± 2, ± 4, and ± 8. We try -1.

$$
\begin{array}{r|rrrrr}
-1 & 1 & -5 & 6 & 4 & -8 \\
   &   & -1 & 6 & -12 & 8 \\
\hline
   & 1 & -6 & 12 & -8 & 0 \\
\end{array}
$$

Now we have $(x+1)(x^3 - 6x^2 + 12x - 8) = 0$. We use synthetic division to determine if 2 is a zero of $x^3 - 6x^2 + 12x - 8 = 0$.

$$
\begin{array}{r|rrrr}
2 & 1 & -6 & 12 & -8 \\
  &   & 2 & -8 & 8 \\
\hline
  & 1 & -4 & 4 & 0 \\
\end{array}
$$

We have $(x+1)(x-2)(x^2 - 4x + 4) = 0$, or $(x+1)(x-2)(x-2)^2 = 0$. Thus the zeros of $f(x)$ are -1 and 2 and the x-intercepts of the graph are $(-1, 0)$ and $(2, 0)$.

3. The zeros divide the x-axis into 3 intervals, $(-\infty, -1)$, $(-1, 2)$, and $(2, \infty)$. We choose a value for x from each interval and find $f(x)$. This tells us the sign of $f(x)$ for all values of x in that interval.

   In $(-\infty, -1)$, test -2:
   $$f(-2) = (-2)^4 - 5(-2)^3 + 6(-2)^2 + 4(-2) - 8 = 64 > 0$$

   In $(-1, 2)$, test 0:
   $$f(0) = 0^4 - 5 \cdot 0^3 + 6 \cdot 0^2 + 4 \cdot 0 - 8 = -8 < 0$$

   In $(2, \infty)$, test 3:
   $$f(3) = 3^4 - 5 \cdot 3^3 + 6 \cdot 3^2 + 4 \cdot 3 - 8 = 4 > 0$$

   Thus the graph lies above the x-axis on $(-\infty, -1)$ and on $(2, \infty)$ and below the x-axis on $(-1, 2)$. We also know that the points $(-2, 64)$, $(0, -8)$, and $(3, 4)$ are on the graph.

5. We find additional points on the graph and then draw the graph.

x	$f(x)$
-1.5	21.4
-0.5	-7.8
1	-2
4	40

$f(x) = x^4 - 5x^3 + 6x^2 + 4x - 8$

6. Checking the graph as described on page 272 in the text, we see that it appears to be correct.

21. $f(1) = 4 \cdot 1^2 - 5 \cdot 1 - 3 = -4$

$f(2) = 4 \cdot 2^2 - 5 \cdot 2 - 3 = 3$

By the intermediate value theorem, since $f(1)$ and $f(2)$ have opposite signs, $f(x)$ has a zero between 1 and 2.

23.

$$
\begin{array}{r}
6x^2 + 16x + 52 \\
x-3 \overline{\smash{\big)}\ 6x^3 - 2x^2 + 4x - 1} \\
\underline{6x^3 - 18x^2} \\
16x^2 + 4x \\
\underline{16x^2 - 48x} \\
52x - 1 \\
\underline{52x - 156} \\
155
\end{array}
$$

$Q(x) = 6x^2 + 16x + 52$; $R(x) = 155$;

$P(x) = (x-3)(6x^2 + 16x + 52) + 155$

25.
$$
\begin{array}{r|rrrr}
5 & 1 & 2 & -13 & 10 \\
  &   & 5 & 35 & 110 \\
\hline
  & 1 & 7 & 22 & \underline{|\,120} \\
\end{array}
$$

The quotient is $x^2 + 7x + 22$; the remainder is 120.

27.
$$
\begin{array}{r|rrrrrr}
-1 & 1 & 0 & 0 & 0 & -2 & 0 \\
   &   & -1 & 1 & -1 & 1 & 1 \\
\hline
   & 1 & -1 & 1 & -1 & -1 & |\,1 \\
\end{array}
$$

The quotient is $x^4 - x^3 + x^2 - x - 1$; the remainder is 1.

29.
$$
\begin{array}{r|rrrrr}
-2 & 1 & 0 & 0 & 0 & -16 \\
   &   & -2 & 4 & -8 & 16 \\
\hline
   & 1 & -2 & 4 & -8 & |\,0 \\
\end{array}
$$

$f(-2) = 0$

31.
$$
\begin{array}{r|rrrr}
-i & 1 & -5 & 1 & -5 \\
   &   & -i & -1+5i & 5 \\
\hline
   & 1 & -5-i & 5i & |\,0 \\
\end{array}
$$

$f(-i) = 0$, so $-i$ is a zero of $f(x)$.

$$
\begin{array}{r|rrrr}
-5 & 1 & -5 & 1 & -5 \\
   &   & -5 & 50 & -255 \\
\hline
   & 1 & -10 & 51 & |\,-260 \\
\end{array}
$$

$f(-5) \neq 0$, so -5 is not a zero of $f(x)$.

33.
$$
\begin{array}{r|rrrr}
\frac{1}{3} & 1 & -\frac{4}{3} & -\frac{5}{3} & \frac{2}{3} \\
   &   & \frac{1}{3} & -\frac{1}{3} & -\frac{2}{3} \\
\hline
   & 1 & -1 & -2 & |\,0 \\
\end{array}
$$

$f\left(\frac{1}{3}\right) = 0$, so $\frac{1}{3}$ is a zero of $f(x)$.

$$
\begin{array}{r|rrrr}
1 & 1 & -\frac{4}{3} & -\frac{5}{3} & \frac{2}{3} \\
  &   & 1 & -\frac{1}{3} & -2 \\
\hline
  & 1 & -\frac{1}{3} & -2 & |\,-\frac{4}{3} \\
\end{array}
$$

$f(1) \neq 0$, so 1 is not a zero of $f(x)$.

35. $f(x) = x^3 + 2x^2 - 7x + 4$

Try $x+1$ and $x+2$. Using synthetic division we find that $f(-1) \neq 0$ and $f(-2) \neq 0$. Thus $x+1$ and $x+2$ are not factors of $f(x)$. Try $x+4$.

$$
\begin{array}{r|rrrr}
-4 & 1 & 2 & -7 & 4 \\
   &   & -4 & 8 & -4 \\
\hline
   & 1 & -2 & 1 & |\,0 \\
\end{array}
$$

Since $f(-4) = 0$, $x+4$ is a factor of $f(x)$. Thus $f(x) = (x+4)(x^2 - 2x + 1) = (x+4)(x-1)^2$.

Now we solve $f(x) = 0$.

$x + 4 = 0 \quad or \quad (x-1)^2 = 0$

$x = -4 \ or \qquad x - 1 = 0$

$x = -4 \ or \qquad\qquad x = 1$

The solutions of $f(x) = 0$ are -4 and 1.

37. $f(x) = x^4 - 4x^3 - 21x^2 + 100x - 100$

Using synthetic division we find that $f(2) = 0$:

$$
\begin{array}{r|rrrrr}
2 & 1 & -4 & -21 & 100 & -100 \\
  &   & 2 & -4 & -50 & 100 \\
\hline
  & 1 & -2 & -25 & 50 & |\,0 \\
\end{array}
$$

Then we have:

$f(x) = (x-2)(x^3 - 2x^2 - 25x + 50)$

$\quad = (x-2)[x^2(x-2) - 25(x-2)]$

$\quad = (x-2)(x-2)(x^2 - 25)$

$\quad = (x-2)^2(x+5)(x-5)$

Now solve $f(x) = 0$.

$(x-2)^2 = 0 \quad or \ x + 5 = 0 \quad or \ x - 5 = 0$

$x - 2 = 0 \ or \qquad x = -5 \ or \qquad x = 5$

$x = 2 \ or \qquad x = -5 \ or \qquad x = 5$

The solutions of $f(x) = 0$ are 2, -5, and 5.

39. A polynomial function of degree 3 with -4, -1, and 2 as zeros has factors $x+4$, $x+1$, and $x-2$ so we have $f(x) = a_n(x+4)(x+1)(x-2)$.

The simplest polynomial is obtained if we let $a_n = 1$.

$f(x) = (x+4)(x+1)(x-2)$

$\quad = (x^2 + 5x + 4)(x-2)$

$\quad = x^3 - 2x^2 + 5x^2 - 10x + 4x - 8$

$\quad = x^3 + 3x^2 - 6x - 8$

41. A polynomial function of degree 3 with $\frac{1}{2}$, $1 - \sqrt{2}$, and $1 + \sqrt{2}$ as zeros has factors $x - \frac{1}{2}$, $x - (1 - \sqrt{2})$, and $x - (1 + \sqrt{2})$ so we have

$f(x) = a_n\left(x - \frac{1}{2}\right)[x - (1 - \sqrt{2})][x - (1 + \sqrt{2})]$.

Let $a_n = 1$.

$$f(x) = \left(x - \frac{1}{2}\right)[x - (1 - \sqrt{2})][x - (1 + \sqrt{2})]$$
$$= \left(x - \frac{1}{2}\right)[(x-1) + \sqrt{2}][(x-1) - \sqrt{2}]$$
$$= \left(x - \frac{1}{2}\right)(x^2 - 2x + 1 - 2)$$
$$= \left(x - \frac{1}{2}\right)(x^2 - 2x - 1)$$
$$= x^3 - 2x^2 - x - \frac{1}{2}x^2 + x + \frac{1}{2}$$
$$= x^3 - \frac{5}{2}x^2 + \frac{1}{2}$$

If we let $a_n = 2$, we obtain $f(x) = 2x^3 - 5x^2 + 1$.

43. A polynomial function of degree 5 has at most 5 real zeros. Since 5 zeros are given, these are all of the zeros of the desired function. We proceed as in Exercise 39 above, letting $a_n = 1$.
$$f(x) = (x+3)^2(x-2)(x-0)^2$$
$$= (x^2 + 6x + 9)(x^3 - 2x^2)$$
$$= x^5 + 6x^4 + 9x^3 - 2x^4 - 12x^3 - 18x^2$$
$$= x^5 + 4x^4 - 3x^3 - 18x^2$$

45. A polynomial function of degree 5 can have at most 5 zeros. Since $f(x)$ has rational coefficients, in addition to the 3 given zeros, the other zeros are the conjugates of $1 + \sqrt{3}$ and $-\sqrt{3}$, or $1 - \sqrt{3}$ and $\sqrt{3}$.

47. $-\sqrt{11}$ is also a zero.
$$f(x) = (x - \sqrt{11})(x + \sqrt{11})$$
$$= x^2 - 11$$

49. $1 - i$ is also a zero.
$$f(x) = (x+1)(x-4)[x - (1+i)][x - (1-i)]$$
$$= (x^2 - 3x - 4)(x^2 - 2x + 2)$$
$$= x^4 - 2x^3 + 2x^2 - 3x^3 + 6x^2 - 6x - 4x^2 + 8x - 8$$
$$= x^4 - 5x^3 + 4x^2 + 2x - 8$$

51. $$f(x) = \left(x - \frac{1}{3}\right)(x - 0)(x + 3)$$
$$= \left(x^2 - \frac{1}{3}x\right)(x + 3)$$
$$= x^3 + \frac{8}{3}x^2 - x$$

53. $g(x) = 3x^4 - x^3 + 5x^2 - x + 1$

$\dfrac{\text{Possibilities for } p}{\text{Possibilities for } q} : \dfrac{\pm 1}{\pm 1, \pm 3}$

Possibilities for p/q: $\pm 1, \pm \dfrac{1}{3}$

55. $f(x) = 3x^5 + 2x^4 - 25x^3 - 28x^2 + 12x$

a) We know that 0 is a zero since
$$f(x) = x(3x^4 + 2x^3 - 25x^2 - 28x + 12).$$
Now consider $g(x) = 3x^4 + 2x^3 - 25x^2 - 28x + 12$.
Possibilities for p/q: $\pm 1, \pm 2, \pm 3, \pm 4, \pm 6, \pm 12$,
$$\pm \frac{1}{3}, \pm \frac{2}{3}, \pm \frac{4}{3}$$

From the graph of $y = 3x^4 + 2x^3 - 25x^2 - 28x + 12$, we see that, of all the possibilities above, only -2, $\dfrac{1}{3}$, $\dfrac{2}{3}$, and 3 might be zeros. We use synthetic division to determine if -2 is a zero.

```
-2 |  3    2   -25   -28    12
   |       -6    8    34   -12
   ----------------------------
      3   -4   -17     6 |   0
```

Now try 3 in the quotient above.

```
 3 |  3   -4   -17    6
   |       9    15   -6
   ---------------------
      3    5    -2 |   0
```

We have $f(x) = (x+2)(x-3)(3x^2 + 5x - 2)$.
We find the other zeros.
$$3x^2 + 5x - 2 = 0$$
$$(3x - 1)(x + 2) = 0$$
$$3x - 1 = 0 \quad or \quad x + 2 = 0$$
$$3x = 1 \quad or \quad \quad\quad x = -2$$
$$x = \frac{1}{3} \quad or \quad \quad\quad x = -2$$

The rational zeros of $g(x) = 3x^4 + 2x^3 - 25x^2 - 28x + 12$ are -2, 3, and $\dfrac{1}{3}$. Since 0 is also a zero of $f(x)$, the zeros of $f(x)$ are -2, 3, $\dfrac{1}{3}$, and 0. (The zero -2 has multiplicity 2.) These are the only zeros.

b) From our work above we see
$$f(x) = x(x+2)(x-3)(3x-1)(x+2), \text{ or }$$
$$x(x+2)^2(x-3)(3x-1).$$

57. $f(x) = x^4 - 6x^3 + 9x^2 + 6x - 10$

a) Possibilities for p/q: $\pm 1, \pm 2, \pm 5, \pm 10$

From the graph of $f(x)$, we see that -1 and 1 might be zeros.

```
-1 |  1   -6    9     6   -10
   |      -1    7   -16    10
   ----------------------------
      1   -7   16   -10 |   0
```

```
 1 |  1   -7    16   -10
   |       1    -6    10
   ----------------------
      1   -6    10 |   0
```

$f(x) = (x+1)(x-1)(x^2 - 6x + 10)$
Using the quadratic formula, we find that the other zeros are $3 \pm i$.

The rational zeros are -1 and 1. The other zeros are $3 \pm i$.

b) $f(x) = (x+1)(x-1)[x - (3+i)][x - (3-i)]$
$$= (x+1)(x-1)(x-3-i)(x-3+i)$$

59. $f(x) = 3x^3 - 8x^2 + 7x - 2$

a) Possibilities for p/q: $\pm 1, \pm 2, \pm \dfrac{1}{3}, \pm \dfrac{2}{3}$

From the graph of $f(x)$, we see that $\dfrac{2}{3}$ and 1 might be zeros.

$$\begin{array}{r|rrrr} 1 & 3 & -8 & 7 & -2 \\ & & 3 & -5 & 2 \\ \hline & 3 & -5 & 2 & 0 \end{array}$$

We have $f(x) = (x-1)(3x^2 - 5x + 2)$.

We find the other zeros.

$$3x^2 - 5x + 2 = 0$$
$$(3x - 2)(x - 1) = 0$$
$$3x - 2 = 0 \quad or \quad x - 1 = 0$$
$$x = \frac{2}{3} \quad or \qquad x = 1$$

The rational zeros are 1 and $\frac{2}{3}$. (The zero 1 has multiplicity 2.) These are the only zeros.

b) $f(x) = (x-1)^2(3x-2)$

61. $f(x) = x^6 + x^5 - 28x^4 - 16x^3 + 192x^2$

a) We know that 0 is a zero since

$$f(x) = x^2(x^4 + x^3 - 28x^2 - 16x + 192).$$

Consider $g(x) = x^4 + x^3 - 28x^2 - 16x + 192$.

Possibilities for p/q: $\pm 1, \pm 2, \pm 3, \pm 4, \pm 6, \pm 8, \pm 12,$
$$\pm 16, \pm 24, \pm 32, \pm 48, \pm 64, \pm 96,$$
$$\pm 192$$

From the graph of $y = g(x)$, we see that -4, 3 and 4 might be zeros.

$$\begin{array}{r|rrrrr} -4 & 1 & 1 & -28 & -16 & 192 \\ & & -4 & 12 & 64 & -192 \\ \hline & 1 & -3 & -16 & 48 & 0 \end{array}$$

We have $f(x) = x^2 \cdot g(x) = $
$x^2(x+4)(x^3 - 3x^2 - 16x + 48)$.

We find the other zeros.

$$x^3 - 3x^2 - 16x + 48 = 0$$
$$x^2(x-3) - 16(x-3) = 0$$
$$(x-3)(x^2 - 16) = 0$$
$$(x-3)(x+4)(x-4) = 0$$
$$x - 3 = 0 \quad or \quad x + 4 = 0 \quad or \quad x - 4 = 0$$
$$x = 3 \quad or \qquad x = -4 \quad or \qquad x = 4$$

The rational zeros are 0, -4, 3, and 4. (The zeros 0 and -4 each have multiplicity 2.) These are the only zeros.

b) $f(x) = x^2(x+4)^2(x-3)(x-4)$

63. $f(x) = 2x^6 - 7x^3 + x^2 - x$

There are 3 variations in sign in $f(x)$, so there are 3 or 1 positive real zeros.

$$f(-x) = 2(-x)^6 - 7(-x)^3 + (-x)^2 - (-x)$$
$$= 2x^6 + 7x^3 + x^2 + x$$

There are no variations in sign in $f(-x)$, so there are no negative real zeros.

65. $g(x) = 5x^5 - 4x^2 + x - 1$

There are 3 variations in sign in $g(x)$, so there are 3 or 1 positive real zeros.

$$g(-x) = 5(-x)^5 - 4(-x)^2 + (-x) - 1$$
$$= -5x^5 - 4x^2 - x - 1$$

There is no variation in sign in $g(-x)$, so there are 0 negative real zeros.

67. $f(x) = \dfrac{5}{(x-2)^2}$

1. The denominator is zero when $x = 2$, so the domain excludes 2. It is $(-\infty, 2) \cup (2, \infty)$. The line $x = 2$ is the vertical asymptote.

2. Because the degree of the numerator is less than the degree of the denominator, the x-axis, or $y = 0$, is the horizontal asymptote. There is no oblique asymptote.

3. The numerator has no zeros, so there is no x-intercept.

4. $f(0) = \dfrac{5}{(0-2)^2} = \dfrac{5}{4}$, so the y-intercept is $\left(0, \dfrac{5}{4}\right)$.

5. Find other function values to determine the shape of the graph and then draw it.

69. $f(x) = \dfrac{x-2}{x^2 - 2x - 15} = \dfrac{x-2}{(x+3)(x-5)}$

1. The denominator is zero when $x = -3$, or $x = 5$, so the domain excludes -3 and 5. It is $(-\infty, -3) \cup (-3, 5) \cup (5, \infty)$. The lines $x = -3$ and $x = 5$ are vertical asymptotes.

2. Because the degree of the numerator is less than the degree of the denominator, the x-axis, or $y = 0$, is the horizontal asymptote. There is no oblique asymptote.

3. The numerator is zero when $x = 2$, so the x-intercept is $(2, 0)$.

4. $f(0) = \dfrac{0-2}{0^2 - 2 \cdot 0 - 15} = \dfrac{2}{15}$, so the y-intercept is $\left(0, \dfrac{2}{15}\right)$.

5. Find other function values to determine the shape of the graph and then draw it.

$$f(x) = \frac{x-2}{x^2 - 2x - 15}$$

71. Answers may vary. The numbers -2 and 3 must be zeros of the denominator, and -3 must be zero of the numerator. In addition, the numerator and denominator must have the same degree and the ratio of the leading coefficients must be 4.

$$f(x) = \frac{4x(x+3)}{(x+2)(x-3)}, \text{ or } f(x) = \frac{4x^2 + 12x}{x^2 - x - 6}$$

73.
$$x^2 - 9 < 0 \quad \text{Polynomial inequality}$$
$$x^2 - 9 = 0 \quad \text{Related equation}$$
$$(x+3)(x-3) = 0 \quad \text{Factoring}$$

The solutions of the related equation are -3 and 3. These numbers divide the x-axis into the intervals $(-\infty, -3)$, $(-3, 3)$, and $(3, \infty)$.

We let $f(x) = (x+3)(x-3)$ and test a value in each interval.

$$(-\infty, -3): f(-4) = 7 > 0$$
$$(-3, 3): f(0) = -9 < 0$$
$$(3, \infty): f(4) = 7 > 0$$

Function values are negative only on $(-3, 3)$. The solution set is $(-3, 3)$.

75. $(1-x)(x+4)(x-2) \le 0$ Polynomial inequality

$(1-x)(x+4)(x-2) = 0$ Related equation

The solutions of the related equation are 1, -4 and 2. These numbers divide the x-axis into the intervals $(-\infty, -4)$, $(-4, 1)$, $(1, 2)$ and $(2, \infty)$.

We let $f(x) = (1-x)(x+4)(x-2)$ and test a value in each interval.

$$(-\infty, -4): \ f(-5) = 42 > 0$$
$$(-4, 1): \ f(0) = -8 < 0$$
$$(1, 2): \ f\left(\frac{3}{2}\right) = \frac{11}{8} > 0$$
$$(2, \infty): \ f(3) = -14 < 0$$

Function values are negative on $(-4, 1)$ and $(2, \infty)$. Since the inequality symbol is $\le$, the endpoints of the intervals must be included in the solution set. It is $[-4, 1] \cup [2, \infty)$.

77. a) We write and solve a polynomial equation.

$$-16t^2 + 80t + 224 = 0$$
$$-16(t^2 - 5t - 14) = 0$$
$$-16(t+2)(t-7) = 0$$

The solutions are $t = -2$ and $t = 7$. Only $t = 7$ has meaning in this application. The rocket reaches the ground at $t = 7$ seconds.

b) We write and solve a polynomial inequality.

$$-16t^2 + 80t + 224 > 320 \quad \text{Polynomial inequality}$$
$$-16t^2 + 80t - 96 > 0 \quad \text{Equivalent inequality}$$
$$-16t^2 + 80t - 96 = 0 \quad \text{Related equation}$$
$$-16(t^2 - 5t + 6) = 0$$
$$-16(t-2)(t-3) = 0$$

The solutions of the related equation are 2 and 3. These numbers divide the t-axis into the intervals $(-\infty, 2)$, $(2, 3)$ and $(3, \infty)$. We restrict our discussion to values of t such that $0 \le t \le 7$ since we know from part (a) the rocket is in the air for 7 sec. We consider the intervals $[0, 2)$, $(2, 3)$ and $(3, 7]$. We let $f(t) = -16t^2 + 80t - 96$ and test a value in each interval.

$$[0, 2): f(1) = -32 < 0$$
$$(2, 3): f\left(\frac{5}{2}\right) = 4 > 0$$
$$(3, 7]: f(4) = -32 < 0$$

Function values are positive on $(2, 3)$. The solution set is $(2, 3)$

79.
$$y = kx$$
$$100 = k \cdot 25$$
$$4 = k \quad \text{Variation constant}$$
Equation of variation: $y = 4x$

81.
$$t = \frac{k}{r}$$
$$35 = \frac{k}{800}$$
$$28{,}000 = k \quad \text{Variation constant}$$
$$t = \frac{28{,}000}{r} \quad \text{Variation equation}$$
$$t = \frac{28{,}000}{1400}$$
$$t = 20 \text{ min}$$

83.
$$N = ka$$
$$87 = k \cdot 28$$
$$\frac{87}{28} = k \quad \text{Variation constant}$$
$$N = \frac{87}{28}a \quad \text{Variation equation}$$
$$N = \frac{87}{28} \cdot 25$$
$$N \approx 78$$

85.
$$y = \frac{kxz^2}{w}$$
$$2 = \frac{k(16)\left(\frac{1}{2}\right)^2}{0.2}$$
$$2 = 20k$$
$$\frac{1}{10} = k$$
$$y = \frac{1}{10} \cdot \frac{xz^2}{w}, \text{ or } \frac{xz^2}{10w}$$

87. $g(x) = \dfrac{x^2 + 2x - 3}{x^2 - 5x + 6} = \dfrac{x^2 + 2x - 3}{(x-2)(x-3)}$

The values of x that make the denominator 0 are 2 and 3, so the domain is $(-\infty, 2) \cup (2, 3) \cup (3, \infty)$. Answer A is correct.

89. $f(x) = x^3 + 3x^2 - 2x - 6$

Graph the function and use the Zero, Maximum and Minimum features.

a) Zeros: $-3, -1.414, 1.414$

b) Relative maximum: 2.303 at $x = -2.291$

c) Relative minimum: -6.303 at $x = 0.291$

d) Domain: all real numbers; range: all real numbers

91. a) Linear: $f(x) = 0.5408695652x - 30.30434783$;
quadratic: $f(x) = 0.0030322581x^2 - 0.5764516129x + 57.53225806$;
cubic: $f(x) = 0.0000247619x^3 - 0.0112857143x^2 + 2.002380952x - 82.14285714$

b) For the linear function, $f(300) \approx 132$; for the quadratic function, $f(300) \approx 158$; for the cubic function, $f(300) \approx 171$. Since the cubic function yields the result that is closest to the actual number of 180, it makes the best prediction.

c) Using the cubic function, we have: $f(350) \approx 298$ and $f(400) \approx 498$.

93. Vertical asymptotes occur at any x-values that make the denominator zero. The graph of a rational function does not cross any vertical asymptotes. Horizontal asymptotes occur when the degree of the numerator is less than or equal to the degree of the denominator. Oblique asymptotes occur when the degree of the numerator is greater than the degree of the denominator. Graphs of rational functions may cross horizontal or oblique asymptotes.

95.
$$\left| 1 - \frac{1}{x^2} \right| < 3$$
$$-3 < 1 - \frac{1}{x^2} < 3$$
$$-3 < \frac{x^2 - 1}{x^2} < 3$$
$$-3 < \frac{(x+1)(x-1)}{x^2} < 3$$
$$-3 < \frac{(x+1)(x-1)}{x^2} \ and \ \frac{(x+1)(x-1)}{x^2} < 3$$

First, solve
$$-3 < \frac{(x+1)(x-1)}{x^2}$$
$$0 < \frac{(x+1)(x-1)}{x^2} + 3$$

The denominator of $f(x) = \dfrac{(x+1)(x-1)}{x^2} + 3$ is zero when $x = 0$, so the function is not defined for this value of x. Solve the related equation.

$$\frac{(x+1)(x-1)}{x^2} + 3 = 0$$
$$(x+1)(x-1) + 3x^2 = 0 \quad \text{Multiplying by } x^2$$
$$x^2 - 1 + 3x^2 = 0$$
$$4x^2 = 1$$
$$x^2 = \frac{1}{4}$$
$$x = \pm\frac{1}{2}$$

The critical values are $-\dfrac{1}{2}$, 0 and $\dfrac{1}{2}$. Test a value in each of the intervals determined by them.

$$\left(-\infty, -\frac{1}{2}\right): \ f(-1) = 3 > 0$$
$$\left(-\frac{1}{2}, 0\right): \ f\left(-\frac{1}{4}\right) = -12 < 0$$
$$\left(0, \frac{1}{2}\right): \ f\left(\frac{1}{4}\right) = -12 < 0$$
$$\left(\frac{1}{2}, \infty\right): \ f(1) = 3 > 0$$

The solution set for this portion of the inequality is $\left(-\infty, -\dfrac{1}{2}\right) \cup \left(\dfrac{1}{2}, \infty\right)$.

Next, solve
$$\frac{(x+1)(x-1)}{x^2} < 3$$
$$\frac{(x+1)(x-1)}{x^2} - 3 < 0$$

The denominator of $f(x) = \dfrac{(x+1)(x-1)}{x^2} - 3$ is zero when $x = 0$, so the function is not defined for this value of x. Now solve the related equation.

$$\frac{(x+1)(x-1)}{x^2} - 3 = 0$$
$$(x+1)(x-1) - 3x^2 = 0 \quad \text{Multiplying by } x^2$$
$$x^2 - 1 - 3x^2 = 0$$
$$2x^2 = -1$$
$$x^2 = -\frac{1}{2}$$

There are no real solutions for this portion of the inequality. The solution set of the original inequality is $\left(-\infty, -\dfrac{1}{2}\right) \cup \left(\dfrac{1}{2}, \infty\right)$.

97. $(x-2)^{-3} < 0$
$$\frac{1}{(x-2)^3} < 0$$

The denominator of $f(x) = \dfrac{1}{(x-2)^3}$ is zero when $x = 2$, so the function is not defined for this value of x. The related equation $\dfrac{1}{(x-2)^3} = 0$ has no solution, so 2 is the only critical point. Test a value in each of the intervals determined by this critical point.

$(-\infty, 2): \ f(1) = -1 < 0$
$(2, \infty): \ f(3) = 1 > 0$

Function values are negative on $(-\infty, 2)$. The solution set is $(-\infty, 2)$.

99. Divide $x^3 + kx^2 + kx - 15$ by $x + 3$.

$$
\begin{array}{r|rrrr}
-3 & 1 & k & k & -15 \\
  &   & -3 & 9-3k & -27+6k \\
\hline
  & 1 & -3+k & 9-2k & -42+6k
\end{array}
$$

Thus $f(-3) = -42 + 6k$.

We know that if $x + 3$ is a factor of $f(x)$, then $f(-3) = 0$. We solve $-42 + 6k = 0$ for k.

$$-42 + 6k = 0$$
$$6k = 42$$
$$k = 7$$

101. $f(x) = \sqrt{x^2 + 3x - 10}$

Since we cannot take the square root of a negative number, then $x^2 + 3x - 10 \geq 0$.

$$x^2 + 3x - 10 \geq 0 \quad \text{Polynomial inequality}$$
$$x^2 + 3x - 10 = 0 \quad \text{Related equation}$$
$$(x + 5)(x - 2) = 0 \quad \text{Factoring}$$

The solutions of the related equation are -5 and 2. These numbers divide the x-axis into the intervals $(-\infty, -5)$, $(-5, 2)$ and $(2, \infty)$.

We let $g(x) = (x + 5)(x - 2)$ and test a value in each interval.

$$(-\infty, -5): \; g(-6) = 8 > 0$$
$$(-5, 2): \; g(0) = -10 < 0$$
$$(2, \infty): \; g(3) = 8 > 0$$

Functions values are positive on $(-\infty, -5)$ and $(2, \infty)$. Since the equality symbol is $\geq$, the endpoints of the intervals must be included in the solution set. It is $(-\infty, -5] \cup [2, \infty)$.

103. $f(x) = \dfrac{1}{\sqrt{5 - |7x + 2|}}$

We cannot take the square root of a negative number; neither can the denominator be zero. Thus we have $5 - |7x + 2| > 0$.

$$5 - |7x + 2| > 0 \quad \text{Polynomial inequality}$$
$$|7x + 2| < 5$$
$$-5 < 7x + 2 < 5$$
$$-7 < 7x < 3$$
$$-1 < x < \frac{3}{7}$$

The solution set is $\left(-1, \dfrac{3}{7}\right)$.

Chapter 3 Test

1. $f(x) = 2x^3 + 6x^2 - x^4 + 11$
$$= -x^4 + 2x^3 + 6x^2 + 11$$

The degree of the polynomial is 4, so the polynomial is quartic. The leading term is $-x^4$ and the leading coefficient is -1.

2. $h(x) = -4.7x + 29$

The degree of the polynomial is 1, so the polynomial is linear. The leading term is $-4.7x$ and the leading coefficient is -4.7.

3. $f(x) = x(3x - 5)(x - 3)^2(x + 1)^3$

The zeros of the function are 0, $\dfrac{5}{3}$, 3, and -1.

The factors x and $3x - 5$ each occur once, so the zeros 0 and $\dfrac{5}{3}$ have multiplicity 1.

The factor $x - 3$ occurs twice, so the zero 3 has multiplicity 2.

The factor $x + 1$ occurs three times, so the zero -1 has multiplicity 3.

4. For 1986, $x = 1986 - 1980 = 6$.

$f(6) = 0.152(6)^4 - 9.786(6)^3 + 173.317(6)^2 - 831.220(6) + 4053.134 \approx 3388$

In 1986, 3388 young people were killed by gunfire.

For 1993, $x = 1993 - 1980 = 13$.

$f(13) = 0.152(13)^4 - 9.786(13)^3 + 173.317(13)^2 - 831.220(13) + 4053.134 \approx 5379$

In 1993, 5379 young people were killed by gunfire.

For 1999, $x = 1999 - 1980 = 19$.

$f(19) = 0.152(19)^4 - 9.786(19)^3 + 173.317(19)^2 - 831.220(19) + 4053.134 \approx 3514$

In 1999, 3514 young people were killed by gunfire.

5. $f(x) = x^3 - x^2 - 6x$

1. The leading term is x^3. The degree, 3, is odd and the leading coefficient, 1, is positive so as $x \to \infty$, $f(x) \to \infty$ and as $x \to -\infty$, $f(x) \to -\infty$.

2. We solve $f(x) = 0$.

$$x^3 - x^2 - 6x = 0$$
$$x(x^2 - x - 6) = 0$$
$$x(x - 3)(x + 2) = 0$$
$$x = 0 \; \text{ or } \; x - 3 = 0 \; \text{ or } \; x + 2 = 0$$
$$x = 0 \; \text{ or } \qquad x = 3 \; \text{ or } \qquad x = -2$$

The zeros of the function are -2, 0, and 3, so the x-intercepts are $(-2, 0)$, $(0, 0)$, and $(3, 0)$.

3. The zeros divide the x-axis in 4 intervals, $(-\infty, -2)$, $(-2, 0)$, $(0, 3)$, and $(3, \infty)$. We choose a value for x in each interval and find $f(x)$. This tells us the sign of $f(x)$ for all values of x in that interval.

In $(-\infty, -2)$, test -3:

$f(-3) = (-3)^3 - (-3)^2 - 6(-3) = -18 < 0$

In $(-2, 0)$, test -1:

$f(-1) = (-1)^3 - (-1)^2 - 6(-1) = 4 > 0$

In $(0, 3)$, test 1:

$f(1) = 1^3 - 1^2 - 6 \cdot 1 = -6 < 0$

In $(3, \infty)$, test 4:

$f(4) = 4^3 - 4^2 - 6 \cdot 4 = 24 > 0$

Thus the graph lies below the x-axis on $(-\infty, -2)$ and on $(0, 3)$ and above the x-axis on $(-2, 0)$ and $(3, \infty)$. We also know the points $(-3, -18)$, $(-1, 4)$, $(1, -6)$, and $(4, 24)$ are on the graph.

4. From Step 2 we know that the y-intercept is $(0, 0)$.

5. We find additional points on the graph and draw the graph.

x	$f(x)$
-1.5	3.375
-0.5	2.625
0.5	-3.125
2.5	-5.625

$f(x) = x^3 - x^2 - 6x$

6. Checking the graph as described on page 272 in the text, we see that it appears to be correct.

6. $f(x) = -2x^4 + x^3 + 11x^2 - 4x - 12$

1. The leading term is $-2x^4$. The degree, 4, is even and the leading coefficient, -2, is negative so $x \to \infty$, $f(x) \to -\infty$ and as $x \to -\infty$, $f(x) \to -\infty$.

2. We solve $f(x) = 0$.

   The possible rational zeros are $\pm 1, \pm 2, \pm 3, \pm 4, \pm 6, \pm 12, \pm\dfrac{1}{2}$, and $\pm\dfrac{3}{2}$. We try -1.

$$\begin{array}{r|rrrrr} -1 & -2 & 1 & 11 & -4 & -12 \\ & & 2 & -3 & -8 & 12 \\ \hline & -2 & 3 & 8 & -12 & 0 \end{array}$$

   We have $f(x) = (x+1)(-2x^3 + 3x^2 + 8x - 12)$. Find the other zeros.

$$-2x^3 + 3x^2 + 8x - 12 = 0$$
$$2x^3 - 3x^2 - 8x + 12 = 0 \quad \text{Multiplying by } -1$$
$$x^2(2x - 3) - 4(2x - 3) = 0$$
$$(2x - 3)(x^2 - 4) = 0$$
$$(2x - 3)(x + 2)(x - 2) = 0$$
$$2x - 3 = 0 \quad \text{or} \quad x + 2 = 0 \quad \text{or} \quad x - 2 = 0$$
$$2x = 3 \quad \text{or} \qquad x = -2 \quad \text{or} \qquad x = 2$$
$$x = \frac{3}{2} \quad \text{or} \qquad x = -2 \quad \text{or} \qquad x = 2$$

   The zeros of the function are -1, $\dfrac{3}{2}$, -2, and 2, so the x-intercepts are $(-2, 0)$, $(-1, 0)$, $\left(\dfrac{3}{2}, 0\right)$, and $(2, 0)$.

3. The zeros divide the x-axis into 5 intervals, $(-\infty, -2)$, $(-2, -1)$, $\left(-1, \dfrac{3}{2}\right)$, $\left(\dfrac{3}{2}, 2\right)$, and $(2, \infty)$. We choose a value for x in each interval and find $f(x)$. This tells us the sign of $f(x)$ for all values of x in that interval.

In $(-\infty, -2)$, test -3: $f(-3) = -90 < 0$

In $(-2, -1)$, test -1.5: $f(-1.5) = 5.25 > 0$

In $\left(-1, \dfrac{3}{2}\right)$, test 0: $f(0) = -12 < 0$

In $\left(\dfrac{3}{2}, 2\right)$, test 1.75: $f(1.75) \approx 1.29 > 0$

In $(2, \infty)$, test 3: $f(3) = -60 < 0$

Thus the graph lies below the x-axis on $(-\infty, -2)$, on $\left(-1, \dfrac{3}{2}\right)$, and on $(2, \infty)$ and above the x-axis on $(-2, -1)$ and on $\left(\dfrac{3}{2}, 2\right)$. We also know the points $(-3, -90)$, $(-1.5, 5.25)$, $(0, -12)$, $(1.75, 1.29)$, and $(3, -60)$ are on the graph.

4. From Step 3 we know that $f(0) = -12$, so the y-intercept is $(0, -12)$.

5. We find additional points on the graph and draw the graph.

x	$f(x)$
-0.5	-7.5
0.5	-11.25
2.5	-15.75

$f(x) = -2x^4 + x^3 + 11x^2 - 4x - 12$

6. Checking the graph as described on page 272 in the text, we see that it appears to be correct.

7. $f(0) = -5 \cdot 0^2 + 3 = 3$

$f(2) = -5 \cdot 2^2 + 3 = -17$

By the intermediate value theorem, since $f(0)$ and $f(2)$ have opposite signs, $f(x)$ has a zero between 0 and 2.

8. $g(-2) = 2(-2)^3 + 6(-2)^2 - 3 = 5$

$g(-1) = 2(-1)^3 + 6(-1)^2 - 3 = 1$

Since both $g(-2)$ and $g(-1)$ are positive, we cannot use the intermediate value theorem to determine if there is a zero between -2 and -1.

9.

$$\begin{array}{r} x^3 + 4x^2 + 4x + 6 \\ x - 1 \overline{\smash{\big)}\ x^4 + 3x^3 + 0x^2 + 2x - 5} \\ \underline{x^4 - x^3} \\ 4x^3 + 0x^2 \\ \underline{4x^3 - 4x^2} \\ 4x^2 + 2x \\ \underline{4x^2 - 4x} \\ 6x - 5 \\ \underline{6x - 6} \\ 1 \end{array}$$

The quotient is $x^3 + 4x^2 + 4x + 6$; the remainder is 1.

$P(x) = (x - 1)(x^3 + 4x^2 + 4x + 6) + 1$

10.

$$\begin{array}{r|rrrr} 5 & 3 & 0 & -12 & 7 \\ & & 15 & 75 & 315 \\ \hline & 3 & 15 & 63 & \underline{|\ 322} \end{array}$$

$Q(x) = 3x^2 + 15x + 63;\ R(x) = 322$

11.

$$\begin{array}{r|rrrr} -3 & 2 & -6 & 1 & -4 \\ & & -6 & 36 & -111 \\ \hline & 2 & -12 & 37 & \underline{|-115} \end{array}$$

$P(-3) = -115$

12.

$$\begin{array}{r|rrrr} -2 & 1 & 4 & 1 & -6 \\ & & -2 & -4 & 6 \\ \hline & 1 & 2 & -3 & \underline{|\ 0} \end{array}$$

$f(-2) = 0$, so -2 is a zero of $f(x)$.

13. The function can be written in the form

$$f(x) = a_n(x+3)^2(x)(x-6).$$

The simplest polynomial is obtained if we let $a_n = 1$.

$$\begin{aligned} f(x) &= (x+3)^2(x)(x-6) \\ &= (x^2 + 6x + 9)(x^2 - 6x) \\ &= x^4 + 6x^3 + 9x^2 - 6x^3 - 36x^2 - 54x \\ &= x^4 - 27x^2 - 54x \end{aligned}$$

14. A polynomial function of degree 5 can have at most 5 zeros. Since $f(x)$ has rational coefficients, in addition to the 3 given zeros, the other zeros are the conjugates of $\sqrt{3}$ and $2 - i$, or $-\sqrt{3}$ and $2 + i$.

15. $-3i$ is also a zero.

$$\begin{aligned} f(x) &= (x+10)(x-3i)(x+3i) \\ &= (x+10)(x^2+9) \\ &= x^3 + 10x^2 + 9x + 90 \end{aligned}$$

16. $\sqrt{3}$ and $1 + i$ are also zeros.

$$\begin{aligned} f(x) &= (x-0)(x+\sqrt{3})(x-\sqrt{3})[x-(1-i)][x-(1+i)] \\ &= x(x^2-3)[(x-1)+i][(x-1)-i] \\ &= (x^3-3x)(x^2-2x+1+1) \\ &= (x^3-3x)(x^2-2x+2) \\ &= x^5 - 2x^4 + 2x^3 - 3x^3 + 6x^2 - 6x \\ &= x^5 - 2x^4 - x^3 + 6x^2 - 6x \end{aligned}$$

17. $f(x) = 2x^3 + x^2 - 2x + 12$

$$\frac{\text{Possibilities for } p}{\text{Possibilities for } q} : \frac{\pm 1, \pm 2, \pm 3, \pm 4, \pm 6, \pm 12}{\pm 1, \pm 2}$$

Possibilities for p/q: $\pm 1, \pm 2, \pm 3, \pm 4, \pm 6, \pm 12, \pm\dfrac{1}{2}, \pm\dfrac{3}{2}$

18. $h(x) = 10x^4 - x^3 + 2x - 5$

$$\frac{\text{Possibilities for } p}{\text{Possibilities for } q} : \frac{\pm 1, \pm 5}{\pm 1, \pm 2, \pm 5, \pm 10}$$

Possibilities for p/q: $\pm 1, \pm 5, \pm\dfrac{1}{2}, \pm\dfrac{5}{2}, \pm\dfrac{1}{5}, \pm\dfrac{1}{10}$

19. $f(x) = x^3 + x^2 - 5x - 5$

a) Possibilities for p/q: $\pm 1, \pm 5$

From the graph of $y = f(x)$, we see that -1 might be a zero.

$$\begin{array}{r|rrrr} -1 & 1 & 1 & -5 & -5 \\ & & -1 & 0 & 5 \\ \hline & 1 & 0 & -5 & \underline{|\ 0} \end{array}$$

We have $f(x) = (x+1)(x^2-5)$. We find the other zeros.

$$\begin{aligned} x^2 - 5 &= 0 \\ x^2 &= 5 \\ x &= \pm\sqrt{5} \end{aligned}$$

The rational zero is -1. The other zeros are $\pm\sqrt{5}$.

b) $f(x) = (x+1)(x-\sqrt{5})(x+\sqrt{5})$

20. $g(x) = 2x^4 - 11x^3 + 16x^2 - x - 6$

a) Possibilities for p/q: $\pm 1, \pm 2, \pm 3, \pm 6, \pm\dfrac{1}{2}, \pm\dfrac{3}{2}$

From the graph of $y = g(x)$, we see that $-\dfrac{1}{2}$, 1, 2, and 3 might be zeros. We try $-\dfrac{1}{2}$.

$$\begin{array}{r|rrrrr} -\frac{1}{2} & 2 & -11 & 16 & -1 & -6 \\ & & -1 & 6 & -11 & 6 \\ \hline & 2 & -12 & 22 & -12 & \underline{|\ 0} \end{array}$$

Now we try 1.

$$\begin{array}{r|rrrr} 1 & 2 & -12 & 22 & -12 \\ & & 2 & -10 & 12 \\ \hline & 2 & -10 & 12 & \underline{|\ 0} \end{array}$$

We have $g(x) = \left(x + \dfrac{1}{2}\right)(x-1)(2x^2 - 10x + 12) = 2\left(x + \dfrac{1}{2}\right)(x-1)(x^2 - 5x + 6)$. We find the other zeros.

$$\begin{aligned} x^2 - 5x + 6 &= 0 \\ (x-2)(x-3) &= 0 \\ x - 2 = 0 \ &or\ x - 3 = 0 \\ x = 2 \ &or\ \quad x = 3 \end{aligned}$$

The rational zeros are $-\dfrac{1}{2}$, 1, 2, and 3. These are the only zeros.

b) $$\begin{aligned} g(x) &= 2\left(x + \dfrac{1}{2}\right)(x-1)(x-2)(x-3) \\ &= (2x+1)(x-1)(x-2)(x-3) \end{aligned}$$

21. $h(x) = x^3 + 4x^2 + 4x + 16$

a) Possibilities for p/q: $\pm 1, \pm 2, \pm 4, \pm 8, \pm 16$

From the graph of $h(x)$, we see that -4 might be a zero.

$$\begin{array}{r|rrrr} -4 & 1 & 4 & 4 & 16 \\ & & -4 & 0 & -16 \\ \hline & 1 & 0 & 4 & \underline{|\ 0} \end{array}$$

We have $h(x) = (x+4)(x^2+4)$. We find the other zeros.

$$x^2 + 4 = 0$$
$$x^2 = -4$$
$$x = \pm 2i$$

The rational zero is -4. The other zeros are $\pm 2i$.

b) $h(x) = (x+4)(x+2i)(x-2i)$

22. $f(x) = 3x^4 - 11x^3 + 15x^2 - 9x + 2$

a) Possibilities for p/q: $\pm 1, \pm 2, \pm \dfrac{1}{3}, \pm \dfrac{2}{3}$

From the graph of $f(x)$, we see that $\dfrac{2}{3}$ and 1 might be zeros. We try $\dfrac{2}{3}$.

$$\begin{array}{r|rrrrr} \tfrac{2}{3} & 3 & -11 & 15 & -9 & 2 \\ & & 2 & -6 & 6 & -2 \\ \hline & 3 & -9 & 9 & -3 & 0 \end{array}$$

Now we try 1.

$$\begin{array}{r|rrrr} 1 & 3 & -9 & 9 & -3 \\ & & 3 & -6 & 3 \\ \hline & 3 & -6 & 3 & 0 \end{array}$$

We have $f(x) = \left(x - \dfrac{2}{3}\right)(x-1)(3x^2 - 6x + 3) =$

$3\left(x - \dfrac{2}{3}\right)(x-1)(x^2 - 2x + 1)$. We find the other zeros.

$$x^2 - 2x + 1 = 0$$
$$(x-1)(x-1) = 0$$
$$x - 1 = 0 \quad or \quad x - 1 = 0$$
$$x = 1 \quad or \quad x = 1$$

The rational zeros are $\dfrac{2}{3}$ and 1. (The zero 1 has multiplicity 3.) These are the only zeros.

b) $f(x) = 3\left(x - \dfrac{2}{3}\right)(x-1)(x-1)(x-1)$
$$= (3x-2)(x-1)^3$$

23. $g(x) = -x^8 + 2x^6 - 4x^3 - 1$

There are 2 variations in sign in $g(x)$, so there are 2 or 0 positive real zeros.

$$g(-x) = -(-x)^8 + 2(-x)^6 - 4(-x)^3 - 1$$
$$= -x^8 + 2x^6 + 4x^3 - 1$$

There are 2 variations in sign in $g(-x)$, so there are 2 or 0 negative real zeros.

24. $f(x) = \dfrac{2}{(x-3)^2}$

1. The denominator is zero when $x = 3$, so the domain excludes 3. it is $(-\infty, 3) \cup (3, \infty)$. The line $x = 3$ is the vertical asymptote.

2. Because the degree of the numerator is less than the degree of the denominator, the x-axis, or $y = 0$, is the horizontal asymptote.

3. The numerator has no zeros, so there is no x-intercept.

4. $f(0) = \dfrac{2}{(0-3)^2} = \dfrac{2}{9}$, so the y-intercept is $\left(0, \dfrac{2}{9}\right)$.

5. Find other function values to determine the shape of the graph and then draw it.

25. $f(x) = \dfrac{x+3}{x^2 - 3x - 4} = \dfrac{x+3}{(x+1)(x-4)}$

1. The denominator is zero when $x = -1$ or $x = 4$, so the domain excludes -1 and 4. It is $(-\infty, -1) \cup (-1, 4) \cup (4, \infty)$. The lines $x = -1$ and $x = 4$ are vertical asymptotes.

2. Because the degree of the numerator is less than the degree of the denominator, the x-axis, or $y = 0$, is the horizontal asymptote.

3. The numerator is zero at $x = -3$, so the x-intercept is $(-3, 0)$.

4. $f(0) = \dfrac{0+3}{(0+1)(0-4)} = -\dfrac{3}{4}$, so the y-intercept is $\left(0, -\dfrac{3}{4}\right)$.

5. Find other function values to determine the shape of the graph and then draw it.

26. Answers may vary. The numbers -1 and 2 must be zeros of the denominator, and -4 must be a zero of the numerator.

$$f(x) = \dfrac{x+4}{(x+1)(x-2)}, \text{ or } f(x) = \dfrac{x+4}{x^2 - x - 2}$$

27.
$$\begin{array}{ll} 2x^2 > 5x + 3 & \text{Polynomial inequality} \\ 2x^2 - 5x - 3 > 0 & \text{Equivalent inequality} \\ 2x^2 - 5x - 3 = 0 & \text{Related equation} \\ (2x+1)(x-3) = 0 & \text{Factoring} \end{array}$$

The solutions of the related equation are $-\dfrac{1}{2}$ and 3. These numbers divide the x-axis into the intervals $\left(-\infty, -\dfrac{1}{2}\right)$, $\left(-\dfrac{1}{2}, 3\right)$, and $(3, \infty)$.

We let $f(x) = (2x + 1)(x - 3)$ and test a value in each interval.

$$\left(-\infty, -\frac{1}{2}\right): \ f(-1) = 4 > 0$$

$$\left(-\frac{1}{2}, 3\right): \ f(0) = -3 < 0$$

$$(3, \infty): \ f(4) = 9 > 0$$

Function values are positive on $\left(-\infty, -\frac{1}{2}\right)$ and $(3, \infty)$.

The solution set is $\left(-\infty, -\frac{1}{2}\right) \cup (3, \infty)$.

28. $\quad \dfrac{x+1}{x-4} \leq 3$ Rational inequality

$\dfrac{x+1}{x-4} - 3 \leq 0$ Equivalent inequality

The denominator of $f(x) = \dfrac{x+1}{x-4} - 3$ is zero when $x = 4$, so the function is not defined for this value of x. We solve the related equation $f(x) = 0$.

$$\frac{x+1}{x-4} - 3 = 0$$

$$(x-4)\left(\frac{x+1}{x-4} - 3\right) = (x-4) \cdot 0$$

$$x + 1 - 3(x-4) = 0$$

$$x + 1 - 3x + 12 = 0$$

$$-2x + 13 = 0$$

$$2x = 13$$

$$x = \frac{13}{2}$$

The critical values are 4 and $\dfrac{13}{2}$. They divide the x-axis into the intervals $(-\infty, 4)$, $\left(4, \dfrac{13}{2}\right)$ and $\left(\dfrac{13}{2}, \infty\right)$. We test a value in each interval.

$$(-\infty, 4): \ f(3) = -7 < 0$$

$$\left(4, \frac{13}{2}\right): \ f(5) = 3 > 0$$

$$\left(\frac{13}{2}, \infty\right): \ f(9) = -1 < 0$$

Function values are negative for $(-\infty, 4)$ and $\left(\dfrac{13}{2}, \infty\right)$. Since the inequality symbol is $\leq$, the endpoint of the interval $\left(\dfrac{13}{2}, \infty\right)$ must be included in the solution set. It is $(-\infty, 4) \cup \left[\dfrac{13}{2}, \infty\right)$.

29. a) We write and solve a polynomial equation.

$$-16t^2 + 64t + 192 = 0$$

$$-16(t^2 - 4t - 12) = 0$$

$$-16(t + 2)(t - 6) = 0$$

The solutions are $t = -2$ and $t = 6$. Only $t = 6$ has meaning in this application. The rocket reaches the ground at $t = 6$ seconds.

b) We write and solve a polynomial inequality.

$$-16t^2 + 64t + 192 > 240$$

$$-16t^2 + 64t - 48 > 0$$

$$-16t^2 + 64t - 48 = 0 \quad \text{Related equation}$$

$$-16(t^2 - 4t + 3) = 0$$

$$-16(t - 1)(t - 3) = 0$$

The solutions of the related equation are 1 and 3. These numbers divide the t-axis into the intervals $(-\infty, 1)$, $(1, 3)$ and $(3, \infty)$. Because the rocket returns to the ground at $t = 6$, we restrict our discussion to values of t such that $0 \leq t \leq 6$. We consider the intervals $[0, 1)$, $(1, 3)$ and $(3, 6]$. We let $f(t) = -16t^2 + 64t - 48$ and test a value in each interval.

$$[0, 1): \ f\left(\frac{1}{2}\right) = -20 < 0$$

$$(1, 3): \ f(2) = 16 > 0$$

$$(3, 6]: \ f(4) = -48 < 0$$

Function values are positive on $(1, 3)$. The solution set is $(1, 3)$.

30. $\quad y = \dfrac{k}{x}$

$$5 = \frac{k}{6}$$

$$30 = k \quad \text{Variation constant}$$

Equation of variation: $y = \dfrac{30}{x}$

31. $\quad d = kr^2$

$$200 = k \cdot 60^2$$

$$\frac{1}{18} = k \quad \text{Variation constant}$$

$$d = \frac{1}{18}r^2 \quad \text{Equation of variation}$$

$$d = \frac{1}{18} \cdot 30^2$$

$$d = 50 \text{ ft}$$

32. $\quad y = \dfrac{kxz^2}{w}$

$$100 = \frac{k(0.1)(10)^2}{5}$$

$$100 = 2k$$

$$50 = k \quad \text{Variation constant}$$

$$y = \frac{50xz^2}{w} \quad \text{Equation of variation}$$

33. $f(x) = \sqrt{x^2 + x - 12}$

Since we cannot take the square root of a negative number, then $x^2 + x - 12 \geq 0$.

$$x^2 + x - 12 \geq 0 \text{ Polynomial inequality}$$

$$x^2 + x - 12 = 0 \text{ Related equation}$$

$$(x + 4)(x - 3) = 0 \text{ Factoring}$$

The solutions of the related equation are -4 and 3. These numbers divide the x-axis into the intervals $(-\infty, -4)$, $(-4, 3)$ and $(3, \infty)$.

We let $f(x) = (x + 4)(x - 3)$ and test a value in each interval.

$(-\infty, -4):\ g(-5) = 8 > 0$

$(-4, 3):\ g(0) = -12 < 0$

$(3, \infty):\ g(4) = 8 > 0$

Functions values are positive on $(-\infty, -4)$ and $(3, \infty)$. Since the inequality symbol is $\geq$, the endpoints of the intervals must be included in the solution set. It is $(-\infty, -4] \cup [3, \infty)$.

Chapter 4

Exponential and Logarithmic Functions

Exercise Set 4.1

1. We interchange the first and second coordinates of each ordered pair to find the inverse of the relation. It is
$$\{(8,7),\ (8,-2),\ (-4,3),\ (-8,8)\}.$$

3. We interchange the first and second coordinates of each ordered pair to find the inverse of the relation. It is
$$\{(-1,-1),\ (4,-3)\}.$$

5. Interchange x and y.
$$y = 4x - 5$$
$$\downarrow \qquad \downarrow$$
$$x = 4y - 5$$

7. Interchange x and y.
$$x^3 y = -5$$
$$\downarrow\downarrow$$
$$y^3 x = -5$$

9. Interchange x and y.
$$x = y^2 - 2y$$
$$\downarrow \qquad \downarrow \quad \downarrow$$
$$y = x^2 - 2x$$

11. Graph $x = y^2 - 3$. Some points on the graph are $(-3,0)$, $(-2,-1)$, $(-2,1)$, $(1,-2)$, and $(1,2)$. Plot these points and draw the curve. Then reflect the graph across the line $y = x$.

13. Graph $y = 3x - 2$. The intercepts are $(0,-2)$ and $\left(\dfrac{2}{3}, 0\right)$.

Plot these points and draw the line. Then reflect the graph across the line $y = x$.

15. Graph $y = |x|$. Some points on the graph are $(0,0)$, $(-2,2)$, $(2,2)$, $(-5,5)$, and $(5,5)$. Plot these points and draw the graph. Then reflect the graph across the line $y = x$.

17. We show that if $f(a) = f(b)$, then $a = b$.
$$\frac{1}{3}a - 6 = \frac{1}{3}b - 6$$
$$\frac{1}{3}a = \frac{1}{3}b \qquad \text{Adding 6}$$
$$a = b \qquad \text{Multiplying by 3}$$
Thus f is one-to-one.

19. We show that if $f(a) = f(b)$, then $a = b$.
$$a^3 + \frac{1}{2} = b^3 + \frac{1}{2}$$
$$a^3 = b^3 \qquad \text{Subtracting } \frac{1}{2}$$
$$a = b \qquad \text{Taking cube roots}$$
Thus f is one-to-one.

21. $g(-1) = 1 - (-1)^2 = 1 - 1 = 0$ and $g(1) = 1 - 1^2 = 1 - 1 = 0$, so $g(-1) = g(1)$ but $-1 \neq 1$. Thus the function is not one-to-one.

23. $f(-2) = (-2)^4 - (-2)^2 = 16 - 4 = 12$ and $f(2) = 2^4 - 2^2 = 16 - 4 = 12$, so $f(-2) = f(2)$ but $-2 \neq 2$. Thus the function is not one-to-one.

25. The function is one-to-one, because no horizontal line crosses the graph more than once.

27. The function is not one-to-one, because there are many horizontal lines that cross the graph more than once.

29. The function is not one-to-one, because there are many horizontal lines that cross the graph more than once.

31. The function is one-to-one, because no horizontal line crosses the graph more than once.

33. The graph of $f(x) = 5x - 8$ is shown below.

Since there is no horizontal line that crosses the graph more than once, the function is one-to-one.

35. The graph of $f(x) = 1 - x^2$ is shown below.

Since there are many horizontal lines that cross the graph more than once, the function is not one-to-one.

37. The graph of $f(x) = |x + 2|$ is shown below.

Since there are many horizontal lines that cross the graph more than once, the function is not one-to-one.

39. The graph of $f(x) = -\dfrac{4}{x}$ is shown below.

Since there is no horizontal line that crosses the graph more than once, the function is one-to-one.

41. The graph of $f(x) = \dfrac{2}{3}$ is shown below.

Since the horizontal line $y = \dfrac{2}{3}$ crosses the graph more than once, the function is not one-to-one.

43. The graph of $f(x) = \sqrt{25 - x^2}$ is shown below.

Since there are many horizontal lines that cross the graph more than once, the function is not one-to-one.

45. a) The graph of $f(x) = x + 4$ is shown below. It passes the horizontal-line test, so it is one-to-one.

b) Replace $f(x)$ with y: $y = x + 4$

Interchange x and y: $x = y + 4$

Solve for y: $x - 4 = y$

Replace y with $f^{-1}(x)$: $f^{-1}(x) = x - 4$

47. a) The graph of $f(x) = 2x - 1$ is shown below. It passes the horizontal-line test, so it is one-to-one.

b) Replace $f(x)$ with y: $y = 2x - 1$

Interchange x and y: $x = 2y - 1$

Solve for y: $\dfrac{x + 1}{2} = y$

Replace y with $f^{-1}(x)$: $f^{-1}(x) = \dfrac{x + 1}{2}$

49. a) The graph of $f(x) = \dfrac{4}{x + 7}$ is shown below. It passes the horizontal-line test, so the function is one-to-one.

b) Replace $f(x)$ with y: $y = \dfrac{4}{x + 7}$

Interchange x and y: $x = \dfrac{4}{y + 7}$

Solve for y: $x(y + 7) = 4$

$$y + 7 = \frac{4}{x}$$

$$y = \frac{4}{x} - 7$$

Replace y with $f^{-1}(x)$: $f^{-1}(x) = \dfrac{4}{x} - 7$

51. a) The graph of $f(x) = \dfrac{x + 4}{x - 3}$ is shown below. It passes the horizontal-line test, so the function is one-to-one.

b) Replace $f(x)$ with y: $y = \dfrac{x + 4}{x - 3}$

Interchange x and y: $x = \dfrac{y + 4}{y - 3}$

Solve for y: $(y - 3)x = y + 4$

$$xy - 3x = y + 4$$

$$xy - y = 3x + 4$$

$$y(x - 1) = 3x + 4$$

$$y = \frac{3x + 4}{x - 1}$$

Replace y with $f^{-1}(x)$: $f^{-1}(x) = \dfrac{3x + 4}{x - 1}$

53. a) The graph of $f(x) = x^3 - 1$ is shown below. It passes the horizontal-line test, so the function is one-to-one.

b) Replace $f(x)$ with y: $y = x^3 - 1$

Interchange x and y: $x = y^3 - 1$

Solve for y: $x + 1 = y^3$

$$\sqrt[3]{x+1} = y$$

Replace y with $f^{-1}(x)$: $f^{-1}(x) = \sqrt[3]{x+1}$

55. a) The graph of $f(x) = x\sqrt{4 - x^2}$ is shown below. Since there are many horizontal lines that cross the graph more than once, the function is not one-to-one and thus does not have an inverse that is a function.

57. a) The graph of $f(x) = 5x^2 - 2$, $x \geq 0$ is shown below. It passes the horizontal-line test, so it is one-to-one.

b) Replace $f(x)$ with y: $y = 5x^2 - 2$

Interchange x and y: $x = 5y^2 - 2$

Solve for y: $x + 2 = 5y^2$

$$\frac{x+2}{5} = y^2$$

$$\sqrt{\frac{x+2}{5}} = y$$

(We take the principal square root, because $x \geq 0$ in the original equation.)

Replace y with $f^{-1}(x)$: $f^{-1}(x) = \sqrt{\dfrac{x+2}{5}}$ for all x

in the range of $f(x)$, or $f^{-1}(x) = \sqrt{\dfrac{x+2}{5}}$, $x \geq -2$.

59. a) The graph of $f(x) = \sqrt{x+1}$ is shown below. It passes the horizontal-line test, so the function is one-to-one.

b) Replace $f(x)$ with y: $y = \sqrt{x+1}$

Interchange x and y: $x = \sqrt{y+1}$

Solve for y: $x^2 = y + 1$

$$x^2 - 1 = y$$

Replace y with $f^{-1}(x)$: $f^{-1}(x) = x^2 - 1$ for all x in the range of $f(x)$, or $f^{-1}(x) = x^2 - 1$, $x \geq 0$.

61. $f(x) = 3x$

The function f multiplies an input by 3. Then to reverse this procedure, f^{-1} would divide each of its inputs by 3. Thus, $f^{-1}(x) = \dfrac{x}{3}$, or $f^{-1}(x) = \dfrac{1}{3}x$.

63. $f(x) = -x$

The outputs of f are the opposites, or additive inverses, of the inputs. Then the outputs of f^{-1} are the opposites of its inputs. Thus, $f^{-1}(x) = -x$.

65. $f(x) = \sqrt[3]{x-5}$

The function f subtracts 5 from each input and then takes the cube root of the result. To reverse this procedure, f^{-1} would raise each input to the third power and then add 5 to the result. Thus, $f^{-1}(x) = x^3 + 5$.

67. We reflect the graph of f across the line $y = x$. The reflections of the labeled points are $(-5, -5)$, $(-3, 0)$, $(1, 2)$, and $(3, 5)$.

69. We reflect the graph of f across the line $y = x$. The reflections of the labeled points are $(-6, -2)$, $(1, -1)$, $(2, 0)$, and $(5.375, 1.5)$.

71. We reflect the graph of f across the line $y = x$.

73. We find $(f^{-1} \circ f)(x)$ and $(f \circ f^{-1})(x)$ and check to see that each is x.

$$(f^{-1} \circ f)(x) = f^{-1}(f(x)) = f^{-1}\left(\frac{7}{8}x\right) =$$

$$\frac{8}{7}\left(\frac{7}{8}x\right) = x$$

$$(f \circ f^{-1})(x) = f(f^{-1}(x)) = f\left(\frac{8}{7}x\right) = \frac{7}{8}\left(\frac{8}{7}x\right) = x$$

75. We find $(f^{-1} \circ f)(x)$ and $(f \circ f^{-1})(x)$ and check to see that each is x.

$$(f^{-1} \circ f)(x) = f^{-1}(f(x)) = f^{-1}\left(\frac{1-x}{x}\right) =$$

$$\frac{1}{\frac{1-x}{x}+1} = \frac{1}{\frac{1-x+x}{x}} = \frac{1}{\frac{1}{x}} = x$$

$$(f \circ f^{-1})(x) = f(f^{-1}(x)) = f\left(\frac{1}{x+1}\right) =$$

$$\frac{1-\frac{1}{x+1}}{\frac{1}{x+1}} = \frac{\frac{x+1-1}{x+1}}{\frac{1}{x+1}} = \frac{\frac{x}{x+1}}{\frac{1}{x+1}} = x$$

77. $(f^{-1} \circ f)(x) = f^{-1}(f(x)) = \dfrac{5\left(\dfrac{2}{5}x+1\right)-5}{2} =$

$$\frac{2x+5-5}{2} = \frac{2x}{2} = x$$

$$(f \circ f^{-1})(x) = f(f^{-1}(x)) = \frac{2}{5}\left(\frac{5x-5}{2}\right) + 1 =$$

$$x - 1 + 1 = x$$

79. Replace $f(x)$ with y: $y = 5x - 3$

Interchange x and y: $x = 5y - 3$

Solve for y: $x + 3 = 5y$

$$\frac{x+3}{5} = y$$

Replace y with $f^{-1}(x)$: $f^{-1}(x) = \dfrac{x+3}{5}$, or $\dfrac{1}{5}x + \dfrac{3}{5}$

The domain and range of f are $(-\infty, \infty)$, so the domain and range of f^{-1} are also $(-\infty, \infty)$.

81. Replace $f(x)$ with y: $y = \dfrac{2}{x}$

Interchange x and y: $x = \dfrac{2}{y}$

Solve for y: $xy = 2$

$$y = \frac{2}{x}$$

Replace y with $f^{-1}(x)$: $f^{-1}(x) = \dfrac{2}{x}$

The domain and range of f are $(-\infty, 0) \cup (0, \infty)$, so the domain and range of f^{-1} are also $(-\infty, 0) \cup (0, \infty)$.

83. Replace $f(x)$ with y: $y = \dfrac{1}{3}x^3 - 2$

Interchange x and y: $x = \dfrac{1}{3}y^3 - 2$

Solve for y: $x + 2 = \dfrac{1}{3}y^3$

$$3x + 6 = y^3$$

$$\sqrt[3]{3x+6} = y$$

Replace y with $f^{-1}(x)$: $f^{-1}(x) = \sqrt[3]{3x+6}$

The domain and range of f are $(-\infty, \infty)$, so the domain and range of f^{-1} are also $(-\infty, \infty)$.

85. Replace $f(x)$ with y: $y = \dfrac{x+1}{x-3}$

Interchange x and y: $x = \dfrac{y+1}{y-3}$

Solve for y: $xy - 3x = y + 1$

$$xy - y = 3x + 1$$
$$y(x - 1) = 3x + 1$$
$$y = \dfrac{3x+1}{x-1}$$

Replace y with $f^{-1}(x)$: $f^{-1}(x) = \dfrac{3x+1}{x-1}$

The domain of f is $(-\infty, 3) \cup (3, \infty)$ and the range of f is $(-\infty, 1) \cup (1, \infty)$. Thus the domain of f^{-1} is $(-\infty, 1) \cup (1, \infty)$ and the range of f^{-1} is $(-\infty, 3) \cup (3, \infty)$.

87. Since $f(f^{-1}(x)) = f^{-1}(f(x)) = x$, then $f(f^{-1}(5)) = 5$ and $f^{-1}(f(a)) = a$.

89. a) $s(x) = \dfrac{2x-3}{2}$

$$s(5) = \dfrac{2 \cdot 5 - 3}{2} = \dfrac{10-3}{2} = \dfrac{7}{2} = 3\dfrac{1}{2}$$

$$s\left(7\dfrac{1}{2}\right) = s(7.5) = \dfrac{2 \cdot 7.5 - 3}{2} = \dfrac{15-3}{2} = \dfrac{12}{2} = 6$$

$$s(8) = \dfrac{2 \cdot 8 - 3}{2} = \dfrac{16-3}{2} = \dfrac{13}{2} = 6\dfrac{1}{2}$$

b) The graph of $s(x)$ passes the horizontal-line test and thus has an inverse that is a function.

Replace $f(x)$ with y: $y = \dfrac{2x-3}{2}$

Interchange x and y: $x = \dfrac{2y-3}{2}$

Solve for y: $2x = 2y - 3$

$$2x + 3 = 2y$$
$$\dfrac{2x+3}{2} = y$$

Replace y with $s^{-1}(x)$: $s^{-1}(x) = \dfrac{2x+3}{2}$

c) $s^{-1}(3) = \dfrac{2 \cdot 3 + 3}{2} = \dfrac{6+3}{2} = \dfrac{9}{2} = 4\dfrac{1}{2}$

$$s^{-1}\left(5\dfrac{1}{2}\right) = s^{-1}(5.5) = \dfrac{2 \cdot 5.5 + 3}{2} = \dfrac{11+3}{2} = \dfrac{14}{2} = 7$$

$$s^{-1}(7) = \dfrac{2 \cdot 7 + 3}{2} = \dfrac{14+3}{2} = \dfrac{17}{2} = 8\dfrac{1}{2}$$

91. $y_1 = 0.8x + 1.7,$

$y_2 = \dfrac{x - 1.7}{0.8}$

Both the domain and the range of f are the set of all real numbers. Then both the domain and the range of f^{-1} are also the set of all real numbers.

93. $y_1 = \dfrac{1}{2}x - 4,$

$y_2 = 2x + 8$

Both the domain and the range of f are the set of all real numbers. Then both the domain and the range of f^{-1} are also the set of all real numbers.

95. $y_1 = \sqrt{x - 3},$

$y_2 = x^2 + 3, x \geqslant 0$

The domain of f is $[3, \infty)$ and the range of f is $[0, \infty)$. Then the domain of f^{-1} is $[0, \infty)$ and the range of f^{-1} is $[3, \infty)$.

97. $y_1 = x^2 - 4, x \geqslant 0; \; y_2 = \sqrt{4 + x}$

Since it is specified that $x \geq 0$, the domain of f is $[0, \infty)$. The range of f is $[-4, \infty)$. Then the domain of f^{-1} is $[-4, \infty)$ and the range of f^{-1} is $[0, \infty)$.

99. $y_1 = (3x - 9)^3$, $y_2 = \dfrac{\sqrt[3]{x} + 9}{3}$

Both the domain and the range of f are the set of all real numbers. Then both the domain and the range of f^{-1} are also the set of all real numbers.

101. a) $D(r) = \dfrac{11r + 5}{10}$

$D(0) = \dfrac{11 \cdot 0 + 5}{10} = \dfrac{5}{10} = 0.5$

$D(10) = \dfrac{11 \cdot 10 + 5}{10} = \dfrac{115}{10} = 11.5$

$D(20) = \dfrac{11 \cdot 20 + 5}{10} = \dfrac{225}{10} = 22.5$

$D(50) = \dfrac{11 \cdot 50 + 5}{10} = \dfrac{555}{10} = 55.5$

$D(65) = \dfrac{11 \cdot 65 + 5}{10} = \dfrac{720}{10} = 72$

b) Replace $D(r)$ with y: $y = \dfrac{11r + 5}{10}$

Interchange r and y: $r = \dfrac{11y + 5}{10}$

Solve for y: $10r = 11y + 5$

$10r - 5 = 11y$

$\dfrac{10r - 5}{11} = y$

Replace y with $D^{-1}(r)$: $D^{-1}(r) = \dfrac{10r - 5}{11}$

$D^{-1}(r)$ represents the speed, in miles per hour, that the car is traveling when the reaction distance is r feet.

c) $y_1 = \dfrac{11x + 5}{10}$, $y_2 = \dfrac{10x - 5}{11}$

103. Discussion and Writing

105. The functions for which the coefficient of x^2 is negative have a maximum value. These are (b), (d), (f), and (h).

107. Since $|2| > 1$ the graph of $f(x) = 2x^2$ can be obtained by stretching the graph of $f(x) = x^2$ vertically. Since $0 < \left|\dfrac{1}{4}\right| < 1$, the graph of $f(x) = \dfrac{1}{4}x^2$ can be obtained by shrinking the graph of $y = x^2$ vertically. Thus the graph of $f(x) = 2x^2$, or (a) is narrower.

109. We can write (f) as $f(x) = -2[x - (-3)]^2 + 1$. Thus the graph of (f) has vertex $(-3, 1)$.

111. The graph of $f(x) = x^2 - 3$ is a parabola with vertex $(0, -3)$. If we consider x-values such that $x \geq 0$, then the graph is the right-hand side of the parabola and it passes the horizontal line test. We find the inverse of $f(x) = x^2 - 3$, $x \geq 0$.

Replace $f(x)$ with y: $y = x^2 - 3$

Interchange x and y: $x = y^2 - 3$

  Solve for y: $x + 3 = y^2$

$\sqrt{x + 3} = y$

(We take the principal square root, because $x \geq 0$ in the original equation.)

Replace y with $f^{-1}(x)$: $f^{-1}(x) = \sqrt{x + 3}$ for all x in the range of $f(x)$, or $f^{-1}(x) = \sqrt{x + 3}$, $x \geq -3$.

Answers may vary. There are other restrictions that also make $f(x)$ one-to-one.

113. Answers may vary. $f(x) = \dfrac{3}{x}$, $f(x) = 1 - x$, $f(x) = x$.

115. If the graph $y = f(x)$ (or the graph of $y = f^{-1}(x)$) is symmetric with respect to the line $y = x$, then $f^{-1}(x) = f(x)$.

Exercise Set 4.2

1. $e^4 \approx 54.5982$

3. $e^{-2.458} \approx 0.0856$

5. $f(x) = -2^x - 1$

$f(0) = -2^0 - 1 = -1 - 1 = -2$

The only graph with y-intercept $(0, -2)$ is (f).

7. $f(x) = e^x + 3$

This is the graph of $f(x) = e^x$ shifted up 3 units. Then (e) is the correct choice.

9. $f(x) = 3^{-x} - 2$

$f(0) = 3^{-0} - 2 = 1 - 2 = -1$

Since the y-intercept is $(0, -1)$, the correct graph is (a) or (c). Check another point on the graph. $f(-1) = 3^{-(-1)} - 2 = 3 - 2 = 1$, so $(-1, 1)$ is on the graph. Thus (a) is the correct choice.

11. Graph $f(x) = 3^x$.

Compute some function values, plot the corresponding points, and connect them with a smooth curve.

x	$y = f(x)$	(x, y)
-3	$\dfrac{1}{27}$	$\left(-3, \dfrac{1}{27}\right)$
-2	$\dfrac{1}{9}$	$\left(-2, \dfrac{1}{9}\right)$
-1	$\dfrac{1}{3}$	$\left(-1, \dfrac{1}{3}\right)$
0	1	$(0, 1)$
1	3	$(1, 3)$
2	9	$(2, 9)$
3	27	$(3, 27)$

13. Graph $f(x) = 6^x$.

Compute some function values, plot the corresponding points, and connect them with a smooth curve.

x	$y = f(x)$	(x, y)
-3	$\dfrac{1}{216}$	$\left(-3, \dfrac{1}{216}\right)$
-2	$\dfrac{1}{36}$	$\left(-2, \dfrac{1}{36}\right)$
-1	$\dfrac{1}{6}$	$\left(-1, \dfrac{1}{6}\right)$
0	1	$(0, 1)$
1	6	$(1, 6)$
2	36	$(2, 36)$
3	216	$(3, 216)$

15. Graph $f(x) = \left(\dfrac{1}{4}\right)^x$.

Compute some function values, plot the corresponding points, and connect them with a smooth curve.

x	$y = f(x)$	(x, y)
-3	64	$(-3, 64)$
-2	16	$(-2, 16)$
-1	4	$(-1, 4)$
0	1	$(0, 1)$
1	$\dfrac{1}{4}$	$\left(1, \dfrac{1}{4}\right)$
2	$\dfrac{1}{16}$	$\left(2, \dfrac{1}{16}\right)$
3	$\dfrac{1}{64}$	$\left(3, \dfrac{1}{64}\right)$

17. Graph $y = -2^x$.

x	y	(x, y)
-3	$-\dfrac{1}{8}$	$\left(-3, -\dfrac{1}{8}\right)$
-2	$-\dfrac{1}{4}$	$\left(-2, -\dfrac{1}{4}\right)$
-1	$-\dfrac{1}{2}$	$\left(-1, -\dfrac{1}{2}\right)$
0	-1	$(0, -1)$
1	-2	$(1, -2)$
2	-4	$(2, -4)$
3	-8	$(3, -8)$

19. Graph $f(x) = -0.25^x + 4$.

x	$y = f(x)$	(x, y)
-3	-60	$(-3, -60)$
-2	-12	$(-2, -12)$
-1	0	$(-1, 0)$
0	3	$(0, 3)$
1	3.75	$(1, 3.75)$
2	3.94	$(2, 3.94)$
3	3.98	$(3, 3.98)$

21. Graph $f(x) = 1 + e^{-x}$.

x	$y = f(x)$	(x, y)
-3	21.1	$(-3, 21.1)$
-2	8.4	$(-2, 8.4)$
-1	3.7	$(-1, 3.7)$
0	2	$(0, 2)$
1	1.4	$(1, 1.4)$
2	1.1	$(2, 1.1)$
3	1.0	$(3, 1.0)$

23. Graph $y = \dfrac{1}{4}e^x$.

Choose values for x and compute the corresponding y-values. Plot the points (x, y) and connect them with a smooth curve.

x	y	(x, y)
-3	0.0124	$(-3, 0.0124)$
-2	0.0338	$(-2, 0.0338)$
-1	0.0920	$(-1, 0.0920)$
0	0.25	$(0, 0.25)$
1	0.6796	$(1, 0.6796)$
2	1.8473	$(2, 1.8473)$
3	5.0214	$(3, 5.0214)$

25. Graph $f(x) = 1 - e^{-x}$.

Compute some function values, plot the corresponding points, and connect them with a smooth curve.

x	y	(x, y)
-3	-19.0855	$(-3, -19.0855)$
-2	-6.3891	$(-2, -6.3891)$
-1	-1.7183	$(-1, -1.7183)$
0	0	$(0, 0)$
1	0.6321	$(1, 0.6321)$
2	0.8647	$(2, 0.8647)$
3	0.9502	$(3, 0.9502)$

27. Shift the graph of $y = 2^x$ left 1 unit.

29. Shift the graph of $y = 2^x$ down 3 units.

31. Reflect the graph of $y = 3^x$ across the y-axis, then across the x-axis, and then shift it up 4 units.

33. Shift the graph of $y = \left(\frac{3}{2}\right)^x$ right 1 unit.

35. Shift the graph of $y = 2^x$ left 3 units and down 5 units.

37. Shrink the graph of $y = e^x$ horizontally.

39. Shift the graph of $y = e^x$ left 1 unit and reflect it across the y-axis.

41. Reflect the graph of $y = e^x$ across the y-axis and then across the x-axis; shift it up 1 unit and then stretch it vertically.

43. a) We use the formula $A = P\left(1 + \dfrac{r}{n}\right)^{nt}$ and substitute 82,000 for P, 0.045 for r, and 4 for n.

$$A(t) = 82,000\left(1 + \frac{0.045}{4}\right)^{4t} = 82,000(1.01125)^{4t}$$

b) $A(0) = 82,000(1.01125)^{4 \cdot 0} = \$82,000$

$A(2) = 82,000(1.01125)^{4 \cdot 2} \approx \$89,677.22$

$A(5) = 82,000(1.01125)^{4 \cdot 5} \approx \$102,561.54$

$A(10) = 82,000(1.01125)^{4 \cdot 10} \approx \$128,278.90$

45. We use the formula $A = P\left(1 + \dfrac{r}{n}\right)^{nt}$ and substitute 3000 for P, 0.05 for r, and 4 for n.

$$A(t) = 3000\left(1 + \frac{0.05}{4}\right)^{4t} = 3000(1.0125)^{4t}$$

On Jacob's sixteenth birthday, $t = 16 - 6 = 10$.

$A(10) = 3000(1.0125)^{4 \cdot 10} = 4930.86$

When the CD matures \$4930.86 will be available.

47. We use the formula $A = P\left(1 + \dfrac{r}{n}\right)^{nt}$ and substitute 3000 for P, 0.04 for r, 2 for n, and 2 for t.

$$A = 3000\left(1 + \frac{0.04}{2}\right)^{2 \cdot 2} \approx \$3247.30$$

49. We use the formula $A = P\left(1 + \dfrac{r}{n}\right)^{nt}$ and substitute 120,000 for P, 0.025 for r, 1 for n, and 10 for t.

$$A = 120,000\left(1 + \frac{0.025}{1}\right)^{1 \cdot 10} \approx \$153,610.15$$

51. We use the formula $A = P\left(1 + \dfrac{r}{n}\right)^{nt}$ and substitute 53,500 for P, 0.055 for r, 4 for n, and 6.5 for t.

$$A = 53,500\left(1 + \frac{0.055}{4}\right)^{4(6.5)} \approx \$76,305.59$$

53. We use the formula $A = P\left(1 + \dfrac{r}{n}\right)^{nt}$ and substitute 17,400 for P, 0.081 for r, 365 for n, and 5 for t.

$$A = 17,400\left(1 + \frac{0.081}{365}\right)^{365 \cdot 5} \approx \$26,086.69$$

55. $C(t) = 1.0283(1.1483)^t$

In 2008, $t = 2008 - 1996 = 12$.

$C(12) = 1.0283(1.1483)^{12} \approx 5.4$ billion gallons

In 2010, $t = 2010 - 1996 = 14$.

$C(14) = 1.0283(1.1483)^{14} \approx 7.1$ billion gallons

57. $G(x) = 433.6(1.5)^x$

In 2009, $x = 2009 - 2004 = 5$.

$G(5) = 433.6(1.5)^5 \approx 3293$ GB

In 2014, $x = 2014 - 2004 = 10$.

$G(10) = 433.6(1.5)^{10} \approx 25,004$ GB

59. In 2007, $x = 2007 - 1996 = 11$.

$E(11) = 139.76(1.194)^{11} \approx \$982.7$ billion

$I(11) = 130.67(1.187)^{11} \approx \$861.3$ billion

In 2012, $x = 2012 - 1996 = 16$.

$E(16) = 139.76(1.194)^{16} \approx \$2384.8$ billion

$I(16) = 130.67(1.187)^{16} \approx \$2029.5$ billion

In 2020, $x = 2020 - 1996 = 24$.

$E(24) = 139.76(1.194)^{24} \approx \$9851.1$ billion

$I(24) = 130.67(1.187)^{24} \approx \$7998.2$ billion

61. $V(0) = 1800(0.8)^0 = \$1800$

$V(1) = 1800(0.8)^1 = \$1440$

$V(2) = 1800(0.8)^2 = \$1152$

$V(5) = 1800(0.8)^5 \approx \$589.82$

$V(10) = 1800(0.8)^{10} \approx \$193.27$

63. $S(t) = 29.0626(1.3438)^t$

In 2000, $t = 2000 - 1999 = 1$.

$S(1) = 29.0626(1.3438)^1 \approx \$39.1$ billion

In 2005, $t = 2005 - 1999 = 6$.

$S(6) = 29.0626(1.3438)^6 \approx \$171.1$ billion

In 2007, $t = 2007 - 1999 = 8$.

$S(8) = 29.0626(1.3438)^8 \approx \$309.0$ billion

65. $f(25) = 100(1 - e^{-0.04(25)}) \approx 63\%$

67. Graph (c) is the graph of $y = 3^x - 3^{-x}$.

69. Graph (a) is the graph of $f(x) = -2.3^x$.

71. Graph (l) is the graph of $y = 2^{-|x|}$.

73. Graph (g) is the graph of $f(x) = (0.58)^x - 1$.

75. Graph (i) is the graph of $g(x) = e^{|x|}$.

77. Graph (k) is the graph of $y = 2^{-x^2}$.

79. Graph (m) is the graph of $g(x) = \dfrac{e^x - e^{-x}}{2}$.

81. Discussion and Writing

83. They are inverses.

85. $(1 - 4i)(7 + 6i) = 7 + 6i - 28i - 24i^2$

$\qquad\qquad\qquad = 7 + 6i - 28i + 24$

$\qquad\qquad\qquad = 31 - 22i$

87. $2x^2 - 13x - 7 = 0 \quad$ Setting $f(x) = 0$

$(2x + 1)(x - 7) = 0$

$2x + 1 = 0 \quad$ or $\quad x - 7 = 0$

$2x = -1 \quad$ or $\qquad x = 7$

$x = -\dfrac{1}{2} \quad$ or $\qquad x = 7$

The zeros of the function are $-\dfrac{1}{2}$ and 7, and the x-intercepts are $\left(-\dfrac{1}{2}, 0\right)$ and $(7, 0)$.

89.
$$x^4 - x^2 = 0 \quad \text{Setting } h(x) = 0$$
$$x^2(x^2 - 1) = 0$$
$$x^2(x + 1)(x - 1) = 0$$
$$x^2 = 0 \quad or \quad x + 1 = 0 \quad or \quad x - 1 = 0$$
$$x = 0 \quad or \qquad x = -1 \quad or \qquad x = 1$$

The zeros of the function are 0, -1, and 1, and the x-intercepts are $(0, 0)$, $(-1, 0)$, and $(1, 0)$.

91.
$$x^3 + 6x^2 - 16x = 0$$
$$x(x^2 + 6x - 16) = 0$$
$$x(x + 8)(x - 2) = 0$$
$$x = 0 \quad or \quad x + 8 = 0 \quad or \quad x - 2 = 0$$
$$x = 0 \quad or \qquad x = -8 \quad or \qquad x = 2$$

The solutions are 0, -8, and 2.

93. $7^\pi \approx 451.8078726$ and $\pi^7 \approx 3020.293228$, so π^7 is larger.
$70^{80} \approx 4.054 \times 10^{147}$ and $80^{70} \approx 1.646 \times 10^{133}$, so 70^{80} is larger.

Exercise Set 4.3

1. Graph $x = 3^y$.

Choose values for y and compute the corresponding x-values. Plot the points (x, y) and connect them with a smooth curve.

x	y	(x, y)
$\frac{1}{27}$	-3	$\left(\frac{1}{27}, -3\right)$
$\frac{1}{9}$	-2	$\left(\frac{1}{9}, -2\right)$
$\frac{1}{3}$	-1	$\left(\frac{1}{3}, -1\right)$
1	0	$(1, 0)$
3	1	$(3, 1)$
9	2	$(9, 2)$
27	3	$(27, 3)$

3. Graph $x = \left(\frac{1}{2}\right)^y$.

Choose values for y and compute the corresponding x-values. Plot the points (x, y) and connect them with a smooth curve.

x	y	(x, y)
8	-3	$(8, -3)$
4	-2	$(4, -2)$
2	-1	$(2, -1)$
1	0	$(1, 0)$
$\frac{1}{2}$	1	$\left(\frac{1}{2}, 1\right)$
$\frac{1}{4}$	2	$\left(\frac{1}{4}, 2\right)$
$\frac{1}{8}$	3	$\left(\frac{1}{8}, 3\right)$

5. Graph $y = \log_3 x$.

The equation $y = \log_3 x$ is equivalent to $x = 3^y$. We can find ordered pairs that are solutions by choosing values for y and computing the corresponding x-values.

For $y = -2$, $x = 3^{-2} = \frac{1}{9}$.

For $y = -1$, $x = 3^{-1} = \frac{1}{3}$.

For $y = 0$, $x = 3^0 = 1$.

For $y = 1$, $x = 3^1 = 3$.

For $y = 2$, $x = 3^2 = 9$.

x, or 3^y	y
$\frac{1}{9}$	-2
$\frac{1}{3}$	-1
1	0
3	1
9	2

7. Graph $f(x) = 2 \ln x$.

Choose values for x and use a calculator to find the corresponding values of $2 \ln x$. Then plot the points and connect them with a smooth curve.

x	$f(x)$
0.1	-4.6
0.5	-1.4
1	0
2	1.4
4	2.8
8	4.2

9. $\log_2 16 = 4$ because the exponent to which we raise 2 to get 16 is 4.

11. $\log_5 125 = 3$, because the exponent to which we raise 5 to get 125 is 3.

13. $\log 0.001 = -3$, because the exponent to which we raise 10 to get 0.001 is -3.

15. $\log_2 \frac{1}{4} = -2$, because the exponent to which we raise 2 to get $\frac{1}{4}$ is -2.

17. $\ln 1 = 0$, because the exponent to which we raise e to get 1 is 0.

19. $\log 10 = 1$, because the exponent to which we raise 10 to get 10 is 1.

21. $\log_5 5^4 = 4$, because the exponent to which we raise 5 to get 5^4 is 4.

23. $\log_3 \sqrt[4]{3} = \log_3 3^{1/4} = \frac{1}{4}$, because the exponent to which we raise 3 to get $3^{1/4}$ is $\frac{1}{4}$.

25. $\log 10^{-7} = -7$, because the exponent to which we raise 10 to get 10^{-7} is -7.

27. $\log_{49} 7 = \frac{1}{2}$, because the exponent to which we raise 49 to get 7 is $\frac{1}{2}$. $(49^{1/2} = \sqrt{49} = 7)$

29. $\ln e^{3/4} = \frac{3}{4}$, because the exponent to which we raise e to get $e^{3/4}$ is $\frac{3}{4}$.

31. $\log_4 1 = 0$, because the exponent to which we raise 4 to get 1 is 0.

33. $\ln \sqrt{e} = \ln e^{1/2} = \frac{1}{2}$, because the exponent to which we raise e to get $e^{1/2}$ is $\frac{1}{2}$.

35. The exponent is the logarithm.

$$10^3 = 1000 \Rightarrow 3 = \log_{10} 1000$$

The base remains the same.

37. The exponent is the logarithm.

$$8^{1/3} = 2 \Rightarrow \log_8 2 = \frac{1}{3}$$

The base remains the same.

39. $e^3 = t \Rightarrow \log_e t = 3$, or $\ln t = 3$

41. $e^2 = 7.3891 \Rightarrow \log_e 7.3891 = 2$, or $\ln 7.3891 = 2$

43. $p^k = 3 \Rightarrow \log_p 3 = k$

45. The logarithm is the exponent.

$$\log_5 5 = 1 \Rightarrow 5^1 = 5$$

The base remains the same.

47. $\log 0.01 = -2$ is equivalent to $\log_{10} 0.01 = -2$.

The logarithm is the exponent.

$$\log_{10} 0.01 = -2 \Rightarrow 10^{-2} = 0.01$$

The base remains the same.

49. $\ln 30 = 3.4012 \Rightarrow e^{3.4012} = 30$

51. $\log_a M = -x \Rightarrow a^{-x} = M$

53. $\log_a T^3 = x \Rightarrow a^x = T^3$

55. $\log 3 \approx 0.4771$

57. $\log 532 \approx 2.7259$

59. $\log 0.57 \approx -0.2441$

61. $\log(-2)$ does not exist. (The calculator gives an error message.)

63. $\ln 2 \approx 0.6931$

65. $\ln 809.3 \approx 6.6962$

67. $\ln(-1.32)$ does not exist. (The calculator gives an error message.)

69. Let $a = 10$, $b = 4$, and $M = 100$ and substitute in the change-of-base formula.

$$\log_4 100 = \frac{\log_{10} 100}{\log_{10} 4} \approx 3.3219$$

71. Let $a = 10$, $b = 100$, and $M = 0.3$ and substitute in the change-of-base formula.

$$\log_{100} 0.3 = \frac{\log_{10} 0.3}{\log_{10} 100} \approx -0.2614$$

73. Let $a = 10$, $b = 200$, and $M = 50$ and substitute in the change-of-base formula.

$$\log_{200} 50 = \frac{\log_{10} 50}{\log_{10} 200} \approx 0.7384$$

75. Let $a = e$, $b = 3$, and $M = 12$ and substitute in the change-of-base formula.

$$\log_3 12 = \frac{\ln 12}{\ln 3} \approx 2.2619$$

77. Let $a = e$, $b = 100$, and $M = 15$ and substitute in the change-of-base formula.

$$\log_{100} 15 = \frac{\ln 15}{\ln 100} \approx 0.5880$$

79. Graph $y = 3^x$ and then reflect this graph across the line $y = x$ to get the graph of $y = \log_3 x$.

81. Graph $y = \log x$ and then reflect this graph across the line $y = x$ to get the graph of $y = 10^x$.

83. Shift the graph of $y = \log_2 x$ left 3 units.

Domain: $(-3, \infty)$

Vertical asymptote: $x = -3$

85. Shift the graph of $y = \log_3 x$ down 1 unit.

Domain: $(0, \infty)$

Vertical asymptote: $x = 0$

87. Stretch the graph of $y = \ln x$ vertically.

Domain: $(0, \infty)$

Vertical asymptote: $x = 0$

89. Reflect the graph of $y = \ln x$ across the x-axis and then shift it up 2 units.

Domain: $(0, \infty)$

Vertical asymptote: $x = 0$

91. a) We substitute 784.242 for P, since P is in thousands.
$$w(784.242) = 0.37 \ln 784.242 + 0.05$$
$$\approx 2.5 \text{ ft/sec}$$

b) We substitute 484.246 for P, since P is in thousands.
$$w(484.246) = 0.37 \ln 484.246 + 0.05$$
$$\approx 2.3 \text{ ft/sec}$$

c) We substitute 276.963 for P, since P is in thousands.
$$w(276.963) = 0.37 \ln 276.963 + 0.05$$
$$\approx 2.1 \text{ ft/sec}$$

d) We substitute 2862.244 for P, since P is in thousands.
$$w(2862.244) = 0.37 \ln 2862.244 + 0.05$$
$$\approx 3.0 \text{ ft/sec}$$

e) We substitute 533.492 for P, since P is in thousands.
$$w(533.492) = 0.37 \ln 533.492 + 0.05$$
$$\approx 2.4 \text{ ft/sec}$$

f) We substitute 304.973 for P, since P is in thousands.
$$w(304.973) = 0.37 \ln 304.973 + 0.05$$
$$\approx 2.2 \text{ ft/sec}$$

g) We substitute 8104.079 for P, since P is in thousands.
$$w(8104.079) = 0.37 \ln 8104.079 + 0.05$$
$$\approx 3.4 \text{ ft/sec}$$

h) We substitute 122.206 for P, since P is in thousands.
$$w(122.206) = 0.37 \ln 122.206 + 0.05$$
$$\approx 1.8 \text{ ft/sec}$$

93. a) $R = \log \dfrac{10^{7.85} \cdot I_0}{I_0} = \log 10^{7.85} = 7.85$

b) $R = \log \dfrac{10^{8.25} \cdot I_0}{I_0} = \log 10^{8.25} = 8.25$

c) $R = \log \dfrac{10^{9.6} \cdot I_0}{I_0} = \log 10^{9.6} = 9.6$

d) $R = \log \dfrac{10^{7.9} \cdot I_0}{I_0} = \log 10^{7.9} = 7.9$

e) $R = \log \dfrac{10^{6.9} \cdot I_0}{I_0} = \log 10^{6.9} = 6.9$

95. a)
$$7 = -\log[H^+]$$
$$-7 = \log[H^+]$$
$$H^+ = 10^{-7} \qquad \text{Using the definition of logarithm}$$

b)
$$5.4 = -\log[H^+]$$
$$-5.4 = \log[H^+]$$
$$H^+ = 10^{-5.4} \qquad \text{Using the definition of logarithm}$$
$$H^+ \approx 4.0 \times 10^{-6}$$

c) $\quad 3.2 = -\log[H^+]$

$\qquad -3.2 = \log[H^+]$

$\qquad\quad H^+ = 10^{-3.2} \qquad$ Using the definition of logarithm

$\qquad\quad H^+ \approx 6.3 \times 10^{-4}$

d) $\quad 4.8 = -\log[H^+]$

$\qquad -4.8 = \log[H^+]$

$\qquad\quad H^+ = 10^{-4.8} \quad$ Using the definition of logarithm

$\qquad\quad H^+ \approx 1.6 \times 10^{-5}$

97. a) $L = 10\log \dfrac{2510 \cdot I_0}{I_0}$

$\qquad = 10\log 2510$

$\qquad \approx 34$ decibels

b) $L = 10\log \dfrac{2,500,000 \cdot I_0}{I_0}$

$\qquad = 10\log 2,500,000$

$\qquad \approx 64$ decibels

c) $L = 10\log \dfrac{10^6 \cdot I_0}{I_0}$

$\qquad = 10\log 10^6$

$\qquad = 60$ decibels

d) $L = 10\log \dfrac{10^9 \cdot I_0}{I_0}$

$\qquad = 10\log 10^9$

$\qquad = 90$ decibels

99. Discussion and Writing

101. $y = 6 = 0 \cdot x + 6$

Slope: 0; y-intercept $(0, 6)$

103. $y = 2x - \dfrac{3}{13}$

Slope: 2, y-intercept: $\left(0, -\dfrac{3}{13}\right)$

105.
$$\begin{array}{r|rrrr} -5 & 1 & -6 & 3 & 10 \\ & & -5 & 55 & -290 \\ \hline & 1 & -11 & 58 & -280 \end{array}$$

The remainder is -280, so $g(-5) = -280$.

107. $\quad f(x) = (x - \sqrt{7})(x + \sqrt{7})(x - 0)$

$\qquad = (x^2 - 7)(x)$

$\qquad = x^3 - 7x$

109. Using the change-of-base formula, we get

$\dfrac{\log_5 8}{\log_5 2} = \log_2 8 = 3.$

111. $f(x) = \log_5 x^3$

x^3 must be positive. Since $x^3 > 0$ for $x > 0$, the domain is $(0, \infty)$.

113. $f(x) = \ln |x|$

$|x|$ must be positive. Since $|x| > 0$ for $x \neq 0$, the domain is $(-\infty, 0) \cup (0, \infty)$.

115. Graph $y = \log_2(2x + 5) = \dfrac{\log(2x + 5)}{\log 2}$. Observe that outputs are negative for inputs between $-\dfrac{5}{2}$ and -2. Thus, the solution set is $\left(-\dfrac{5}{2}, -2\right)$.

117. Graph (d) is the graph of $f(x) = \ln |x|$.

119. Graph (b) is the graph of $f(x) = \ln x^2$.

Exercise Set 4.4

1. Use the product rule.

$\log_3(81 \cdot 27) = \log_3 81 + \log_3 27 = 4 + 3 = 7$

3. Use the product rule.

$\log_5(5 \cdot 125) = \log_5 5 + \log_5 125 = 1 + 3 = 4$

5. Use the product rule.

$\log_t 8Y = \log_t 8 + \log_t Y$

7. Use the product rule.

$\ln xy = \ln x + \ln y$

9. Use the power rule.

$\log_b t^3 = 3\log_b t$

11. Use the power rule.

$\log y^8 = 8\log y$

13. Use the power rule.

$\log_c K^{-6} = -6\log_c K$

15. Use the power rule.

$\ln \sqrt[3]{4} = \ln 4^{1/3} = \dfrac{1}{3}\ln 4$

17. Use the quotient rule.

$\log_t \dfrac{M}{8} = \log_t M - \log_t 8$

19. Use the quotient rule.

$\log \dfrac{x}{y} = \log x - \log y$

21. Use the quotient rule.

$\ln \dfrac{r}{s} = \ln r - \ln s$

23. $\quad \log_a 6xy^5z^4$

$\qquad = \log_a 6 + \log_a x + \log_a y^5 + \log_a z^4$

$\qquad\qquad\qquad\qquad\qquad$ Product rule

$\qquad = \log_a 6 + \log_a x + 5\log_a y + 4\log_a z$

$\qquad\qquad\qquad\qquad\qquad$ Power rule

25. $\log_b \dfrac{p^2 q^5}{m^4 b^9}$

$= \log_b p^2 q^5 - \log_b m^4 b^9$ Quotient rule

$= \log_b p^2 + \log_b q^5 - (\log_b m^4 + \log_b b^9)$

                           Product rule

$= \log_b p^2 + \log_b q^5 - \log_b m^4 - \log_b b^9$

$= \log_b p^2 + \log_b q^5 - \log_b m^4 - 9$ $(\log_b b^9 = 9)$

$= 2\log_b p + 5\log_b q - 4\log_b m - 9$ Power rule

27. $\ln \dfrac{2}{3x^3 y}$

$= \ln 2 - \ln 3x^3 y$ Quotient rule

$= \ln 2 - (\ln 3 + \ln x^3 + \ln y)$ Product rule

$= \ln 2 - \ln 3 - \ln x^3 - \ln y$

$= \ln 2 - \ln 3 - 3\ln x - \ln y$ Power rule

29. $\log \sqrt{r^3 t}$

$= \log (r^3 t)^{1/2}$

$= \dfrac{1}{2} \log r^3 t$ Power rule

$= \dfrac{1}{2} (\log r^3 + \log t)$ Product rule

$= \dfrac{1}{2} (3\log r + \log t)$ Power rule

$= \dfrac{3}{2} \log r + \dfrac{1}{2} \log t$

31. $\dfrac{1}{2} \log_a \dfrac{x^6}{p^5 q^8}$

$= \dfrac{1}{2} [\log_a x^6 - \log_a (p^5 q^8)]$ Quotient rule

$= \dfrac{1}{2} [\log_a x^6 - (\log_a p^5 + \log_a q^8)]$ Product rule

$= \dfrac{1}{2} (\log_a x^6 - \log_a p^5 - \log_a q^8)$

$= \dfrac{1}{2} (6\log_a x - 5\log_a p - 8\log_a q)$ Power rule

$= 3\log_a x - \dfrac{5}{2} \log_a p - 4\log_a q$

33. $\log_a \sqrt[4]{\dfrac{m^8 n^{12}}{a^3 b^5}}$

$= \dfrac{1}{4} \log_a \dfrac{m^8 n^{12}}{a^3 b^5}$ Power rule

$= \dfrac{1}{4} (\log_a m^8 n^{12} - \log_a a^3 b^5)$ Quotient rule

$= \dfrac{1}{4} [\log_a m^8 + \log_a n^{12} - (\log_a a^3 + \log_a b^5)]$

                          Product rule

$= \dfrac{1}{4} (\log_a m^8 + \log_a n^{12} - \log_a a^3 - \log_a b^5)$

$= \dfrac{1}{4} (\log_a m^8 + \log_a n^{12} - 3 - \log_a b^5)$

                       $(\log_a a^3 = 3)$

$= \dfrac{1}{4} (8\log_a m + 12\log_a n - 3 - 5\log_a b)$

                         Power rule

$= 2\log_a m + 3\log_a n - \dfrac{3}{4} - \dfrac{5}{4} \log_a b$

35. $\log_a 75 + \log_a 2$

$= \log_a (75 \cdot 2)$ Product rule

$= \log_a 150$

37. $\log 10,000 - \log 100$

$= \log \dfrac{10,000}{100}$ Quotient rule

$= \log 100$

$= 2$

39. $\dfrac{1}{2} \log n + 3\log m$

$= \log n^{1/2} + \log m^3$ Power rule

$= \log n^{1/2} m^3$, or   Product rule

   $\log m^3 \sqrt{n}$ $n^{1/2} = \sqrt{n}$

41. $\dfrac{1}{2} \log_a x + 4\log_a y - 3\log_a x$

$= \log_a x^{1/2} + \log_a y^4 - \log_a x^3$ Power rule

$= \log_a x^{1/2} y^4 - \log_a x^3$ Product rule

$= \log_a \dfrac{x^{1/2} y^4}{x^3}$ Quotient rule

$= \log_a x^{-5/2} y^4$, or $\log_a \dfrac{y^4}{x^{5/2}}$ Simplifying

43. $\ln x^2 - 2\ln \sqrt{x}$

$= \ln x^2 - \ln (\sqrt{x})^2$ Power rule

$= \ln x^2 - \ln x$ $[(\sqrt{x})^2 = x]$

$= \ln \dfrac{x^2}{x}$ Quotient rule

$= \ln x$

45. $\ln (x^2 - 4) - \ln (x + 2)$

$= \ln \dfrac{x^2 - 4}{x + 2}$ Quotient rule

$= \ln \dfrac{(x + 2)(x - 2)}{x + 2}$ Factoring

$= \ln (x - 2)$ Removing a factor of 1

47. $\log(x^2 - 5x - 14) - \log(x^2 - 4)$

$= \log \dfrac{x^2 - 5x - 14}{x^2 - 4}$ Quotient rule

$= \log \dfrac{(x+2)(x-7)}{(x+2)(x-2)}$ Factoring

$= \log \dfrac{x-7}{x-2}$ Removing a factor of 1

49. $\ln x - 3[\ln(x-5) + \ln(x+5)]$

$= \ln x - 3\ln[(x-5)(x+5)]$ Product rule

$= \ln x - 3\ln(x^2 - 25)$

$= \ln x - \ln(x^2 - 25)^3$ Power rule

$= \ln \dfrac{x}{(x^2 - 25)^3}$ Quotient rule

51. $\dfrac{3}{2}\ln 4x^6 - \dfrac{4}{5}\ln 2y^{10}$

$= \dfrac{3}{2}\ln 2^2 x^6 - \dfrac{4}{5}\ln 2y^{10}$ Writing 4 as 2^2

$= \ln(2^2 x^6)^{3/2} - \ln(2y^{10})^{4/5}$ Power rule

$= \ln(2^3 x^9) - \ln(2^{4/5}y^8)$

$= \ln \dfrac{2^3 x^9}{2^{4/5}y^8}$ Quotient rule

$= \ln \dfrac{2^{11/5} x^9}{y^8}$

53. $\log_a \dfrac{2}{11} = \log_a 2 - \log_a 11$ Quotient rule

$\approx 0.301 - 1.041$

≈ -0.74

55. $\log_a 98 = \log_a(7^2 \cdot 2)$

$= \log_a 7^2 + \log_a 2$ Product rule

$= 2\log_a 7 + \log_a 2$ Power rule

$\approx 2(0.845) + 0.301$

≈ 1.991

57. $\dfrac{\log_a 2}{\log_a 7} \approx \dfrac{0.301}{0.845} \approx 0.356$

59. $\log_b 125 = \log_b 5^3$

$= 3\log_b 5$ Power rule

$\approx 3(1.609)$

≈ 4.827

61. $\log_b \dfrac{1}{6} = \log_b 1 - \log_b 6$ Quotient rule

$= \log_b 1 - \log_b(2 \cdot 3)$

$= \log_b 1 - (\log_b 2 + \log_b 3)$ Product rule

$= \log_b 1 - \log_b 2 - \log_b 3$

$\approx 0 - 0.693 - 1.099$

≈ -1.792

63. $\log_b \dfrac{3}{b} = \log_b 3 - \log_b b$ Quotient rule

$\approx 1.099 - 1$

≈ 0.099

65. $\log_p p^3 = 3$ $(\log_a a^x = x)$

67. $\log_e e^{|x-4|} = |x-4|$ $(\log_a a^x = x)$

69. $3^{\log_3 4x} = 4x$ $(a^{\log_a x} = x)$

71. $10^{\log w} = w$ $(a^{\log_a x} = x)$

73. $\ln e^{8t} = 8t$ $(\log_a a^x = x)$

75. $\log_b \sqrt{b} = \log_b b^{1/2}$

$= \dfrac{1}{2}\log_b b$ Power rule

$= \dfrac{1}{2} \cdot 1$ $(\log_b b = 1)$

$= \dfrac{1}{2}$

77. Discussion and Writing

79. The degree of $f(x) = 5 - x^2 + x^4$ is 4, so the function is quartic.

81. $f(x) = -\dfrac{3}{4}$ is of the form $f(x) = mx + b$ $\left(\text{with } m = 0 \text{ and } b = -\dfrac{3}{4}\right)$, so it is a linear function. In fact, it is a constant function.

83. $f(x) = -\dfrac{3}{x}$ is of the form $f(x) = \dfrac{p(x)}{q(x)}$ where $p(x)$ and $q(x)$ are polynomials and $q(x)$ is not the zero polynomial, so $f(x)$ is a rational function.

85. The degree of $f(x) = -\dfrac{1}{3}x^3 - 4x^2 + 6x + 42$ is 3, so the function is cubic.

87. $f(x) = \dfrac{1}{2}x + 3$ is of the form $f(x) = mx + b$, so it is a linear function.

89. $5^{\log_5 8} = 2x$

$8 = 2x$ $(a^{\log_a x} = x)$

$4 = x$

The solution is 4.

91. $\log_a(x^2 + xy + y^2) + \log_a(x - y)$

$= \log_a[(x^2 + xy + y^2)(x - y)]$ Product rule

$= \log_a(x^3 - y^3)$ Multiplying

93. $\log_a \dfrac{x - y}{\sqrt{x^2 - y^2}}$

$= \log_a \dfrac{x - y}{(x^2 - y^2)^{1/2}}$

$= \log_a(x - y) - \log_a(x^2 - y^2)^{1/2}$ Quotient rule

$= \log_a(x - y) - \dfrac{1}{2}\log_a(x^2 - y^2)$ Power rule

$= \log_a(x - y) - \dfrac{1}{2}\log_a[(x+y)(x-y)]$

$= \log_a(x - y) - \dfrac{1}{2}[\log_a(x+y) + \log_a(x-y)]$

                                Product rule

$= \log_a(x - y) - \dfrac{1}{2}\log_a(x+y) - \dfrac{1}{2}\log_a(x-y)$

$= \dfrac{1}{2}\log_a(x - y) - \dfrac{1}{2}\log_a(x+y)$

95.
$$\log_a \frac{\sqrt[4]{y^2 z^5}}{\sqrt[4]{x^3 z^{-2}}}$$

$$= \log_a \sqrt[4]{\frac{y^2 z^5}{x^3 z^{-2}}}$$

$$= \log_a \sqrt[4]{\frac{y^2 z^7}{x^3}}$$

$$= \log_a \left(\frac{y^2 z^7}{x^3}\right)^{1/4}$$

$$= \frac{1}{4} \log_a \left(\frac{y^2 z^7}{x^3}\right) \qquad \text{Power rule}$$

$$= \frac{1}{4} (\log_a y^2 z^7 - \log_a x^3) \qquad \text{Quotient rule}$$

$$= \frac{1}{4} (\log_a y^2 + \log_a z^7 - \log_a x^3) \qquad \text{Product rule}$$

$$= \frac{1}{4} (2 \log_a y + 7 \log_a z - 3 \log_a x) \quad \text{Power rule}$$

$$= \frac{1}{4} (2 \cdot 3 + 7 \cdot 4 - 3 \cdot 2)$$

$$= \frac{1}{4} \cdot 28$$

$$= 7$$

97. $\log_a M - \log_a N = \log_a \dfrac{M}{N}$

This is the quotient rule, so it is true.

99. $\dfrac{\log_a M}{x} = \dfrac{1}{x} \log_a M = \log_a M^{1/x}$. The statement is true by the power rule.

101. $\log_a 8x = \log_a 8 + \log_a x = \log_a x + \log_a 8$. The statement is true by the product rule and the commutative property of addition.

103. $\log_a \left(\dfrac{1}{x}\right) = \log_a x^{-1} = -1 \cdot \log_a x = -1 \cdot 2 = -2$

105. We use the change-of-base formula.

$$\log_{10} 11 \cdot \log_{11} 12 \cdot \log_{12} 13 \cdots$$
$$\log_{998} 999 \cdot \log_{999} 1000$$

$$= \log_{10} 11 \cdot \frac{\log_{10} 12}{\log_{10} 11} \cdot \frac{\log_{10} 13}{\log_{10} 12} \cdots$$
$$\frac{\log_{10} 999}{\log_{10} 998} \cdot \frac{\log_{10} 1000}{\log_{10} 999}$$

$$= \frac{\log_{10} 11}{\log_{10} 11} \cdot \frac{\log_{10} 12}{\log_{10} 12} \cdots \frac{\log_{10} 999}{\log_{10} 999} \cdot \log_{10} 1000$$

$$= \log_{10} 1000$$

$$= 3$$

107. $\ln a - \ln b + xy = 0$

$$\ln a - \ln b = -xy$$

$$\ln \frac{a}{b} = -xy$$

Then, using the definition of a logarithm, we have
$e^{-xy} = \dfrac{a}{b}$, or $e^{xy} = \dfrac{b}{a}$.

109.
$$\log_a \left(\frac{x + \sqrt{x^2 - 5}}{5}\right)$$

$$= \log_a \left(\frac{x + \sqrt{x^2 - 5}}{5} \cdot \frac{x - \sqrt{x^2 - 5}}{x - \sqrt{x^2 - 5}}\right)$$

$$= \log_a \left(\frac{5}{5(x - \sqrt{x^2 - 5})}\right) = \log_a \left(\frac{1}{x - \sqrt{x^2 - 5}}\right)$$

$$= \log_a 1 - \log_a (x - \sqrt{x^2 - 5})$$

$$= -\log_a (x - \sqrt{x^2 - 5})$$

Exercise Set 4.5

1. $3^x = 81$

$$3^x = 3^4$$

$$x = 4 \qquad \text{The exponents are the same.}$$

The solution is 4.

3. $2^{2x} = 8$

$$2^{2x} = 2^3$$

$$2x = 3 \qquad \text{The exponents are the same.}$$

$$x = \frac{3}{2}$$

The solution is $\dfrac{3}{2}$.

5. $\qquad 2^x = 33$

$$\log 2^x = \log 33 \qquad \text{Taking the common logarithm on both sides}$$

$$x \log 2 = \log 33 \qquad \text{Power rule}$$

$$x = \frac{\log 33}{\log 2}$$

$$x \approx \frac{1.5185}{0.3010}$$

$$x \approx 5.044$$

The solution is 5.044.

7. $5^{4x-7} = 125$

$$5^{4x-7} = 5^3$$

$$4x - 7 = 3$$

$$4x = 10$$

$$x = \frac{10}{4} = \frac{5}{2}$$

The solution is $\dfrac{5}{2}$.

9. $27 = 3^{5x} \cdot 9^{x^2}$

$$3^3 = 3^{5x} \cdot (3^2)^{x^2}$$

$$3^3 = 3^{5x} \cdot 3^{2x^2}$$

$$3^3 = 3^{5x + 2x^2}$$

$$3 = 5x + 2x^2$$

$$0 = 2x^2 + 5x - 3$$

$$0 = (2x - 1)(x + 3)$$

$$x = \frac{1}{2} \quad \text{or} \quad x = -3$$

The solutions are -3 and $\dfrac{1}{2}$.

11. $84^x = 70$

$\log 84^x = \log 70$

$x \log 84 = \log 70$

$x = \dfrac{\log 70}{\log 84}$

$x \approx \dfrac{1.8451}{1.9243}$

$x \approx 0.959$

The solution is 0.959.

13. $e^{-c} = 5^{2c}$

$\ln e^{-c} = \ln 5^{2c}$

$-c = 2c \ln 5$

$0 = c + 2c \ln 5$

$0 = c(1 + 2 \ln 5)$

$0 = c$ Dividing by $1 + 2 \ln 5$

The solution is 0.

15. $e^t = 1000$

$\ln e^t = \ln 1000$

$t = \ln 1000$ Using $\log_a a^x = x$

$t \approx 6.908$

The solution is 6.908.

17. $e^{-0.03t} = 0.08$

$\ln e^{-0.03t} = \ln 0.08$

$-0.03t = \ln 0.08$

$t = \dfrac{\ln 0.08}{-0.03}$

$t \approx \dfrac{-2.5257}{-0.03}$

$t \approx 84.191$

The solution is 84.191.

19. $3^x = 2^{x-1}$

$\ln 3^x = \ln 2^{x-1}$

$x \ln 3 = (x - 1) \ln 2$

$x \ln 3 = x \ln 2 - \ln 2$

$\ln 2 = x \ln 2 - x \ln 3$

$\ln 2 = x(\ln 2 - \ln 3)$

$\dfrac{\ln 2}{\ln 2 - \ln 3} = x$

$\dfrac{0.6931}{0.6931 - 1.0986} \approx x$

$-1.710 \approx x$

The solution is -1.710.

21. $(3.9)^x = 48$

$\log(3.9)^x = \log 48$

$x \log 3.9 = \log 48$

$x = \dfrac{\log 48}{\log 3.9}$

$x \approx \dfrac{1.6812}{0.5911}$

$x \approx 2.844$

The solution is 2.844.

23. $e^x + e^{-x} = 5$

$e^{2x} + 1 = 5e^x$ Multiplying by e^x

$e^{2x} - 5e^x + 1 = 0$ This equation is quadratic in e^x.

$e^x = \dfrac{5 \pm \sqrt{21}}{2}$

$x = \ln\left(\dfrac{5 \pm \sqrt{21}}{2}\right) \approx \pm 1.567$

The solutions are -1.567 and 1.567.

25. $\dfrac{e^x + e^{-x}}{e^x - e^{-x}} = 3$

$e^x + e^{-x} = 3e^x - 3e^{-x}$ Multiplying by $e^x - e^{-x}$

$4e^{-x} = 2e^x$ Subtracting e^x and adding e^{-x}

$2e^{-x} = e^x$

$2 = e^{2x}$ Multiplying by e^x

$\ln 2 = \ln e^{2x}$

$\ln 2 = 2x$

$\dfrac{\ln 2}{2} = x$

$0.347 \approx x$

The solution is 0.347.

27. $\log_5 x = 4$

$x = 5^4$ Writing an equivalent exponential equation

$x = 625$

The solution is 625.

29. $\log x = -4$ The base is 10.

$x = 10^{-4}$, or 0.0001

The solution is 0.0001.

31. $\log_{64} \dfrac{1}{4} = x$

$\dfrac{1}{4} = 64^x$

$\dfrac{1}{4} = (4^3)^x$

$4^{-1} = 4^{3x}$

$-1 = 3x$

$-\dfrac{1}{3} = x$

The solution is $-\dfrac{1}{3}$.

33. $\ln x = 1$ The base is e.

$x = e^1 = e$

The solution is e.

35. $\log_2(10 + 3x) = 5$

$$2^5 = 10 + 3x$$
$$32 = 10 + 3x$$
$$22 = 3x$$
$$\frac{22}{3} = x$$

The answer checks. The solution is $\frac{22}{3}$.

37. $\log x + \log(x - 9) = 1$ The base is 10.

$$\log_{10}[x(x - 9)] = 1$$
$$x(x - 9) = 10^1$$
$$x^2 - 9x = 10$$
$$x^2 - 9x - 10 = 0$$
$$(x - 10)(x + 1) = 0$$
$$x = 10 \ \text{ or } \ x = -1$$

Check: For 10:

$$\log x + \log(x - 9) = 1$$

$\log 10 + \log(10 - 9)$? 1
$\log 10 + \log 1$
$1 + 0$
1 | 1 TRUE

For -1:

$$\log x + \log(x - 9) = 1$$

$\log(-1) + \log(-1 - 9)$? 1

The number -1 does not check, because negative numbers do not have logarithms. The solution is 10.

39. $\log_2(x + 20) - \log_2(x + 2) = \log_2 x$

$$\log_2 \frac{x + 20}{x + 2} = \log_2 x$$
$$\frac{x + 20}{x + 2} = x \quad \text{Using the property of}$$
$$\text{logarithmic equality}$$
$$x + 20 = x^2 + 2x \quad \text{Multiplying by}$$
$$x + 2$$
$$0 = x^2 + x - 20$$
$$0 = (x + 5)(x - 4)$$

$x + 5 = 0 \quad or \quad x - 4 = 0$
$x = -5 \quad or \quad\quad x = 4$

Check: For -5:

$$\log_2(x + 20) - \log_2(x + 2) = \log_2 x$$

$\log_2(-5 + 20) - \log_2(-5 + 2)$? $\log_2(-5)$

The number -5 does not check, because negative numbers do not have logarithms.

For 4:

$$\log_2(x + 20) - \log_2(x + 2) = \log_2 x$$

$\log_2(4 + 20) - \log_2(4 + 2)$? $\log_2 4$
$\log_2 24 - \log_2 6$
$\log_2 \dfrac{24}{6}$
$\log_2 4$ | $\log_2 4$ TRUE

The solution is 4.

41. $\log(x + 8) - \log(x + 1) = \log 6$

$$\log \frac{x + 8}{x + 1} = \log 6 \quad \text{Quotient rule}$$
$$\frac{x + 8}{x + 1} = 6 \quad \text{Using the property of}$$
$$\text{logarithmic equality}$$
$$x + 8 = 6x + 6 \quad \text{Multiplying by } x + 1$$
$$2 = 5x$$
$$\frac{2}{5} = x$$

The answer checks. The solution is $\frac{2}{5}$.

43. $\log_8(x + 1) - \log_8 x = 2$

$$\log_8 \left(\frac{x + 1}{x}\right) = 2 \quad \text{Quotient rule}$$
$$\frac{x + 1}{x} = 8^2$$
$$\frac{x + 1}{x} = 64$$
$$x + 1 = 64x$$
$$1 = 63x$$
$$\frac{1}{63} = x$$

The answer checks. The solution is $\frac{1}{63}$.

45. $\log x + \log(x + 4) = \log 12$

$$\log x(x + 4) = \log 12$$
$$x(x + 4) = 12 \quad \text{Using the property of}$$
$$\text{logarithmic equality}$$
$$x^2 + 4x = 12$$
$$x^2 + 4x - 12 = 0$$
$$(x + 6)(x - 2) = 0$$

$x + 6 = 0 \quad or \quad x - 2 = 0$
$x = -6 \quad or \quad\quad x = 2$

Check: For -6:

$$\log x + \log(x + 4) = \log 12$$

$\log(-6) + \log(-6 + 4)$? $\log 12$

The number -6 does not check, because negative numbers do not have logarithms.

For 2:

$$\log x + \log(x + 4) = \log 12$$

$$\frac{}{\log 2 + \log(2 + 4) \ ? \ \log 12}$$

$$\log 2 + \log 6 \ \Big|$$

$$\log(2 \cdot 6) \ \Big|$$

$$\log 12 \ \Big| \ \log 12 \qquad \text{TRUE}$$

The solution is 2.

47. $\log_4(x + 3) + \log_4(x - 3) = 2$

$$\log_4[(x + 3)(x - 3)] = 2 \qquad \text{Product rule}$$

$$(x + 3)(x - 3) = 4^2$$

$$x^2 - 9 = 16$$

$$x^2 = 25$$

$$x = \pm 5$$

The number 5 checks, but -5 does not. The solution is 5.

49. $\log(2x + 1) - \log(x - 2) = 1$

$$\log\left(\frac{2x + 1}{x - 2}\right) = 1 \qquad \text{Quotient rule}$$

$$\frac{2x + 1}{x - 2} = 10^1 = 10$$

$$2x + 1 = 10x - 20$$

$$\qquad\qquad \text{Multiplying by } x - 2$$

$$21 = 8x$$

$$\frac{21}{8} = x$$

The answer checks. The solution is $\dfrac{21}{8}$.

51. $\ln(x + 8) + \ln(x - 1) = 2\ln x$

$$\ln(x + 8)(x - 1) = \ln x^2$$

$$(x + 8)(x - 1) = x^2 \qquad \text{Using the property of}$$
$$\qquad\qquad\qquad\qquad\qquad \text{logarithmic equality}$$

$$x^2 + 7x - 8 = x^2$$

$$7x - 8 = 0$$

$$7x = 8$$

$$x = \frac{8}{7}$$

The answer checks. The solution is $\dfrac{8}{7}$.

53.

$$\log_6 x = 1 - \log_6(x - 5)$$

$$\log_6 x + \log_6(x - 5) = 1$$

$$\log_6 x(x - 5) = 1$$

$$6^1 = x(x - 5)$$

$$6 = x^2 - 5x$$

$$0 = x^2 - 5x - 6$$

$$0 = (x + 1)(x - 6)$$

$$x + 1 = 0 \quad or \quad x - 6 = 0$$

$$x = -1 \quad or \qquad x = 6$$

The number -1 does not check, but 6 does. The answer is 6.

55.

$$9^{x-1} = 100(3^x)$$

$$(3^2)^{x-1} = 100(3^x)$$

$$3^{2x-2} = 100(3^x)$$

$$\frac{3^{2x-2}}{3^x} = 100$$

$$3^{x-2} = 100$$

$$\log 3^{x-2} = \log 100$$

$$(x - 2)\log 3 = 2$$

$$x - 2 = \frac{2}{\log 3}$$

$$x = 2 + \frac{2}{\log 3}$$

$$x \approx 6.192$$

The solution is 6.192.

57.

$$e^x - 2 = -e^{-x}$$

$$e^x - 2 = -\frac{1}{e^x}$$

$$e^{2x} - 2e^x = -1 \quad \text{Multiplying by } e^x$$

$$e^{2x} - 2e^x + 1 = 0$$

Let $u = e^x$.

$$u^2 - 2u + 1 = 0$$

$$(u - 1)(u - 1) = 0$$

$$u - 1 = 0 \quad or \quad u - 1 = 0$$

$$u = 1 \quad or \qquad u = 1$$

$$e^x = 1 \quad or \qquad e^x = 1 \quad \text{Replacing } u \text{ with } e^x$$

$$x = 0 \quad or \qquad x = 0$$

The solution is 0.

59. $e^{7.2x} = 14.009$

Graph $y_1 = e^{7.2x}$ and $y_2 = 14.009$ and find the first coordinate of the point of intersection using the Intersect feature. The solution is 0.367.

61. $xe^{3x} - 1 = 3$

Graph $y_1 = xe^{3x} - 1$ and $y_2 = 3$ and find the first coordinate of the point of intersection using the Intersect feature. The solution is 0.621.

63. $4\ln(x + 3.4) = 2.5$

Graph $y_1 = 4\ln(x + 3.4)$ and $y_2 = 2.5$ and find the first coordinate of the point of intersection using the Intersect feature. The solution is -1.532.

65. $\log_8 x + \log_8(x + 2) = 2$

Graph $y_1 = \dfrac{\log x}{\log 8} + \dfrac{\log(x + 2)}{\log 8}$ and $y_2 = 2$ and find the first coordinate of the point of intersection using the intersect feature. The solution is 7.062.

67. $\log_5(x + 7) - \log_5(2x - 3) = 1$

Graph $y_1 = \dfrac{\log(x + 7)}{\log 5} - \dfrac{\log(2x - 3)}{\log 5}$ and $y_2 = 1$ and find the first coordinate of the point of intersection using the Intersect feature. The solution is 2.444.

69. Solving the first equation for y, we get
$y = \dfrac{12.4 - 2.3x}{3.8}$. Graph $y_1 = \dfrac{12.4 - 2.3x}{3.8}$ and
$y_2 = 1.1 \ln(x - 2.05)$ and use the Intersect feature to find
the point of intersection. It is $(4.093, 0.786)$.

71. Graph $y_1 = 2.3 \ln(x + 10.7)$ and $y_2 = 10e^{-0.007x^2}$ and use
the Intersect feature to find the point of intersection. It is
$(7.586, 6.684)$.

73. Discussion and Writing

75. $g(x) = x^2 - 6$

a) $-\dfrac{b}{2a} = -\dfrac{0}{2 \cdot 1} = 0$

$g(0) = 0^2 - 6 = -6$

The vertex is $(0, -6)$.

b) The axis of symmetry is $x = 0$.

c) Since the coefficient of the x^2-term is positive, the
function has a minimum value. It is the second coordinate of the vertex, -6, and it occurs when $x = 0$.

77. $G(x) = -2x^2 - 4x - 7$

a) $-\dfrac{b}{2a} = -\dfrac{-4}{2(-2)} = -1$

$G(-1) = -2(-1)^2 - 4(-1) - 7 = -5$

The vertex is $(-1, -5)$.

b) The axis of symmetry is $x = -1$.

c) Since the coefficient of the x^2-term is negative, the
function has a maximum value. It is the second
coordinate of the vertex, -5, and it occurs when
$x = -1$.

79. $\ln(\log x) = 0$

$\log x = e^0$

$\log x = 1$

$x = 10^1 = 10$

The answer checks. The solution is 10.

81. $\sqrt{\ln x} = \ln \sqrt{x}$

$\sqrt{\ln x} = \dfrac{1}{2} \ln x$ \qquad Power rule

$\ln x = \dfrac{1}{4}(\ln x)^2$ \qquad Squaring both sides

$0 = \dfrac{1}{4}(\ln x)^2 - \ln x$

Let $u = \ln x$ and substitute.

$\dfrac{1}{4}u^2 - u = 0$

$u\left(\dfrac{1}{4}u - 1\right) = 0$

$u = 0 \qquad or \qquad \dfrac{1}{4}u - 1 = 0$

$u = 0 \qquad or \qquad \dfrac{1}{4}u = 1$

$u = 0 \qquad or \qquad u = 4$

$\ln x = 0 \qquad or \qquad \ln x = 4$

$x = e^0 = 1 \qquad or \qquad x = e^4 \approx 54.598$

Both answers check. The solutions are 1 and e^4, or 1 and
54.598.

83. $\qquad (\log_3 x)^2 - \log_3 x^2 = 3$

$(\log_3 x)^2 - 2\log_3 x - 3 = 0$

Let $u = \log_3 x$ and substitute:

$u^2 - 2u - 3 = 0$

$(u - 3)(u + 1) = 0$

$u = 3 \qquad or \qquad u = -1$

$\log_3 x = 3 \quad or \quad \log_3 x = -1$

$x = 3^3 \quad or \qquad x = 3^{-1}$

$x = 27 \quad or \qquad x = \dfrac{1}{3}$

Both answers check. The solutions are $\dfrac{1}{3}$ and 27.

85. $\ln x^2 = (\ln x)^2$

$2 \ln x = (\ln x)^2$

$0 = (\ln x)^2 - 2 \ln x$

Let $u = \ln x$ and substitute.

$0 = u^2 - 2u$

$0 = u(u - 2)$

$u = 0 \quad or \qquad u = 2$

$\ln x = 0 \quad or \quad \ln x = 2$

$x = 1 \quad or \qquad x = e^2 \approx 7.389$

Both answers check. The solutions are 1 and e^2, or 1 and
7.389.

87. $5^{2x} - 3 \cdot 5^x + 2 = 0$

$(5^x - 1)(5^x - 2) = 0$ \quad This equation is
\qquad\qquad\qquad\qquad\qquad quadratic in 5^x.

$5^x = 1 \qquad or \qquad 5^x = 2$

$\log 5^x = \log 1 \quad or \quad \log 5^x = \log 2$

$x \log 5 = 0 \qquad or \quad x \log 5 = \log 2$

$x = 0 \qquad or \qquad x = \dfrac{\log 2}{\log 5} \approx 0.431$

The solutions are 0 and 0.431.

89. $\log_3 |x| = 2$

$|x| = 3^2$

$|x| = 9$

$x = -9 \ or \ x = 9$

Both answers check. The solutions are -9 and 9.

91. $\ln x^{\ln x} = 4$

$\ln x \cdot \ln x = 4$

$(\ln x)^2 = 4$

$\ln x = \pm 2$

$\ln x = -2 \qquad or \quad \ln x = 2$

$x = e^{-2} \qquad or \qquad x = e^2$

$x \approx 0.135 \quad or \qquad x \approx 7.389$

Both answers check. The solutions are e^{-2} and e^2, or 0.135
and 7.389.

93.

$$\frac{\sqrt{(e^{2x} \cdot e^{-5x})^{-4}}}{e^x \div e^{-x}} = e^7$$

$$\frac{\sqrt{e^{12x}}}{e^{x-(-x)}} = e^7$$

$$\frac{e^{6x}}{e^{2x}} = e^7$$

$$e^{4x} = e^7$$

$$4x = 7$$

$$x = \frac{7}{4}$$

The solution is $\frac{7}{4}$.

95. $|\log_5 x| + 3\log_5 |x| = 4$

Note that we must have $x > 0$. First consider the case when $0 < x < 1$. When $0 < x < 1$, then $\log_5 x < 0$, so $|\log_5 x| = -\log_5 x$ and $|x| = x$. Thus we have:

$$-\log_5 x + 3\log_5 x = 4$$

$$2\log_5 x = 4$$

$$\log_5 x^2 = 4$$

$$x^2 = 5^4$$

$$x = 5^2$$

$$x = 25 \quad \text{(Recall that } x > 0.\text{)}$$

25 cannot be a solution since we assumed $0 < x < 1$.

Now consider the case when $x > 1$. In this case $\log_5 x > 0$, so $|\log_5 x| = \log_5 x$ and $|x| = x$. Thus we have:

$$\log_5 x + 3\log_5 x = 4$$

$$4\log_5 x = 4$$

$$\log_5 x = 1$$

$$x = 5$$

This answer checks. The solution is 5.

97. $a = \log_8 225$, so $8^a = 225 = 15^2$.

$b = \log_2 15$, so $2^b = 15$.

$$\text{Then} \quad 8^a = (2^b)^2$$

$$(2^3)^a = 2^{2b}$$

$$2^{3a} = 2^{2b}$$

$$3a = 2b$$

$$a = \frac{2}{3}b.$$

99. $\log_2[\log_3(\log_4 x)] = 0$ yields $x = 64$.

$\log_3[\log_2(\log_4 y)] = 0$ yields $y = 16$.

$\log_4[\log_3(\log_2 z)] = 0$ yields $z = 8$.

Then $x + y + z = 64 + 16 + 8 = 88$.

Exercise Set 4.6

1. a) Substitute 6.4 for P_0 and 0.012 for k in $P(t) = P_0 e^{kt}$. We have:

$P(t) = 6.4e^{0.012t}$, where $P(t)$ is in billions and t is the number of years after 2005.

b) In 2008, $t = 2008 - 2005 = 3$.

$P(3) = 6.4e^{0.012(3)} \approx 6.6$ billion

In 2010, $t = 2010 - 2005 = 5$.

$P(5) = 6.4e^{0.012(5)} \approx 6.8$ billion

c) Substitute 8 for $P(t)$ and solve for t.

$$8 = 6.4e^{0.012t}$$

$$1.25 = e^{0.012t}$$

$$\ln 1.25 = \ln e^{0.012t}$$

$$\ln 1.25 = 0.012t$$

$$\frac{\ln 1.25}{0.012} = t$$

$$18.6 \approx t$$

The world population will be 8 billion about 18.6 yr after 2005.

d) $T = \dfrac{\ln 2}{0.012} \approx 57.8$ yr

3. a) $T = \dfrac{\ln 2}{0.024} \approx 28.9$ yr

b) $k = \dfrac{\ln 2}{63} \approx 1.1\%$ per yr

c) $T = \dfrac{\ln 2}{0.014} \approx 49.5$ yr

d) $T = \dfrac{\ln 2}{0.002} \approx 346.6$ yr

e) $k = \dfrac{\ln 2}{38.5} \approx 1.8\%$ per yr

f) $k = \dfrac{\ln 2}{36.5} \approx 1.9\%$ per yr

g) $k = \dfrac{\ln 2}{0.009} \approx 77.0$ yr

h) $k = \dfrac{\ln 2}{69.3} \approx 1.0\%$ per yr

i) $k = \dfrac{\ln 2}{0.012} \approx 57.8$ yr

j) $k = \dfrac{\ln 2}{0.026} \approx 26.7$ yr

5.
$$P(t) = P_0 e^{kt}$$

$$24,839,654,400 = 6,276,883 e^{0.013t}$$

$$\frac{24,839,654,400}{6,276,883} = e^{0.013t}$$

$$\ln\left(\frac{24,839,654,400}{6,276,883}\right) = \ln e^{0.013t}$$

$$\ln\left(\frac{24,839,654,400}{6,276,883}\right) = 0.013t$$

$$\frac{\ln\left(\dfrac{24,839,654,400}{6,276,883}\right)}{0.013} = t$$

$$637 \approx t$$

There will be one person for every square yard of land about 637 yr after 2005.

7. a) Substitute 10,000 for P_0 and 5.4%, or 0.054 for k.

$P(t) = 10,000 e^{0.054t}$

b) $P(1) = 10,000e^{0.054(1)} \approx \$10,555$

$P(2) = 10,000e^{0.054(2)} \approx \$11,140$

$P(5) = 10,000e^{0.054(5)} \approx \$13,100$

$P(10) = 10,000e^{0.054(10)} \approx \$17,160$

c) $T = \dfrac{\ln 2}{0.054} \approx 12.8 \text{ yr}$

9. We use the function found in Example 5. If the mummy has lost 46% of its carbon-14 from an initial amount P_0, then $54\% P_0$, or $0.54 P_0$ remains. We substitute in the function.

$$0.54 P_0 = P_0 e^{-0.00012t}$$

$$0.54 = e^{-0.00012t}$$

$$\ln 0.54 = \ln e^{-0.00012t}$$

$$\ln 0.54 = -0.00012t$$

$$\frac{\ln 0.54}{-0.00012} = t$$

$$5135 \approx t$$

The mummy is about 5135 years old.

11. a) $K = \dfrac{\ln 2}{3} \approx 0.231$, or 23.1% per min

b) $k = \dfrac{\ln 2}{22} \approx 0.0315$, or 3.15% per yr

c) $T = \dfrac{\ln 2}{0.096} \approx 7.2$ days

d) $T = \dfrac{\ln 2}{0.063} \approx 11$ yr

e) $k = \dfrac{\ln 2}{25} \approx 0.028$, or 2.8% per yr

f) $k = \dfrac{\ln 2}{4560} \approx 0.00015$, or 0.015% per yr

g) $k = \dfrac{\ln 2}{23,105} \approx 0.00003$, or 0.003% per yr

13. a)
$$N(t) = N_0 e^{-kt}$$

$$4.8 = 125.0 e^{-k \cdot 20}$$

$$\frac{4.8}{125.0} = e^{-20k}$$

$$\ln \frac{4.8}{125.0} = \ln e^{-20k}$$

$$\ln \frac{4.8}{125.0} = -20k$$

$$\frac{\ln \dfrac{4.8}{125.0}}{-20} = k$$

$$0.1630 \approx k$$

We have $N(t) = 125.0 e^{-0.1630t}$, where t is the number of years since 1983.

b) In 1995, $t = 1995 - 1983 = 12$.

$N(12) = 125.0 e^{-0.1630(12)} \approx 17.7$ students per computer

In 2007, $t = 2007 - 1983 = 24$.

$N(24) = 125.0 e^{-0.1630(24)} \approx 2.5$ students per computer

c) Substitute 1 for $N(t)$ and solve for t.

$$1 = 125.0 e^{-0.1630t}$$

$$0.008 = e^{-0.1630t}$$

$$\ln 0.008 = \ln e^{-0.1630t}$$

$$\ln 0.008 = -0.1630t$$

$$\frac{\ln 0.008}{-0.1630} = t$$

$$30 \approx t$$

There will be a computer for each student about 30 yr after 1983, or in 2013.

15. a) $N(0) = \dfrac{3500}{1 + 19.9 e^{-0.6(0)}} \approx 167$

b) $N(2) = \dfrac{3500}{1 + 19.9 e^{-0.6(2)}} \approx 500$

$N(5) = \dfrac{3500}{1 + 19.9 e^{-0.6(5)}} \approx 1758$

$N(8) = \dfrac{3500}{1 + 19.9 e^{-0.6(8)}} \approx 3007$

$N(12) = \dfrac{3500}{1 + 19.9 e^{-0.6(12)}} \approx 3449$

$N(16) = \dfrac{3500}{1 + 19.9 e^{-0.6(16)}} \approx 3495$

c) As $t \to \infty$, $N(t) \to 3500$; the number of people infected approaches 3500 but never actually reaches it.

17. To find k we substitute 105 for T_1, 0 for T_0, 5 for t, and 70 for $T(t)$ and solve for k.

$$70 = 0 + (105 - 0)e^{-5k}$$

$$70 = 105 e^{-5k}$$

$$\frac{70}{105} = e^{-5k}$$

$$\ln \frac{70}{105} = \ln e^{-5k}$$

$$\ln \frac{70}{105} = -5k$$

$$\frac{\ln \dfrac{70}{105}}{-5} = k$$

$$0.081 \approx k$$

The function is $T(t) = 105 e^{-0.081t}$.

Now we find $T(10)$.

$T(10) = 105 e^{-0.081(10)} \approx 46.7 \,°\text{F}$

19. To find k we substitute 43 for T_1, 68 for T_0, 12 for t, and 55 for $T(t)$ and solve for k.

$$55 = 68 + (43 - 68)e^{-12k}$$

$$-13 = -25 e^{-12k}$$

$$0.52 = e^{-12k}$$

$$\ln 0.52 = \ln e^{-12k}$$

$$\ln 0.52 = -12k$$

$$0.0545 \approx k$$

The function is $T(t) = 68 - 25e^{-0.0545t}$.

Now we find $T(20)$.

$T(20) = 68 - 25e^{-0.0545(20)} \approx 59.6°F$

21. a) $y = 0.1367470864(1.024068023)^x$, where x is the number of years after 1900.

Since $r^2 \approx 0.9688$, the function is a good fit.

b)

c) In 2007, $x = 2007 - 1900 = 107$.

$y = 0.1367470864(1.024068023)^{107} \approx 1.7\%$

In 2015, $x = 2015 - 1900 = 115$.

$y = 0.1367470864(1.024068023)^{115} \approx 2.1\%$

In 2020, $x = 2020 - 1900 = 120$.

$y = 0.1367470864(1.024068023)^{120} \approx 2.4\%$

23. a)

b) Linear: $y = 0.3920710572x + 0.695407279$; $r^2 \approx 0.9465$

Quadratic: $y = 0.0072218083x^2 + 0.1172668587x + 1.973805022$; $r^2 \approx 0.9851$

Exponential: $y = 2.062091236(1.059437743)^x$; $r^2 \approx 0.9775$

The value of r^2 is greatest for the quadratic function, so it provides the best fit.

c)

d) In 2008, $x = 2008 - 1967 = 41$.

For the linear function, when $x = 41$, $y \approx 16.8\%$.

For the quadratic function, when $x = 41$, $y \approx 18.9\%$.

For the exponential function, when $x = 41$, $y \approx 22.0\%$.

Given the trend shown in the table, the quadratic model seems the most realistic. Answers may vary.

25. a) $y = 255.0890581(1.277632801)^x$, where x is the number of years after 1995.

b) In 2005, $x = 2005 - 1995 = 10$.

$y = 255.0890581(1.277632801)^{10} \approx 2956$ surgeries

In 2009, $x = 2009 - 1995 = 14$.

$y = 255.0890581(1.277632801)^{14} \approx 7877$ surgeries

c) Graph $y_1 = 255.0890581(1.277632801)^x$ and $y_2 = 12,000$ and find the first coordinate of the point of intersection of the graphs. It is approximately 16, so there will be 12,000 surgeries about 16 yr after 1995, or in 2011.

27. Discussion and Writing

29. Multiplication principle for inequalities

31. Principle of zero products

33. Power rule

35. $480e^{-0.003p} = 150e^{0.004p}$

$$\frac{480}{150} = \frac{e^{0.004p}}{e^{-0.003p}}$$

$$3.2 = e^{0.007p}$$

$$\ln 3.2 = \ln e^{0.007p}$$

$$\ln 3.2 = 0.007p$$

$$\frac{\ln 3.2}{0.007} = p$$

$$\$166.16 \approx p$$

37. $P(t) = P_0 e^{kt}$

$$50,000 = P_0 e^{0.07(18)}$$

$$\frac{50,000}{e^{0.07(18)}} = P_0$$

$$\$14,182.70 \approx P_0$$

39. $$i = \frac{V}{R}\left[1 - e^{-(R/L)t}\right]$$

$$\frac{iR}{V} = 1 - e^{-(R/L)t}$$

$$e^{-(R/L)t} = 1 - \frac{iR}{V}$$

$$\ln e^{-(R/L)t} = \ln\left(1 - \frac{iR}{V}\right)$$

$$-\frac{R}{L}t = \ln\left(1 - \frac{iR}{V}\right)$$

$$t = -\frac{L}{R}\left[\ln\left(1 - \frac{iR}{V}\right)\right]$$

41. $y = ae^x$

$\ln y = \ln(ae^x)$

$\ln y = \ln a + \ln e^x$

$\ln y = \ln a + x$

$Y = x + \ln a$

This function is of the form $y = mx + b$, so it is linear.

Chapter 4 Review Exercises

1. The statement is true. See page 351 in the text.

3. The graph of f^{-1} is a reflection of the graph of f across $y = x$, so the statement is false.

5. The domain of all logarithmic functions is $(0, \infty)$, so the statement is false.

7. We interchange the first and second coordinates of each pair to find the inverse of the relation. It is $\{(-2.7, 1.3), (-3, 8), (3, -5), (-3, 6), (-5, 7)\}$.

9. The graph of $f(x) = -|x| + 3$ is shown below. The function is not one-to-one, because there are many horizontal lines that cross the graph more than once.

11. The graph of $f(x) = 2x - \dfrac{3}{4}$ is shown below. The function is one-to-one, because no horizontal line crosses the graph more than once.

13. a) The graph of $f(x) = 2 - 3x$ is shown below. It passes the horizontal-line test, so the function is one-to-one.

b) Replace $f(x)$ with y: $y = 2 - 3x$

Interchange x and y: $x = 2 - 3y$

Solve for y: $y = \dfrac{-x + 2}{3}$

Replace y with $f^{-1}(x)$: $f^{-1}(x) = \dfrac{-x + 2}{3}$

15. a) The graph of $f(x) = \sqrt{x - 6}$ is shown below. It passes the horizontal-line test, so the function is one-to-one.

b) Replace $f(x)$ with y: $y = \sqrt{x - 6}$

Interchange x and y: $x = \sqrt{y - 6}$

Solve for y: $\quad x^2 = y - 6$

$\qquad\qquad x^2 + 6 = y$

Replace y with $f^{-1}(x)$: $f^{-1}(x) = x^2 + 6$, for all x in the range of $f(x)$, or $f^{-1}(x) = x^2 + 6$, $x \geq 0$.

17. a) The graph of $f(x) = 3x^2 + 2x - 1$ is shown below. It is not one-to-one since there are many horizontal lines that cross the graph more than once. The function does not have an inverse that is a function.

19. We find $(f^{-1} \circ f)(x)$ and $(f \circ f^{-1})(x)$ and check to see that each is x.

$$(f^{-1} \circ f)(x) = f^{-1}(f(x)) = f^{-1}(6x - 5) =$$
$$\frac{(6x - 5) + 5}{6} = \frac{6x}{6} = x$$

$$(f \circ f^{-1})(x) = f(f^{-1}(x)) = f\left(\frac{x + 5}{6}\right) =$$
$$6\left(\frac{x + 5}{6}\right) - 5 = x + 5 - 5 = x$$

21. Replace $f(x)$ with y: $y = 2 - 5x$

Interchange x and y: $x = 2 - 5y$

Solve for y: $y = \dfrac{2 - x}{5}$

Replace y with $f^{-1}(x)$: $f^{-1}(x) = \dfrac{2 - x}{5}$

The domain and range of f are $(-\infty, \infty)$, so the domain and range of f^{-1} are also $(-\infty, \infty)$.

23. Since $f(f^{-1}(x)) = x$, then $f(f^{-1}(657)) = 657$.

25.

27.

29.

31. $f(x) = e^{x-3}$

This is the graph of $f(x) = e^x$ shifted right 3 units. The correct choice is (c).

33. $f(x) = -\log_3(x + 1)$

This is the graph of $\log_3 x$ shifted left 1 unit and reflected across the y-axis. The correct choice is (b).

35. $f(x) = 3(1 - e^{-x})$, $x \geq 0$

This is the graph of $f(x) = e^x$ reflected across the y-axis, reflected across the x-axis, shifted up 1 unit, and stretched by a factor of 3. The correct choice is (e).

37. $\log_5 125 = 3$ because the exponent to which we raise 5 to get 125 is 3.

39. $\ln e = 1$ because the exponent to which we raise e to get e is 1.

41. $\log 10^{1/4} = \dfrac{1}{4}$ because the exponent to which we raise 10 to get $10^{1/4}$ is $\dfrac{1}{4}$.

43. $\log 1 = 0$ because the exponent to which we raise 10 to get 1 is 0.

45. $\log_2 \sqrt[3]{2} = \log_2 2^{1/3} = \dfrac{1}{3}$ because the exponent to which we raise 2 to get $2^{1/3}$ is $\dfrac{1}{3}$.

47. $\log_4 x = 2 \Rightarrow 4^2 = x$

49. $4^{-3} = \dfrac{1}{64} \Rightarrow \log_4 \dfrac{1}{64} = -3$

51. $\log 11 \approx 1.0414$

53. $\ln 3 \approx 1.0986$

55. $\log(-3)$ does not exist. (The calculator gives an error message.)

57. $\log_5 24 = \dfrac{\log 24}{\log 5} \approx 1.9746$

59.
$$3\log_b x - 4\log_b y + \dfrac{1}{2}\log_b z$$
$$= \log_b x^3 - \log_b y^4 + \log_b z^{1/2}$$
$$= \log_b \dfrac{x^3 z^{1/2}}{y^4}, \text{ or } \log_b \dfrac{x^3 \sqrt{z}}{y^4}$$

61.
$$\ln \sqrt[4]{wr^2} = \ln(wr^2)^{1/4}$$
$$= \dfrac{1}{4}\ln wr^2$$
$$= \dfrac{1}{4}(\ln w + \ln r^2)$$
$$= \dfrac{1}{4}(\ln w + 2\ln r)$$
$$= \dfrac{1}{4}\ln w + \dfrac{1}{2}\ln r$$

63.
$$\log_a 3 = \log_a \left(\dfrac{6}{2}\right)$$
$$= \log_a 6 - \log_a 2$$
$$\approx 0.778 - 0.301$$
$$\approx 0.477$$

65.
$$\log_a \dfrac{1}{5} = \log_a 5^{-1}$$
$$= -\log_a 5$$
$$\approx -0.699$$

67. $\ln e^{-5k} = -5k$ $(\log_a a^x = x)$

69.
$$\log_4 x = 2$$
$$x = 4^2 = 16$$

The solution is 16.

71. $e^x = 80$

$\ln e^x = \ln 80$

$x = \ln 80$

$x \approx 4.382$

The solution is 4.382.

73. $\log_{16} 4 = x$

$16^x = 4$

$(4^2)^x = 4^1$

$4^{2x} = 4^1$

$2x = 1$

$x = \dfrac{1}{2}$

The solution is $\dfrac{1}{2}$.

75. $\log_2 x + \log_2(x-2) = 3$

$\log_2 x(x-2) = 3$

$x(x-2) = 2^3$

$x^2 - 2x = 8$

$x^2 - 2x - 8 = 0$

$(x+2)(x-4) = 0$

$x + 2 = 0 \quad or \quad x - 4 = 0$

$x = -2 \ or \quad\quad x = 4$

The number 4 checks, but -2 does not. The solution is 4.

77. $\log x^2 = \log x$

$x^2 = x$

$x^2 - x = 0$

$x(x-1) = 0$

$x = 0 \ or \ x - 1 = 0$

$x = 0 \ or \quad\quad x = 1$

The number 1 checks, but 0 does not. The solution is 1.

79. a) $A(t) = 16,000(1.0105)^{4t}$

b) $A(0) = 16,000(1.0105)^{4\cdot 0} = \$16,000$

$A(6) = 16,000(1.0105)^{4\cdot 6} \approx \$20,558.51$

$A(12) = 16,000(1.0105)^{4\cdot 12} \approx \$26,415.77$

$A(18) = 16,000(1.0105)^{4\cdot 18} \approx \$33,941.80$

81. $T = \dfrac{\ln 2}{0.086} \approx 8.1$ years

83. $P(t) = P_0 e^{kt}$

$0.73 P_0 = P_0 e^{-0.00012t}$

$0.73 = e^{-0.00012t}$

$\ln 0.73 = \ln e^{-0.00012t}$

$\ln 0.73 = -0.00012t$

$\dfrac{\ln 0.73}{-0.00012} = t$

$2623 \approx t$

The skeleton is about 2623 years old.

85. $R = \log \dfrac{10^{6.3} \cdot I_0}{I_0} = \log 10^{6.3} = 6.3$

87. a) We substitute 353.823 for P, since P is in thousands.

$W(353.823) = 0.37 \ln 353.823 + 0.05$

≈ 2.2 ft/sec

b) We substitute 3.4 for W and solve for P.

$3.4 = 3.7 \ln P + 0.05$

$3.35 = 0.37 \ln P$

$\dfrac{3.35}{0.37} = \ln P$

$e^{3.35/0.37} = P$

$P \approx 8553.143$

The population is about 8553.143 thousand, or 8,553,143. (Answers may vary due to rounding differences.)

89. a) $P(t) = 3.039 e^{0.013t}$, where $P(t)$ is in millions and t is the number of years after 2005.

b) In 2009, $t = 2009 - 2005 = 4$.

$P(4) = 3.039 e^{0.013(4)} \approx 3.201$ million

In 2015, $t = 2015 - 2005 = 10$.

$P(10) = 3.039 e^{0.013(10)} \approx 3.461$ million

c) $10 = 3.039 e^{0.013t}$

$\dfrac{10}{3.039} = e^{0.013t}$

$\ln\left(\dfrac{10}{3.039}\right) = \ln e^{0.013t}$

$\ln\left(\dfrac{10}{3.039}\right) = 0.013t$

$\dfrac{\ln\left(\dfrac{10}{3.039}\right)}{0.013} = t$

$92 \approx t$

The population will be 10 million about 92 yr after 2005.

d) $T = \dfrac{\ln 2}{0.013} \approx 53.3$ yr

91. We must have $2x - 3 > 0$, or $x > \dfrac{3}{2}$, so answer A is correct.

93. We graph $y_1 = \dfrac{4+3x}{x-2}$, $y_2 = \dfrac{x+4}{x-3}$, and $y_3 = x$ and observe that the graphs of y_1 and y_2 are not reflections of each other across the third line, $y = x$. Thus $f(x)$ and $g(x)$ are not inverses.

95. By the product rule, $\log_2 x + \log_2 5 = \log_2 5x$, not $\log_2(x+5)$. Also, substituting various numbers for x shows that the two sides of the inequality are indeed unequal. You could also graph each side and show that the graphs do not coincide.

97. $|\log_4 x| = 3$

$\log_4 x = -3 \quad or \quad \log_4 x = 3$

$x = 4^{-3} \quad or \qquad x = 4^3$

$x = \dfrac{1}{64} \quad or \qquad x = 64$

Both answers check. The solutions are $\dfrac{1}{64}$ and 64.

99. $5^{\sqrt{x}} = 625$

$5^{\sqrt{x}} = 5^4$

$\sqrt{x} = 4$

$x = 16$

Chapter 4 Test

1. We interchange the first and second coordinates of each pair to find the inverse of the relation. It is $\{(5, -2), (3, 4)(-1, 0), (-3, -6)\}$.

2. The function is not one-to-one, because there are many horizontal lines that cross the graph more than once.

3. The function is one-to-one, because no horizontal line crosses the graph more than once.

4. a) The graph of $f(x) = x^3 + 1$ is shown below. It passes the horizontal-line test, so the function is one-to-one.

b) Replace $f(x)$ with y: $y = x^3 + 1$

Interchange x and y: $x = y^3 + 1$

Solve for y: $y^3 = x - 1$

$y = \sqrt[3]{x - 1}$

Replace y with $f^{-1}(x)$: $f^{-1}(x) = \sqrt[3]{x - 1}$

5. a) The graph of $f(x) = 1 - x$ is shown below. It passes the horizontal-line test, so the function is one-to-one.

b) Replace $f(x)$ with y: $y = 1 - x$

Interchange x and y: $x = 1 - y$

Solve for y: $y = 1 - x$

Replace y with $f^{-1}(x)$: $f^{-1}(x) = 1 - x$

6. a) The graph of $f(x) = \dfrac{x}{2 - x}$ is shown below. It passes the horizontal-line test, so the function is one-to-one.

b) Replace $f(x)$ with y: $y = \dfrac{x}{2 - x}$

Interchange x and y: $x = \dfrac{y}{2 - y}$

Solve for y: $(2 - y)x = y$

$2x - xy = y$

$xy + y = 2x$

$y(x + 1) = 2x$

$y = \dfrac{2x}{x + 1}$

Replace y with $f^{-1}(x)$: $f^{-1}(x) = \dfrac{2x}{x + 1}$

7. a) The graph of $f(x) = x^2 + x - 3$ is shown below. It is not one-to-one since there are many horizontal lines that cross the graph more than once. The function does not have an inverse that is a function.

8. We find $(f^{-1} \circ f)(x)$ and $(f \circ f^{-1})(x)$ and check to see that each is x.

$$(f^{-1} \circ f)(x) = f^{-1}(f(x)) = f^{-1}(-4x + 3) =$$

$$\frac{3 - (-4x + 3)}{4} = \frac{3 + 4x - 3}{4} = \frac{4x}{4} = x$$

$$(f \circ f^{-1})(x) = f(f^{-1}(x)) = f\left(\frac{3 - x}{4}\right) =$$

$$-4\left(\frac{3 - x}{4}\right) + 3 = -3 + x + 3 = x$$

9. Replace $f(x)$ with y: $y = \dfrac{1}{x - 4}$

Interchange x and y: $x = \dfrac{1}{y - 4}$

Solve for y: $x(y - 4) = 1$

$$xy - 4x = 1$$

$$xy = 4x + 1$$

$$y = \frac{4x + 1}{x}$$

Replace y with $f^{-1}(x)$: $f^{-1}(x) = \dfrac{4x + 1}{x}$

The domain of $f(x)$ is $(-\infty, 4) \cup (4, \infty)$ and the range of $f(x)$ is $(-\infty, 0) \cup (0, \infty)$. Thus, the domain of f^{-1} is $(-\infty, 0) \cup (0, \infty)$ and the range of f^{-1} is $(-\infty, 4) \cup (4, \infty)$.

10.

11.

12.

13.

14. $\log 0.00001 = -5$ because the exponent to which we raise 10 to get 0.00001 is -5.

15. $\ln e = 1$ because the exponent to which we raise e to get e is 1.

16. $\ln 1 = 0$ because the exponent to which we raise e to get 1 is 0.

17. $\log_4 \sqrt[5]{4} = \log_4 4^{1/5} = \dfrac{1}{5}$ because the exponent to which we raise 4 to get $4^{1/5}$ is $\dfrac{1}{5}$.

18. $\ln x = 4 \Rightarrow x = e^4$

19. $3^x = 5.4 \Rightarrow x = \log_3 5.4$

20. $\ln 16 \approx 2.7726$

21. $\log 0.293 \approx -0.5331$

22. $\log_6 10 = \dfrac{\log 10}{\log 6} \approx 1.2851$

23. $\quad 2 \log_a x - \log_a y + \dfrac{1}{2} \log_a z$

$$= \log_a x^2 - \log_a y + \log_a z^{1/2}$$

$$= \log_a \frac{x^2 z^{1/2}}{y}, \text{ or } \log_a \frac{x^2 \sqrt{z}}{y}$$

24. $\ln \sqrt[5]{x^2 y} = \ln(x^2 y)^{1/5}$

$\qquad = \dfrac{1}{5} \ln x^2 y$

$\qquad = \dfrac{1}{5}(\ln x^2 + \ln y)$

$\qquad = \dfrac{1}{5}(2\ln x + \ln y)$

$\qquad = \dfrac{2}{5}\ln x + \dfrac{1}{5}\ln y$

25. $\log_a 4 = \log_a \left(\dfrac{8}{2}\right)$

$\qquad = \log_a 8 - \log_a 2$

$\qquad \approx 0.984 - 0.328$

$\qquad \approx 0.656$

26. $\ln e^{-4t} = -4t \quad (\log_a a^x = x)$

27. $\log_{25} 5 = x$

$\qquad 25^x = 5$

$\qquad (5^2)^x = 5^1$

$\qquad 5^{2x} = 5^1$

$\qquad 2x = 1$

$\qquad x = \dfrac{1}{2}$

The solution is $\dfrac{1}{2}$.

28. $\log_3 x + \log_3 (x+8) = 2$

$\qquad \log_3 x(x+8) = 2$

$\qquad x(x+8) = 3^2$

$\qquad x^2 + 8x = 9$

$\qquad x^2 + 8x - 9 = 0$

$\qquad (x+9)(x-1) = 0$

$x = -9 \ \ or \ \ x = 1$

The number 1 checks, but -9 does not. The solution is 1.

29. $3^{4-x} = 27^x$

$\qquad 3^{4-x} = (3^3)^x$

$\qquad 3^{4-x} = 3^{3x}$

$\qquad 4 - x = 3x$

$\qquad 4 = 4x$

$\qquad x = 1$

The solution is 1.

30. $e^x = 65$

$\qquad \ln e^x = \ln 65$

$\qquad x = \ln 65$

$\qquad x \approx 4.174$

The solution is 4.174.

31. $R = \log \dfrac{10^{6.6} \cdot I_0}{I_0} = \log 10^{6.6} = 6.6$

32. $k = \dfrac{\ln 2}{45} \approx 0.0154 \approx 1.54\%$

33. a) $\qquad 1144.54 = 1000e^{3k}$

$\qquad 1.14454 = e^{3k}$

$\qquad \ln 1.14454 = \ln e^{3k}$

$\qquad \ln 1.14454 = 3k$

$\qquad \dfrac{\ln 1.14454}{3} = k$

$\qquad 0.045 \approx k$

The interest rate is about 4.5%.

b) $P(t) = 1000e^{0.045t}$

c) $P(8) = 1000e^{0.045 \cdot 8} \approx \$1433.33$

d) $T = \dfrac{\ln 2}{0.045} \approx 15.4 \text{ yr}$

34. $\qquad 4^{\sqrt[3]{x}} = 8$

$\qquad (2^2)^{\sqrt[3]{x}} = 2^3$

$\qquad 2^{2\sqrt[3]{x}} = 2^3$

$\qquad 2\sqrt[3]{x} = 3$

$\qquad \sqrt[3]{x} = \dfrac{3}{2}$

$\qquad x = \left(\dfrac{3}{2}\right)^3$

$\qquad x = \dfrac{27}{8}$

The solution is $\dfrac{27}{8}$.

Chapter 5

Systems of Equations and Matrices

Exercise Set 5.1

1. Graph (c) is the graph of this system.

3. Graph (f) is the graph of this system.

5. Graph (b) is the graph of this system.

7. Graph $x + y = 2$ and $3x + y = 0$ and find the coordinates of the point of intersection.

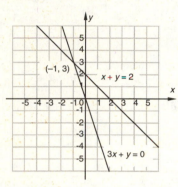

The solution is $(-1, 3)$.

9. Graph $x + 2y = 1$ and $x + 4y = 3$ and find the coordinates of the point of intersection.

The solution is $(-1, 1)$.

11. Graph $y + 1 = 2x$ and $y - 1 = 2x$ and find the coordinates of the point of intersection.

The graphs do not intersect, so there is no solution.

13. Graph $x - y = -6$ and $y = -2x$ and find the coordinates of the point of intersection.

The solution is $(-2, 4)$.

15. Graph $2y = x - 1$ and $3x = 6y + 3$ and find the coordinates of the point of intersection.

The graphs coincide so there are infinitely many solutions. Solving either equation for y, we get $y = \dfrac{x-1}{2}$, so the solutions can be expressed as $\left(x, \dfrac{x-1}{2}\right)$. Similarly, solving either equation for x, we get $x = 2y + 1$, so the solutions can also be expressed as $(2y + 1, y)$.

17. $x +\ y = 9,$ (1)

$2x - 3y = -2$ (2)

Solve equation (1) for either x or y. We choose to solve for y.

$y = 9 - x$

Then substitute $9 - x$ for y in equation (2) and solve the resulting equation.

$2x - 3(9 - x) = -2$

$2x - 27 + 3x = -2$

$5x - 27 = -2$

$5x = 25$

$x = 5$

Now substitute 5 for x in either equation (1) or (2) and solve for y.

$5 + y = 9$ Using equation (1)

$y = 4$

The solution is $(5, 4)$.

19. $x - 2y = 7,$ (1)

$x = y + 4$ (2)

Use equation (2) and substitute $y+4$ for x in equation (1). Then solve for y.

$y + 4 - 2y = 7$

$-y + 4 = 7$

$-y = 3$

$y = -3$

Substitute -3 for y in equation (2) to find x.

$x = -3 + 4 = 1$

The solution is $(1, -3)$.

21. $y = 2x - 6,$ (1)

$5x - 3y = 16$ (2)

Use equation (1) and substitute $2x - 6$ for y in equation (2). Then solve for x.

$5x - 3(2x - 6) = 16$

$5x - 6x + 18 = 16$

$-x + 18 = 16$

$-x = -2$

$x = 2$

Substitute 2 for x in equation (1) to find y.

$y = 2 \cdot 2 - 6 = 4 - 6 = -2$

The solution is $(2, -2)$.

23. $x - 5y = 4,$ (1)

$y = 7 - 2x$ (2)

Use equation (2) and substitute $7 - 2x$ for y in equation (1). Then solve for x.

$x - 5(7 - 2x) = 4$

$x - 35 + 10x = 4$

$11x - 35 = 4$

$11x = 39$

$x = \dfrac{39}{11}$

Substitute $\dfrac{39}{11}$ for x in equation (2) to find y.

$y = 7 - 2 \cdot \dfrac{39}{11} = 7 - \dfrac{78}{11} = -\dfrac{1}{11}$

The solution is $\left(\dfrac{39}{11}, -\dfrac{1}{11} \right)$.

25. $2x - 3y = 5,$ (1)

$5x + 4y = 1$ (2)

Solve one equation for either x or y. We choose to solve equation (1) for x.

$2x - 3y = 5$

$2x = 3y + 5$

$x = \dfrac{3}{2}y + \dfrac{5}{2}$

Substitute $\dfrac{3}{2}y + \dfrac{5}{2}$ for x in equation (2) and solve for y.

$5\left(\dfrac{3}{2}y + \dfrac{5}{2} \right) + 4y = 1$

$\dfrac{15}{2}y + \dfrac{25}{2} + 4y = 1$

$\dfrac{23}{2}y + \dfrac{25}{2} = 1$

$\dfrac{23}{2}y = -\dfrac{23}{2}$

$y = -1$

Substitute -1 for y in either equation (1) or (2) and solve for x.

$2x - 3(-1) = 5$ Using equation (1)

$2x + 3 = 5$

$2x = 2$

$x = 1$

The solution is $(1, -1)$.

27. $x + 2y = 2,$ (1)

$4x + 4y = 5$ (2)

Solve one equation for either x or y. We choose to solve equation (1) for x since x has a coefficient of 1 in that equation.

$x + 2y = 2$

$x = -2y + 2$

Substitute $-2y + 2$ for x in equation (2) and solve for y.

$4(-2y + 2) + 4y = 5$

$-8y + 8 + 4y = 5$

$-4y + 8 = 5$

$-4y = -3$

$y = \dfrac{3}{4}$

Substitute $\frac{3}{4}$ for y in either equation (1) or equation (2) and solve for x.

$$x + 2y = 2 \quad \text{Using equation (1)}$$
$$x + 2 \cdot \frac{3}{4} = 2$$
$$x + \frac{3}{2} = 2$$
$$x = \frac{1}{2}$$

The solution is $\left(\frac{1}{2}, \frac{3}{4} \right)$.

29. $\quad x + 2y = 7, \quad (1)$
$\quad x - 2y = -5 \quad (2)$

We add the equations to eliminate y.

$$\begin{array}{r} x + 2y = 7 \\ \underline{x - 2y = -5} \\ 2x \quad\quad = 2 \quad \text{Adding} \\ x = 1 \end{array}$$

Back-substitute in either equation and solve for y.

$$1 + 2y = 7 \quad \text{Using equation (1)}$$
$$2y = 6$$
$$y = 3$$

The solution is $(1, 3)$. Since the system of equations has exactly one solution it is consistent and independent.

31. $\quad x - 3y = 2, \quad (1)$
$\quad 6x + 5y = -34 \quad (2)$

Multiply equation (1) by -6 and add it to equation (2) to eliminate x.

$$\begin{array}{r} -6x + 18y = -12 \\ \underline{6x + \ 5y = -34} \\ 23y = -46 \\ y = -2 \end{array}$$

Back-substitute to find x.

$$x - 3(-2) = 2 \quad \text{Using equation (1)}$$
$$x + 6 = 2$$
$$x = -4$$

The solution is $(-4, -2)$. Since the system of equations has exactly one solution it is consistent and independent.

33. $\quad 3x - 12y = 6, \quad (1)$
$\quad 2x - \ 8y = 4 \quad (2)$

Multiply equation (1) by 2 and equation (2) by -3 and add.

$$\begin{array}{r} 6x - 24y = 12 \\ \underline{-6x + 24y = -12} \\ 0 = 0 \end{array}$$

The equation $0 = 0$ is true for all values of x and y. Thus, the system of equations has infinitely many solutions. Solving either equation for y, we can write $y = \frac{1}{4}x - \frac{1}{2}$ so the solutions are ordered pairs of the form $\left(x, \frac{1}{4}x - \frac{1}{2} \right)$.

Equivalently, if we solve either equation for x we get $x = 4y + 2$ so the solutions can also be expressed as $(4y + 2, y)$. Since there are infinitely many solutions, the system of equations is consistent and dependent.

35. $\quad 2x = 5 - 3y, \quad (1)$
$\quad 4x = 11 - 7y \quad (2)$

We rewrite the equations.

$$2x + 3y = 5, \quad (1a)$$
$$4x + 7y = 11 \quad (2a)$$

Multiply equation (2a) by -2 and add to eliminate x.

$$\begin{array}{r} -4x - 6y = -10 \\ \underline{4x + 7y = 11} \\ y = 1 \end{array}$$

Back-substitute to find x.

$$2x = 5 - 3 \cdot 1 \quad \text{Using equation (1)}$$
$$2x = 2$$
$$x = 1$$

The solution is $(1, 1)$. Since the system of equations has exactly one solution it is consistent and independent.

37. $\quad 0.3x - 0.2y = -0.9,$
$\quad 0.2x - 0.3y = -0.6$

First, multiply each equation by 10 to clear the decimals.

$$3x - 2y = -9 \quad (1)$$
$$2x - 3y = -6 \quad (2)$$

Now multiply equation (1) by 3 and equation (2) by -2 and add to eliminate y.

$$\begin{array}{r} 9x - 6y = -27 \\ \underline{-4x + 6y = 12} \\ 5x \quad\quad = -15 \\ x = -3 \end{array}$$

Back-substitute to find y.

$$3(-3) - 2y = -9 \quad \text{Using equation (1)}$$
$$-9 - 2y = -9$$
$$-2y = 0$$
$$y = 0$$

The solution is $(-3, 0)$. Since the system of equations has exactly one solution it is consistent and independent.

39. $\quad \frac{1}{5}x + \frac{1}{2}y = 6, \quad (1)$
$\quad \frac{3}{5}x - \frac{1}{2}y = 2 \quad (2)$

We could multiply both equations by 10 to clear fractions, but since the y-coefficients differ only by sign we will just add to eliminate y.

$$\begin{array}{r} \frac{1}{5}x + \frac{1}{2}y = 6 \\ \underline{\frac{3}{5}x - \frac{1}{2}y = 2} \\ \frac{4}{5}x \quad\quad = 8 \\ x = 10 \end{array}$$

Back-substitute to find y.

$$\frac{1}{5} \cdot 10 + \frac{1}{2}y = 6 \quad \text{Using equation (1)}$$

$$2 + \frac{1}{2}y = 6$$

$$\frac{1}{2}y = 4$$

$$y = 8$$

The solution is $(10, 8)$. Since the system of equations has exactly one solution it is consistent and independent.

41. Familiarize. Let $x =$ the number of injuries that occur in snowboarding each winter and $y =$ the number of injuries that occur in skiing.

Translate. The total number of snowboarding and skiing injuries is 288,400 so we have one equation:

$$x + y = 288,400$$

Skiing accounts for 400 more injuries than snowboarding, so we have a second equation:

$$y = x + 400$$

Carry out. We solve the system of equations

$$x + y = 288,400, \quad (1)$$
$$y = x + 400. \quad (2)$$

Substitute $x + 400$ for y in equation (1) and solve for x.

$$x + (x + 400) = 288,400$$
$$2x + 400 = 288,400$$
$$2x = 288,000$$
$$x = 144,000$$

Back-substitute in equation (2) to find y.

$$y = 144,000 + 400 = 144,400$$

Check. $144,400 + 144,000 = 288,400$ injuries and $144,400$ is 400 more than $144,000$. The answer checks.

State. Snowboarding accounts for 144,000 injuries each winter and skiing accounts for about 144,400 injuries.

43. Familiarize. Let $m =$ the number of streets named Main Street and $s =$ the number of streets named Second Street.

Translate. The total number of streets bearing one of these names is 15,684, so we have one equation:

$$m + s = 15,684$$

A second equation comes from the fact that there are 260 more streets named Second Street than Main Street:

$$s = m + 260$$

Carry out. We solve the system of equations

$$m + s = 15,684, \quad (1)$$
$$s = m + 260. \quad (2)$$

Substitute $m + 260$ for s in equation (1) and solve for m.

$$m + (m + 260) = 15,684$$
$$2m + 260 = 15,684$$
$$2m = 15,424$$
$$m = 7712$$

Back-substitute in equation (2) to find s.

$$s = 7712 + 260 = 7972$$

Check. The total number of streets is $7712 + 7972$, or $15,684$ and 7972 is 260 more than 7712. The answer checks.

State. There are 7712 streets named Main Street and 7972 streets named Second Street.

45. Familiarize. Let $w =$ the number of visitors hosted by the website and $m =$ the number hosted by the museums, in millions.

Translate. The total number of visitors was 142 million, so we have one equation:

$$m + w = 142$$

A second equation comes from the fact that the museums had 94 million fewer visitors than the website:

$$m = w - 94$$

Carry out. We solve the system of equations

$$m + w = 142, \quad (1)$$
$$m = w - 94. \quad (2)$$

Substitute $w - 94$ for m in equation (1) and solve for w.

$$(w - 94) + w = 142$$
$$2w - 94 = 142$$
$$2w = 236$$
$$w = 118$$

Back-substitute in equation (2) to find m.

$$m = 118 - 94 = 24$$

Check. The total number of visitors is $24 + 118$, or 142 million, and 24 million is 94 million less than 118 million. The answer checks

State. The Smithsonian Institution hosted 24 million visitors and the Smithsonian website hosted 118 million visitors.

47. Familiarize. Let $x =$ the number of video rentals and $y =$ the number of boxes of popcorn given away. Then the total cost of the rentals is $1 \cdot x$, or x, and the total cost of the popcorn is $2y$.

Translate. The number of new members is the same as the number of incentives so we have one equation:

$$x + y = 48.$$

The total cost of the incentive was \$86. This gives us another equation:

$$x + 2y = 86.$$

Carry out. We solve the system of equations

$$x + y = 48, \quad (1)$$
$$x + 2y = 86. \quad (2)$$

Multiply equation (1) by -1 and add.

$$\begin{array}{r} -x - y = -48 \\ x + 2y = 86 \\ \hline y = 38 \end{array}$$

Back-substitute to find x.

$x + 38 = 48$ Using equation (1)

$\qquad x = 10$

Check. When 10 rentals and 38 boxes of popcorn are given away, a total of 48 new members have signed up. Ten rentals cost the store $10 and 38 boxes of popcorn cost $2 \cdot 38$, or \$76, so the total cost of the incentives was $10 + 76$, or \$86. The solution checks.

State. Ten rentals and 38 boxes of popcorn were given away.

49. Familiarize and Translate. We use the system of equations given in the problem.

$$y = 70 + 2x \qquad (1)$$
$$y = 175 - 5x, \quad (2)$$

Carry out. Substitute $175 - 5x$ for y in equation (1) and solve for x.

$175 - 5x = 70 + 2x$

$\qquad 105 = 7x$ Adding $5x$ and subtracting 70

$\qquad 15 = x$

Back-substitute in either equation to find y. We choose equation (1).

$y = 70 + 2 \cdot 15 = 70 + 30 = 100$

Check. Substituting 15 for x and 100 for y in both of the original equations yields true equations, so the solution checks. We could also check graphically by finding the point of intersection of equations (1) and (2).

State. The equilibrium point is $(15, \$100)$.

51. Familiarize and Translate. We find the value of x for which $C = R$, where

$C = 14x + 350,$

$R = 16.5x.$

Carry out. When $C = R$ we have:

$14x + 350 = 16.5x$

$\qquad 350 = 2.5x$

$\qquad 140 = x$

Check. When $x = 140$, $C = 14 \cdot 140 + 350$, or 2310 and $R = 16.5(140)$, or 2310. Since $C = R$, the solution checks.

State. 140 units must be produced and sold in order to break even.

53. Familiarize and Translate. We find the value of x for which $C = R$, where

$C = 15x + 12,000,$

$R = 18x - 6000.$

Carry out. When $C = R$ we have:

$15x + 12,000 = 18x - 6000$

$\qquad 18,000 = 3x$ Subtracting $15x$ and

$\qquad\qquad\qquad\qquad$ adding 6000

$\qquad 6000 = x$

Check. When $x = 6000$, $C = 15 \cdot 6000 + 12,000$, or 102,000 and $R = 18 \cdot 6000 - 6000$, or 102,000. Since $C = R$, the solution checks.

State. 6000 units must be produced and sold in order to break even.

55. Familiarize. Let $x =$ the number of servings of spaghetti and meatballs required and $y =$ the number of servings of iceberg lettuce required. Then x servings of spaghetti contain $260x$ Cal and $32x$ g of carbohydrates; y servings of lettuce contain $5y$ Cal and $1 \cdot y$ or y, g of carbohydrates.

Translate. One equation comes from the fact that 400 Cal are desired:

$260x + 5y = 400.$

A second equation comes from the fact that 50g of carbohydrates are required:

$32x + y = 50.$

Carry out. We solve the system

$260x + 5y = 400, \quad (1)$

$32x + \ y = 50. \quad (2)$

Multiply equation (2) by -5 and add.

$\qquad 260x + 5y = 400$

$\underline{-160x - 5y = -250}$

$\qquad 100x \qquad\ = 150$

$\qquad\qquad x = 1.5$

Back-substitute to find y.

$32(1.5) + y = 50$ Using equation (2)

$\qquad 48 + y = 50$

$\qquad\qquad y = 2$

Check. 1.5 servings of spaghetti contain $260(1.5)$, or 390 Cal and $32(1.5)$, or 48 g of carbohydrates; 2 servings of lettuce contain $5 \cdot 2$, or 10 Cal and $1 \cdot 2$, or 2 g of carbohydrates. Together they contain $390 + 10$, or 400 Cal and $48 + 2$, or 50 g of carbohydrates. The solution checks.

State. 1.5 servings of spaghetti and meatballs and 2 servings of iceberg lettuce are required.

57. Familiarize. It helps to make a drawing. Then organize the information in a table. Let $x =$ the speed of the boat and $y =$ the speed of the stream. The speed upstream is $x - y$. The speed downstream is $x + y$.

$\quad$ 46 km $\qquad$ 2 hr $\quad (x + y)$ km/h

Downstream

$\quad$ 51 km $\qquad$ 3 hr $\quad (x - y)$ km/h

Upstream

	Distance	Speed	Time
Downstream	46	$x + y$	2
Upstream	51	$x - y$	3

Translate. Using $d = rt$ in each row of the table, we get a system of equations.

$46 = (x + y)2 \qquad\qquad x + y = 23, \quad (1)$

$\qquad\qquad\qquad$ or

$51 = (x - y)3 \qquad\qquad x - y = 17 \quad (2)$

Carry out. We begin by adding equations (1) and (2).

$$x + y = 23$$
$$\underline{x - y = 17}$$
$$2x \quad\;\; = 40$$
$$x = 20$$

Back-substitute to find y.

$$20 + y = 23 \quad \text{Using equation (1)}$$
$$y = 3$$

Check. The speed downstream is $20+3$, or 23 km/h. The distance traveled downstream in 2 hr is $23 \cdot 2$, or 46 km. The speed upstream is $20 - 3$, or 17 km/h. The distance traveled upstream in 3 hr is $17 \cdot 3$, or 51 km. The solution checks.

State. The speed of the boat is 20 km/h. The speed of the stream is 3 km/h.

59. Familiarize. Let $x =$ the amount invested at 7% and $y =$ the amount invested at 9%. Then the interest from the investments is 7%x and 9%y, or $0.07x$ and $0.09y$.

Translate.

The total investment is $15,000.

$$x + y = 15,000$$

The total interest is $1230.

$$0.07x + 0.09y = 1230$$

We have a system of equations:

$$x + \quad\;\; y = 15,000,$$
$$0.07x + 0.09y = 1230$$

Multiplying the second equation by 100 to clear the decimals, we have:

$$x + \;\; y = 15,000, \quad (1)$$
$$7x + 9y = 123,000. \quad (2)$$

Carry out. We begin by multiplying equation (1) by -7 and adding.

$$-7x - 7y = -105,000$$
$$\underline{7x + 9y = 123,000}$$
$$2y = 18,000$$
$$y = 9000$$

Back-substitute to find x.

$$x + 9000 = 15,000 \quad \text{Using equation (1)}$$
$$x = 6000$$

Check. The total investment is $6000+$9000, or $15,000. The total interest is $0.07($6000$) + 0.09($9000$)$, or $420 + $810, or $1230. The solution checks.

State. $6000 was invested at 7% and $9000 was invested at 9%.

61. Familiarize. Let $x =$ the number of pounds of French roast coffee used and $y =$ the number of pounds of Kenyan coffee. We organize the information in a table.

	French roast	Kenyan	Mixture
Amount	x	y	10 lb
Price per pound	$9.00	$7.50	$8.40
Total cost	$9x$	$7.50y$	$8.40(10), or $84

Translate. The first and third rows of the table give us a system of equations.

$$x + \quad\; y = 10,$$
$$9x + 7.5y = 84$$

Multiply the second equation by 10 to clear the decimals.

$$x + \;\; y = 10, \quad (1)$$
$$90x + 75y = 840 \quad (2)$$

Carry out. Begin by multiplying equation (1) by -75 and adding.

$$-75x - 75y = -750$$
$$\underline{90x + 75y = 840}$$
$$15x \qquad\;\; = 90$$
$$x = 6$$

Back-substitute to find y.

$$6 + y = 10 \quad \text{Using equation (1)}$$
$$y = 4$$

Check. The total amount of coffee in the mixture is $6+4$, or 10 lb. The total value of the mixture is $6($9$)+4($7.50$)$, or $54 + $30, or $84. The solution checks.

State. 6 lb of French roast coffee and 4 lb of Kenyan coffee should be used.

63. Familiarize. Let $x =$ the speed of the plane and $y =$ the speed of the wind. We organize the information in a table.

	Distance	Speed	Time
LA to NY	3000	$x + y$	5
NY to LA	3000	$x - y$	6

Translate. Using $d = rt$ in each row of the table, we get a system of equations.

$$3000 = (x + y)5 \qquad x + y = 600, \quad (1)$$
$$\text{or}$$
$$3000 = (x - y)6 \qquad x - y = 500 \quad (2)$$

Carry out. We begin by adding equations (1) and (2).

$$x + y = 600$$
$$\underline{x - y = 500}$$
$$2x \quad\;\; = 1100$$
$$x = 550$$

Back-substitute to find y.

$$550 + y = 600$$
$$y = 50$$

Check. The speed with the wind is $550 + 50$, or 600 mph. Then the distance traveled with a tailwind is $600 \cdot 5$, or 3000 mi. The speed against the wind is $550 - 50$, or 500 mi.

The distance traveled against the wind is 500·6, or 3000 mi. The distances are both 3000 mi, so the answer checks.

State. The speed of the plane is 550 mph, and the speed of the wind is 50 mph.

65. Left to the student

67. a) $b(x) = -0.0657894737x + 63.95394737$;

$c(x) = 1.186842105x + 48.24078947$

b) Graph $y_1 = b(x)$ and $y_2 = c(x)$ and find the first coordinate of the point of intersection of the graphs. It is approximately 12.5, so we estimate that chicken consumption will equal beef consumption about 12.5 yr after 1995.

69. Discussion and Writing

71. *Familiarize*. Let h = the number of hardback books sold, in millions. Then $h + 40$ = the number of paperback books sold.

Translate.

Hardback sales plus paperback sales is total sales.

$$h + (h + 40) = 110$$

Carry out. We solve the equation.

$$h + (h + 40) = 110$$
$$2h + 40 = 110$$
$$2h = 70$$
$$h = 35$$

If $h = 35$, then $h + 40 = 35 + 40 = 75$.

Check. Total sales were $35 + 75$, or 110 million, and 75 million is 40 million more than 35 million. The answer checks.

State. 35 million hardback books and 75 million paperback books were sold.

73. Substituting 15 for $f(x)$, we solve the following equation.

$$15 = x^2 - 4x + 3$$
$$0 = x^2 - 4x - 12$$
$$0 = (x - 6)(x + 2)$$
$$x - 6 = 0 \quad or \quad x + 2 = 0$$
$$x = 6 \quad or \quad x = -2$$

If the output is 15, the input is 6 or -2.

75. $f(-2) = (-2)^2 - 4(-2) + 3 = 15$

77. *Familiarize*. Let x = the time spent jogging and y = the time spent walking. Then Nancy jogs $8x$ km and walks $4y$ km.

Translate.

The total time is 1 hr.

$$x + y = 1$$

The total distance is 6 km.

$$8x + 4y = 6$$

Carry out. Solve the system

$$x + y = 1, \quad (1)$$
$$8x + 4y = 6. \quad (2)$$

Multiply equation (1) by -4 and add.

$$-4x - 4y = -4$$
$$\underline{8x + 4y = 6}$$
$$4x = 2$$
$$x = \frac{1}{2}$$

This is the time we need to find the distance spent jogging, so we could stop here. However, we will not be able to check the solution unless we find y also so we continue. We back-substitute.

$$\frac{1}{2} + y = 1 \quad \text{Using equation (1)}$$
$$y = \frac{1}{2}$$

Then the distance jogged is $8 \cdot \frac{1}{2}$, or 4 km, and the distance walked is $4 \cdot \frac{1}{2}$, or 2 km.

Check. The total time is $\frac{1}{2}$ hr $+ \frac{1}{2}$ hr, or 1 hr. The total distance is 4 km $+$ 2 km, or 6 km. The solution checks.

State. Nancy jogged 4 km on each trip.

79. *Familiarize and Translate*. We let x and y represent the speeds of the trains. Organize the information in a table. Using $d = rt$, we let $3x$, $2y$, $1.5x$, and $3y$ represent the distances the trains travel.

First situation:

3 hours	x km/h		y km/h	2 hours
Union		•		Central

|—————————— 216 km ——————————|

Second situation:

1.5 hours	x km/h		y km/h	3 hours
Union		•		Central

|—————————— 216 km ——————————|

	Distance traveled in first situation	Distance traveled in second situation
Train$_1$ (from Union to Central)	$3x$	$1.5x$
Train$_2$ (from Central to Union)	$2y$	$3y$
Total	216	216

The total distance in each situation is 216 km. Thus, we have a system of equations.

$$3x + 2y = 216, \quad (1)$$
$$1.5x + 3y = 216 \quad (2)$$

Carry out. Multiply equation (2) by -2 and add.

$$3x + 2y = 216$$
$$\underline{-3x - 6y = -432}$$
$$-4y = -216$$
$$y = 54$$

Back-substitute to find x.

$$3x + 2 \cdot 54 = 216 \quad \text{Using equation (1)}$$
$$3x + 108 = 216$$
$$3x = 108$$
$$x = 36$$

Check. If $x = 36$ and $y = 54$, the total distance the trains travel in the first situation is $3 \cdot 36 + 2 \cdot 54$, or 216 km. The total distance they travel in the second situation is $1.5 \cdot 36 + 3 \cdot 54$, or 216 km. The solution checks.

State. The speed of the first train is 36 km/h. The speed of the second train is 54 km/h.

81. Substitute the given solutions in the equation $Ax + By = 1$ to get a system of equations.

$$3A - B = 1, \quad (1)$$
$$-4A - 2B = 1 \quad (2)$$

Multiply equation (1) by -2 and add.

$$-6A + 2B = -2$$
$$\underline{-4A - 2B = 1}$$
$$-10A = -1$$
$$A = \frac{1}{10}$$

Back-substitute to find B.

$$3\left(\frac{1}{10}\right) - B = 1 \qquad \text{Using equation (1)}$$
$$\frac{3}{10} - B = 1$$
$$-B = \frac{7}{10}$$
$$B = -\frac{7}{10}$$

We have $A = \frac{1}{10}$ and $B = -\frac{7}{10}$.

83. Familiarize. Let x and y represent the number of gallons of gasoline used in city driving and in highway driving, respectively. Then $49x$ and $51y$ represent the number of miles driven in the city and on the highway, respectively.

Translate. The fact that 9 gal of gasoline were used gives us one equation:

$$x + y = 9.$$

A second equation comes from the fact that the car is driven 447 mile:

$$49x + 51y = 447.$$

Carry out. We solve the system of equations

$$x + y = 9, \quad (1)$$
$$49x + 51y = 447. \quad (2)$$

Multiply equation (1) by -49 and add.

$$-49x - 49y = -441$$
$$\underline{49x + 51y = 447}$$
$$2y = 6$$
$$y = 3$$

Back-substitute to find x.

$$x + 3 = 9$$
$$x = 6$$

Then in the city the car is driven 49(6), or 294 mi; on the highway it is driven 51(3), or 153 mi.

Check. The number of gallons of gasoline used is $6 + 3$, or 9. The number of miles driven is $294 + 153 = 447$. The answer checks.

State. The car was driven 294 mi in the city and 153 mi on the highway.

Exercise Set 5.2

1.
$$x + y + z = 2, \quad (1)$$
$$6x - 4y + 5z = 31, \quad (2)$$
$$5x + 2y + 2z = 13 \quad (3)$$

Multiply equation (1) by -6 and add it to equation (2). We also multiply equation (1) by -5 and add it to equation (3).

$$x + y + z = 2 \quad (1)$$
$$-10y - z = 19 \quad (4)$$
$$-3y - 3z = 3 \quad (5)$$

Multiply the last equation by 10 to make the y-coefficient a multiple of the y-coefficient in equation (4).

$$x + y + z = 2 \quad (1)$$
$$-10y - z = 19 \quad (4)$$
$$-30y - 30z = 30 \quad (6)$$

Multiply equation (4) by -3 and add it to equation (6).

$$x + y + z = 2 \quad (1)$$
$$-10y - z = 19 \quad (4)$$
$$-27z = -27 \quad (7)$$

Solve equation (7) for z.

$$-27z = -27$$
$$z = 1$$

Back-substitute 1 for z in equation (4) and solve for y.

$$-10y - 1 = 19$$
$$-10y = 20$$
$$y = -2$$

Back-substitute 1 for z for -2 and y in equation (1) and solve for x.

$$x + (-2) + 1 = 2$$
$$x - 1 = 2$$
$$x = 3$$

The solution is $(3, -2, 1)$.

3.
$$x - y + 2z = -3 \quad (1)$$
$$x + 2y + 3z = 4 \quad (2)$$
$$2x + y + z = -3 \quad (3)$$

Multiply equation (1) by -1 and add it to equation (2). We also multiply equation (1) by -2 and add it to equation (3).
$$x - y + 2z = -3 \quad (1)$$
$$3y + z = 7 \quad (4)$$
$$3y - 3z = 3 \quad (5)$$

Multiply equation (4) by -1 and add it to equation (5).
$$x - y + 2z = -3 \quad (1)$$
$$3y + z = 7 \quad (4)$$
$$- 4z = -4 \quad (6)$$

Solve equation (6) for z.
$$-4z = -4$$
$$z = 1$$

Back-substitute 1 for z in equation (4) and solve for y.
$$3y + 1 = 7$$
$$3y = 6$$
$$y = 2$$

Back-substitute 1 for z and 2 for y in equation (1) and solve for x.
$$x - 2 + 2 \cdot 1 = -3$$
$$x = -3$$

The solution is $(-3, 2, 1)$.

5.
$$x + 2y - z = 5, \quad (1)$$
$$2x - 4y + z = 0, \quad (2)$$
$$3x + 2y + 2z = 3 \quad (3)$$

Multiply equation (1) by -2 and add it to equation (2). Also, multiply equation (1) by -3 and add it to equation (3).
$$x + 2y - z = 5, \quad (1)$$
$$- 8y + 3z = -10, \quad (4)$$
$$- 4y + 5z = -12 \quad (5)$$

Multiply equation (5) by 2 to make the y-coefficient a multiple of the y-coefficient of equation (4).
$$x + 2y - z = 5, \quad (1)$$
$$- 8y + 3z = -10, \quad (4)$$
$$- 8y + 10z = -24 \quad (6)$$

Multiply equation (4) by -1 and add it to equation (6).
$$x + 2y - z = 5, \quad (1)$$
$$- 8y + 3z = -10, \quad (4)$$
$$7z = -14 \quad (7)$$

Solve equation (7) for z.
$$7z = -14$$
$$z = -2$$

Back-substitute -2 for z in equation (4) and solve for y.

$$-8y + 3(-2) = -10$$
$$-8y - 6 = -10$$
$$-8y = -4$$
$$y = \frac{1}{2}$$

Back-substitute $\frac{1}{2}$ for y and -2 for z in equation (1) and solve for x.
$$x + 2 \cdot \frac{1}{2} - (-2) = 5$$
$$x + 1 + 2 = 5$$
$$x = 2$$

The solution is $\left(2, \frac{1}{2}, -2\right)$.

7.
$$x + 2y - z = -8, \quad (1)$$
$$2x - y + z = 4, \quad (2)$$
$$8x + y + z = 2 \quad (3)$$

Multiply equation (1) by -2 and add it to equation (2). Also, multiply equation (1) by -8 and add it to equation (3).
$$x + 2y - z = -8, \quad (1)$$
$$- 5y + 3z = 20, \quad (4)$$
$$- 15y + 9z = 66 \quad (5)$$

Multiply equation (4) by -3 and add it to equation (5).
$$x + 2y - z = -8, \quad (1)$$
$$- 5y + 3z = 20, \quad (4)$$
$$0 = 6 \quad (6)$$

Equation (6) is false, so the system of equations has no solution.

9.
$$2x + y - 3z = 1, \quad (1)$$
$$x - 4y + z = 6, \quad (2)$$
$$4x - 7y - z = 13 \quad (3)$$

Interchange equations (1) and (2).
$$x - 4y + z = 6, \quad (2)$$
$$2x + y - 3z = 1, \quad (1)$$
$$4x - 7y - z = 13 \quad (3)$$

Multiply equation (2) by -2 and add it to equation (1). Also, multiply equation (2) by -4 and add it to equation (3).
$$x - 4y + z = 6, \quad (2)$$
$$9y - 5z = -11, \quad (4)$$
$$9y - 5z = -11 \quad (5)$$

Multiply equation (4) by -1 and add it to equation (5).
$$x - 4y + z = 6, \quad (1)$$
$$9y - 5z = -11, \quad (4)$$
$$0 = 0 \quad (6)$$

The equation $0 = 0$ tells us that equation (3) of the original system is dependent on the first two equations. The system of equations has infinitely many solutions and is equivalent to

$$2x + y - 3z = 1, \quad (1)$$
$$x - 4y + z = 6. \quad (2)$$

To find an expression for the solutions, we first solve equation (4) for either y or z. We choose to solve for z.

$$9y - 5z = -11$$
$$-5z = -9y - 11$$
$$z = \frac{9y + 11}{5}$$

Back-substitute in equation (2) to find an expression for x in terms of y.

$$x - 4y + \frac{9y + 11}{5} = 6$$
$$x - 4y + \frac{9}{5}y + \frac{11}{5} = 6$$
$$x = \frac{11}{5}y + \frac{19}{5} = \frac{11y + 19}{5}$$

The solutions are given by $\left(\dfrac{11y + 19}{5}, y, \dfrac{9y + 11}{5} \right)$, where y is any real number.

11.
$$4a + 9b \qquad = 8, \quad (1)$$
$$8a \qquad + 6c = -1, \quad (2)$$
$$6b + 6c = -1 \quad (3)$$

Multiply equation (1) by -2 and add it to equation (2).

$$4a + 9b \qquad = 8, \quad (1)$$
$$- 18b + 6c = -17, \quad (4)$$
$$6b + 6c = -1 \quad (3)$$

Multiply equation (3) by 3 to make the b-coefficient a multiple of the b-coefficient in equation (4).

$$4a + 9b \qquad = 8, \quad (1)$$
$$- 18b + 6c = -17, \quad (4)$$
$$18b + 18c = -3 \quad (5)$$

Add equation (4) to equation (5).

$$4a + 9b \qquad = 8, \quad (1)$$
$$- 18b + 6c = -17, \quad (4)$$
$$24c = -20 \quad (6)$$

Solve equation (6) for c.

$$24c = -20$$
$$c = -\frac{20}{24} = -\frac{5}{6}$$

Back-substitute $-\dfrac{5}{6}$ for c in equation (4) and solve for b.

$$-18b + 6c = -17$$
$$-18b + 6\left(-\frac{5}{6} \right) = -17$$
$$-18b - 5 = -17$$
$$-18b = -12$$
$$b = \frac{12}{18} = \frac{2}{3}$$

Back-substitute $\dfrac{2}{3}$ for b in equation (1) and solve for a.

$$4a + 9b = 8$$
$$4a + 9 \cdot \frac{2}{3} = 8$$
$$4a + 6 = 8$$
$$4a = 2$$
$$a = \frac{1}{2}$$

The solution is $\left(\dfrac{1}{2}, \dfrac{2}{3}, -\dfrac{5}{6} \right)$.

13.
$$2x \qquad + z = 1, \quad (1)$$
$$3y - 2z = 6, \quad (2)$$
$$x - 2y \qquad = -9 \quad (3)$$

Interchange equations (1) and (3).

$$x - 2y \qquad = -9, \quad (3)$$
$$3y - 2z = 6, \quad (2)$$
$$2x \qquad + z = 1 \quad (1)$$

Multiply equation (3) by -2 and add it to equation (1).

$$x - 2y \qquad = -9, \quad (3)$$
$$3y - 2z = 6, \quad (2)$$
$$4y + z = 19 \quad (4)$$

Multiply equation (4) by 3 to make the y-coefficient a multiple of the y-coefficient in equation (2).

$$x - 2y \qquad = -9, \quad (3)$$
$$3y - 2z = 6, \quad (2)$$
$$12y + 3z = 57 \quad (5)$$

Multiply equation (2) by -4 and add it to equation (5).

$$x - 2y \qquad = -9, \quad (3)$$
$$3y - 2z = 6, \quad (2)$$
$$11z = 33 \quad (6)$$

Solve equation (6) for z.

$$11z = 33$$
$$z = 3$$

Back-substitute 3 for z in equation (2) and solve for y.

$$3y - 2z = 6$$
$$3y - 2 \cdot 3 = 6$$
$$3y - 6 = 6$$
$$3y = 12$$
$$y = 4$$

Back-substitute 4 for y in equation (3) and solve for x.

$$x - 2y = -9$$
$$x - 2 \cdot 4 = -9$$
$$x - 8 = -9$$
$$x = -1$$

The solution is $(-1, 4, 3)$.

15.
$$w + x + y + z = 2 \quad (1)$$
$$w + 2x + 2y + 4z = 1 \quad (2)$$
$$-w + x - y - z = -6 \quad (3)$$
$$-w + 3x + y - z = -2 \quad (4)$$

Multiply equation (1) by -1 and add to equation (2). Add equation (1) to equation (3) and to equation (4).

$$w + x + y + z = 2 \quad (1)$$
$$x + y + 3z = -1 \quad (5)$$
$$2x = -4 \quad (6)$$
$$4x + 2y = 0 \quad (7)$$

Solve equation (6) for x.

$$2x = -4$$
$$x = -2$$

Back-substitute -2 for x in equation (7) and solve for y.

$$4(-2) + 2y = 0$$
$$-8 + 2y = 0$$
$$2y = 8$$
$$y = 4$$

Back-substitute -2 for x and 4 for y in equation (5) and solve for z.

$$-2 + 4 + 3z = -1$$
$$3z = -3$$
$$z = -1$$

Back-substitute -2 for x, 4 for y, and -1 for z in equation (1) and solve for w.

$$w - 2 + 4 - 1 = 2$$
$$w = 1$$

The solution is $(1, -2, 4, -1)$.

17. *Familiarize.* Let c, r, and w represent the amount spent on cross-training shoes, running shoes, and walking shoes in 2004, respectively, in billions of dollars.

Translate. Total sales were \$6.7 billion.

$$c + r + w = 6.7$$

The amount spent on walking shoes was \$0.3 billion more than the total amount spent on cross-training and running shoes.

$$w = c + r + 0.3$$

The amount spent on running shoes was \$0.4 billion more than the amount spent on cross-training shoes.

$$r = c + 0.4$$

We have a system of equations

$$c + r + w = 6.7, \qquad c + r + w = 6.7,$$
$$w = c + r + 0.3, \text{ or } -c - r + w = 0.3,$$
$$r = c + 0.4 \qquad -c + r = 0.4$$

Carry out. Solving the system of equations, we get $(1.4, 1.8, 3.5)$.

Check. Total sales were $1.4 + 1.8 + 3.5$, or \$6.7 billion. Also, \$3.5 billion is \$0.3 billion more than the total of \$1.4

billion and \$1.8 billion, and \$1.8 billion is \$0.4 billion more than \$1.4 billion. The answer checks.

State. \$1.4 billion was spent on cross-training shoes, \$1.8 billion was spent on running shoes, and \$3.5 billion was spent on walking shoes.

19. *Familiarize.* Let x y, and z represents the number of milligrams of caffeine in an 8-oz serving of brewed coffee, Red Bull energy drink, and Mountain Dew, respectively.

Translate. The total amount of caffeine in one serving of each beverage is 197 mg.

$$x + y + z = 197$$

One serving of brewed coffee has 6 mg more caffeine than two servings of Mountain Dew.

$$x = 2z + 6$$

One serving of Red Bull contains 37 mg less caffeine than one serving each of brewed coffee and Mountain Dew.

$$y = x + z - 37$$

We have a system of equations

$$x + y + z = 197, \qquad x + y + z = 197,$$
$$x = 2z + 6, \qquad \text{or} \qquad x - 2z = 6,$$
$$y = x + z - 37 \qquad -x + y - z = -37$$

Carry out. Solving the system of equations, we get $(80, 80, 37)$.

Check. The total amount of caffeine is $80 + 80 + 37$, or 197 mg. Also, 80 mg is 6 mg more than twice 37 mg, and 80 mg is 37 mg less than the total of 80 mg and 37 mg, or 117 mg. The answer checks.

State. One serving each of brewed coffee, Red Bull energy drink, and Mountain Dew contains 80 mg, 80 mg, and 37 mg of caffeine, respectively.

21. *Familiarize.* Let x, y, and z represent the number of grams of carbohydrates in 1 cup of raw lettuce, 6 raw asparagus spears, and 1 cup of raw tomatoes, respectively.

Translate. Together, 1 cup of raw lettuce, 6 raw asparagus spears, and 1 cup of raw tomatoes contain 12 grams of carbohydrates.

$$x + y + z = 12$$

One cup of raw lettuce and 6 raw asparagus spears have one-half the carbohydrates of 1 cup of raw tomatoes.

$$x + y = \frac{1}{2}z$$

One cup each of raw lettuce and raw tomatoes have 3 times the carbohydrates of 6 raw asparagus spears.

$$x + z = 3y$$

We have

$$x + y + z = 12,$$
$$x + y = \frac{1}{2}z,$$
$$x + z = 3y$$

or

$$x + y + z = 12,$$
$$x + y - \frac{1}{2}z = 0,$$
$$x - 3y + z = 0.$$

Carry out. Solving the system of equations, we get $(1, 3, 8)$.

Check. The total carbohydrate content is $1+3+8$, or 12 g. One cup of raw lettuce and 6 raw asparagus spears contain $1+3$, or 4 g, of carbohydrates. This is one-half of 8 g, the carbohydrate content of 1 cup of raw tomatoes. One cup each of raw lettuce and raw tomatoes contain $1+8$, or 9 g, of carbohydrates. This is 3 times the 3 g of carbohydrates in 6 raw asparagus spears. The solution checks.

State. One cup of raw lettuce contains 1 g of carbohydrates, 6 raw asparagus spears contain 3 g of carbohydrates, and 1 cup of raw tomatoes contains 8 g of carbohydrates.

23. *Familiarize*. Let $x =$ the number of orders under 10 lb, $y =$ the number of orders from 10 lb up to 15 lb, and $z =$ the number of orders of 15 lb or more. Then the total shipping charges for each category of order are $\$3x$, $\$5y$, and $\$7.50z$.

Translate.

The total number of orders was 150.

$$x + y + z = 150$$

Total shipping charges were $680.

$$3x + 5y + 7.5z = 680$$

The number of orders under 10 lb was three times the number of orders weighing 15 lb or more.

$$x = 3z$$

We have a system of equations

$$
\begin{aligned}
x + y + z &= 150, \\
3x + 5y + 7.5z &= 680, \\
x &= 3z
\end{aligned}
\quad \text{or} \quad
\begin{aligned}
x + y + z &= 150, \\
30x + 50y + 75z &= 6800, \\
x - 3z &= 0
\end{aligned}
$$

Carry out. Solving the system of equations, we get $(60, 70, 20)$.

Check. The total number of orders is $60 + 70 + 20$, or 150. The total shipping charges are $\$3 \cdot 60 + \$5 \cdot 70 + \$7.50(20)$, or $680. The number of orders under 10 lb, 60, is three times the number of orders weighing over 15 lb, 20. The solution checks.

State. There were 60 packages under 10 lb, 70 packages from 10 lb up to 15 lb, and 20 packages weighing 15 lb or more.

25. *Familiarize*. Let x, y, and z represent the number of servings of ground beef, baked potato, and strawberries required, respectively. One serving of ground beef contains $245x$ Cal, $0x$ or 0 g of carbohydrates, and $9x$ mg of calcium. One baked potato contains $145y$ Cal, $34y$ g of carbohydrates, and $8y$ mg of calcium. One serving of strawberries contains $45z$ Cal, $10z$ g of carbohydrates, and $21z$ mg of calcium.

Translate.

The total number of calories is 485.

$$245x + 145y + 45z = 485$$

A total of 41.5 g of carbohydrates is required.

$$34y + 10z = 41.5$$

A total of 35 mg of calcium is required.

$$9x + 8y + 21z = 35$$

We have a system of equations.

$$
\begin{aligned}
245x + 145y + 45z &= 485, \\
34y + 10z &= 41.5, \\
9x + 8y + 21z &= 35
\end{aligned}
$$

Carry out. Solving the system of equations, we get $(1.25, 1, 0.75)$.

Check. 1.25 servings of ground beef contains 306.25 Cal, no carbohydrates, and 11.25 mg of calcium; 1 baked potato contains 145 Cal, 34 g of carbohydrates, and 8 mg of calcium; 0.75 servings of strawberries contains 33.75 Cal, 7.5 g of carbohydrates, and 15.75 mg of calcium. Thus, there are a total of $306.25 + 145 + 33.75$, or 485 Cal, $34 + 7.5$, or 41.5 g of carbohydrates, and $11.25 + 8 + 15.75$, or 35 mg of calcium. The solution checks.

State. 1.25 servings of ground beef, 1 baked potato, and 0.75 serving of strawberries are required.

27. *Familiarize*. Let x, y, and z represent the amounts invested at 3%, 4%, and 6%, respectively. Then the annual interest from the investments is $3\%x$, $4\%y$, and $6\%z$, or $0.03x$, $0.04y$, and $0.06z$.

Translate.

A total of $5000 was invested.

$$x + y + z = 5000$$

The total interest is $243.

$$0.03x + 0.04y + 0.06z = 243$$

The amount invested at 6% is $1500 more than the amount invested at 3%.

$$z = x + 1500$$

We have a system of equations.

$$
\begin{aligned}
x + y + z &= 5000, \\
0.03x + 0.04y + 0.06z &= 243, \\
z &= x + 1500
\end{aligned}
$$

or

$$
\begin{aligned}
x + y + z &= 5000, \\
3x + 4y + 6z &= 24,300, \\
-x + z &= 1500
\end{aligned}
$$

Carry out. Solving the system of equations, we get $(1300, 900, 2800)$.

Check. The total investment was $\$1300 + \$900 + \$2800$, or $5000. The total interest was $0.03(\$1300) + 0.04(\$900) + 0.06(\$2800) = \$39 + \$36 + \$168$, or $243. The amount invested at 6%, $2800, is $1500 more than the amount invested at 3%, $1300. The solution checks.

State. $1300 was invested at 3%, $900 at 4%, and $2800 at 6%.

29. *Familiarize*. Let x, y, and z represent the prices of orange juice, a raisin bagel, and a cup of coffee, respectively. The new price for orange juice is $x + 50\%x$, or $x + 0.5x$, or $1.5x$; the new price of a bagel is $y + 20\%y$, or $y + 0.2y$, or $1.2y$.

Translate.

Orange juice, a raisin bagel, and a cup of coffee cost $4.10.

$$x + y + z = 4.1$$

After the price increase, orange juice, a raisin bagel, and a cup of coffee will cost $5.10.

$$1.5x + 1.2y + z = 5.1$$

After the price increases, orange juice will cost twice as much as coffee.

$$1.5x = 2z$$

We have a system of equations.

$$x + y + z = 4.1, \qquad 10x + 10y + 10z = 41,$$
$$1.5x + 1.2y + z = 5.1, \text{ or } 15x + 12y + 10z = 51,$$
$$1.5x = 2z \qquad 15x \qquad -20z = 0$$

Carry out. Solving the system of equations, we get $(1.2, 2, 0.9)$.

Check. If orange juice costs $1.20, a bagel costs $2, and a cup of coffee costs $0.90, then together they cost $1.20 + $2 + $0.90, or $4.10. After the price increases orange juice will cost 1.5($1.20), or $1.80 and a bagel will cost 1.2($2) or $2.40. Then orange juice, a bagel, and coffee will cost $1.80 + $2.40 + $0.90, or $5.10. After the price increase the price of orange juice, $1.80, will be twice the price of coffee, $0.90. The solution checks.

State. Before the increase orange juice cost $1.20, a raisin bagel cost $2, and a cup of coffee cost $0.90.

31. Familiarize. Let x, y, and z represent the number of par-3, par-4, and par-5 holes, respectively. A golfer who shoots par on every hole has a score of $3x$ from the par-3 holes, $4y$ from the par-4 holes, and $5z$ from the par-5 holes.

Translate.

The total number of holes is 18.

$$x + y + z = 18$$

A golfer who shoots par on every hole has a score of 72.

$$3x + 4y + 5z = 72$$

The sum of the number of par-3 holes and the number of par-5 holes is 8.

$$x + z = 8$$

We have a system of equations.

$$x + y + z = 18,$$
$$3x + 4y + 5z = 72,$$
$$x \qquad + z = 8$$

Carry out. We solve the system. The solution is $(4, 10, 4)$.

Check. The total number of holes is $4 + 10 + 4$, or 18. A golfer who shoots par on every hole has a score of $3 \cdot 4 + 4 \cdot 10 + 5 \cdot 4$, or 72. The sum of the number of par-3 holes and the number of par-5 holes is $4 + 4$, or 8. The solution checks.

State. There are 4 par-3 holes, 10 par-4 holes, and 4 par-5 holes.

33. a) Substitute the data points $(0, 279)$, $(2, 9741)$, and $(3, 4459)$ in the function $f(x) = ax^2 + bx + c$.

$$279 = a \cdot 0^2 + b \cdot 0 + c$$
$$9741 = a \cdot 2^2 + b \cdot 2 + c$$
$$4459 = a \cdot 3^2 + b \cdot 3 + c$$

We have a system of equations.

$$c = 279,$$
$$4a + 2b + c = 9741,$$
$$9a + 3b + c = 4459$$

Solving the system of equations, we get
$$\left(-\frac{10,013}{3}, \frac{34,219}{3}, 279 \right), \text{ so}$$
$$f(x) = -\frac{10,013}{3}x^2 + \frac{34,219}{3}x + 279, \text{ where } x \text{ is the}$$
number of years after 2002.

b) In 2003, $x = 1$.
$$f(1) = -\frac{10,013}{3} \cdot 1^2 + \frac{34,219}{3} \cdot 1 + 279 \approx 8348 \text{ fines}$$

35. a) Substitute the data points $(0, 36)$, $(10, 42)$, and $(14, 36)$ in the function $f(x) = ax^2 + bx + c$.

$$36 = a \cdot 0^2 + b \cdot 0 + c$$
$$42 = a \cdot 10^2 + b \cdot 10 + c$$
$$36 = a \cdot 14^2 + b \cdot 14 + c$$

We have a system of equations.

$$c = 36,$$
$$100a + 10b + c = 42,$$
$$196a + 14b + c = 36$$

Solving the system of equations, we get

$(-0.15, 2.1, 36)$, so $f(x) = -0.15x^2 + 2.1x + 36$, where $f(x)$ is in thousands and x is the number of years after 1990.

b) In 2008, $x = 2008 - 1990$, or 18.

$$f(18) = -0.15(18)^2 + 2.1(18) + 36 \approx 25$$

There were about 25 thousand marriages in Arkansas in 2008.

37. a) Using the quadratic regression feature on a graphing calculator, we have $f(x) = 0.1822373078x^2 - 11.23256978x + 468.7226133$, where x is the number of years after 1920.

b) In 2008, $x = 2008 - 1920$, or 88; $f(88) \approx 892$, so we estimate that there will be about 892 morning newspapers in 2008.

39. Discussion and Writing

41. Perpendicular

43. A vertical line

45. A rational function

47. A vertical asymptote

49.
$$\frac{2}{x} - \frac{1}{y} - \frac{3}{z} = -1,$$
$$\frac{2}{x} - \frac{1}{y} + \frac{1}{z} = -9,$$
$$\frac{1}{x} + \frac{2}{y} - \frac{4}{z} = 17$$

First substitute u for $\frac{1}{x}$, v for $\frac{1}{y}$, and w for $\frac{1}{z}$ and solve for u, v, and w.

$$2u - v - 3w = -1,$$
$$2u - v + w = -9,$$
$$u + 2v - 4w = 17$$

Solving this system we get $(-1, 5, -2)$.

If $u = -1$, and $u = \frac{1}{x}$, then $-1 = \frac{1}{x}$, or $x = -1$.

If $v = 5$ and $v = \frac{1}{y}$, then $5 = \frac{1}{y}$, or $y = \frac{1}{5}$.

If $w = -2$ and $w = \frac{1}{z}$, then $-2 = \frac{1}{z}$, or $z = -\frac{1}{2}$.

The solution of the original system is $\left(-1, \frac{1}{5}, -\frac{1}{2}\right)$.

51. Label the angle measures at the tips of the stars a, b, c, d, and e. Also label the angles of the pentagon p, q, r, s, and t.

Using the geometric fact that the sum of the angle measures of a triangle is $180°$, we get 5 equations.

$$p + b + d = 180$$
$$q + c + e = 180$$
$$r + a + d = 180$$
$$s + b + e = 180$$
$$t + a + c = 180$$

Adding these equations, we get

$$(p + q + r + s + t) + 2a + 2b + 2c + 2d + 2e = 5(180).$$

The sum of the angle measures of any convex polygon with n sides is given by the formula $S = (n-2)180$. Thus $p + q + r + s + t = (5-2)180$, or 540. We substitute and solve for $a + b + c + d + e$.

$$540 + 2(a + b + c + d + e) = 900$$
$$2(a + b + c + d + e) = 360$$
$$a + b + c + d + e = 180$$

The sum of the angle measures at the tips of the star is $180°$.

53. Substituting, we get

$$A + \frac{3}{4}B + 3C = 12,$$
$$\frac{4}{3}A + B + 2C = 12,$$
$$2A + B + C = 12, \text{ or}$$

$$4A + 3B + 12C = 48,$$
$$4A + 3B + 6C = 36, \quad \text{Clearing fractions}$$
$$2A + B + C = 12.$$

Solving the system of equations, we get $(3, 4, 2)$. The equation is $3x + 4y + 2z = 12$.

55. Substituting, we get

$$59 = a(-2)^3 + b(-2)^2 + c(-2) + d,$$
$$13 = a(-1)^3 + b(-1)^2 + c(-1) + d,$$
$$-1 = a \cdot 1^3 + b \cdot 1^2 + c \cdot 1 + d,$$
$$-17 = a \cdot 2^3 + b \cdot 2^2 + c \cdot 2 + d, \text{ or}$$

$$-8a + 4b - 2c + d = 59,$$
$$-a + b - c + d = 13,$$
$$a + b + c + d = -1,$$
$$8a + 4b + 2c + d = -17.$$

Solving the system of equations, we get $(-4, 5, -3, 1)$, so $y = -4x^3 + 5x^2 - 3x + 1$.

57. ***Familiarize and Translate***. Let a, s, and c represent the number of adults, students, and children in attendance, respectively.

The total attendance was 100.

$$a + s + c = 100$$

The total amount of money taken in was $100.

(Express 50 cents as $\frac{1}{2}$ dollar.)

$$10a + 3s + \frac{1}{2}c = 100$$

The resulting system is

$$a + s + c = 100,$$
$$10a + 3s + \frac{1}{2}c = 100.$$

Carry out. Multiply the first equation by -3 and add it to the second equation to obtain

$7a - \frac{5}{2}c = -200$ or $a = \frac{5}{14}(c - 80)$ where $(c - 80)$ is a positive multiple of 14 (because a must be a positive integer). That is $(c - 80) = k \cdot 14$ or $c = 80 + k \cdot 14$, where k is a positive integer. If $k > 1$, then $c > 100$. This is impossible since the total attendance is 100. Thus $k = 1$, so $c = 80 + 1 \cdot 14 = 94$. Then $a = \frac{5}{14}(94 - 80) = \frac{5}{14} \cdot 14 = 5$, and $5 + s + 94 = 100$, or $s = 1$.

Check. The total attendance is $5 + 1 + 94$, or 100.

The total amount of money taken in was

$\$10 \cdot 5 + \$3 \cdot 1 + \$\frac{1}{2} \cdot 94 = \$100$. The result checks.

State. There were 5 adults, 1 student, and 94 children in attendance.

Exercise Set 5.3

1. The matrix has 3 rows and 2 columns, so its order is 3×2.

3. The matrix has 1 row and 4 columns, so its order is 1×4.

5. The matrix has 3 rows and 3 columns, so its order is 3×3.

7. We omit the variables and replace the equals signs with a vertical line.

$$\left[\begin{array}{rr|r} 2 & -1 & 7 \\ 1 & 4 & -5 \end{array}\right]$$

9. We omit the variables, writing zeros for the missing terms, and replace the equals signs with a vertical line.

$$\left[\begin{array}{rrr|r} 1 & -2 & 3 & 12 \\ 2 & 0 & -4 & 8 \\ 0 & 3 & 1 & 7 \end{array}\right]$$

11. Insert variables and replace the vertical line with equals signs.

$$3x - 5y = 1,$$
$$x + 4y = -2$$

13. Insert variables and replace the vertical line with equals signs.

$$2x + y - 4z = 12,$$
$$3x \qquad + 5z = -1,$$
$$x - y + z = 2$$

15. $4x + 2y = 11,$
$\quad 3x - y = 2$

Write the augmented matrix. We will use Gaussian elimination.

$$\left[\begin{array}{rr|r} 4 & 2 & 11 \\ 3 & -1 & 2 \end{array}\right]$$

Multiply row 2 by 4 to make the first number in row 2 a multiple of 4.

$$\left[\begin{array}{rr|r} 4 & 2 & 11 \\ 12 & -4 & 8 \end{array}\right]$$

Multiply row 1 by -3 and add it to row 2.

$$\left[\begin{array}{rr|r} 4 & 2 & 11 \\ 0 & -10 & -25 \end{array}\right]$$

Multiply row 1 by $\frac{1}{4}$ and row 2 by $-\frac{1}{10}$.

$$\left[\begin{array}{rr|r} 1 & \frac{1}{2} & \frac{11}{4} \\ 0 & 1 & \frac{5}{2} \end{array}\right]$$

Write the system of equations that corresponds to the last matrix.

$$x + \frac{1}{2}y = \frac{11}{4}, \quad (1)$$
$$y = \frac{5}{2} \quad (2)$$

Back-substitute in equation (1) and solve for x.

$$x + \frac{1}{2} \cdot \frac{5}{2} = \frac{11}{4}$$
$$x + \frac{5}{4} = \frac{11}{4}$$
$$x = \frac{6}{4} = \frac{3}{2}$$

The solution is $\left(\frac{3}{2}, \frac{5}{2}\right)$.

17. $5x - 2y = -3,$
$\quad 2x + 5y = -24$

Write the augmented matrix. We will use Gaussian elimination.

$$\left[\begin{array}{rr|r} 5 & -2 & -3 \\ 2 & 5 & -24 \end{array}\right]$$

Multiply row 2 by 5 to make the first number in row 2 a multiple of 5.

$$\left[\begin{array}{rr|r} 5 & -2 & -3 \\ 10 & 25 & -120 \end{array}\right]$$

Multiply row 1 by -2 and add it to row 2.

$$\left[\begin{array}{rr|r} 5 & -2 & -3 \\ 0 & 29 & -114 \end{array}\right]$$

Multiply row 1 by $\frac{1}{5}$ and row 2 by $\frac{1}{29}$.

$$\left[\begin{array}{rr|r} 1 & -\frac{2}{5} & -\frac{3}{5} \\ 0 & 1 & -\frac{114}{29} \end{array}\right]$$

Write the system of equations that corresponds to the last matrix.

$$x - \frac{2}{5}y = -\frac{3}{5}, \quad (1)$$
$$y = -\frac{114}{29} \quad (2)$$

Back-substitute in equation (1) and solve for x.

$$x - \frac{2}{5}\left(-\frac{114}{29}\right) = -\frac{3}{5}$$

$$x + \frac{228}{145} = -\frac{3}{5}$$

$$x = -\frac{315}{145} = -\frac{63}{29}$$

The solution is $\left(-\frac{63}{29}, -\frac{114}{29}\right)$.

19. $3x + 4y = 7,$
$-5x + 2y = 10$

Write the augmented matrix. We will use Gaussian elimination.

$$\begin{bmatrix} 3 & 4 & | & 7 \\ -5 & 2 & | & 10 \end{bmatrix}$$

Multiply row 2 by 3 to make the first number in row 2 a multiple of 3.

$$\begin{bmatrix} 3 & 4 & | & 7 \\ -15 & 6 & | & 30 \end{bmatrix}$$

Multiply row 1 by 5 and add it to row 2.

$$\begin{bmatrix} 3 & 4 & | & 7 \\ 0 & 26 & | & 65 \end{bmatrix}$$

Multiply row 1 by $\frac{1}{3}$ and row 2 by $\frac{1}{26}$.

$$\begin{bmatrix} 1 & \frac{4}{3} & | & \frac{7}{3} \\ 0 & 1 & | & \frac{5}{2} \end{bmatrix}$$

Write the system of equations that corresponds to the last matrix.

$$x + \frac{4}{3}y = \frac{7}{3}, \quad (1)$$
$$y = \frac{5}{2} \quad (2)$$

Back-substitute in equation (1) and solve for x.

$$x + \frac{4}{3} \cdot \frac{5}{2} = \frac{7}{3}$$
$$x + \frac{10}{3} = \frac{7}{3}$$
$$x = -\frac{3}{3} = -1$$

The solution is $\left(-1, \frac{5}{2}\right)$.

21. $3x + 2y = 6,$
$2x - 3y = -9$

Write the augmented matrix. We will use Gauss-Jordan elimination.

$$\begin{bmatrix} 3 & 2 & | & 6 \\ 2 & -3 & | & -9 \end{bmatrix}$$

Multiply row 2 by 3 to make the first number in row 2 a multiple of 3.

$$\begin{bmatrix} 3 & 2 & | & 6 \\ 6 & -9 & | & -27 \end{bmatrix}$$

Multiply row 1 by -2 and add it to row 2.

$$\begin{bmatrix} 3 & 2 & | & 6 \\ 0 & -13 & | & -39 \end{bmatrix}$$

Multiply row 2 by $-\frac{1}{13}$.

$$\begin{bmatrix} 3 & 2 & | & 6 \\ 0 & 1 & | & 3 \end{bmatrix}$$

Multiply row 2 by -2 and add it to row 1.

$$\begin{bmatrix} 3 & 0 & | & 0 \\ 0 & 1 & | & 3 \end{bmatrix}$$

Multiply row 1 by $\frac{1}{3}$.

$$\begin{bmatrix} 1 & 0 & | & 0 \\ 0 & 1 & | & 3 \end{bmatrix}$$

We have $x = 0$, $y = 3$. The solution is $(0, 3)$.

23. $x - 3y = 8,$
$2x - 6y = 3$

Write the augmented matrix.

$$\begin{bmatrix} 1 & -3 & | & 8 \\ 2 & -6 & | & 3 \end{bmatrix}$$

Multiply row 1 by -2 and add it to row 2.

$$\begin{bmatrix} 1 & -3 & | & 8 \\ 0 & 0 & | & -13 \end{bmatrix}$$

The last row corresponds to the false equation $0 = -13$, so there is no solution.

25. $-2x + 6y = 4,$
$3x - 9y = -6$

Write the augmented matrix.

$$\begin{bmatrix} -2 & 6 & | & 4 \\ 3 & -9 & | & -6 \end{bmatrix}$$

Multiply row 1 by $-\frac{1}{2}$.

$$\begin{bmatrix} 1 & -3 & \big| & -2 \\ 3 & -9 & \big| & -6 \end{bmatrix}$$

Multiply row 1 by -3 and add it to row 2.

$$\begin{bmatrix} 1 & -3 & \big| & -2 \\ 0 & 0 & \big| & 0 \end{bmatrix}$$

The last row corresponds to the equation $0 = 0$ which is true for all values of x and y. Thus, the system of equations is dependent and is equivalent to the first equation $-2x + 6y = 4$, or $x - 3y = -2$. Solving for x, we get $x = 3y - 2$. Then the solutions are of the form $(3y - 2, y)$, where y is any real number.

27. $\quad x + 2y - 3z = 9,$
$\quad 2x - y + 2z = -8,$
$\quad 3x - y - 4z = 3$

Write the augmented matrix. We will use Gauss-Jordan elimination.

$$\begin{bmatrix} 1 & 2 & -3 & \big| & 9 \\ 2 & -1 & 2 & \big| & -8 \\ 3 & -1 & -4 & \big| & 3 \end{bmatrix}$$

Multiply row 1 by -2 and add it to row 2. Also, multiply row 1 by -3 and add it to row 3.

$$\begin{bmatrix} 1 & 2 & -3 & \big| & 9 \\ 0 & -5 & 8 & \big| & -26 \\ 0 & -7 & 5 & \big| & -24 \end{bmatrix}$$

Multiply row 2 by $-\frac{1}{5}$ to get a 1 in the second row, second column.

$$\begin{bmatrix} 1 & 2 & -3 & \big| & 9 \\ 0 & 1 & -\frac{8}{5} & \big| & \frac{26}{5} \\ 0 & -7 & 5 & \big| & -24 \end{bmatrix}$$

Multiply row 2 by -2 and add it to row 1. Also, multiply row 2 by 7 and add it to row 3.

$$\begin{bmatrix} 1 & 0 & \frac{1}{5} & \big| & -\frac{7}{5} \\ 0 & 1 & -\frac{8}{5} & \big| & \frac{26}{5} \\ 0 & 0 & -\frac{31}{5} & \big| & \frac{62}{5} \end{bmatrix}$$

Multiply row 3 by $-\frac{5}{31}$ to get a 1 in the third row, third column.

$$\begin{bmatrix} 1 & 0 & \frac{1}{5} & \big| & -\frac{7}{5} \\ 0 & 1 & -\frac{8}{5} & \big| & \frac{26}{5} \\ 0 & 0 & 1 & \big| & -2 \end{bmatrix}$$

Multiply row 3 by $-\frac{1}{5}$ and add it to row 1. Also, multiply row 3 by $\frac{8}{5}$ and add it to row 2.

$$\begin{bmatrix} 1 & 0 & 0 & \big| & -1 \\ 0 & 1 & 0 & \big| & 2 \\ 0 & 0 & 1 & \big| & -2 \end{bmatrix}$$

We have $x = -1$, $y = 2$, $z = -2$. The solution is $(-1, 2, -2)$.

29. $\quad 4x - y - 3z = 1,$
$\quad 8x + y - z = 5,$
$\quad 2x + y + 2z = 5$

Write the augmented matrix. We will use Gauss-Jordan elimination.

$$\begin{bmatrix} 4 & -1 & -3 & \big| & 1 \\ 8 & 1 & -1 & \big| & 5 \\ 2 & 1 & 2 & \big| & 5 \end{bmatrix}$$

First interchange rows 1 and 3 so that each number below the first number in the first row is a multiple of that number.

$$\begin{bmatrix} 2 & 1 & 2 & \big| & 5 \\ 8 & 1 & -1 & \big| & 5 \\ 4 & -1 & -3 & \big| & 1 \end{bmatrix}$$

Multiply row 1 by -4 and add it to row 2. Also, multiply row 1 by -2 and add it to row 3.

$$\begin{bmatrix} 2 & 1 & 2 & \big| & 5 \\ 0 & -3 & -9 & \big| & -15 \\ 0 & -3 & -7 & \big| & -9 \end{bmatrix}$$

Multiply row 2 by -1 and add it to row 3.

$$\begin{bmatrix} 2 & 1 & 2 & \big| & 5 \\ 0 & -3 & -9 & \big| & -15 \\ 0 & 0 & 2 & \big| & 6 \end{bmatrix}$$

Multiply row 2 by $-\frac{1}{3}$ to get a 1 in the second row, second column.

$$\begin{bmatrix} 2 & 1 & 2 & \big| & 5 \\ 0 & 1 & 3 & \big| & 5 \\ 0 & 0 & 2 & \big| & 6 \end{bmatrix}$$

Multiply row 2 by -1 and add it to row 1.

$$\begin{bmatrix} 2 & 0 & -1 & \big| & 0 \\ 0 & 1 & 3 & \big| & 5 \\ 0 & 0 & 2 & \big| & 6 \end{bmatrix}$$

Multiply row 3 by $\frac{1}{2}$ to get a 1 in the third row, third column.

$$\begin{bmatrix} 2 & 0 & -1 & | & 0 \\ 0 & 1 & 3 & | & 5 \\ 0 & 0 & 1 & | & 3 \end{bmatrix}$$

Add row 3 to row 1. Also multiply row 3 by -3 and add it to row 2.

$$\begin{bmatrix} 2 & 0 & 0 & | & 3 \\ 0 & 1 & 0 & | & -4 \\ 0 & 0 & 1 & | & 3 \end{bmatrix}$$

Finally, multiply row 1 by $\frac{1}{2}$.

$$\begin{bmatrix} 1 & 0 & 0 & | & \frac{3}{2} \\ 0 & 1 & 0 & | & -4 \\ 0 & 0 & 1 & | & 3 \end{bmatrix}$$

We have $x = \frac{3}{2}$, $y = -4$, $z = 3$. The solution is $\left(\frac{3}{2}, -4, 3\right)$.

31. $x - 2y + 3z = 4,$
$\quad 3x + y - z = 0,$
$\quad 2x + 3y - 5z = 1$

Write the augmented matrix. We will use Gaussian elimination.

$$\begin{bmatrix} 1 & -2 & 3 & | & -4 \\ 3 & 1 & -1 & | & 0 \\ 2 & 3 & -5 & | & 1 \end{bmatrix}$$

Multiply row 1 by -3 and add it to row 2. Also, multiply row 1 by -2 and add it to row 3.

$$\begin{bmatrix} 1 & -2 & 3 & | & -4 \\ 0 & 7 & -10 & | & 12 \\ 0 & 7 & -11 & | & 9 \end{bmatrix}$$

Multiply row 2 by -1 and add it to row 3.

$$\begin{bmatrix} 1 & -2 & 3 & | & -4 \\ 0 & 7 & -10 & | & 12 \\ 0 & 0 & -1 & | & -3 \end{bmatrix}$$

Multiply row 2 by $\frac{1}{7}$ and multiply row 3 by -1.

$$\begin{bmatrix} 1 & -2 & 3 & | & -4 \\ 0 & 1 & -\frac{10}{7} & | & \frac{12}{7} \\ 0 & 0 & 1 & | & 3 \end{bmatrix}$$

Now write the system of equations that corresponds to the last matrix.

$$x - 2y + 3z = -4, \quad (1)$$
$$y - \frac{10}{7}z = \frac{12}{7}, \quad (2)$$
$$z = 3 \quad (3)$$

Back-substitute 3 for z in equation (2) and solve for y.

$$y - \frac{10}{7} \cdot 3 = \frac{12}{7}$$
$$y - \frac{30}{7} = \frac{12}{7}$$
$$y = \frac{42}{7} = 6$$

Back-substitute 6 for y and 3 for z in equation (1) and solve for x.

$$x - 2 \cdot 6 + 3 \cdot 3 = -4$$
$$x - 3 = -4$$
$$x = -1$$

The solution is $(-1, 6, 3)$.

33. $2x - 4y - 3z = 3,$
$\quad x + 3y + z = -1,$
$\quad 5x + y - 2z = 2$

Write the augmented matrix.

$$\begin{bmatrix} 2 & -4 & -3 & | & 3 \\ 1 & 3 & 1 & | & -1 \\ 5 & 1 & -2 & | & 2 \end{bmatrix}$$

Interchange the first two rows to get a 1 in the first row, first column.

$$\begin{bmatrix} 1 & 3 & 1 & | & -1 \\ 2 & -4 & -3 & | & 3 \\ 5 & 1 & -2 & | & 2 \end{bmatrix}$$

Multiply row 1 by -2 and add it to row 2. Also, multiply row 1 by -5 and add it to row 3.

$$\begin{bmatrix} 1 & 3 & 1 & | & -1 \\ 0 & -10 & -5 & | & 5 \\ 0 & -14 & -7 & | & 7 \end{bmatrix}$$

Multiply row 2 by $-\frac{1}{10}$ to get a 1 in the second row, second column.

$$\begin{bmatrix} 1 & 3 & 1 & | & -1 \\ 0 & 1 & \frac{1}{2} & | & -\frac{1}{2} \\ 0 & -14 & -7 & | & 7 \end{bmatrix}$$

Multiply row 2 by 14 and add it to row 3.

$$\begin{bmatrix} 1 & 3 & 1 & | & -1 \\ 0 & 1 & \frac{1}{2} & | & -\frac{1}{2} \\ 0 & 0 & 0 & | & 0 \end{bmatrix}$$

The last row corresponds to the equation $0 = 0$. This indicates that the system of equations is dependent. It is equivalent to

$$x + 3y + z = -1,$$
$$y + \frac{1}{2}z = -\frac{1}{2}$$

We solve the second equation for y.

$$y = -\frac{1}{2}z - \frac{1}{2}$$

Substitute for y in the first equation and solve for x.

$$x + 3\left(-\frac{1}{2}z - \frac{1}{2}\right) + z = -1$$

$$x - \frac{3}{2}z - \frac{3}{2} + z = -1$$

$$x = \frac{1}{2}z + \frac{1}{2}$$

The solution is $\left(\frac{1}{2}z + \frac{1}{2}, -\frac{1}{2}z - \frac{1}{2}, z\right)$, where z is any real number.

35.
$$\begin{aligned} p + q + r &= 1, \\ p + 2q + 3r &= 4, \\ 4p + 5q + 6r &= 7 \end{aligned}$$

Write the augmented matrix.

$$\left[\begin{array}{ccc|c} 1 & 1 & 1 & 1 \\ 1 & 2 & 3 & 4 \\ 4 & 5 & 6 & 7 \end{array}\right]$$

Multiply row 1 by -1 and add it to row 2. Also, multiply row 1 by -4 and add it to row 3.

$$\left[\begin{array}{ccc|c} 1 & 1 & 1 & 1 \\ 0 & 1 & 2 & 3 \\ 0 & 1 & 2 & 3 \end{array}\right]$$

Multiply row 2 by -1 and add it to row 3.

$$\left[\begin{array}{ccc|c} 1 & 1 & 1 & 1 \\ 0 & 1 & 2 & 3 \\ 0 & 0 & 0 & 0 \end{array}\right]$$

The last row corresponds to the equation $0 = 0$. This indicates that the system of equations is dependent. It is equivalent to

$$\begin{aligned} p + q + r &= 1, \\ q + 2r &= 3. \end{aligned}$$

We solve the second equation for q.

$$q = -2r + 3$$

Substitute for y in the first equation and solve for p.

$$p - 2r + 3 + r = 1$$

$$p - r + 3 = 1$$

$$p = r - 2$$

The solution is $(r - 2, -2r + 3, r)$, where r is any real number.

37.
$$\begin{aligned} a + b - c &= 7, \\ a - b + c &= 5, \\ 3a + b - c &= -1 \end{aligned}$$

Write the augmented matrix.

$$\left[\begin{array}{ccc|c} 1 & 1 & -1 & 7 \\ 1 & -1 & 1 & 5 \\ 3 & 1 & -1 & -1 \end{array}\right]$$

Multiply row 1 by -1 and add it to row 2. Also, multiply row 1 by -3 and add it to row 3.

$$\left[\begin{array}{ccc|c} 1 & 1 & -1 & 7 \\ 0 & -2 & 2 & -2 \\ 0 & -2 & 2 & -22 \end{array}\right]$$

Multiply row 2 by -1 and add it to row 3.

$$\left[\begin{array}{ccc|c} 1 & 1 & -1 & 7 \\ 0 & -2 & 2 & -2 \\ 0 & 0 & 0 & -20 \end{array}\right]$$

The last row corresponds to the false equation $0 = -20$. Thus, the system of equations has no solution.

39.
$$\begin{aligned} -2w + 2x + 2y - 2z &= -10, \\ w + x + y + z &= -5, \\ 3w + x - y + 4z &= -2, \\ w + 3x - 2y + 2z &= -6 \end{aligned}$$

Write the augmented matrix. We will use Gaussian elimination.

$$\left[\begin{array}{cccc|c} -2 & 2 & 2 & -2 & -10 \\ 1 & 1 & 1 & 1 & -5 \\ 3 & 1 & -1 & 4 & -2 \\ 1 & 3 & -2 & 2 & -6 \end{array}\right]$$

Interchange rows 1 and 2.

$$\left[\begin{array}{cccc|c} 1 & 1 & 1 & 1 & -5 \\ -2 & 2 & 2 & -2 & -10 \\ 3 & 1 & -1 & 4 & -2 \\ 1 & 3 & -2 & 2 & -6 \end{array}\right]$$

Multiply row 1 by 2 and add it to row 2. Multiply row 1 by -3 and add it to row 3. Multiply row 1 by -1 and add it to row 4.

$$\left[\begin{array}{cccc|c} 1 & 1 & 1 & 1 & -5 \\ 0 & 4 & 4 & 0 & -20 \\ 0 & -2 & -4 & 1 & 13 \\ 0 & 2 & -3 & 1 & -1 \end{array}\right]$$

Interchange rows 2 and 3.

$$\left[\begin{array}{cccc|c} 1 & 1 & 1 & 1 & -5 \\ 0 & -2 & -4 & 1 & 13 \\ 0 & 4 & 4 & 0 & -20 \\ 0 & 2 & -3 & 1 & -1 \end{array}\right]$$

Multiply row 2 by 2 and add it to row 3. Add row 2 to row 4.

$$\begin{bmatrix} 1 & 1 & 1 & 1 & | & -5 \\ 0 & -2 & -4 & 1 & | & 13 \\ 0 & 0 & -4 & 2 & | & 6 \\ 0 & 0 & -7 & 2 & | & 12 \end{bmatrix}$$

Multiply row 4 by 4.

$$\begin{bmatrix} 1 & 1 & 1 & 1 & | & -5 \\ 0 & -2 & -4 & 1 & | & 13 \\ 0 & 0 & -4 & 2 & | & 6 \\ 0 & 0 & -28 & 8 & | & 48 \end{bmatrix}$$

Multiply row 3 by -7 and add it to row 4.

$$\begin{bmatrix} 1 & 1 & 1 & 1 & | & -5 \\ 0 & -2 & -4 & 1 & | & 13 \\ 0 & 0 & -4 & 2 & | & 6 \\ 0 & 0 & 0 & -6 & | & 6 \end{bmatrix}$$

Multiply row 2 by $-\dfrac{1}{2}$, row 3 by $-\dfrac{1}{4}$, and row 6 by $-\dfrac{1}{6}$.

$$\begin{bmatrix} 1 & 1 & 1 & 1 & | & -5 \\ 0 & 1 & 2 & -\dfrac{1}{2} & | & -\dfrac{13}{2} \\ 0 & 0 & 1 & -\dfrac{1}{2} & | & -\dfrac{3}{2} \\ 0 & 0 & 0 & 1 & | & -1 \end{bmatrix}$$

Write the system of equations that corresponds to the last matrix.

$$\begin{aligned} w + x + y + z &= -5, & (1) \\ x + 2y - \tfrac{1}{2}z &= -\tfrac{13}{2}, & (2) \\ y - \tfrac{1}{2}z &= -\tfrac{3}{2}, & (3) \\ z &= -1 & (4) \end{aligned}$$

Back-substitute in equation (3) and solve for y.

$$y - \tfrac{1}{2}(-1) = -\tfrac{3}{2}$$
$$y + \tfrac{1}{2} = -\tfrac{3}{2}$$
$$y = -2$$

Back-substitute in equation (2) and solve for x.

$$x + 2(-2) - \tfrac{1}{2}(-1) = -\tfrac{13}{2}$$
$$x - 4 + \tfrac{1}{2} = -\tfrac{13}{2}$$
$$x = -3$$

Back-substitute in equation (1) and solve for w.

$$w - 3 - 2 - 1 = -5$$
$$w = 1$$

The solution is $(1, -3, -2, -1)$.

41. Familiarize. Let $x =$ the number of hours the Houlihans were out before 11 P.M. and $y =$ the number of hours after 11 P.M. Then they pay the babysitter $\$5x$ before 11 P.M. and $\$7.50y$ after 11 P.M.

Translate.

The Houlihans were out for a total of 5 hr.

$$x + y = 5$$

They paid the sitter a total of \$30.

$$5x + 7.5y = 30$$

Carry out. Use Gaussian elimination or Gauss-Jordan elimination to solve the system of equations.

$$\begin{aligned} x + \;\;\;\; y &= 5, \\ 5x + 7.5y &= 30. \end{aligned}$$

The solution is $(3, 2)$. The coordinate $y = 2$ indicates that the Houlihans were out 2 hr after 11 P.M., so they came home at 1 A.M.

Check. The total time is $3 + 2$, or 5 hr. The total pay is $\$5 \cdot 3 + \$7.50(2)$, or $\$15 + \$15$, or \$30. The solution checks.

State. The Houlihans came home at 1 A.M.

43. Familiarize. Let x, y, and z represent the amounts borrowed at 8%, 10%, and 12%, respectively. Then the annual interest is $8\%x$, $10\%y$, and $12\%z$, or $0.08x$, $0.1y$, and $0.12z$.

Translate.

The total amount borrowed was \$30,000.

$$x + y + z = 30,000$$

The total annual interest was \$3040.

$$0.08x + 0.1y + 0.12z = 3040$$

The total amount borrowed at 8% and 10% was twice the amount borrowed at 12%.

$$x + y = 2z$$

We have a system of equations.

$$\begin{aligned} x + y + z &= 30,000, \\ 0.08x + 0.1y + 0.12z &= 3040, \\ x + y &= 2z, \text{ or} \end{aligned}$$

$$\begin{aligned} x + \;\;\;\; y + \;\;\;\;\;\; z &= 30,000, \\ 0.08x + 0.1y + 0.12z &= 3040, \\ x + \;\;\;\; y - \;\;\;\; 2z &= 0. \end{aligned}$$

Carry out. Using Gaussian elimination or Gauss-Jordan elimination, we find that the solution is $(8000, 12,000, 10,000)$.

Check. The total amount borrowed was $\$8000 + \$12,000 + \$10,000$, or \$30,000. The total annual interest was $0.08(\$8000) + 0.1(\$12,000) + 0.12(\$10,000)$, or $\$640 + \$1200 + \$1200$, or \$3040. The total amount borrowed at 8% and 10%, $\$8000 + \$12,000$ or \$20,000, was twice the amount borrowed at 12%, \$10,000. The solution checks.

State. The amounts borrowed at 8%, 10%, and 12% were \$8000, \$12,000 and \$10,000, respectively.

45. Discussion and Writing

47. The function has a variable in the exponent, so it is an exponential function.

49. The function is the quotient of two polynomials, so it is a rational function.

51. The function is of the form $f(x) = \log_a x$, so it is logarithmic.

53. The function is of the form $f(x) = mx + b$, so it is linear.

55. Substitute to find three equations.
$$12 = a(-3)^2 + b(-3) + c$$
$$-7 = a(-1)^2 + b(-1) + c$$
$$-2 = a \cdot 1^2 + b \cdot 1 + c$$

We have a system of equations.
$$9a - 3b + c = 12,$$
$$a - b + c = -7,$$
$$a + b + c = -2$$

Write the augmented matrix. We will use Gaussian elimination.

$$\begin{bmatrix} 9 & -3 & 1 & | & 12 \\ 1 & -1 & 1 & | & -7 \\ 1 & 1 & 1 & | & -2 \end{bmatrix}$$

Interchange the first two rows.

$$\begin{bmatrix} 1 & -1 & 1 & | & -7 \\ 9 & -3 & 1 & | & 12 \\ 1 & 1 & 1 & | & -2 \end{bmatrix}$$

Multiply row 1 by -9 and add it to row 2. Also, multiply row 1 by -1 and add it to row 3.

$$\begin{bmatrix} 1 & -1 & 1 & | & -7 \\ 0 & 6 & -8 & | & 75 \\ 0 & 2 & 0 & | & 5 \end{bmatrix}$$

Interchange row 2 and row 3.

$$\begin{bmatrix} 1 & -1 & 1 & | & -7 \\ 0 & 2 & 0 & | & 5 \\ 0 & 6 & -8 & | & 75 \end{bmatrix}$$

Multiply row 2 by -3 and add it to row 3.

$$\begin{bmatrix} 1 & -1 & 1 & | & -7 \\ 0 & 2 & 0 & | & 5 \\ 0 & 0 & -8 & | & 60 \end{bmatrix}$$

Multiply row 2 by $\frac{1}{2}$ and row 3 by $-\frac{1}{8}$.

$$\begin{bmatrix} 1 & -1 & 1 & | & -7 \\ 0 & 1 & 0 & | & \frac{5}{2} \\ 0 & 0 & 1 & | & -\frac{15}{2} \end{bmatrix}$$

Write the system of equations that corresponds to the last matrix.
$$x - y + z = -7,$$
$$y = \frac{5}{2},$$
$$z = -\frac{15}{2}$$

Back-substitute $\frac{5}{2}$ for y and $-\frac{15}{2}$ for z in the first equation and solve for x.
$$x - \frac{5}{2} - \frac{15}{2} = -7$$
$$x - 10 = -7$$
$$x = 3$$

The solution is $\left(3, \frac{5}{2}, -\frac{15}{2}\right)$, so the equation is

$$y = 3x^2 + \frac{5}{2}x - \frac{15}{2}.$$

57. $\begin{bmatrix} 1 & 5 \\ 3 & 2 \end{bmatrix}$

Multiply row 1 by -3 and add it to row 2.

$$\begin{bmatrix} 1 & 5 \\ 0 & -13 \end{bmatrix}$$

Multiply row 2 by $-\frac{1}{13}$.

$$\begin{bmatrix} 1 & 5 \\ 0 & 1 \end{bmatrix} \quad \text{Row-echelon form}$$

Multiply row 2 by -5 and add it to row 1.

$$\begin{bmatrix} 1 & 0 \\ 0 & 1 \end{bmatrix} \quad \text{Reduced row-echelon form}$$

59. $y = x + z,$
$$3y + 5z = 4,$$
$$x + 4 = y + 3z, \text{ or}$$
$$x - y + z = 0,$$
$$3y + 5z = 4,$$
$$x - y - 3z = -4$$

Write the augmented matrix. We will use Gauss-Jordan elimination.

$$\begin{bmatrix} 1 & -1 & 1 & | & 0 \\ 0 & 3 & 5 & | & 4 \\ 1 & -1 & -3 & | & -4 \end{bmatrix}$$

Multiply row 1 by -1 and add it to row 3.

$$\begin{bmatrix} 1 & -1 & 1 & | & 0 \\ 0 & 3 & 5 & | & 4 \\ 0 & 0 & -4 & | & -4 \end{bmatrix}$$

Multiply row 3 by $-\dfrac{1}{4}$.

$$\begin{bmatrix} 1 & -1 & 1 & | & 0 \\ 0 & 3 & 5 & | & 4 \\ 0 & 0 & 1 & | & 1 \end{bmatrix}$$

Multiply row 3 by -1 and add it to row 1. Also, multiply row 3 by -5 and add it to row 2.

$$\begin{bmatrix} 1 & -1 & 0 & | & -1 \\ 0 & 3 & 0 & | & -1 \\ 0 & 0 & 1 & | & 1 \end{bmatrix}$$

Multiply row 2 by $\dfrac{1}{3}$.

$$\begin{bmatrix} 1 & -1 & 0 & | & -1 \\ 0 & 1 & 0 & | & -\dfrac{1}{3} \\ 0 & 0 & 1 & | & 1 \end{bmatrix}$$

Add row 2 to row 1.

$$\begin{bmatrix} 1 & 0 & 0 & | & -\dfrac{4}{3} \\ 0 & 1 & 0 & | & -\dfrac{1}{3} \\ 0 & 0 & 1 & | & 1 \end{bmatrix}$$

Read the solution from the last matrix. It is
$\left(-\dfrac{4}{3}, -\dfrac{1}{3}, 1 \right)$.

61. $x - 4y + 2z = 7,$
$3x + y + 3z = -5$

Write the augmented matrix.

$$\begin{bmatrix} 1 & -4 & 2 & | & 7 \\ 3 & 1 & 3 & | & -5 \end{bmatrix}$$

Multiply row 1 by -3 and add it to row 2.

$$\begin{bmatrix} 1 & -4 & 2 & | & 7 \\ 0 & 13 & -3 & | & -26 \end{bmatrix}$$

Multiply row 2 by $\dfrac{1}{13}$.

$$\begin{bmatrix} 1 & -4 & 2 & | & 7 \\ 0 & 1 & -\dfrac{3}{13} & | & -2 \end{bmatrix}$$

Write the system of equations that corresponds to the last matrix.
$$x - 4y + 2z = 7,$$
$$y - \dfrac{3}{13}z = -2$$

Solve the second equation for y.

$$y = \dfrac{3}{13}z - 2$$

Substitute in the first equation and solve for x.

$$x - 4\left(\dfrac{3}{13}z - 2 \right) + 2z = 7$$
$$x - \dfrac{12}{13}z + 8 + 2z = 7$$
$$x = -\dfrac{14}{13}z - 1$$

The solution is $\left(-\dfrac{14}{13}z - 1, \ \dfrac{3}{13}z - 2, z \right)$, where z is any real number.

63. $4x + 5y = 3,$
$-2x + y = 9,$
$3x - 2y = -15$

Write the augmented matrix.

$$\begin{bmatrix} 4 & 5 & | & 3 \\ -2 & 1 & | & 9 \\ 3 & -2 & | & -15 \end{bmatrix}$$

Multiply row 2 by 2 and row 3 by 4.

$$\begin{bmatrix} 4 & 5 & | & 3 \\ -4 & 2 & | & 18 \\ 12 & -8 & | & -60 \end{bmatrix}$$

Add row 1 to row 2. Also, multiply row 1 by -3 and add it to row 3.

$$\begin{bmatrix} 4 & 5 & | & 3 \\ 0 & 7 & | & 21 \\ 0 & -23 & | & -69 \end{bmatrix}$$

Multiply row 2 by $\dfrac{1}{7}$ and row 3 by $-\dfrac{1}{23}$.

$$\begin{bmatrix} 4 & 5 & | & 3 \\ 0 & 1 & | & 3 \\ 0 & 1 & | & 3 \end{bmatrix}$$

Multiply row 2 by -1 and add it to row 3.

$$\begin{bmatrix} 4 & 5 & | & 3 \\ 0 & 1 & | & 3 \\ 0 & 0 & | & 0 \end{bmatrix}$$

The last row corresponds to the equation $0 = 0$. Thus we have a dependent system that is equivalent to
$$4x + 5y = 3, \quad (1)$$
$$y = 3. \quad (2)$$
Back-substitute in equation (1) to find x.
$$4x + 5 \cdot 3 = 3$$
$$4x + 15 = 3$$
$$4x = -12$$
$$x = -3$$
The solution is $(-3, 3)$.

Exercise Set 5.4

1. $\begin{bmatrix} 5 & x \end{bmatrix} = \begin{bmatrix} y & -3 \end{bmatrix}$

Corresponding entries of the two matrices must be equal. Thus we have $5 = y$ and $x = -3$.

3. $\begin{bmatrix} 3 & 2x \\ y & -8 \end{bmatrix} = \begin{bmatrix} 3 & -2 \\ 1 & -8 \end{bmatrix}$

Corresponding entries of the two matrices must be equal. Thus, we have:

$2x = -2$ and $y = 1$

$x = -1$ and $y = 1$

5. $\mathbf{A} + \mathbf{B} = \begin{bmatrix} 1 & 2 \\ 4 & 3 \end{bmatrix} + \begin{bmatrix} -3 & 5 \\ 2 & -1 \end{bmatrix}$

$= \begin{bmatrix} 1 + (-3) & 2 + 5 \\ 4 + 2 & 3 + (-1) \end{bmatrix}$

$= \begin{bmatrix} -2 & 7 \\ 6 & 2 \end{bmatrix}$

7. $\mathbf{E} + \mathbf{0} = \begin{bmatrix} 1 & 3 \\ 2 & 6 \end{bmatrix} + \begin{bmatrix} 0 & 0 \\ 0 & 0 \end{bmatrix}$

$= \begin{bmatrix} 1 + 0 & 3 + 0 \\ 2 + 0 & 6 + 0 \end{bmatrix}$

$= \begin{bmatrix} 1 & 3 \\ 2 & 6 \end{bmatrix}$

9. $3\mathbf{F} = 3 \begin{bmatrix} 3 & 3 \\ -1 & -1 \end{bmatrix}$

$= \begin{bmatrix} 3 \cdot 3 & 3 \cdot 3 \\ 3 \cdot (-1) & 3 \cdot (-1) \end{bmatrix}$

$= \begin{bmatrix} 9 & 9 \\ -3 & -3 \end{bmatrix}$

11. $3\mathbf{F} = 3 \begin{bmatrix} 3 & 3 \\ -1 & -1 \end{bmatrix} = \begin{bmatrix} 9 & 9 \\ -3 & -3 \end{bmatrix},$

$2\mathbf{A} = 2 \begin{bmatrix} 1 & 2 \\ 4 & 3 \end{bmatrix} = \begin{bmatrix} 2 & 4 \\ 8 & 6 \end{bmatrix}$

$3\mathbf{F} + 2\mathbf{A} = \begin{bmatrix} 9 & 9 \\ -3 & -3 \end{bmatrix} + \begin{bmatrix} 2 & 4 \\ 8 & 6 \end{bmatrix}$

$= \begin{bmatrix} 9 + 2 & 9 + 4 \\ -3 + 8 & -3 + 6 \end{bmatrix}$

$= \begin{bmatrix} 11 & 13 \\ 5 & 3 \end{bmatrix}$

13. $\mathbf{B} - \mathbf{A} = \begin{bmatrix} -3 & 5 \\ 2 & -1 \end{bmatrix} - \begin{bmatrix} 1 & 2 \\ 4 & 3 \end{bmatrix}$

$= \begin{bmatrix} -3 & 5 \\ 2 & -1 \end{bmatrix} + \begin{bmatrix} -1 & -2 \\ -4 & -3 \end{bmatrix}$

$[\mathbf{B} - \mathbf{A} = \mathbf{B} + (-\mathbf{A})]$

$= \begin{bmatrix} -3 + (-1) & 5 + (-2) \\ 2 + (-4) & -1 + (-3) \end{bmatrix}$

$= \begin{bmatrix} -4 & 3 \\ -2 & -4 \end{bmatrix}$

15. $\mathbf{BA} = \begin{bmatrix} -3 & 5 \\ 2 & -1 \end{bmatrix} \begin{bmatrix} 1 & 2 \\ 4 & 3 \end{bmatrix}$

$= \begin{bmatrix} -3 \cdot 1 + 5 \cdot 4 & -3 \cdot 2 + 5 \cdot 3 \\ 2 \cdot 1 + (-1)4 & 2 \cdot 2 + (-1)3 \end{bmatrix}$

$= \begin{bmatrix} 17 & 9 \\ -2 & 1 \end{bmatrix}$

17. $\mathbf{CD} = \begin{bmatrix} 1 & -1 \\ -1 & 1 \end{bmatrix} \begin{bmatrix} 1 & 1 \\ 1 & 1 \end{bmatrix}$.

$= \begin{bmatrix} 1 \cdot 1 + (-1) \cdot 1 & 1 \cdot 1 + (-1) \cdot 1 \\ -1 \cdot 1 + 1 \cdot 1 & -1 \cdot 1 + 1 \cdot 1 \end{bmatrix}$

$= \begin{bmatrix} 0 & 0 \\ 0 & 0 \end{bmatrix}$

19. $\mathbf{AI} = \begin{bmatrix} 1 & 2 \\ 4 & 3 \end{bmatrix} \begin{bmatrix} 1 & 0 \\ 0 & 1 \end{bmatrix}$

$= \begin{bmatrix} 1 \cdot 1 + 2 \cdot 0 & 1 \cdot 0 + 2 \cdot 1 \\ 4 \cdot 1 + 3 \cdot 0 & 4 \cdot 0 + 3 \cdot 1 \end{bmatrix}$

$= \begin{bmatrix} 1 & 2 \\ 4 & 3 \end{bmatrix}$

21. $\begin{bmatrix} -1 & 0 & 7 \\ 3 & -5 & 2 \end{bmatrix} \begin{bmatrix} 6 \\ -4 \\ 1 \end{bmatrix}$

$= \begin{bmatrix} -1 \cdot 6 + 0(-4) + 7 \cdot 1 \\ 3 \cdot 6 + (-5)(-4) + 2 \cdot 1 \end{bmatrix}$

$= \begin{bmatrix} 1 \\ 40 \end{bmatrix}$

23. $\begin{bmatrix} -2 & 4 \\ 5 & 1 \\ -1 & -3 \end{bmatrix} \begin{bmatrix} 3 & -6 \\ -1 & 4 \end{bmatrix}$

$= \begin{bmatrix} -2 \cdot 3 + 4(-1) & -2(-6) + 4 \cdot 4 \\ 5 \cdot 3 + 1(-1) & 5(-6) + 1 \cdot 4 \\ -1 \cdot 3 + (-3)(-1) & -1(-6) + (-3) \cdot 4 \end{bmatrix}$

$= \begin{bmatrix} -10 & 28 \\ 14 & -26 \\ 0 & -6 \end{bmatrix}$

25. $\begin{bmatrix} 1 \\ -5 \\ 3 \end{bmatrix} \begin{bmatrix} -6 & 5 & 8 \\ 0 & 4 & -1 \end{bmatrix}$

This product is not defined because the number of columns of the first matrix, 1, is not equal to the number of rows of the second matrix, 2.

27. $\begin{bmatrix} 1 & -4 & 3 \\ 0 & 8 & 0 \\ -2 & -1 & 5 \end{bmatrix} \begin{bmatrix} 3 & 0 & 0 \\ 0 & -4 & 0 \\ 0 & 0 & 1 \end{bmatrix}$

$= \begin{bmatrix} 3 + 0 + 0 & 0 + 16 + 0 & 0 + 0 + 3 \\ 0 + 0 + 0 & 0 - 32 + 0 & 0 + 0 + 0 \\ -6 + 0 + 0 & 0 + 4 + 0 & 0 + 0 + 5 \end{bmatrix}$

$= \begin{bmatrix} 3 & 16 & 3 \\ 0 & -32 & 0 \\ -6 & 4 & 5 \end{bmatrix}$

29. a) $\mathbf{B} = \begin{bmatrix} 150 & 80 & 40 \end{bmatrix}$

b) $\$150 + 5\% \cdot \$150 = 1.05(\$150) = \$157.50$

$\$80 + 5\% \cdot \$80 = 1.05(\$80) = \$84$

$\$40 + 5\% \cdot \$40 = 1.05(\$40) = \$42$

We write the matrix that corresponds to these amounts.

$\mathbf{R} = \begin{bmatrix} 157.5 & 84 & 42 \end{bmatrix}$

c) $\mathbf{B+R} = \begin{bmatrix} 150 & 80 & 40 \end{bmatrix} + \begin{bmatrix} 157.5 & 84 & 42 \end{bmatrix}$

$= \begin{bmatrix} 307.5 & 164 & 82 \end{bmatrix}$

The entries represent the total budget for each type of expenditure for June and July.

31. a) $\mathbf{C} = \begin{bmatrix} 140 & 27 & 3 & 13 & 64 \end{bmatrix}$

$\mathbf{P} = \begin{bmatrix} 180 & 4 & 11 & 24 & 662 \end{bmatrix}$

$\mathbf{B} = \begin{bmatrix} 50 & 5 & 1 & 82 & 20 \end{bmatrix}$

b) $\mathbf{C} + 2\mathbf{P} + 3\mathbf{B}$

$= \begin{bmatrix} 140 & 27 & 3 & 13 & 64 \end{bmatrix} +$

$\begin{bmatrix} 360 & 8 & 22 & 48 & 1324 \end{bmatrix} +$

$\begin{bmatrix} 150 & 15 & 3 & 246 & 60 \end{bmatrix}$

$= \begin{bmatrix} 650 & 50 & 28 & 307 & 1448 \end{bmatrix}$

The entries represent the total nutritional value of 1 serving of chicken, 1 cup of potato salad, and 3 broccoli spears.

33. a) $\mathbf{M} = \begin{bmatrix} 45.29 & 6.63 & 10.94 & 7.42 & 8.01 \\ 53.78 & 4.95 & 9.83 & 6.16 & 12.56 \\ 47.13 & 8.47 & 12.66 & 8.29 & 9.43 \\ 51.64 & 7.12 & 11.57 & 9.35 & 10.72 \end{bmatrix}$

b) $\mathbf{N} = \begin{bmatrix} 65 & 48 & 93 & 57 \end{bmatrix}$

c) $\mathbf{NM} =$

$\begin{bmatrix} 12,851.86 & 1862.1 & 3019.81 & 2081.9 & 2611.56 \end{bmatrix}$

d) The entries of $\mathbf{NM}$ represent the total cost, in cents, of each item for the day's meals.

35. a) $\mathbf{S} = \begin{bmatrix} 8 & 15 \\ 6 & 10 \\ 4 & 3 \end{bmatrix}$

b) $\mathbf{C} = \begin{bmatrix} 3 & 1.5 & 2 \end{bmatrix}$

c) $\mathbf{CS} = \begin{bmatrix} 41 & 66 \end{bmatrix}$

d) The entries of $\mathbf{CS}$ represent the total cost, in dollars, of ingredients for each coffee shop.

37. a) $\mathbf{P} = \begin{bmatrix} 6 & 4.5 & 5.2 \end{bmatrix}$

b) $\mathbf{PS} = \begin{bmatrix} 6 & 4.5 & 5.2 \end{bmatrix} \begin{bmatrix} 8 & 15 \\ 6 & 10 \\ 4 & 3 \end{bmatrix}$

$= \begin{bmatrix} 95.8 & 150.6 \end{bmatrix}$

The profit from Mugsey's Coffee Shop is \$95.80, and the profit from The Coffee Club is \$150.60.

39. $2x - 3y = 7,$

$x + 5y = -6$

Write the coefficients on the left in a matrix. Then write the product of that matrix and the column matrix containing the variables, and set the result equal to the column matrix containing the constants on the right.

$\begin{bmatrix} 2 & -3 \\ 1 & 5 \end{bmatrix} \begin{bmatrix} x \\ y \end{bmatrix} = \begin{bmatrix} 7 \\ -6 \end{bmatrix}$

41. $x + y - 2z = 6,$

$3x - y + z = 7,$

$2x + 5y - 3z = 8$

Write the coefficients on the left in a matrix. Then write the product of that matrix and the column matrix containing the variables, and set the result equal to the column matrix containing the constants on the right.

$\begin{bmatrix} 1 & 1 & -2 \\ 3 & -1 & 1 \\ 2 & 5 & -3 \end{bmatrix} \begin{bmatrix} x \\ y \\ z \end{bmatrix} = \begin{bmatrix} 6 \\ 7 \\ 8 \end{bmatrix}$

43. $3x - 2y + 4z = 17,$

$2x + y - 5z = 13$

Write the coefficients on the left in a matrix. Then write the product of that matrix and the column matrix containing the variables, and set the result equal to the column matrix containing the constants on the right.

$\begin{bmatrix} 3 & -2 & 4 \\ 2 & 1 & -5 \end{bmatrix} \begin{bmatrix} x \\ y \\ z \end{bmatrix} = \begin{bmatrix} 17 \\ 13 \end{bmatrix}$

45. $-4w + x - y + 2z = 12,$

$w + 2x - y - z = 0,$

$-w + x + 4y - 3z = 1,$

$2w + 3x + 5y - 7z = 9$

Write the coefficients on the left in a matrix. Then write the product of that matrix and the column matrix containing the variables, and set the result equal to the column matrix containing the constants on the right.

$\begin{bmatrix} -4 & 1 & -1 & 2 \\ 1 & 2 & -1 & -1 \\ -1 & 1 & 4 & -3 \\ 2 & 3 & 5 & -7 \end{bmatrix} \begin{bmatrix} w \\ x \\ y \\ z \end{bmatrix} = \begin{bmatrix} 12 \\ 0 \\ 1 \\ 9 \end{bmatrix}$

47. Left to the student

49. Discussion and Writing

51. $f(x) = x^2 - x - 6$

a) $-\dfrac{b}{2a} = -\dfrac{-1}{2 \cdot 1} = \dfrac{1}{2}$

$f\left(\dfrac{1}{2}\right) = \left(\dfrac{1}{2}\right)^2 - \dfrac{1}{2} - 6 = -\dfrac{25}{4}$

The vertex is $\left(\dfrac{1}{2}, -\dfrac{25}{4}\right)$.

b) The axis of symmetry is $x = \dfrac{1}{2}$.

c) Since the coefficient of x^2 is positive, the function has a minimum value. It is the second coordinate of the vertex, $-\dfrac{25}{4}$.

d) Plot some points and draw the graph of the function.

$$f(x) = x^2 - x - 6$$

53. $f(x) = -x^2 - 3x + 2$

a) $-\dfrac{b}{2a} = -\dfrac{-3}{2(-1)} = -\dfrac{3}{2}$

$f\left(-\dfrac{3}{2}\right) = -\left(-\dfrac{3}{2}\right)^2 - 3\left(-\dfrac{3}{2}\right) + 2 = \dfrac{17}{4}$

The vertex is $\left(-\dfrac{3}{2}, \dfrac{17}{4}\right)$.

b) The axis of symmetry is $x = -\dfrac{3}{2}$.

c) Since the coefficient of x^2 is negative, the function has a maximum value. It is the second coordinate of the vertex, $\dfrac{17}{4}$.

d) Plot some points and draw the graph of the function.

$$f(x) = -x^2 - 3x + 2$$

55. $\mathbf{A} = \begin{bmatrix} -1 & 0 \\ 2 & 1 \end{bmatrix}$, $\mathbf{B} = \begin{bmatrix} 1 & -1 \\ 0 & 2 \end{bmatrix}$

$(\mathbf{A} + \mathbf{B})(\mathbf{A} - \mathbf{B}) = \begin{bmatrix} 0 & -1 \\ 2 & 3 \end{bmatrix} \begin{bmatrix} -2 & 1 \\ 2 & -1 \end{bmatrix}$

$\phantom{(\mathbf{A} + \mathbf{B})(\mathbf{A} - \mathbf{B})} = \begin{bmatrix} -2 & 1 \\ 2 & -1 \end{bmatrix}$

$\mathbf{A}^2 - \mathbf{B}^2$

$= \begin{bmatrix} -1 & 0 \\ 2 & 1 \end{bmatrix} \begin{bmatrix} -1 & 0 \\ 2 & 1 \end{bmatrix} - \begin{bmatrix} 1 & -1 \\ 0 & 2 \end{bmatrix} \begin{bmatrix} 1 & -1 \\ 0 & 2 \end{bmatrix}$

$= \begin{bmatrix} 1 & 0 \\ 0 & 1 \end{bmatrix} - \begin{bmatrix} 1 & -3 \\ 0 & 4 \end{bmatrix}$

$= \begin{bmatrix} 0 & 3 \\ 0 & -3 \end{bmatrix}$

Thus $(\mathbf{A} + \mathbf{B})(\mathbf{A} - \mathbf{B}) \neq \mathbf{A}^2 - \mathbf{B}^2$.

57. In Exercise 55 we found that $(\mathbf{A} + \mathbf{B})(\mathbf{A} - \mathbf{B}) = \begin{bmatrix} -2 & 1 \\ 2 & -1 \end{bmatrix}$

and we also found $\mathbf{A}^2$ and $\mathbf{B}^2$.

$\mathbf{BA} = \begin{bmatrix} 1 & -1 \\ 0 & 2 \end{bmatrix} \begin{bmatrix} -1 & 0 \\ 2 & 1 \end{bmatrix} = \begin{bmatrix} -3 & -1 \\ 4 & 2 \end{bmatrix}$

$\mathbf{AB} = \begin{bmatrix} -1 & 0 \\ 2 & 1 \end{bmatrix} \begin{bmatrix} 1 & -1 \\ 0 & 2 \end{bmatrix} = \begin{bmatrix} -1 & 1 \\ 2 & 0 \end{bmatrix}$

$\mathbf{A}^2 + \mathbf{BA} - \mathbf{AB} - \mathbf{B}^2$

$= \begin{bmatrix} 1 & 0 \\ 0 & 1 \end{bmatrix} + \begin{bmatrix} -3 & -1 \\ 4 & 2 \end{bmatrix} - \begin{bmatrix} -1 & 1 \\ 2 & 0 \end{bmatrix} - \begin{bmatrix} 1 & -3 \\ 0 & 4 \end{bmatrix}$

$= \begin{bmatrix} -2 & 1 \\ 2 & -1 \end{bmatrix}$

Thus $(\mathbf{A} + \mathbf{B})(\mathbf{A} - \mathbf{B}) = \mathbf{A}^2 + \mathbf{BA} - \mathbf{AB} - \mathbf{B}^2$.

59. See the answer section in the text.

61. See the answer section in the text.

63. See the answer section in the text.

Exercise Set 5.5

1. $\mathbf{BA} = \begin{bmatrix} 7 & 3 \\ 2 & 1 \end{bmatrix} \begin{bmatrix} 1 & -3 \\ -2 & 7 \end{bmatrix} = \begin{bmatrix} 1 & 0 \\ 0 & 1 \end{bmatrix}$

$\mathbf{AB} = \begin{bmatrix} 1 & -3 \\ -2 & 7 \end{bmatrix} \begin{bmatrix} 7 & 3 \\ 2 & 1 \end{bmatrix} = \begin{bmatrix} 1 & 0 \\ 0 & 1 \end{bmatrix}$

Since $\mathbf{BA} = \mathbf{I} = \mathbf{AB}$, $\mathbf{B}$ is the inverse of $\mathbf{A}$.

3. $\mathbf{BA} = \begin{bmatrix} 2 & 3 & 2 \\ 3 & 3 & 4 \\ 1 & 1 & 1 \end{bmatrix} \begin{bmatrix} -1 & -1 & 6 \\ 1 & 0 & -2 \\ 1 & 0 & -3 \end{bmatrix} =$

$\begin{bmatrix} 3 & -2 & 0 \\ 4 & -3 & 0 \\ 1 & -1 & 1 \end{bmatrix}$

Since $\mathbf{BA} \neq \mathbf{I}$, $\mathbf{B}$ is not the inverse of $\mathbf{A}$.

5. $\mathbf{A} = \begin{bmatrix} 3 & 2 \\ 5 & 3 \end{bmatrix}$

Write the augmented matrix.

$\begin{bmatrix} 3 & 2 & | & 1 & 0 \\ 5 & 3 & | & 0 & 1 \end{bmatrix}$

Multiply row 2 by 3.

$\begin{bmatrix} 3 & 2 & | & 1 & 0 \\ 15 & 9 & | & 0 & 3 \end{bmatrix}$

Multiply row 1 by -5 and add it to row 2.

$\begin{bmatrix} 3 & 2 & | & 1 & 0 \\ 0 & -1 & | & -5 & 3 \end{bmatrix}$

Multiply row 2 by 2 and add it to row 1.

$\begin{bmatrix} 3 & 0 & | & -9 & 6 \\ 0 & -1 & | & -5 & 3 \end{bmatrix}$

Multiply row 1 by $\frac{1}{3}$ and row 2 by -1.

$$\begin{bmatrix} 1 & 0 & | & -3 & 2 \\ 0 & 1 & | & 5 & -3 \end{bmatrix}$$

Then $\mathbf{A}^{-1} = \begin{bmatrix} -3 & 2 \\ 5 & -3 \end{bmatrix}$.

7. $\mathbf{A} = \begin{bmatrix} 6 & 9 \\ 4 & 6 \end{bmatrix}$

Write the augmented matrix.

$$\begin{bmatrix} 6 & 9 & | & 1 & 0 \\ 4 & 6 & | & 0 & 1 \end{bmatrix}$$

Multiply row 2 by 3.

$$\begin{bmatrix} 6 & 9 & | & 1 & 0 \\ 12 & 18 & | & 0 & 3 \end{bmatrix}$$

Multiply row 1 by -2 and add it to row 2.

$$\begin{bmatrix} 6 & 9 & | & 1 & 0 \\ 0 & 0 & | & -2 & 3 \end{bmatrix}$$

We cannot obtain the identity matrix on the left since the second row contains only zeros to the left of the vertical line. Thus, $\mathbf{A}^{-1}$ does not exist.

9. $\mathbf{A} = \begin{bmatrix} 4 & -3 \\ 1 & -2 \end{bmatrix}$

Write the augmented matrix.

$$\begin{bmatrix} 4 & -3 & | & 1 & 0 \\ 1 & -2 & | & 0 & 1 \end{bmatrix}$$

Interchange the rows.

$$\begin{bmatrix} 1 & -2 & | & 0 & 1 \\ 4 & -3 & | & 1 & 0 \end{bmatrix}$$

Multiply row 1 by -4 and add it to row 2.

$$\begin{bmatrix} 1 & -2 & | & 0 & 1 \\ 0 & 5 & | & 1 & -4 \end{bmatrix}$$

Multiply row 2 by $\frac{1}{5}$.

$$\begin{bmatrix} 1 & -2 & | & 0 & 1 \\ 0 & 1 & | & \frac{1}{5} & -\frac{4}{5} \end{bmatrix}$$

Multiply row 2 by 2 and add it to row 1.

$$\begin{bmatrix} 1 & 0 & | & \frac{2}{5} & -\frac{3}{5} \\ 0 & 1 & | & \frac{1}{5} & -\frac{4}{5} \end{bmatrix}$$

Then $\mathbf{A}^{-1} = \begin{bmatrix} \frac{2}{5} & -\frac{3}{5} \\ \frac{1}{5} & -\frac{4}{5} \end{bmatrix}$.

11. $\mathbf{A} = \begin{bmatrix} 3 & 1 & 0 \\ 1 & 1 & 1 \\ 1 & -1 & 2 \end{bmatrix}$

Write the augmented matrix.

$$\begin{bmatrix} 3 & 1 & 0 & | & 1 & 0 & 0 \\ 1 & 1 & 1 & | & 0 & 1 & 0 \\ 1 & -1 & 2 & | & 0 & 0 & 1 \end{bmatrix}$$

Interchange the first two rows.

$$\begin{bmatrix} 1 & 1 & 1 & | & 0 & 1 & 0 \\ 3 & 1 & 0 & | & 1 & 0 & 0 \\ 1 & -1 & 2 & | & 0 & 0 & 1 \end{bmatrix}$$

Multiply row 1 by -3 and add it to row 2. Also, multiply row 1 by -1 and add it to row 3.

$$\begin{bmatrix} 1 & 1 & 1 & | & 0 & 1 & 0 \\ 0 & -2 & -3 & | & 1 & -3 & 0 \\ 0 & -2 & 1 & | & 0 & -1 & 1 \end{bmatrix}$$

Multiply row 2 by $-\frac{1}{2}$.

$$\begin{bmatrix} 1 & 1 & 1 & | & 0 & 1 & 0 \\ 0 & 1 & \frac{3}{2} & | & -\frac{1}{2} & \frac{3}{2} & 0 \\ 0 & -2 & 1 & | & 0 & -1 & 1 \end{bmatrix}$$

Multiply row 2 by -1 and add it to row 1. Also, multiply row 2 by 2 and add it to row 3.

$$\begin{bmatrix} 1 & 0 & -\frac{1}{2} & | & \frac{1}{2} & -\frac{1}{2} & 0 \\ 0 & 1 & \frac{3}{2} & | & -\frac{1}{2} & \frac{3}{2} & 0 \\ 0 & 0 & 4 & | & -1 & 2 & 1 \end{bmatrix}$$

Multiply row 3 by $\frac{1}{4}$.

$$\begin{bmatrix} 1 & 0 & -\frac{1}{2} & | & \frac{1}{2} & -\frac{1}{2} & 0 \\ 0 & 1 & \frac{3}{2} & | & -\frac{1}{2} & \frac{3}{2} & 0 \\ 0 & 0 & 1 & | & -\frac{1}{4} & \frac{1}{2} & \frac{1}{4} \end{bmatrix}$$

Multiply row 3 by $\frac{1}{2}$ and add it to row 1. Also, multiply row 3 by $-\frac{3}{2}$ and add it to row 2.

$$\begin{bmatrix} 1 & 0 & 0 & | & \frac{3}{8} & -\frac{1}{4} & \frac{1}{8} \\ 0 & 1 & 0 & | & -\frac{1}{8} & \frac{3}{4} & -\frac{3}{8} \\ 0 & 0 & 1 & | & -\frac{1}{4} & \frac{1}{2} & \frac{1}{4} \end{bmatrix}$$

Then $\mathbf{A}^{-1} = \begin{bmatrix} \frac{3}{8} & -\frac{1}{4} & \frac{1}{8} \\ -\frac{1}{8} & \frac{3}{4} & -\frac{3}{8} \\ -\frac{1}{4} & \frac{1}{2} & \frac{1}{4} \end{bmatrix}$.

13. $\mathbf{A} = \begin{bmatrix} 1 & -4 & 8 \\ 1 & -3 & 2 \\ 2 & -7 & 10 \end{bmatrix}$

Write the augmented matrix.

$$\begin{bmatrix} 1 & -4 & 8 & | & 1 & 0 & 0 \\ 1 & -3 & 2 & | & 0 & 1 & 0 \\ 2 & -7 & 10 & | & 0 & 0 & 1 \end{bmatrix}$$

Multiply row 1 by -1 and add it to row 2. Also, multiply row 1 by -2 and add it to row 3.

$$\begin{bmatrix} 1 & -4 & 8 & | & 1 & 0 & 0 \\ 0 & 1 & -6 & | & -1 & 1 & 0 \\ 0 & 1 & -6 & | & -2 & 0 & 1 \end{bmatrix}$$

Since the second and third rows are identical left of the vertical line, it will not be possible to obtain the identity matrix on the left side. Thus, $\mathbf{A}^{-1}$ does not exist.

15. $\mathbf{A} = \begin{bmatrix} 2 & 3 & 2 \\ 3 & 3 & 4 \\ -1 & -1 & -1 \end{bmatrix}$

Write the augmented matrix.

$\left[\begin{array}{ccc|ccc} 2 & 3 & 2 & 1 & 0 & 0 \\ 3 & 3 & 4 & 0 & 1 & 0 \\ -1 & -1 & -1 & 0 & 0 & 1 \end{array}\right]$

Interchange rows 1 and 3.

$\left[\begin{array}{ccc|ccc} -1 & -1 & -1 & 0 & 0 & 1 \\ 3 & 3 & 4 & 0 & 1 & 0 \\ 2 & 3 & 2 & 1 & 0 & 0 \end{array}\right]$

Multiply row 1 by 3 and add it to row 2. Also, multiply row 1 by 2 and add it to row 3.

$\left[\begin{array}{ccc|ccc} -1 & -1 & -1 & 0 & 0 & 1 \\ 0 & 0 & 1 & 0 & 1 & 3 \\ 0 & 1 & 0 & 1 & 0 & 2 \end{array}\right]$

Multiply row 1 by -1.

$\left[\begin{array}{ccc|ccc} 1 & 1 & 1 & 0 & 0 & -1 \\ 0 & 0 & 1 & 0 & 1 & 3 \\ 0 & 1 & 0 & 1 & 0 & 2 \end{array}\right]$

Interchange rows 2 and 3.

$\left[\begin{array}{ccc|ccc} 1 & 1 & 1 & 0 & 0 & -1 \\ 0 & 1 & 0 & 1 & 0 & 2 \\ 0 & 0 & 1 & 0 & 1 & 3 \end{array}\right]$

Multiply row 2 by -1 and add it to row 1.

$\left[\begin{array}{ccc|ccc} 1 & 0 & 1 & -1 & 0 & -3 \\ 0 & 1 & 0 & 1 & 0 & 2 \\ 0 & 0 & 1 & 0 & 1 & 3 \end{array}\right]$

Multiply row 3 by -1 and add it to row 1.

$\left[\begin{array}{ccc|ccc} 1 & 0 & 0 & -1 & -1 & -6 \\ 0 & 1 & 0 & 1 & 0 & 2 \\ 0 & 0 & 1 & 0 & 1 & 3 \end{array}\right]$

Then $\mathbf{A}^{-1} = \begin{bmatrix} -1 & -1 & -6 \\ 1 & 0 & 2 \\ 0 & 1 & 3 \end{bmatrix}$.

17. $\mathbf{A} = \begin{bmatrix} 1 & 2 & -1 \\ -2 & 0 & 1 \\ 1 & -1 & 0 \end{bmatrix}$

Write the augmented matrix.

$\left[\begin{array}{ccc|ccc} 1 & 2 & -1 & 1 & 0 & 0 \\ -2 & 0 & 1 & 0 & 1 & 0 \\ 1 & -1 & 0 & 0 & 0 & 1 \end{array}\right]$

Multiply row 1 by 2 and add it to row 2. Also, multiply row 1 by -1 and add it to row 3.

$\left[\begin{array}{ccc|ccc} 1 & 2 & -1 & 1 & 0 & 0 \\ 0 & 4 & -1 & 2 & 1 & 0 \\ 0 & -3 & 1 & -1 & 0 & 1 \end{array}\right]$

Add row 3 to row 1 and also to row 2.

$\left[\begin{array}{ccc|ccc} 1 & -1 & 0 & 0 & 0 & 1 \\ 0 & 1 & 0 & 1 & 1 & 1 \\ 0 & -3 & 1 & -1 & 0 & 1 \end{array}\right]$

Add row 2 to row 1. Also, multiply row 2 by 3 and add it to row 3.

$\left[\begin{array}{ccc|ccc} 1 & 0 & 0 & 1 & 1 & 2 \\ 0 & 1 & 0 & 1 & 1 & 1 \\ 0 & 0 & 1 & 2 & 3 & 4 \end{array}\right]$

Then $\mathbf{A}^{-1} = \begin{bmatrix} 1 & 1 & 2 \\ 1 & 1 & 1 \\ 2 & 3 & 4 \end{bmatrix}$.

19. $\mathbf{A} = \begin{bmatrix} 1 & 3 & -1 \\ 0 & 2 & -1 \\ 1 & 1 & 0 \end{bmatrix}$

Write the augmented matrix.

$\left[\begin{array}{ccc|ccc} 1 & 3 & -1 & 1 & 0 & 0 \\ 0 & 2 & -1 & 0 & 1 & 0 \\ 1 & 1 & 0 & 0 & 0 & 1 \end{array}\right]$

Multiply row 1 by -1 and add it to row 3.

$\left[\begin{array}{ccc|ccc} 1 & 3 & -1 & 1 & 0 & 0 \\ 0 & 2 & -1 & 0 & 1 & 0 \\ 0 & -2 & 1 & -1 & 0 & 1 \end{array}\right]$

Add row 3 to row 1 and also to row 2.

$\left[\begin{array}{ccc|ccc} 1 & 1 & 0 & 0 & 0 & 0 \\ 0 & 0 & 0 & -1 & 1 & 1 \\ 0 & -2 & 1 & -1 & 0 & 1 \end{array}\right]$

Since the second row consists only of zeros to the left of the vertical line, it will not be possible to obtain the identity matrix on the left side. Thus, $\mathbf{A}^{-1}$ does not exist.

21. $\mathbf{A} = \begin{bmatrix} 1 & 2 & 3 & 4 \\ 0 & 1 & 3 & -5 \\ 0 & 0 & 1 & -2 \\ 0 & 0 & 0 & -1 \end{bmatrix}$

Write the augmented matrix.

$\left[\begin{array}{cccc|cccc} 1 & 2 & 3 & 4 & 1 & 0 & 0 & 0 \\ 0 & 1 & 3 & -5 & 0 & 1 & 0 & 0 \\ 0 & 0 & 1 & -2 & 0 & 0 & 1 & 0 \\ 0 & 0 & 0 & -1 & 0 & 0 & 0 & 1 \end{array}\right]$

Multiply row 4 by -1.

$\left[\begin{array}{cccc|cccc} 1 & 2 & 3 & 4 & 1 & 0 & 0 & 0 \\ 0 & 1 & 3 & -5 & 0 & 1 & 0 & 0 \\ 0 & 0 & 1 & -2 & 0 & 0 & 1 & 0 \\ 0 & 0 & 0 & 1 & 0 & 0 & 0 & -1 \end{array}\right]$

Multiply row 4 by -4 and add it to row 1. Multiply row 4 by 5 and add it to row 2. Also, multiply row 4 by 2 and add it to row 3.

$\left[\begin{array}{cccc|cccc} 1 & 2 & 3 & 0 & 1 & 0 & 0 & 4 \\ 0 & 1 & 3 & 0 & 0 & 1 & 0 & -5 \\ 0 & 0 & 1 & 0 & 0 & 0 & 1 & -2 \\ 0 & 0 & 0 & 1 & 0 & 0 & 0 & -1 \end{array}\right]$

Multiply row 3 by -3 and add it to row 1 and to row 2.

$\left[\begin{array}{cccc|cccc} 1 & 2 & 0 & 0 & 1 & 0 & -3 & 10 \\ 0 & 1 & 0 & 0 & 0 & 1 & -3 & 1 \\ 0 & 0 & 1 & 0 & 0 & 0 & 1 & -2 \\ 0 & 0 & 0 & 1 & 0 & 0 & 0 & -1 \end{array}\right]$

Multiply row 2 by -2 and add it to row 1.

$$\left[\begin{array}{cccc|cccc} 1 & 0 & 0 & 0 & 1 & -2 & 3 & 8 \\ 0 & 1 & 0 & 0 & 0 & 1 & -3 & 1 \\ 0 & 0 & 1 & 0 & 0 & 0 & 1 & -2 \\ 0 & 0 & 0 & 1 & 0 & 0 & 0 & -1 \end{array}\right]$$

Then $\mathbf{A}^{-1} = \left[\begin{array}{cccc} 1 & -2 & 3 & 8 \\ 0 & 1 & -3 & 1 \\ 0 & 0 & 1 & -2 \\ 0 & 0 & 0 & -1 \end{array}\right]$.

23. $\mathbf{A} = \left[\begin{array}{cccc} 1 & -14 & 7 & 38 \\ -1 & 2 & 1 & -2 \\ 1 & 2 & -1 & -6 \\ 1 & -2 & 3 & 6 \end{array}\right]$

Write the augmented matrix.

$$\left[\begin{array}{cccc|cccc} 1 & -14 & 7 & 38 & 1 & 0 & 0 & 0 \\ -1 & 2 & 1 & -2 & 0 & 1 & 0 & 0 \\ 1 & 2 & -1 & -6 & 0 & 0 & 1 & 0 \\ 1 & -2 & 3 & 6 & 0 & 0 & 0 & 1 \end{array}\right]$$

Add row 1 to row 2. Also, multiply row 1 by -1 and add it to row 3 and to row 4.

$$\left[\begin{array}{cccc|cccc} 1 & -14 & 7 & 38 & 1 & 0 & 0 & 0 \\ 0 & -12 & 8 & 36 & 1 & 1 & 0 & 0 \\ 0 & 16 & -8 & -44 & -1 & 0 & 1 & 0 \\ 0 & 12 & -4 & -32 & -1 & 0 & 0 & 1 \end{array}\right]$$

Add row 2 to row 4.

$$\left[\begin{array}{cccc|cccc} 1 & -14 & 7 & 38 & 1 & 0 & 0 & 0 \\ 0 & -12 & 8 & 36 & 1 & 1 & 0 & 0 \\ 0 & 16 & -8 & -44 & -1 & 0 & 1 & 0 \\ 0 & 0 & 4 & 4 & 0 & 1 & 0 & 1 \end{array}\right]$$

Multiply row 4 by $\frac{1}{4}$.

$$\left[\begin{array}{cccc|cccc} 1 & -14 & 7 & 38 & 1 & 0 & 0 & 0 \\ 0 & -12 & 8 & 36 & 1 & 1 & 0 & 0 \\ 0 & 16 & -8 & -44 & -1 & 0 & 1 & 0 \\ 0 & 0 & 1 & 1 & 0 & \frac{1}{4} & 0 & \frac{1}{4} \end{array}\right]$$

Multiply row 4 by -38 and add it to row 1. Multiply row 4 by -36 and add it to row 2. Also, multiply row 4 by 44 and add it to row 3.

$$\left[\begin{array}{cccc|cccc} 1 & -14 & -31 & 0 & 1 & -\frac{19}{2} & 0 & -\frac{19}{2} \\ 0 & -12 & -28 & 0 & 1 & -8 & 0 & -9 \\ 0 & 16 & 36 & 0 & -1 & 11 & 1 & 11 \\ 0 & 0 & 1 & 1 & 0 & \frac{1}{4} & 0 & \frac{1}{4} \end{array}\right]$$

Multiply row 3 by $\frac{1}{36}$.

$$\left[\begin{array}{cccc|cccc} 1 & -14 & -31 & 0 & 1 & -\frac{19}{2} & 0 & -\frac{19}{2} \\ 0 & -12 & -28 & 0 & 1 & -8 & 0 & -9 \\ 0 & \frac{4}{9} & 1 & 0 & -\frac{1}{36} & \frac{11}{36} & \frac{1}{36} & \frac{11}{36} \\ 0 & 0 & 1 & 1 & 0 & \frac{1}{4} & 0 & \frac{1}{4} \end{array}\right]$$

Multiply row 3 by 31 and add it to row 1. Multiply row 3 by 28 and add it to row 2. Also, multiply row 3 by -1 and add it to row 4.

$$\left[\begin{array}{cccc|cccc} 1 & -\frac{2}{9} & 0 & 0 & \frac{5}{36} & -\frac{1}{36} & \frac{31}{36} & -\frac{1}{36} \\ 0 & \frac{4}{9} & 0 & 0 & \frac{2}{9} & \frac{5}{9} & \frac{7}{9} & -\frac{4}{9} \\ 0 & \frac{4}{9} & 1 & 0 & -\frac{1}{36} & \frac{11}{36} & \frac{1}{36} & \frac{11}{36} \\ 0 & -\frac{4}{9} & 0 & 1 & \frac{1}{36} & -\frac{1}{18} & -\frac{1}{36} & -\frac{1}{18} \end{array}\right]$$

Multiply row 2 by $\frac{1}{2}$ and add it to row 1. Also, multiply row 2 by -1 and add it to row 3. Add row 2 to row 4.

$$\left[\begin{array}{cccc|cccc} 1 & 0 & 0 & 0 & \frac{1}{4} & \frac{1}{4} & \frac{5}{4} & -\frac{1}{4} \\ 0 & \frac{4}{9} & 0 & 0 & \frac{2}{9} & \frac{5}{9} & \frac{7}{9} & -\frac{4}{9} \\ 0 & 0 & 1 & 0 & -\frac{1}{4} & -\frac{1}{4} & -\frac{3}{4} & \frac{3}{4} \\ 0 & 0 & 0 & 1 & \frac{1}{4} & \frac{1}{2} & \frac{3}{4} & -\frac{1}{2} \end{array}\right]$$

Multiply row 2 by $\frac{9}{4}$.

$$\left[\begin{array}{cccc|cccc} 1 & 0 & 0 & 0 & \frac{1}{4} & \frac{1}{4} & \frac{5}{4} & -\frac{1}{4} \\ 0 & 1 & 0 & 0 & \frac{1}{2} & \frac{5}{4} & \frac{7}{4} & -1 \\ 0 & 0 & 1 & 0 & -\frac{1}{4} & -\frac{1}{4} & -\frac{3}{4} & \frac{3}{4} \\ 0 & 0 & 0 & 1 & \frac{1}{4} & \frac{1}{2} & \frac{3}{4} & -\frac{1}{2} \end{array}\right]$$

Then $\mathbf{A}^{-1} = \left[\begin{array}{cccc} \frac{1}{4} & \frac{1}{4} & \frac{5}{4} & -\frac{1}{4} \\ \frac{1}{2} & \frac{5}{4} & \frac{7}{4} & -1 \\ -\frac{1}{4} & -\frac{1}{4} & -\frac{3}{4} & \frac{3}{4} \\ \frac{1}{4} & \frac{1}{2} & \frac{3}{4} & -\frac{1}{2} \end{array}\right]$, or

$$\left[\begin{array}{cccc} 0.25 & 0.25 & 1.25 & -0.25 \\ 0.5 & 1.25 & 1.75 & -1 \\ -0.25 & -0.25 & -0.75 & 0.75 \\ 0.25 & 0.5 & 0.75 & -0.5 \end{array}\right].$$

25. Write an equivalent matrix equation, $\mathbf{AX} = \mathbf{B}$.

$$\left[\begin{array}{cc} 11 & 3 \\ 7 & 2 \end{array}\right]\left[\begin{array}{c} x \\ y \end{array}\right] = \left[\begin{array}{c} -4 \\ 5 \end{array}\right]$$

Then we have $\mathbf{X} = \mathbf{A}^{-1}\mathbf{B}$.

$$\left[\begin{array}{c} x \\ y \end{array}\right] = \left[\begin{array}{cc} 2 & -3 \\ -7 & 11 \end{array}\right]\left[\begin{array}{c} -4 \\ 5 \end{array}\right] = \left[\begin{array}{c} -23 \\ 83 \end{array}\right]$$

The solution is $(-23, 83)$.

27. Write an equivalent matrix equation, $\mathbf{AX} = \mathbf{B}$.

$$\left[\begin{array}{ccc} 3 & 1 & 0 \\ 2 & -1 & 2 \\ 1 & 1 & 1 \end{array}\right]\left[\begin{array}{c} x \\ y \\ z \end{array}\right] = \left[\begin{array}{c} 2 \\ -5 \\ 5 \end{array}\right]$$

Then we have $\mathbf{X} = \mathbf{A}^{-1}\mathbf{B}$.

$$\begin{bmatrix} x \\ y \\ z \end{bmatrix} = \frac{1}{9} \begin{bmatrix} 3 & 1 & -2 \\ 0 & -3 & 6 \\ -3 & 2 & 5 \end{bmatrix} \begin{bmatrix} 2 \\ -5 \\ 5 \end{bmatrix} = \frac{1}{9} \begin{bmatrix} -9 \\ 45 \\ 9 \end{bmatrix} = \begin{bmatrix} -1 \\ 5 \\ 1 \end{bmatrix}$$

The solution is $(-1, 5, 1)$.

29. $4x + 3y = 2,$
$\quad\;\; x - 2y = 6$

Write an equivalent matrix equation, $\mathbf{AX} = \mathbf{B}$.

$$\begin{bmatrix} 4 & 3 \\ 1 & -2 \end{bmatrix} \begin{bmatrix} x \\ y \end{bmatrix} = \begin{bmatrix} 2 \\ 6 \end{bmatrix}$$

Then $\mathbf{X} = \mathbf{A}^{-1}\mathbf{B} = \begin{bmatrix} \dfrac{2}{11} & \dfrac{3}{11} \\ \dfrac{1}{11} & -\dfrac{4}{11} \end{bmatrix} \begin{bmatrix} 2 \\ 6 \end{bmatrix} = \begin{bmatrix} 2 \\ -2 \end{bmatrix}.$

The solution is $(2, -2)$.

31. $5x + \;\; y = 2,$
$\quad\; 3x - 2y = -4$

Write an equivalent matrix equation, $\mathbf{AX} = \mathbf{B}$.

$$\begin{bmatrix} 5 & 1 \\ 3 & -2 \end{bmatrix} \begin{bmatrix} x \\ y \end{bmatrix} = \begin{bmatrix} \dfrac{2}{13} & \dfrac{1}{13} \\ \dfrac{3}{13} & -\dfrac{5}{13} \end{bmatrix} \begin{bmatrix} 2 \\ -4 \end{bmatrix} = \begin{bmatrix} 2 \\ -4 \end{bmatrix}$$

Then $\mathbf{X} = \mathbf{A}^{-1}\mathbf{B} = \begin{bmatrix} 0 \\ 2 \end{bmatrix}.$

The solution is $(0, 2)$.

33. $\;\; x \quad\;\;\; + z = 1,$
$\;\; 2x + y \quad\;\;\; = 3,$
$\;\;\; x - y + z = 4$

Write an equivalent matrix equation, $\mathbf{AX} = \mathbf{B}$.

$$\begin{bmatrix} 1 & 0 & 1 \\ 2 & 1 & 0 \\ 1 & -1 & 1 \end{bmatrix} \begin{bmatrix} x \\ y \\ z \end{bmatrix} = \begin{bmatrix} 1 \\ 3 \\ 4 \end{bmatrix}$$

Then $\mathbf{X} = \mathbf{A}^{-1}\mathbf{B} = \begin{bmatrix} -\dfrac{1}{2} & \dfrac{1}{2} & \dfrac{1}{2} \\ 1 & 0 & -1 \\ \dfrac{3}{2} & -\dfrac{1}{2} & -\dfrac{1}{2} \end{bmatrix} \begin{bmatrix} 1 \\ 3 \\ 4 \end{bmatrix} = \begin{bmatrix} 3 \\ -3 \\ -2 \end{bmatrix}.$

The solution is $(3, -3, -2)$.

35. $2x + 3y + 4z = 2,$
$\quad\; x - 4y + 3z = 2,$
$\quad\; 5x + \;\; y + \;\; z = -4$

Write an equivalent matrix equation, $\mathbf{AX} = \mathbf{B}$.

$$\begin{bmatrix} 2 & 3 & 4 \\ 1 & -4 & 3 \\ 5 & 1 & 1 \end{bmatrix} \begin{bmatrix} x \\ y \\ z \end{bmatrix} = \begin{bmatrix} 2 \\ 2 \\ -4 \end{bmatrix}$$

Then $\mathbf{X} = \mathbf{A}^{-1}\mathbf{B} = \begin{bmatrix} -\dfrac{1}{16} & \dfrac{1}{112} & \dfrac{25}{112} \\ \dfrac{1}{8} & -\dfrac{9}{56} & -\dfrac{1}{56} \\ \dfrac{3}{16} & \dfrac{13}{112} & -\dfrac{11}{112} \end{bmatrix} \begin{bmatrix} 2 \\ 2 \\ -4 \end{bmatrix} = \begin{bmatrix} -1 \\ 0 \\ 1 \end{bmatrix}.$

The solution is $(-1, 0, 1)$.

37. $2w - 3x + 4y - 5z = 0,$
$\quad 3w - 2x + 7y - 3z = 2,$
$\quad\;\; w + \;\; x - \;\; y + \;\; z = 1,$
$\quad -w - 3x - 6y + 4z = 6$

Write an equivalent matrix equation, $\mathbf{AX} = \mathbf{B}$.

$$\begin{bmatrix} 2 & -3 & 4 & -5 \\ 3 & -2 & 7 & -3 \\ 1 & 1 & -1 & 1 \\ -1 & -3 & -6 & 4 \end{bmatrix} \begin{bmatrix} w \\ x \\ y \\ z \end{bmatrix} = \begin{bmatrix} 0 \\ 2 \\ 1 \\ 6 \end{bmatrix}$$

Then $\mathbf{X} = \mathbf{A}^{-1}\mathbf{B} = \dfrac{1}{203} \begin{bmatrix} 26 & 11 & 127 & 9 \\ -8 & -19 & 39 & -34 \\ -37 & 39 & -48 & -5 \\ -55 & 47 & -11 & 20 \end{bmatrix} \begin{bmatrix} 0 \\ 2 \\ 1 \\ 6 \end{bmatrix} = \begin{bmatrix} 1 \\ -1 \\ 0 \\ 1 \end{bmatrix}.$

The solution is $(1, -1, 0, 1)$.

39. *Familiarize.* Let $x =$ the number of hot dogs sold and $y =$ the number of sausages.

Translate.

The total number of items sold was 145.

$\quad x + y = 145$

The number of hot dogs sold is 45 more than the number of sausages.

$\quad x = y + 45$

We have a system of equations:

$\quad\; x + y = 145, \qquad x + y = 145,$
$\qquad\qquad\quad\; \text{or}$
$\quad\; x = y + 45, \qquad\; x - y = 45.$

Carry out. Write an equivalent matrix equation, $\mathbf{AX} = \mathbf{B}$.

$$\begin{bmatrix} 1 & 1 \\ 1 & -1 \end{bmatrix} \begin{bmatrix} x \\ y \end{bmatrix} = \begin{bmatrix} 145 \\ 45 \end{bmatrix}$$

Then $\mathbf{X} = \mathbf{A}^{-1}\mathbf{B} = \begin{bmatrix} \dfrac{1}{2} & \dfrac{1}{2} \\ \dfrac{1}{2} & -\dfrac{1}{2} \end{bmatrix} \begin{bmatrix} 145 \\ 45 \end{bmatrix} = \begin{bmatrix} 95 \\ 50 \end{bmatrix},$ so the

solution is $(95, 50)$.

Check. The total number of items is $95 + 50$, or 145. The number of hot dogs, 95, is 45 more than the number of sausages. The solution checks.

State. Stefan sold 95 hot dogs and 50 Italian sausages.

41. *Familiarize.* Let x, y, and z represent the prices of one ton of topsoil, mulch, and pea gravel, respectively.

Translate.

Four tons of topsoil, 3 tons of mulch, and 6 tons of pea gravel costs $2825.

$$4x + 3y + 6z = 2825$$

Five tons of topsoil, 2 tons of mulch, and 5 tons of pea gravel costs $2663.

$$5x + 2y + 5z = 2663$$

Pea gravel costs $17 less per ton than topsoil.

$$z = x - 17$$

We have a system of equations.

$$4x + 3y + 6z = 2825,$$
$$5x + 2y + 5z = 2663,$$
$$z = x - 17, \text{ or}$$

$$4x + 3y + 6z = 2825,$$
$$5x + 2y + 5z = 2663,$$
$$x \qquad - z = 17$$

Carry out. Write an equivalent matrix equation, $\mathbf{AX} = \mathbf{B}$.

$$\begin{bmatrix} 4 & 3 & 6 \\ 5 & 2 & 5 \\ 1 & 0 & -1 \end{bmatrix} \begin{bmatrix} x \\ y \\ z \end{bmatrix} = \begin{bmatrix} 2825 \\ 2663 \\ 17 \end{bmatrix}$$

Then $\mathbf{X} = \mathbf{A}^{-1}\mathbf{B} = \begin{bmatrix} -\dfrac{1}{5} & \dfrac{3}{10} & \dfrac{3}{10} \\ 1 & -1 & 1 \\ -\dfrac{1}{5} & \dfrac{3}{10} & -\dfrac{7}{10} \end{bmatrix} \begin{bmatrix} 2825 \\ 2663 \\ 17 \end{bmatrix} =$

$\begin{bmatrix} 239 \\ 179 \\ 222 \end{bmatrix}$, so the solution is $(239, 179, 222)$.

Check. Four tons of topsoil, 3 tons of mulch, and 6 tons of pea gravel costs $4 \cdot \$239 + 3 \cdot \$179 + 6 \cdot \$222$, or $\$956 + \$537 + \$1332$, or $\$2825$. Five tons of topsoil, 2 tons of mulch, and 5 tons of pea gravel costs $5 \cdot \$239 + 2 \cdot \$179 + 5 \cdot \$222$, or $\$1195 + \$358 + \$1110$, or $\$2663$. The price of pea gravel, $\$222$, is $\$17$ less than the price of topsoil, $\$239$. The solution checks.

State. The price of topsoil is $239 per ton, of mulch is $179 per ton, and of pea gravel is $222 per ton.

43. Left to the student

45. Left to the student

47. Discussion and Writing

49.

$$\begin{array}{r|rrrr} -2 & 1 & -6 & 4 & -8 \\ & & -2 & 16 & -40 \\ \hline & 1 & -8 & 20 & -48 \end{array}$$

$$f(-2) = -48$$

51.
$$2x^2 + x = 7$$
$$2x^2 + x - 7 = 0$$
$$a = 2, \, b = 1, \, c = -7$$
$$x = \frac{-b \pm \sqrt{b^2 - 4ac}}{2a}$$
$$= \frac{-1 \pm \sqrt{1^2 - 4 \cdot 2 \cdot (-7)}}{2 \cdot 2} = \frac{-1 \pm \sqrt{1 + 56}}{4}$$
$$= \frac{-1 \pm \sqrt{57}}{4}$$

The solutions are $\dfrac{-1 + \sqrt{57}}{4}$ and $\dfrac{-1 - \sqrt{57}}{4}$, or $\dfrac{-1 \pm \sqrt{57}}{4}$.

53.
$$\sqrt{2x + 1} - 1 = \sqrt{2x - 4}$$
$$(\sqrt{2x + 1} - 1)^2 = (\sqrt{2x - 4})^2 \quad \text{Squaring both sides}$$
$$2x + 1 - 2\sqrt{2x + 1} + 1 = 2x - 4$$
$$2x + 2 - 2\sqrt{2x + 1} = 2x - 4$$
$$2 - 2\sqrt{2x + 1} = -4 \quad \text{Subtracting } 2x$$
$$-2\sqrt{2x + 1} = -6 \quad \text{Subtracting } 2$$
$$\sqrt{2x + 1} = 3 \quad \text{Dividing by } -2$$
$$(\sqrt{2x + 1})^2 = 3^2 \quad \text{Squaring both sides}$$
$$2x + 1 = 9$$
$$2x = 8$$
$$x = 4$$

The number 4 checks. It is the solution.

55. $f(x) = x^3 - 3x^2 - 6x + 8$

We use synthetic division to find one factor. We first try $x - 1$.

$$\begin{array}{r|rrrr} 1 & 1 & -3 & -6 & 8 \\ & & 1 & -2 & -8 \\ \hline & 1 & -2 & -8 & 0 \end{array}$$

Since $f(1) = 0$, $x - 1$ is a factor of $f(x)$. We have $f(x) = (x-1)(x^2 - 2x - 8)$. Factoring the trinomial we get $f(x) = (x - 1)(x - 4)(x + 2)$.

57. $\mathbf{A} = [x]$

Write the augmented matrix.

$$[\, x \mid 1 \,]$$

Multiply by $\dfrac{1}{x}$.

$$\left[\, 1 \mid \dfrac{1}{x} \,\right]$$

Then $\mathbf{A}^{-1}$ exists if and only if $x \neq 0$. $\mathbf{A}^{-1} = \left[\dfrac{1}{x}\right]$.

59. $\mathbf{A} = \begin{bmatrix} 0 & 0 & x \\ 0 & y & 0 \\ z & 0 & 0 \end{bmatrix}$

Write the augmented matrix.

$$\left[\begin{array}{ccc|ccc} 0 & 0 & x & 1 & 0 & 0 \\ 0 & y & 0 & 0 & 1 & 0 \\ z & 0 & 0 & 0 & 0 & 1 \end{array}\right]$$

Interchange row 1 and row 3.

$$\left[\begin{array}{ccc|ccc} z & 0 & 0 & 0 & 0 & 1 \\ 0 & y & 0 & 0 & 1 & 0 \\ 0 & 0 & x & 1 & 0 & 0 \end{array}\right]$$

Multiply row 1 by $\dfrac{1}{z}$, row 2 by $\dfrac{1}{y}$, and row 3 by $\dfrac{1}{x}$.

$$\left[\begin{array}{ccc|ccc} 1 & 0 & 0 & 0 & 0 & \frac{1}{z} \\ 0 & 1 & 0 & 0 & \frac{1}{y} & 0 \\ 0 & 0 & 1 & \frac{1}{x} & 0 & 0 \end{array}\right]$$

Then $\mathbf{A}^{-1}$ exists if and only if $x \neq 0$ and $y \neq 0$ and $z \neq 0$, or if and only if $xyz \neq 0$.

$$\mathbf{A}^{-1} = \begin{bmatrix} 0 & 0 & \frac{1}{z} \\ 0 & \frac{1}{y} & 0 \\ \frac{1}{x} & 0 & 0 \end{bmatrix}$$

Exercise Set 5.6

1. $\begin{vmatrix} 5 & 3 \\ -2 & -4 \end{vmatrix} = 5(-4) - (-2) \cdot 3 = -20 + 6 = -14$

3. $\begin{vmatrix} 4 & -7 \\ -2 & 3 \end{vmatrix} = 4 \cdot 3 - (-2)(-7) = 12 - 14 = -2$

5. $\begin{vmatrix} -2 & -\sqrt{5} \\ -\sqrt{5} & 3 \end{vmatrix} = -2 \cdot 3 - (-\sqrt{5})(-\sqrt{5}) = -6 - 5 = -11$

7. $\begin{vmatrix} x & 4 \\ x & x^2 \end{vmatrix} = x \cdot x^2 - x \cdot 4 = x^3 - 4x$

9. $\mathbf{A} = \begin{bmatrix} 7 & -4 & -6 \\ 2 & 0 & -3 \\ 1 & 2 & -5 \end{bmatrix}$

M_{11} is the determinant of the matrix formed by deleting the first row and first column of $\mathbf{A}$:

$M_{11} = \begin{vmatrix} 0 & -3 \\ 2 & -5 \end{vmatrix} = 0(-5) - 2(-3) = 0 + 6 = 6$

M_{32} is the determinant of the matrix formed by deleting the third row and second column of $\mathbf{A}$:

$M_{32} = \begin{vmatrix} 7 & -6 \\ 2 & -3 \end{vmatrix} = 7(-3) - 2(-6) = -21 + 12 = -9$

M_{22} is the determinant of the matrix formed by deleting the second row and second column of $\mathbf{A}$:

$M_{22} = \begin{vmatrix} 7 & -6 \\ 1 & -5 \end{vmatrix} = 7(-5) - 1(-6) = -35 + 6 = -29$

11. In Exercise 9 we found that $M_{11} = 6$.

$A_{11} = (-1)^{1+1} M_{11} = 1 \cdot 6 = 6$

In Exercise 9 we found that $M_{32} = -9$.

$A_{32} = (-1)^{3+2} M_{32} = -1(-9) = 9$

In Exercise 9 we found that $M_{22} = -29$.

$A_{22} = (-1)^{2+2}(-29) = 1(-29) = -29$

13. $\mathbf{A} = \begin{bmatrix} 7 & -4 & -6 \\ 2 & 0 & -3 \\ 1 & 2 & -5 \end{bmatrix}$

$|\mathbf{A}|$

$= 2A_{21} + 0A_{22} + (-3)A_{23}$

$= 2(-1)^{2+1} \begin{vmatrix} -4 & -6 \\ 2 & -5 \end{vmatrix} + 0 + (-3)(-1)^{2+3} \begin{vmatrix} 7 & -4 \\ 1 & 2 \end{vmatrix}$

$= 2(-1)[-4(-5) - 2(-6)] + 0 +$
$\qquad (-3)(-1)[7 \cdot 2 - 1(-4)]$

$= -2(32) + 0 + 3(18) = -64 + 0 + 54$

$= -10$

15. $\mathbf{A} = \begin{bmatrix} 7 & -4 & -6 \\ 2 & 0 & -3 \\ 1 & 2 & -5 \end{bmatrix}$

$|\mathbf{A}|$

$= -6A_{13} + (-3)A_{23} + (-5)A_{33}$

$= -6(-1)^{1+3} \begin{vmatrix} 2 & 0 \\ 1 & 2 \end{vmatrix} + (-3)(-1)^{2+3} \begin{vmatrix} 7 & -4 \\ 1 & 2 \end{vmatrix} +$

$\qquad (-5)(-1)^{3+3} \begin{vmatrix} 7 & -4 \\ 2 & 0 \end{vmatrix}$

$= -6 \cdot 1(2 \cdot 2 - 1 \cdot 0) + (-3)(-1)[7 \cdot 2 - 1(-4)] +$
$\qquad -5 \cdot 1(7 \cdot 0 - 2(-4))$

$= -6(4) + 3(18) - 5(8) = -24 + 54 - 40$

$= -10$

17. $\mathbf{A} = \begin{bmatrix} 1 & 0 & 0 & -2 \\ 4 & 1 & 0 & 0 \\ 5 & 6 & 7 & 8 \\ -2 & -3 & -1 & 0 \end{bmatrix}$

M_{41} is the determinant of the matrix formed by deleting the fourth row and the first column of $\mathbf{A}$.

$M_{41} = \begin{bmatrix} 0 & 0 & -2 \\ 1 & 0 & 0 \\ 6 & 7 & 8 \end{bmatrix}$

We will expand M_{41} across the first row.

$M_{41} = 0(-1)^{1+1} \begin{vmatrix} 0 & 0 \\ 7 & 8 \end{vmatrix} + 0(-1)^{1+2} \begin{vmatrix} 1 & 0 \\ 6 & 8 \end{vmatrix} +$

$\qquad (-2)(-1)^{3+1} \begin{vmatrix} 1 & 0 \\ 6 & 7 \end{vmatrix}$

$\qquad = 0 + 0 + (-2)(1)(1 \cdot 7 - 6 \cdot 0)$

$\qquad = 0 + 0 - 2(7) = -14$

M_{33} is the determinant of the matrix formed by deleting the third row and the third column of $\mathbf{A}$.

$M_{33} = \begin{bmatrix} 1 & 0 & -2 \\ 4 & 1 & 0 \\ -2 & -3 & 0 \end{bmatrix}$

We will expand M_{33} down the third column.

$$M_{33} = -2(-1)^{1+3}\begin{vmatrix} 4 & 1 \\ -2 & -3 \end{vmatrix} + 0(-1)^{2+3}\begin{vmatrix} 1 & 0 \\ -2 & -3 \end{vmatrix} +$$

$$0(-1)^{3+3}\begin{vmatrix} 1 & 0 \\ 4 & 1 \end{vmatrix}$$

$$= -2(1)[4(-3) - (-2)(1)] + 0 + 0$$

$$= -2(-10) + 0 + 0 = 20$$

19. $\mathbf{A} = \begin{bmatrix} 1 & 0 & 0 & -2 \\ 4 & 1 & 0 & 0 \\ 5 & 6 & 7 & 8 \\ -2 & -3 & -1 & 0 \end{bmatrix}$

$$A_{24} = (-1)^{2+4}M_{24} = 1 \cdot M_{24} = M_{24}$$

$$= \begin{vmatrix} 1 & 0 & 0 \\ 5 & 6 & 7 \\ -2 & -3 & -1 \end{vmatrix}$$

We will expand across the first row.

$$\begin{vmatrix} 1 & 0 & 0 \\ 5 & 6 & 7 \\ -2 & -3 & -1 \end{vmatrix}$$

$$= 1(-1)^{1+1}\begin{vmatrix} 6 & 7 \\ -3 & -1 \end{vmatrix} + 0(-1)^{1+2}\begin{vmatrix} 5 & 7 \\ -2 & -1 \end{vmatrix} +$$

$$0(-1)^{1+3}\begin{vmatrix} 5 & 6 \\ -2 & -3 \end{vmatrix}$$

$$= 1 \cdot 1[6(-1) - (-3)(7)] + 0 + 0$$

$$= 1(15) = 15$$

$$A_{43} = (-1)^{4+3}M_{43} = -1 \cdot M_{43}$$

$$= -1 \cdot \begin{vmatrix} 1 & 0 & -2 \\ 4 & 1 & 0 \\ 5 & 6 & 8 \end{vmatrix}$$

We will expand across the first row.

$$-1 \cdot \begin{vmatrix} 1 & 0 & -2 \\ 4 & 1 & 0 \\ 5 & 6 & 8 \end{vmatrix}$$

$$= -1\left[1(-1)^{1+1}\begin{vmatrix} 1 & 0 \\ 6 & 8 \end{vmatrix} + 0(-1)^{1+2}\begin{vmatrix} 4 & 0 \\ 5 & 8 \end{vmatrix} +\right.$$

$$\left.(-2)(-1)^{1+3}\begin{vmatrix} 4 & 1 \\ 5 & 6 \end{vmatrix}\right]$$

$$= -1[1 \cdot 1(1 \cdot 8 - 6 \cdot 0) + 0 + (-2) \cdot 1(4 \cdot 6 - 5 \cdot 1)]$$

$$= -1[1(8) + 0 - 2(19)]$$

$$= -1(8 - 38) = -1(-30)$$

$$= 30$$

21. $\mathbf{A} = \begin{bmatrix} 1 & 0 & 0 & -2 \\ 4 & 1 & 0 & 0 \\ 5 & 6 & 7 & 8 \\ -2 & -3 & -1 & 0 \end{bmatrix}$

$$|\mathbf{A}|$$

$$= 1 \cdot A_{11} + 0 \cdot A_{12} + 0 \cdot A_{13} + (-2)A_{14}$$

$$= A_{11} + (-2)A_{14}$$

$$= (-1)^{1+1}\begin{vmatrix} 1 & 0 & 0 \\ 6 & 7 & 8 \\ -3 & -1 & 0 \end{vmatrix} +$$

$$(-2)(-1)^{1+4}\begin{vmatrix} 4 & 1 & 0 \\ 5 & 6 & 7 \\ -2 & -3 & -1 \end{vmatrix}$$

$$= \begin{vmatrix} 1 & 0 & 0 \\ 6 & 7 & 8 \\ -3 & -1 & 0 \end{vmatrix} + 2\begin{vmatrix} 4 & 1 & 0 \\ 5 & 6 & 7 \\ -2 & -3 & -1 \end{vmatrix}$$

We will expand each determinant across the first row. We have:

$$1(-1)^{1+1}\begin{vmatrix} 7 & 8 \\ -1 & 0 \end{vmatrix} + 0 + 0 +$$

$$2\left[4(-1)^{1+1}\begin{vmatrix} 6 & 7 \\ -3 & -1 \end{vmatrix} + 1(-1)^{1+2}\begin{vmatrix} 5 & 7 \\ -2 & -1 \end{vmatrix} + 0\right]$$

$$= 1 \cdot 1[7 \cdot 0 - (-1)8] + 2[4 \cdot 1[6(-1) - (-3) \cdot 7] +$$

$$1(-1)[5(-1) - (-2) \cdot 7]$$

$$= 1(8) + 2[4(15) - 1(9)] = 8 + 2(51)$$

$$= 8 + 102 = 110$$

23. We will expand across the first row. We could have chosen any other row or column just as well.

$$\begin{vmatrix} 3 & 1 & 2 \\ -2 & 3 & 1 \\ 3 & 4 & -6 \end{vmatrix}$$

$$= 3(-1)^{1+1}\begin{vmatrix} 3 & 1 \\ 4 & -6 \end{vmatrix} + 1 \cdot (-1)^{1+2}\begin{vmatrix} -2 & 1 \\ 3 & -6 \end{vmatrix} +$$

$$2(-1)^{1+3}\begin{vmatrix} -2 & 3 \\ 3 & 4 \end{vmatrix}$$

$$= 3 \cdot 1[3(-6) - 4 \cdot 1] + 1(-1)[-2(-6) - 3 \cdot 1] +$$

$$2 \cdot 1(-2 \cdot 4 - 3 \cdot 3)$$

$$= 3(-22) - (9) + 2(-17)$$

$$= -109$$

25. We will expand down the second column. We could have chosen any other row or column just as well.

$$\begin{vmatrix} x & 0 & -1 \\ 2 & x & x^2 \\ -3 & x & 1 \end{vmatrix}$$

$$= 0(-1)^{1+2}\begin{vmatrix} 2 & x^2 \\ -3 & 1 \end{vmatrix} + x(-1)^{2+2}\begin{vmatrix} x & -1 \\ -3 & 1 \end{vmatrix} +$$

$$\qquad x(-1)^{3+2}\begin{vmatrix} x & -1 \\ 2 & x^2 \end{vmatrix}$$

$$= 0(-1)[2\cdot 1-(-3)x^2]+x\cdot 1[x\cdot 1-(-3)(-1)]+$$
$$\qquad x(-1)[x\cdot x^2-2(-1)]$$

$$= 0+x(x-3)-x(x^3+2)$$

$$= x^2-3x-x^4-2x=-x^4+x^2-5x$$

27. $-2x+4y=3,$
$\quad 3x-7y=1$

$$D = \begin{vmatrix} -2 & 4 \\ 3 & -7 \end{vmatrix} = -2(-7)-3\cdot 4 = 14-12 = 2$$

$$D_x = \begin{vmatrix} 3 & 4 \\ 1 & -7 \end{vmatrix} = 3(-7)-1\cdot 4 = -21-4 = -25$$

$$D_y = \begin{vmatrix} -2 & 3 \\ 3 & 1 \end{vmatrix} = -2\cdot 1-3\cdot 3 = -2-9 = -11$$

$$x = \frac{D_x}{D} = \frac{-25}{2} = -\frac{25}{2}$$

$$y = \frac{D_y}{D} = \frac{-11}{2} = -\frac{11}{2}$$

The solution is $\left(-\dfrac{25}{2}, -\dfrac{11}{2}\right)$.

29. $2x-\ y=5,$
$\quad x-2y=1$

$$D = \begin{vmatrix} 2 & -1 \\ 1 & -2 \end{vmatrix} = 2(-2)-1(-1) = -4+1 = -3$$

$$D_x = \begin{vmatrix} 5 & -1 \\ 1 & -2 \end{vmatrix} = 5(-2)-1(-1) = -10+1 = -9$$

$$D_y = \begin{vmatrix} 2 & 5 \\ 1 & 1 \end{vmatrix} = 2\cdot 1-1\cdot 5 = 2-5 = -3$$

$$x = \frac{D_x}{D} = \frac{-9}{-3} = 3$$

$$y = \frac{D_y}{D} = \frac{-3}{-3} = 1$$

The solution is $(3,1)$.

31. $2x+9y=-2,$
$\quad 4x-3y=3$

$$D = \begin{vmatrix} 2 & 9 \\ 4 & -3 \end{vmatrix} = 2(-3)-4\cdot 9 = -6-36 = -42$$

$$D_x = \begin{vmatrix} -2 & 9 \\ 3 & -3 \end{vmatrix} = -2(-3)-3\cdot 9 = 6-27 = -21$$

$$D_y = \begin{vmatrix} 2 & -2 \\ 4 & 3 \end{vmatrix} = 2\cdot 3-4(-2) = 6+8 = 14$$

$$x = \frac{D_x}{D} = \frac{-21}{-42} = \frac{1}{2}$$

$$y = \frac{D_y}{D} = \frac{14}{-42} = -\frac{1}{3}$$

The solution is $\left(\dfrac{1}{2}, -\dfrac{1}{3}\right)$.

33. $2x+5y=7,$
$\quad 3x-2y=1$

$$D = \begin{vmatrix} 2 & 5 \\ 3 & -2 \end{vmatrix} = 2(-2)-3\cdot 5 = -4-15 = -19$$

$$D_x = \begin{vmatrix} 7 & 5 \\ 1 & -2 \end{vmatrix} = 7(-2)-1\cdot 5 = -14-5 = -19$$

$$D_y = \begin{vmatrix} 2 & 7 \\ 3 & 1 \end{vmatrix} = 2\cdot 1-3\cdot 7 = 2-21 = -19$$

$$x = \frac{D_x}{D} = \frac{-19}{-19} = 1$$

$$y = \frac{D_y}{D} = \frac{-19}{-19} = 1$$

The solution is $(1,1)$.

35. $3x+2y-z=4,$
$\quad 3x-2y+z=5,$
$\quad 4x-5y-z=-1$

$$D = \begin{vmatrix} 3 & 2 & -1 \\ 3 & -2 & 1 \\ 4 & -5 & -1 \end{vmatrix} = 42$$

$$D_x = \begin{vmatrix} 4 & 2 & -1 \\ 5 & -2 & 1 \\ -1 & -5 & -1 \end{vmatrix} = 63$$

$$D_y = \begin{vmatrix} 3 & 4 & -1 \\ 3 & 5 & 1 \\ 4 & -1 & -1 \end{vmatrix} = 39$$

$$D_z = \begin{vmatrix} 3 & 2 & 4 \\ 3 & -2 & 5 \\ 4 & -5 & -1 \end{vmatrix} = 99$$

$$x = \frac{D_x}{D} = \frac{63}{42} = \frac{3}{2}$$

$$y = \frac{D_y}{D} = \frac{39}{42} = \frac{13}{14}$$

$$z = \frac{D_z}{D} = \frac{99}{42} = \frac{33}{14}$$

The solution is $\left(\dfrac{3}{2}, \dfrac{13}{14}, \dfrac{33}{14}\right)$.

(Note that we could have used Cramer's rule to find only two of the values and then used substitution to find the remaining value.)

37. $3x + 5y - z = -2,$
$\quad x - 4y + 2z = 13,$
$\quad 2x + 4y + 3z = 1$

$$D = \begin{vmatrix} 3 & 5 & -1 \\ 1 & -4 & 2 \\ 2 & 4 & 3 \end{vmatrix} = -67$$

$$D_x = \begin{vmatrix} -2 & 5 & -1 \\ 13 & -4 & 2 \\ 1 & 4 & 3 \end{vmatrix} = -201$$

$$D_y = \begin{vmatrix} 3 & -2 & -1 \\ 1 & 13 & 2 \\ 2 & 1 & 3 \end{vmatrix} = 134$$

$$D_z = \begin{vmatrix} 3 & 5 & -2 \\ 1 & -4 & 13 \\ 2 & 4 & 1 \end{vmatrix} = -67$$

$$x = \frac{D_x}{D} = \frac{-201}{-67} = 3$$

$$y = \frac{D_y}{D} = \frac{134}{-67} = -2$$

$$z = \frac{D_z}{D} = \frac{-67}{-67} = 1$$

The solution is $(3, -2, 1)$.

(Note that we could have used Cramer's rule to find only two of the values and then used substitution to find the remaining value.)

39. $x - 3y - 7z = 6,$
$\quad 2x + 3y + z = 9,$
$\quad 4x + y = 7$

$$D = \begin{vmatrix} 1 & -3 & -7 \\ 2 & 3 & 1 \\ 4 & 1 & 0 \end{vmatrix} = 57$$

$$D_x = \begin{vmatrix} 6 & -3 & -7 \\ 9 & 3 & 1 \\ 7 & 1 & 0 \end{vmatrix} = 57$$

$$D_y = \begin{vmatrix} 1 & 6 & -7 \\ 2 & 9 & 1 \\ 4 & 7 & 0 \end{vmatrix} = 171$$

$$D_z = \begin{vmatrix} 1 & -3 & 6 \\ 2 & 3 & 9 \\ 4 & 1 & 7 \end{vmatrix} = -114$$

$$x = \frac{D_x}{D} = \frac{57}{57} = 1$$

$$y = \frac{D_y}{D} = \frac{171}{57} = 3$$

$$z = \frac{D_z}{D} = \frac{-114}{57} = -2$$

The solution is $(1, 3, -2)$.

(Note that we could have used Cramer's rule to find only two of the values and then used substitution to find the remaining value.)

41. $\qquad 6y + 6z = -1,$
$\quad 8x \qquad + 6z = -1,$
$\quad 4x + 9y \qquad = 8$

$$D = \begin{vmatrix} 0 & 6 & 6 \\ 8 & 0 & 6 \\ 4 & 9 & 0 \end{vmatrix} = 576$$

$$D_x = \begin{vmatrix} -1 & 6 & 6 \\ -1 & 0 & 6 \\ 8 & 9 & 0 \end{vmatrix} = 288$$

$$D_y = \begin{vmatrix} 0 & -1 & 6 \\ 8 & -1 & 6 \\ 4 & 8 & 0 \end{vmatrix} = 384$$

$$D_z = \begin{vmatrix} 0 & 6 & -1 \\ 8 & 0 & -1 \\ 4 & 9 & 8 \end{vmatrix} = -480$$

$$x = \frac{D_x}{D} = \frac{288}{576} = \frac{1}{2}$$

$$y = \frac{D_y}{D} = \frac{384}{576} = \frac{2}{3}$$

$$z = \frac{D_z}{D} = \frac{-480}{576} = -\frac{5}{6}$$

The solution is $\left(\frac{1}{2}, \frac{2}{3}, -\frac{5}{6}\right)$.

(Note that we could have used Cramer's rule to find only two of the values and then used substitution to find the remaining value.)

43. Left to the student

45. Enter matrix **A** and use the "det(" option from the MATRIX MATH menu.

$|\mathbf{A}| = 110$

47. Left to the student

49. Discussion and Writing

51. The graph of $f(x) = 3x + 2$ is shown below. Since it passes the horizontal-line test, the function is one-to-one.

We find a formula for $f^{-1}(x)$.

Replace $f(x)$ with y: $y = 3x + 2$

Interchange x and y: $x = 3y + 2$

Solve for y: $y = \dfrac{x - 2}{3}$

Replace y with $f^{-1}(x)$: $f^{-1}(x) = \dfrac{x - 2}{3}$

53. The graph of $f(x) = |x| + 3$ is shown below. It fails the horizontal-line test, so it is not one-to-one.

55. $(3 - 4i) - (-2 - i) = 3 - 4i + 2 + i =$
$(3 + 2) + (-4 + 1)i = 5 - 3i$

57. $(1 - 2i)(6 + 2i) = 6 + 2i - 12i - 4i^2 =$
$6 + 2i - 12i + 4 = 10 - 10i$

59. $\begin{vmatrix} x & 5 \\ -4 & x \end{vmatrix} = 24$

$x \cdot x - (-4)(5) = 24$ Evaluating the determinant

$x^2 + 20 = 24$

$x^2 = 4$

$x = \pm 2$

The solutions are -2 and 2.

61. $\begin{vmatrix} x & -3 \\ -1 & x \end{vmatrix} \geq 0$

$x \cdot x - (-1)(-3) \geq 0$

$x^2 - 3 \geq 0$

We solve the related equation.

$x^2 - 3 = 0$

$x^2 = 3$

$x = \pm\sqrt{3}$

The numbers $-\sqrt{3}$ and $\sqrt{3}$ divide the x-axis into three intervals. We let $f(x) = x^2 - 3$ and test a value in each interval.

$(-\infty, -\sqrt{3}): \ f(-2) = (-2)^2 - 3 = 1 > 0$

$(-\sqrt{3}, \sqrt{3}): \ f(0) = 0^2 - 3 = -3 < 0$

$(\sqrt{3}, \infty): \ f(2) = 2^2 - 3 = 1 > 0$

The function is positive in $(-\infty, -\sqrt{3})$, and $(\sqrt{3}, \infty)$. We also include the endpoints of the intervals since the inequality symbol is $\geq$. The solution set is $\{x | x \leq -\sqrt{3} \text{ or } x \geq \sqrt{3}\}$, or $(-\infty, -\sqrt{3}] \cup [\sqrt{3}, \infty)$.

63. $\begin{vmatrix} x + 3 & 4 \\ x - 3 & 5 \end{vmatrix} = -7$

$(x + 3)(5) - (x - 3)(4) = -7$

$5x + 15 - 4x + 12 = -7$

$x + 27 = -7$

$x = -34$

The solution is -34.

65. $\begin{vmatrix} 2 & x & 1 \\ 1 & 2 & -1 \\ 3 & 4 & -2 \end{vmatrix} = -6$

$-x - 2 = -6$ Evaluating the determinant

$-x = -4$

$x = 4$

The solution is 4.

67. Answers may vary.

$\begin{vmatrix} L & -W \\ 2 & 2 \end{vmatrix}$

69. Answers may vary.

$\begin{vmatrix} a & b \\ -b & a \end{vmatrix}$

71. Answers may vary.

$\begin{vmatrix} 2\pi r & 2\pi r \\ -h & r \end{vmatrix}$

Exercise Set 5.7

1. Graph (f) is the graph of $y > x$.

3. Graph (h) is the graph of $y \leq x - 3$.

5. Graph (g) is the graph of $2x + y < 4$.

7. Graph (b) is the graph of $2x - 5y > 10$.

9. Graph: $y > 2x$

1. We first graph the related equation $y = 2x$. We draw the line dashed since the inequality symbol is $>$.

2. To determine which half-plane to shade, test a point not on the line. We try $(1, 1)$ and substitute:

$$\begin{array}{c} y > 2x \\ \hline 1 \ ? \ 2 \cdot 1 \\ \hline 1 \ | \ 2 \quad \text{FALSE} \end{array}$$

Since $1 > 2$ is false, $(1, 1)$ is not a solution, nor are any points in the half-plane containing $(1, 1)$. The points in the opposite half-plane are solutions, so we shade that half-plane and obtain the graph.

11. Graph: $y + x \geq 0$

1. First graph the related equation $y + x = 0$. Draw the line solid since the inequality is $\geq$.

2. Next determine which half-plane to shade by testing a point not on the line. Here we use $(2, 2)$ as a check.

$$\frac{y + x \geq 0}{2 + 2 \ ? \ 0}$$
$$4 \ \big| \ 0 \quad \text{TRUE}$$

Since $4 \geq 0$ is true, $(2, 2)$ is a solution. Thus shade the half-plane containing $(2, 2)$.

13. Graph: $y > x - 3$

1. We first graph the related equation $y = x - 3$. Draw the line dashed since the inequality symbol is $>$.

2. To determine which half-plane to shade, test a point not on the line. We try $(0, 0)$.

$$\frac{y > x - 3}{0 \ ? \ 0 - 3}$$
$$0 \ \big| \ -3 \quad \text{TRUE}$$

Since $0 > -3$ is true, $(0, 0)$ is a solution. Thus we shade the half-plane containing $(0, 0)$.

15. Graph: $x + y < 4$

1. First graph the related equation $x + y = 4$. Draw the line dashed since the inequality is $<$.

2. To determine which half-plane to shade, test a point not on the line. We try $(0, 0)$.

$$\frac{x + y < 4}{0 + 0 \ ? \ 4}$$
$$0 \ \big| \ 4 \quad \text{TRUE}$$

Since $0 < 4$ is true, $(0, 0)$ is a solution. Thus shade the half-plane containing $(0, 0)$.

17. Graph: $3x - 2y \leq 6$

1. First graph the related equation $3x - 2y = 6$. Draw the line solid since the inequality is $\leq$.

2. To determine which half-plane to shade, test a point not on the line. We try $(0, 0)$.

$$\frac{3x - 2y \leq 6}{3(0) - 2(0) \ ? \ 6}$$
$$0 \ \big| \ 6 \quad \text{TRUE}$$

Since $0 \leq 6$ is true, $(0, 0)$ is a solution. Thus shade the half-plane containing $(0, 0)$.

19. Graph: $3y + 2x \geq 6$

1. First graph the related equation $3y + 2x = 6$. Draw the line solid since the inequality is $\geq$.

2. To determine which half-plane to shade, test a point not on the line. We try $(0, 0)$.

$$\frac{3y + 2x \geq 6}{3 \cdot 0 + 2 \cdot 0 \ ? \ 6}$$
$$0 \ \big| \ 6 \quad \text{FALSE}$$

Since $0 \geq 6$ is false, $(0, 0)$ is not a solution. We shade the half-plane which does not contain $(0, 0)$.

$3y + 2x \geq 6$

21. Graph: $3x - 2 \leq 5x + y$

$$-2 \leq 2x + y \quad \text{Adding } -3x$$

1. First graph the related equation $2x + y = -2$. Draw the line solid since the inequality is $\leq$.

2. To determine which half-plane to shade, test a point not on the line. We try $(0, 0)$.

$$\frac{2x + y \geq -2}{2(0) + 0 \; ? \; -2}$$
$$0 \; \Big| \; -2 \quad \text{TRUE}$$

Since $0 \geq -2$ is true, $(0, 0)$ is a solution. Thus shade the half-plane containing the origin.

$3x - 2 \leq 5x + y$

23. Graph: $x < -4$

1. We first graph the related equation $x = -4$. Draw the line dashed since the inequality is $<$.

2. To determine which half-plane to shade, test a point not on the line. We try $(0, 0)$.

$$\frac{x < -4}{0 \; ? \; -4 \quad \text{FALSE}}$$

Since $0 < -4$ is false, $(0, 0)$ is not a solution. Thus, we shade the half-plane which does not contain the origin.

$x < -4$

25. Graph: $y > -3$

1. First we graph the related equation $y = -3$. Draw the line dashed since the inequality is $>$.

2. To determine which half-plane to shade we test a point not on the line. We try $(0, 0)$.

$$\frac{y > -3}{0 \; ? \; -3 \quad \text{TRUE}}$$

Since $0 > -3$ is true, $(0, 0)$ is a solution. We shade the half-plane containing $(0, 0)$.

$y > -3$

27. Graph: $-4 < y < -1$

This is a conjunction of two inequalities

$$-4 < y \quad \text{and} \quad y < -1.$$

We can graph $-4 < y$ and $y < -1$ separately and then graph the intersection, or region in both solution sets.

29. Graph: $y \geq |x|$

1. Graph the related equation $y = |x|$. Draw the line solid since the inequality symbol is $\geq$.

2. To determine the region to shade, observe that the solution set consists of all ordered pairs (x, y) where the second coordinate is greater than or equal to the absolute value of the first coordinate. We see that the solutions are the points on or above the graph of $y = |x|$.

31. Graph (f) is the correct graph.

33. Graph (a) is the correct graph.

35. Graph (b) is the correct graph.

37. First we find the related equations. One line goes through $(0, 4)$ and $(4, 0)$. We find its slope:

$$m = \frac{0 - 4}{4 - 0} = \frac{-4}{4} = -1$$

This line has slope -1 and y-intercept $(0, 4)$, so its equation is $y = -x + 4$.

The other line goes through $(0, 0)$ and $(1, 3)$. We find the slope.

$$m = \frac{3 - 0}{1 - 0} = 3$$

This line has slope 3 and y-intercept $(0, 0)$, so its equation is $y = 3x + 0$, or $y = 3x$.

Observing the shading on the graph and the fact that the lines are solid, we can write the system of inequalities as

$$y \leq -x + 4,$$
$$y \leq 3x.$$

Answers may vary.

39. The equation of the vertical line is $x = 2$ and the equation of the horizontal line is $y = -1$. The lines are dashed and the shaded area is to the left of the vertical line and above the horizontal line, so the system of inequalities can be written

$$x < 2,$$
$$y > -1.$$

41. First we find the related equations. One line goes through $(0, 3)$ and $(3, 0)$. We find its slope:

$$m = \frac{0 - 3}{3 - 0} = \frac{-3}{3} = -1$$

This line has slope -1 and y-intercept $(0, 3)$, so its equation is $y = -x + 3$.

The other line goes through $(0, 1)$ and $(1, 2)$. We find its slope:

$$m = \frac{2 - 1}{1 - 0} = \frac{1}{1} = 1$$

This line has slope 1 and y-intercept $(0, 1)$, so its equation is $y = x + 1$.

Observe that both lines are solid and that the shading lies below both lines, to the right of the y-axis, and above the x-axis. We can write this system of inequalities as

$$y \leq -x + 3,$$
$$y \leq x + 1,$$
$$x \geq 0,$$
$$y \geq 0.$$

43.   Graph: $y \leq x,$
$$y \geq 3 - x$$

We graph the related equations $y = x$ and $y = 3 - x$ using solid lines. The arrows at the ends of the lines indicate the half-plane containing the solution set for each inequality. Shade the region common to both solution sets.

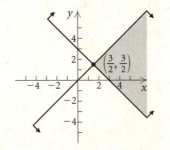

We find the vertex $\left(\dfrac{3}{2}, \dfrac{3}{2}\right)$ by solving the system

$$y = x,$$
$$y = 3 - x.$$

45.   Graph: $y \geq x,$
$$y \leq x - 4$$

We graph the related equations $y = x$ and $y = x - 4$ using solid lines. The arrows at the ends of the lines indicate the half-plane containing the solution set for each inequality. There is not a region common to both solution sets, so there are no vertices.

47. Graph: $y \geq -3$,

$\qquad x \geq 1$

We graph the related equations $y = -3$ and $x = 1$ using solid lines. The arrows at the ends of the lines indicate the half-plane containing the solution set for each inequality. Shade the region common to both solution sets.

We find the vertex $(1, -3)$ by solving the system

$\qquad y = -3$,

$\qquad x = 1$.

49. Graph: $x \leq 3$,

$\qquad y \geq 2 - 3x$

We graph the related equations $x = 3$ and $y = 2 - 3x$ using solid lines. The arrows at the ends of the lines indicate the half-plane containing the solution set for each inequality. Shade the region common to both solution sets.

We find the vertex $(3, -7)$ by solving the system

$\qquad x = 3$,

$\qquad y = 2 - 3x$.

51. Graph: $x + y \leq 1$,

$\qquad x - y \leq 2$

We graph the related equations $x + y = 1$ and $x - y = 2$ using solid lines. The arrows at the ends of the lines

indicate the half-plane containing the solution set for each inequality. Shade the region common to both solution sets.

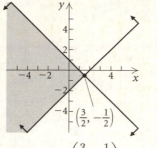

We find the vertex $\left(\dfrac{3}{2}, -\dfrac{1}{2}\right)$ by solving the system

$\qquad x + y = 1$,

$\qquad x - y = 2$.

53. Graph: $2y - x \leq 2$,

$\qquad y + 3x \geq -1$

We graph the related equations $2y - x = 2$ and $y + 3x = -1$ using solid lines. The arrows at the ends of the lines indicate the half-plane containing the solution set for each inequality. Shade the region common to both solution sets.

We find the vertex $\left(-\dfrac{4}{7}, \dfrac{5}{7}\right)$ by solving the system

$\qquad 2y - x = 2$,

$\qquad y + 3x = -1$.

55. Graph: $x - y \leq 2$,

$\qquad x + 2y \geq 8$,

$\qquad y \leq 4$

We graph the related equations $x - y = 2$, $x + 2y = 8$, and $y = 4$ using solid lines. The arrows at the ends of the lines indicate the half-plane containing the solution set for each inequality. Shade the region common to all three solution sets.

We find the vertex $(0, 4)$ by solving the system

$$x + 2y = 8,$$
$$y = 4.$$

We find the vertex $(6, 4)$ by solving the system

$$x - y = 2,$$
$$y = 4.$$

We find the vertex $(4, 2)$ by solving the system

$$x - y = 2,$$
$$x + 2y = 8.$$

57. Graph: $4x - 3y \geq -12,$
$$4x + 3y \geq -36,$$
$$y \leq 0,$$
$$x \leq 0$$

Shade the intersection of the graphs of the four inequalities.

We find the vertex $(-12, 0)$ by solving the system

$$4y + 3x = -36,$$
$$y = 0.$$

We find the vertex $(0, 0)$ by solving the system

$$y = 0,$$
$$x = 0.$$

We find the vertex $(0, -3)$ by solving the system

$$4y - 3x = -12,$$
$$x = 0.$$

We find the vertex $(-4, -6)$ by solving the system

$$4y - 3x = -12,$$
$$4y + 3x = -36.$$

59. Graph: $3x + 4y \geq 12,$
$$5x + 6y \leq 30,$$
$$1 \leq x \leq 3$$

Shade the intersection of the graphs of the given inequalities.

We find the vertex $\left(1, \dfrac{25}{6}\right)$ by solving the system

$$5x + 6y = 30,$$
$$x = 1.$$

We find the vertex $\left(3, \dfrac{5}{2}\right)$ by solving the system

$$5x + 6y = 30,$$
$$x = 3.$$

We find the vertex $\left(3, \dfrac{3}{4}\right)$ by solving the system

$$3x + 4y = 12,$$
$$x = 3.$$

We find the vertex $\left(1, \dfrac{9}{4}\right)$ by solving the system

$$3x + 4y = 12,$$
$$x = 1.$$

61. Find the maximum and minimum values of

$$P = 17x - 3y + 60, \text{ subject to}$$
$$6x + 8y \leq 48,$$
$$0 \leq y \leq 4,$$
$$0 \leq x \leq 7.$$

Graph the system of inequalities and determine the vertices.

Vertex A: $(0, 0)$

Vertex B:
We solve the system $x = 0$ and $y = 4$. The coordinates of point B are $(0, 4)$.

Vertex C:
We solve the system $6x + 8y = 48$ and $y = 4$. The coordinates of point C are $\left(\dfrac{8}{3}, 4\right)$.

Vertex D:
We solve the system $6x + 8y = 48$ and $x = 7$. The coordinates of point D are $\left(7, \dfrac{3}{4}\right)$.

Vertex E:
We solve the system $x = 7$ and $y = 0$. The coordinates of point E are $(7, 0)$.

Evaluate the objective function P at each vertex.

Vertex	$P = 17x - 3y + 60$
$A(0,0)$	$17 \cdot 0 - 3 \cdot 0 + 60 = 60$
$B(0,4)$	$17 \cdot 0 - 3 \cdot 4 + 60 = 48$
$C\left(\dfrac{8}{3}, 4\right)$	$17 \cdot \dfrac{8}{3} - 3 \cdot 4 + 60 = 66\dfrac{2}{3}$
$D\left(7, \dfrac{3}{4}\right)$	$17 \cdot 7 - 3 \cdot \dfrac{3}{4} + 60 = 176\dfrac{3}{4}$
$E(7,0)$	$17 \cdot 7 - 3 \cdot 0 + 60 = 179$

The maximum value of P is 179 when $x = 7$ and $y = 0$.

The minimum value of P is 48 when $x = 0$ and $y = 4$.

63. Find the maximum and minimum values of

$F = 5x + 36y$, subject to

$$5x + 3y \le 34,$$
$$3x + 5y \le 30,$$
$$x \ge 0,$$
$$y \ge 0.$$

Graph the system of inequalities and find the vertices.

Vertex A: $(0,0)$

Vertex B:

We solve the system $3x + 5y = 30$ and $x = 0$. The coordinates of point B are $(0, 6)$.

Vertex C:

We solve the system $5x + 3y = 34$ and $3x + 5y = 30$. The coordinates of point C are $(5, 3)$.

Vertex D:

We solve the system $5x + 3y = 34$ and $y = 0$. The coordinates of point D are $\left(\dfrac{34}{5}, 0\right)$.

Evaluate the objective function F at each vertex.

Vertex	$F = 5x + 36y$
$A(0,0)$	$5 \cdot 0 + 36 \cdot 0 + = 0$
$B(0,6)$	$5 \cdot 0 + 36 \cdot 6 = 216$
$C(5,3)$	$5 \cdot 5 + 39 \cdot 3 = 133$
$D\left(\dfrac{34}{5}, 0\right)$	$5 \cdot \dfrac{34}{5} + 36 \cdot 0 = 34$

The maximum value of F is 216 when $x = 0$ and $y = 6$.

The minimum value of F is 0 when $x = 0$ and $y = 0$.

65. Let x = the number of jumbo biscuits and y = the number of regular biscuits to be made per day. The income I is given by

$$I = 0.10x + 0.08y$$

subject to the constraints

$$x + y \le 200,$$
$$2x + y \le 300,$$
$$x \ge 0,$$
$$y \ge 0.$$

We graph the system of inequalities, determine the vertices, and find the value if I at each vertex.

Vertex	$I = 0.10x + 0.08y$
$(0,0)$	$0.10(0) + 0.08(0) = 0$
$(0, 200)$	$0.10(0) + 0.08(200) = 16$
$(100, 100)$	$0.10(100) + 0.08(100) = 18$
$(150, 0)$	$0.10(150) + 0.08(0) = 15$

The company will have a maximum income of $18 when 100 of each type of biscuit is made.

67. Let x = the number of units of lumber and y = the number of units of plywood produced per week. The profit P is given by

$$P = 20x + 30y$$

subject to the constraints

$$x + y \le 400,$$
$$x \ge 100,$$
$$y \ge 150.$$

We graph the system of inequalities, determine the vertices and find the value of P at each vertex.

Vertex	$P = 20x + 30y$
$(100, 150)$	$20 \cdot 100 + 30 \cdot 150 = 6500$
$(100, 300)$	$20 \cdot 100 + 30 \cdot 300 = 11,000$
$(250, 150)$	$20 \cdot 250 + 30 \cdot 150 = 9500$

The maximum profit of $11,000 is achieved by producing 100 units of lumber and 300 units of plywood.

69. Let x = the number of sacks of soybean meal to be used and y = the number of sacks of oats. The minimum cost is given by

$$C = 15x + 5y$$

subject to the constraints

$$50x + 15y \geq 120,$$
$$8x + 5y \geq 24,$$
$$5x + y \geq 10,$$
$$x \geq 0,$$
$$y \geq 0.$$

Graph the system of inequalities, determine the vertices, and find the value of C at each vertex.

Vertex	$C = 15x + 5y$
$A(0, 10)$	$15 \cdot 0 + 5 \cdot 10 = 50$
$B\left(\dfrac{6}{5}, 4\right)$	$15 \cdot \dfrac{6}{5} + 5 \cdot 4 = 38$
$C\left(\dfrac{24}{13}, \dfrac{24}{13}\right)$	$15 \cdot \dfrac{24}{13} + 5 \cdot \dfrac{24}{13} = 36\dfrac{12}{13}$
$D(3, 0)$	$15 \cdot 3 + 5 \cdot 0 = 45$

The minimum cost of $\$36\dfrac{12}{13}$ is achieved by using $\dfrac{24}{13}$, or $1\dfrac{11}{13}$ sacks of soybean meal and $\dfrac{24}{13}$, or $1\dfrac{11}{13}$ sacks of oats.

71. Let $x =$ the amount invested in corporate bonds and $y =$ the amount invested in municipal bonds. The income I is given by

$$I = 0.08x + 0.075y$$

subject to the constraints

$$x + y \leq 40,000,$$
$$6000 \leq x \leq 22,000,$$
$$y \leq 30,000.$$

We graph the system of inequalities, determine the vertices, and find the value of I at each vertex.

Vertex	$I = 0.08x + 0.075y$
$(6000, 0)$	480
$(6000, 30,000)$	2730
$(10,000, 30,000)$	3050
$(22,000, 18,000)$	3110
$(22,000, 0)$	1760

The maximum income of $\$3110$ occurs when $\$22,000$ is invested in corporate bonds and $\$18,000$ is invested in municipal bonds.

73. Let $x =$ the number of P_1 airplanes and $y =$ the number of P_2 airplanes to be used. The operating cost C, in thousands of dollars, is given by

$$C = 12x + 10y$$

subject to the constraints

$$40x + 80y \geq 2000,$$
$$40x + 30y \geq 1500,$$
$$120x + 40y \geq 2400,$$
$$x \geq 0,$$
$$y \geq 0.$$

Graph the system of inequalities, determine the vertices, and find the value of C at each vertex.

Vertex	$C = 12x + 10y$
$(0, 60)$	$12 \cdot 0 + 10 \cdot 60 = 600$
$(6, 42)$	$12 \cdot 6 + 10 \cdot 42 = 492$
$(30, 10)$	$12 \cdot 30 + 10 \cdot 10 = 460$
$(50, 0)$	$12 \cdot 50 + 10 \cdot 0 = 600$

The minimum cost of $\$460$ thousand is achieved using 30 P_1's and 10 P_2's.

75. Let $x =$ the number of knit suits and $y =$ the number of worsted suits made. The profit is given by

$$P = 34x + 31y$$

subject to

$$2x + 4y \leq 20,$$
$$4x + 2y \leq 16,$$
$$x \geq 0,$$
$$y \geq 0.$$

Graph the system of inequalities, determine the vertices, and find the value of P at each vertex.

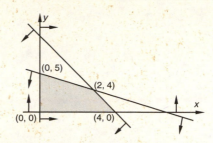

Vertex	$P = 34x + 31y$
$(0,0)$	$34 \cdot 0 + 31 \cdot 0 = 0$
$(0,5)$	$34 \cdot 0 + 31 \cdot 5 = 155$
$(2,4)$	$34 \cdot 2 + 31 \cdot 4 = 192$
$(4,0)$	$34 \cdot 4 + 31 \cdot 0 = 136$

The maximum profit per day is \$192 when 2 knit suits and 4 worsted suits are made.

77. Let $x =$ the number of pounds of meat and $y =$ the number of pounds of cheese in the diet in a week. The cost is given by

$$C = 3.50x + 4.60y$$

subject to

$$2x + 3y \geq 12,$$
$$2x + y \geq 6,$$
$$x \geq 0,$$
$$y \geq 0.$$

Graph the system of inequalities, determine the vertices, and find the value of C at each vertex.

Vertex	$C = 3.50x + 4.60y$
$(0,6)$	$3.50(0) + 4.60(6) = 27.60$
$(1.5,3)$	$3.50(1.5) + 4.60(3) = 19.05$
$(6,0)$	$3.50(6) + 4.60(0) = 21.00$

The minimum weekly cost of \$19.05 is achieved when 1.5 lb of meat and 3 lb of cheese are used.

79. Let $x =$ the number of animal A and $y =$ the number of animal B. The total number of animals is given by

$$T = x + y$$

subject to

$$x + 0.2y \leq 600,$$
$$0.5x + y \leq 525,$$
$$x \geq 0,$$
$$y \geq 0.$$

Graph the system of inequalities, determine the vertices, and find the value of T at each vertex.

Vertex	$T = x + y$
$(0,0)$	$0 + 0 = 0$
$(0,525)$	$0 + 525 = 525$
$(550,250)$	$550 + 250 = 800$
$(600,0)$	$600 + 0 = 600$

The maximum total number of 800 is achieved when there are 550 of A and 250 of B.

81. Left to the student

83. Discussion and Writing

85. $-5 \leq x + 2 < 4$

$-7 \leq x < 2$ Subtracting 2

The solution set is $\{x| -7 \leq x < 2\}$, or $[-7, 2)$.

87. $x^2 - 2x \leq 3$ Polynomial inequality

$x^2 - 2x - 3 \leq 0$

$x^2 - 2x - 3 = 0$ Related equation

$(x + 1)(x - 3) = 0$ Factoring

Using the principle of zero products or by observing the graph of $y = x^2 - 2x - 3$, we see that the solutions of the related equation are -1 and 3. These numbers divide the x-axis into the intervals $(-\infty, -1)$, $(-1, 3)$, and $(3, \infty)$. We let $f(x) = x^2 - 2x - 3$ and test a value in each interval.

$(-\infty, -1)$: $f(-2) = 5 > 0$

$(-1, 3)$: $f(0) = -3 < 0$

$(3, \infty)$: $f(4) = 5 > 0$

Function values are negative on $(-1, 3)$. This can also be determined from the graph of $y = x^2 - 2x - 3$. Since the inequality symbol is $\leq$, the endpoints of the interval must be included in the solution set. It is $\{x| -1 \leq x \leq 3\}$ or $[-1, 3]$.

89. Graph: $y \geq x^2 - 2$,
$ y \leq 2 - x^2$

First graph the related equations $y = x^2 - 2$ and $y = 2 - x^2$ using solid lines. The solution set consists of the region above the graph of $y = x^2 - 2$ and below the graph of $y = 2 - x^2$.

91. See the answer section in the text.

93. See the answer section in the text.

95. Let $x =$ the number of less expensive speaker assemblies and $y =$ the number of more expensive assemblies. The income is given by

$$I = 350x + 600y$$

subject to

$$y \leq 44$$
$$x + y \leq 60,$$
$$x + 2y \leq 90,$$
$$x \geq 0,$$
$$y \geq 0.$$

Graph the system of inequalities, determine the vertices, and find the value of I at each vertex.

Vertex	$I = 350x + 600y$
$(0,0)$	$350 \cdot 0 + 600 \cdot 0 = 0$
$(0,44)$	$350 \cdot 0 + 600 \cdot 44 = 26,400$
$(2,44)$	$350 \cdot 2 + 600 \cdot 44 = 27,100$
$(30,30)$	$350 \cdot 30 + 600 \cdot 30 = 28,500$
$(60,0)$	$350 \cdot 60 + 600 \cdot 0 = 21,000$

The maximum income of \$28,500 is achieved when 30 less expensive and 30 more expensive assemblies are made.

Exercise Set 5.8

1. $\dfrac{x+7}{(x-3)(x+2)} = \dfrac{A}{x-3} + \dfrac{B}{x+2}$

$\dfrac{x+7}{(x-3)(x+2)} = \dfrac{A(x+2) + B(x-3)}{(x-3)(x+2)}$ Adding

Equate the numerators:

$x + 7 = A(x+2) + B(x-3)$

Let $x + 2 = 0$, or $x = -2$. Then we get

$-2 + 7 = 0 + B(-2-3)$
$5 = -5B$
$-1 = B$

Next let $x - 3 = 0$, or $x = 3$. Then we get

$3 + 7 = A(3+2) + 0$
$10 = 5A$
$2 = A$

The decomposition is as follows:

$$\frac{2}{x-3} - \frac{1}{x+2}$$

3. $\dfrac{7x-1}{6x^2 - 5x + 1}$

$= \dfrac{7x-1}{(3x-1)(2x-1)}$ Factoring the denominator

$= \dfrac{A}{3x-1} + \dfrac{B}{2x-1}$

$= \dfrac{A(2x-1) + B(3x-1)}{(3x-1)(2x-1)}$ Adding

Equate the numerators:

$7x - 1 = A(2x-1) + B(3x-1)$

Let $2x - 1 = 0$, or $x = \dfrac{1}{2}$. Then we get

$7\left(\dfrac{1}{2}\right) - 1 = 0 + B\left(3 \cdot \dfrac{1}{2} - 1\right)$

$\dfrac{5}{2} = \dfrac{1}{2}B$

$5 = B$

Next let $3x - 1 = 0$, or $x = \dfrac{1}{3}$. We get

$7\left(\dfrac{1}{3}\right) - 1 = A\left(2 \cdot \dfrac{1}{3} - 1\right) + 0$

$\dfrac{7}{3} - 1 = A\left(\dfrac{2}{3} - 1\right)$

$\dfrac{4}{3} = -\dfrac{1}{3}A$

$-4 = A$

The decomposition is as follows:

$$-\frac{4}{3x-1} + \frac{5}{2x-1}$$

5. $\dfrac{3x^2 - 11x - 26}{(x^2 - 4)(x+1)}$

$= \dfrac{3x^2 - 11x - 26}{(x+2)(x-2)(x+1)}$ Factoring the denominator

$= \dfrac{A}{x+2} + \dfrac{B}{x-2} + \dfrac{C}{x+1}$

$= \dfrac{A(x-2)(x+1) + B(x+2)(x+1) + C(x+2)(x-2)}{(x+2)(x-2)(x+1)}$

Adding

Equate the numerators:

$3x^2 - 11x - 26 = A(x-2)(x+1) +$
$\qquad\qquad\qquad B(x+2)(x+1) + C(x+2)(x-2)$

Let $x + 2 = 0$ or $x = -2$. Then we get
$$3(-2)^2 - 11(-2) - 26 = A(-2-2)(-2+1)+0+0$$
$$12 + 22 - 26 = A(-4)(-1)$$
$$8 = 4A$$
$$2 = A$$

Next let $x - 2 = 0$, or $x = 2$. Then, we get
$$3 \cdot 2^2 - 11 \cdot 2 - 26 = 0 + B(2+2)(2+1) + 0$$
$$12 - 22 - 26 = B \cdot 4 \cdot 3$$
$$-36 = 12B$$
$$-3 = B$$

Finally let $x + 1 = 0$, or $x = -1$. We get
$$3(-1)^2 - 11(-1) - 26 = 0 + 0 + C(-1+2)(-1-2)$$
$$3 + 11 - 26 = C(1)(-3)$$
$$-12 = -3C$$
$$4 = C$$

The decomposition is as follows:
$$\frac{2}{x+2} - \frac{3}{x-2} + \frac{4}{x+1}$$

7.
$$\frac{9}{(x+2)^2(x-1)}$$
$$= \frac{A}{x+2} + \frac{B}{(x+2)^2} + \frac{C}{x-1}$$
$$= \frac{A(x+2)(x-1) + B(x-1) + C(x+2)^2}{(x+2)^2(x-1)}$$
$$\text{Adding}$$

Equate the numerators:
$$9 = A(x+2)(x-1) + B(x-1) + C(x+2)^2 \quad (1)$$

Let $x - 1 = 0$, or $x = 1$. Then, we get
$$9 = 0 + 0 + C(1+2)^2$$
$$9 = 9C$$
$$1 = C$$

Next let $x + 2 = 0$, or $x = -2$. Then, we get
$$9 = 0 + B(-2-1) + 0$$
$$9 = -3B$$
$$-3 = B$$

To find A we first simplify equation (1).
$$9 = A(x^2 + x - 2) + B(x-1) + C(x^2 + 4x + 4)$$
$$= Ax^2 + Ax - 2A + Bx - B + Cx^2 + 4Cx + 4C$$
$$= (A+C)x^2 + (A+B+4C)x + (-2A - B + 4C)$$

Then we equate the coefficients of x^2.
$$0 = A + C$$
$$0 = A + 1 \quad \text{Substituting 1 for } C$$
$$-1 = A$$

The decomposition is as follows:
$$-\frac{1}{x+2} - \frac{3}{(x+2)^2} + \frac{1}{x-1}$$

9.
$$\frac{2x^2 + 3x + 1}{(x^2 - 1)(2x - 1)}$$
$$= \frac{2x^2 + 3x + 1}{(x+1)(x-1)(2x-1)} \quad \begin{array}{l}\text{Factoring the}\\\text{denominator}\end{array}$$
$$= \frac{A}{x+1} + \frac{B}{x-1} + \frac{C}{2x-1}$$
$$= \frac{A(x-1)(2x-1) + B(x+1)(2x-1) + C(x+1)(x-1)}{(x+1)(x-1)(2x-1)}$$
$$\text{Adding}$$

Equate the numerators:
$$2x^2 + 3x + 1 = A(x-1)(2x-1) +$$
$$B(x+1)(2x-1) + C(x+1)(x-1)$$

Let $x + 1 = 0$, or $x = -1$. Then, we get
$$2(-1)^2 + 3(-1) + 1 = A(-1-1)[2(-1)-1]+0+0$$
$$2 - 3 + 1 = A(-2)(-3)$$
$$0 = 6A$$
$$0 = A$$

Next let $x - 1 = 0$, or $x = 1$. Then, we get
$$2 \cdot 1^2 + 3 \cdot 1 + 1 = 0 + B(1+1)(2 \cdot 1 - 1) + 0$$
$$2 + 3 + 1 = B \cdot 2 \cdot 1$$
$$6 = 2B$$
$$3 = B$$

Finally we let $2x - 1 = 0$, or $x = \frac{1}{2}$. We get
$$2\left(\frac{1}{2}\right)^2 + 3\left(\frac{1}{2}\right) + 1 = 0 + 0 + C\left(\frac{1}{2}+1\right)\left(\frac{1}{2}-1\right)$$
$$\frac{1}{2} + \frac{3}{2} + 1 = C \cdot \frac{3}{2} \cdot \left(-\frac{1}{2}\right)$$
$$3 = -\frac{3}{4}C$$
$$-4 = C$$

The decomposition is as follows:
$$\frac{3}{x-1} - \frac{4}{2x-1}$$

11.
$$\frac{x^4 - 3x^3 - 3x^2 + 10}{(x+1)^2(x-3)}$$
$$= \frac{x^4 - 3x^3 - 3x^2 + 10}{x^3 - x^2 - 5x - 3} \quad \begin{array}{l}\text{Multiplying the}\\\text{denominator}\end{array}$$

Since the degree of the numerator is greater than the degree of the denominator, we divide.

$$\begin{array}{r}
x - 2 \\
x^3 - x^2 - 5x - 3 \overline{)x^4 - 3x^3 - 3x^2 + 0x + 10} \\
\underline{x^4 - x^3 - 5x^2 - 3x} \\
-2x^3 + 2x^2 + 3x + 10 \\
\underline{-2x^3 + 2x^2 + 10x + 6} \\
- 7x + 4
\end{array}$$

The original expression is thus equivalent to the following:
$$x - 2 + \frac{-7x + 4}{x^3 - x^2 - 5x - 3}$$

We proceed to decompose the fraction.

$$\frac{-7x+4}{(x+1)^2(x-3)}$$

$$= \frac{A}{x+1} + \frac{B}{(x+1)^2} + \frac{C}{x-3}$$

$$= \frac{A(x+1)(x-3) + B(x-3) + C(x+1)^2}{(x+1)^2(x-3)}$$

Adding

Equate the numerators:

$$-7x + 4 = A(x+1)(x-3) + B(x-3) +$$
$$C(x+1)^2 \qquad (1)$$

Let $x - 3 = 0$, or $x = 3$. Then, we get

$$-7 \cdot 3 + 4 = 0 + 0 + C(3+1)^2$$
$$-17 = 16C$$
$$-\frac{17}{16} = C$$

Let $x + 1 = 0$, or $x = -1$. Then, we get

$$-7(-1) + 4 = 0 + B(-1-3) + 0$$
$$11 = -4B$$
$$-\frac{11}{4} = B$$

To find A we first simplify equation (1).

$$-7x + 4$$
$$= A(x^2 - 2x - 3) + B(x-3) + C(x^2 + 2x + 1)$$
$$= Ax^2 - 2Ax - 3A + Bx - 3B + Cx^2 - 2Cx + C$$
$$= (A+C)x^2 + (-2A+B-2C)x + (-3A-3B+C)$$

Then equate the coefficients of x^2.

$$0 = A + C$$

Substituting $-\frac{17}{16}$ for C, we get $A = \frac{17}{16}$.

The decomposition is as follows:

$$\frac{17/16}{x+1} - \frac{11/4}{(x+1)^2} - \frac{17/16}{x-3}$$

The original expression is equivalent to the following:

$$x - 2 + \frac{17/16}{x+1} - \frac{11/4}{(x+1)^2} - \frac{17/16}{x-3}$$

13. $$\frac{-x^2 + 2x - 13}{(x^2+2)(x-1)}$$

$$= \frac{Ax + B}{x^2 + 2} + \frac{C}{x-1}$$

$$= \frac{(Ax+B)(x-1) + C(x^2+2)}{(x^2+2)(x-1)}$$

Adding

Equate the numerators:

$$-x^2 + 2x - 13 = (Ax+B)(x-1) + C(x^2+2) \quad (1)$$

Let $x - 1 = 0$, or $x = 1$. Then we get

$$-1^2 + 2 \cdot 1 - 13 = 0 + C(1^2 + 2)$$
$$-1 + 2 - 13 = C(1+2)$$
$$-12 = 3C$$
$$-4 = C$$

To find A and B we first simplify equation (1).

$$-x^2 + 2x - 13$$
$$= Ax^2 - Ax + Bx - B + Cx^2 + 2C$$
$$= (A+C)x^2 + (-A+B)x + (-B+2C)$$

Equate the coefficients of x^2:

$$-1 = A + C$$

Substituting -4 for C, we get $A = 3$.

Equate the constant terms:

$$-13 = -B + 2C$$

Substituting -4 for C, we get $B = 5$.

The decomposition is as follows:

$$\frac{3x+5}{x^2+2} - \frac{4}{x-1}$$

15. $$\frac{6 + 26x - x^2}{(2x-1)(x+2)^2}$$

$$= \frac{A}{2x-1} + \frac{B}{x+2} + \frac{C}{(x+2)^2}$$

$$= \frac{A(x+2)^2 + B(2x-1)(x+2) + C(2x-1)}{(2x-1)(x+2)^2}$$

Adding

Equate the numerators:

$$6 + 26x - x^2 = A(x+2)^2 + B(2x-1)(x+2) +$$
$$C(2x-1) \qquad (1)$$

Let $2x - 1 = 0$, or $x = \frac{1}{2}$. Then, we get

$$6 + 26 \cdot \frac{1}{2} - \left(\frac{1}{2}\right)^2 = A\left(\frac{1}{2} + 2\right)^2 + 0 + 0$$
$$6 + 13 - \frac{1}{4} = A\left(\frac{5}{2}\right)^2$$
$$\frac{75}{4} = \frac{25}{4}A$$
$$3 = A$$

Let $x + 2 = 0$, or $x = -2$. We get

$$6 + 26(-2) - (-2)^2 = 0 + 0 + C[2(-2) - 1]$$
$$6 - 52 - 4 = -5C$$
$$-50 = -5C$$
$$10 = C$$

To find B we first simplify equation (1).

$$6 + 26x - x^2$$
$$= A(x^2 + 4x + 4) + B(2x^2 + 3x - 2) + C(2x-1)$$
$$= Ax^2 + 4Ax + 4A + 2Bx^2 + 3Bx - 2B + 2Cx - C$$
$$= (A+2B)x^2 + (4A+3B+2C)x + (4A-2B-C)$$

Equate the coefficients of x^2:

$$-1 = A + 2B$$

Substituting 3 for A, we obtain $B = -2$.

The decomposition is as follows:

$$\frac{3}{2x-1} - \frac{2}{x+2} + \frac{10}{(x+2)^2}$$

17. $\dfrac{6x^3 + 5x^2 + 6x - 2}{2x^2 + x - 1}$

Since the degree of the numerator is greater than the degree of the denominator, we divide.

$$
\begin{array}{r}
3x+1 \\
2x^2 + x - 1\overline{)6x^3+5x^2+6x-2} \\
\underline{6x^3+3x^2-3x} \\
2x^2+9x-2 \\
\underline{2x^2+x-1} \\
8x-1
\end{array}
$$

The original expression is equivalent to

$$3x + 1 + \dfrac{8x - 1}{2x^2 + x - 1}$$

We proceed to decompose the fraction.

$$\dfrac{8x-1}{2x^2+x-1} = \dfrac{8x-1}{(2x-1)(x+1)} \quad \text{Factoring the denominator}$$

$$= \dfrac{A}{2x-1} + \dfrac{B}{x+1}$$

$$= \dfrac{A(x+1) + B(2x-1)}{(2x-1)(x+1)} \quad \text{Adding}$$

Equate the numerators:

$$8x - 1 = A(x+1) + B(2x-1)$$

Let $x + 1 = 0$, or $x = -1$. Then we get

$$8(-1) - 1 = 0 + B[2(-1) - 1]$$
$$-8 - 1 = B(-2 - 1)$$
$$-9 = -3B$$
$$3 = B$$

Next let $2x - 1 = 0$, or $x = \dfrac{1}{2}$. We get

$$8\left(\dfrac{1}{2}\right) - 1 = A\left(\dfrac{1}{2} + 1\right) + 0$$
$$4 - 1 = A\left(\dfrac{3}{2}\right)$$
$$3 = \dfrac{3}{2}A$$
$$2 = A$$

The decomposition is

$$\dfrac{2}{2x-1} + \dfrac{3}{x+1}.$$

The original expression is equivalent to

$$3x + 1 + \dfrac{2}{2x-1} + \dfrac{3}{x+1}.$$

19. $\dfrac{2x^2 - 11x + 5}{(x-3)(x^2 + 2x - 5)}$

$$= \dfrac{A}{x-3} + \dfrac{Bx + C}{x^2 + 2x - 5}$$

$$= \dfrac{A(x^2+2x-5) + (Bx+C)(x-3)}{(x-3)(x^2+2x-5)} \quad \text{Adding}$$

Equate the numerators:

$$2x^2 - 11x + 5 = A(x^2 + 2x - 5) +$$
$$(Bx + C)(x - 3) \qquad (1)$$

Let $x - 3 = 0$, or $x = 3$. Then, we get

$$2 \cdot 3^2 - 11 \cdot 3 + 5 = A(3^2 + 2 \cdot 3 - 5) + 0$$
$$18 - 33 + 5 = A(9 + 6 - 5)$$
$$-10 = 10A$$
$$-1 = A$$

To find B and C, we first simplify equation (1).

$$2x^2 - 11x + 5 = Ax^2 + 2Ax - 5A + Bx^2 - 3Bx +$$
$$Cx - 3C$$
$$= (A + B)x^2 + (2A - 3B + C)x +$$
$$(-5A - 3C)$$

Equate the coefficients of x^2:

$$2 = A + B$$

Substituting -1 for A, we get $B = 3$.

Equate the constant terms:

$$5 = -5A - 3C$$

Substituting -1 for A, we get $C = 0$.

The decomposition is as follows:

$$-\dfrac{1}{x-3} + \dfrac{3x}{x^2 + 2x - 5}$$

21. $\dfrac{-4x^2 - 2x + 10}{(3x+5)(x+1)^2}$

The decomposition looks like

$$\dfrac{A}{3x+5} + \dfrac{B}{x+1} + \dfrac{C}{(x+1)^2}.$$

Add and equate the numerators.

$$-4x^2 - 2x + 10$$
$$= A(x+1)^2 + B(3x+5)(x+1) + C(3x+5)$$
$$= A(x^2 + 2x + 1) + B(3x^2 + 8x + 5) + C(3x+5)$$

or

$$-4x^2 - 2x + 10$$
$$= (A + 3B)x^2 + (2A + 8B + 3C)x + (A + 5B + 5C)$$

Then equate corresponding coefficients.

$-4 = A + 3B$	Coefficients of x^2-terms
$-2 = 2A + 8B + 3C$	Coefficients of x-terms
$10 = A + 5B + 5C$	Constant terms

We solve this system of three equations and find $A = 5$, $B = -3$, $C = 4$.

The decomposition is

$$\dfrac{5}{3x+5} - \dfrac{3}{x+1} + \dfrac{4}{(x+1)^2}.$$

23. $\dfrac{36x + 1}{12x^2 - 7x - 10} = \dfrac{36x + 1}{(4x - 5)(3x + 2)}$

The decomposition looks like

$$\dfrac{A}{4x-5} + \dfrac{B}{3x+2}.$$

Add and equate the numerators.

$$36x + 1 = A(3x + 2) + B(4x - 5)$$
$$\text{or } 36x + 1 = (3A + 4B)x + (2A - 5B)$$

Then equate corresponding coefficients.

$36 = 3A + 4B$ Coefficients of x-terms

$1 = 2A - 5B$ Constant terms

We solve this system of equations and find

$A = 8$ and $B = 3$.

The decomposition is

$$\frac{8}{4x - 5} + \frac{3}{3x + 2}.$$

25. $\dfrac{-4x^2 - 9x + 8}{(3x^2 + 1)(x - 2)}$

The decomposition looks like

$$\frac{Ax + B}{3x^2 + 1} + \frac{C}{x - 2}.$$

Add and equate the numerators.

$-4x^2 - 9x + 8$

$= (Ax + B)(x - 2) + C(3x^2 + 1)$

$= Ax^2 - 2Ax + Bx - 2B + 3Cx^2 + C$

or

$-4x^2 - 9x + 8$

$= (A + 3C)x^2 + (-2A + B)x + (-2B + C)$

Then equate corresponding coefficients.

$-4 = A + 3C$ Coefficients of x^2-terms

$-9 = -2A + B$ Coefficients of x-terms

$8 = -2B + C$ Constant terms

We solve this system of equations and find

$A = 2$, $B = -5$, $C = -2$.

The decomposition is

$$\frac{2x - 5}{3x^2 + 1} - \frac{2}{x - 2}.$$

27. Discussion and Writing

29. The degree of the numerator is equal to the degree of the denominator, so the first step should be to divide the numerator by the denominator in order to express the fraction as a quotient + remainder/denominator.

31. $f(x) = x^3 + x^2 - 3x - 2$

We use synthetic division to factor the polynomial. Using the possibilities found by the rational zeros theorem we find that $x + 2$ is a factor:

$$\begin{array}{r|rrrr} -2 & 1 & 1 & -3 & -2 \\ & & -2 & 2 & 2 \\ \hline & 1 & -1 & -1\,| & 0 \end{array}$$

We have $x^3 + x^2 - 3x - 2 = (x + 2)(x^2 - x - 1)$.

$x^3 + x^2 - 3x - 2 = 0$

$(x + 2)(x^2 - x - 1) = 0$

$x + 2 = 0$ or $x^2 - x - 1 = 0$

The solution of the first equation is -2. We use the quadratic formula to solve the second equation.

$$x = \frac{-b \pm \sqrt{b^2 - 4ac}}{2a}$$

$$= \frac{-(-1) \pm \sqrt{(-1)^2 - 4 \cdot 1 \cdot (-1)}}{2 \cdot 1}$$

$$= \frac{1 \pm \sqrt{5}}{2}$$

The solutions are -2, $\dfrac{1 + \sqrt{5}}{2}$ and $\dfrac{1 - \sqrt{5}}{2}$.

33. $f(x) = x^3 + 5x^2 + 5x - 3$

$$\begin{array}{r|rrrr} -3 & 1 & 5 & 5 & -3 \\ & & -3 & -6 & 3 \\ \hline & 1 & 2 & -1\,| & 0 \end{array}$$

$x^3 + 5x^2 + 5x - 3 = 0$

$(x + 3)(x^2 + 2x - 1) = 0$

$x + 3 = 0$ or $x^2 + 2x - 1 = 0$

The solution of the first equation is -3. We use the quadratic formula to solve the second equation.

$$x = \frac{-b \pm \sqrt{b^2 - 4ac}}{2a}$$

$$= \frac{-2 \pm \sqrt{2^2 - 4 \cdot 1 \cdot (-1)}}{2 \cdot 1} = \frac{-2 \pm \sqrt{8}}{2}$$

$$= \frac{-2 \pm 2\sqrt{2}}{2} = \frac{2(-1 \pm \sqrt{2})}{2}$$

$$= -1 \pm \sqrt{2}$$

The solutions are -3, $-1 + \sqrt{2}$, and $-1 - \sqrt{2}$.

35. $\dfrac{x}{x^4 - a^4}$

$$= \frac{x}{(x^2 + a^2)(x + a)(x - a)} \quad \text{Factoring the denominator}$$

$$= \frac{Ax + B}{x^2 + a^2} + \frac{C}{x + a} + \frac{D}{x - a}$$

$$= [(Ax + B)(x + a)(x - a) + C(x^2 + a^2)(x - a) + D(x^2 + a^2)(x + a)] / [(x^2 + a^2)(x + a)(x - a)]$$

Equate the numerators:

$x = (Ax + B)(x + a)(x - a) + C(x^2 + a^2)(x - a) +$
$\qquad D(x^2 + a^2)(x + a)$

Let $x - a = 0$, or $x = a$. Then, we get

$a = 0 + 0 + D(a^2 + a^2)(a + a)$

$a = D(2a^2)(2a)$

$a = 4a^3 D$

$\dfrac{1}{4a^2} = D$

Let $x + a = 0$, or $x = -a$. We get

$-a = 0 + C[(-a)^2 + a^2](-a - a) + 0$

$-a = C(2a^2)(-2a)$

$-a = -4a^3 C$

$\dfrac{1}{4a^2} = C$

Equate the coefficients of x^3:

$0 = A + C + D$

Substituting $\dfrac{1}{4a^2}$ for C and for D, we get

$$A = -\dfrac{1}{2a^2}.$$

Equate the constant terms:

$$0 = -Ba^2 - Ca^3 + Da^3$$

Substitute $\dfrac{1}{4a^2}$ for C and for D. Then solve for B.

$$0 = -Ba^2 - \dfrac{1}{4a^2} \cdot a^3 + \dfrac{1}{4a^2} \cdot a^3$$

$$0 = -Ba^2$$

$$0 = B$$

The decomposition is as follows:

$$-\dfrac{\frac{1}{2a^2}x}{x^2 + a^2} + \dfrac{\frac{1}{4a^2}}{x + a} + \dfrac{\frac{1}{4a^2}}{x - a}$$

37. $\dfrac{1 + \ln x^2}{(\ln x + 2)(\ln x - 3)^2} = \dfrac{1 + 2\ln x}{(\ln x + 2)(\ln x - 3)^2}$

Let $u = \ln x$. Then we have:

$$\dfrac{1 + 2u}{(u + 2)(u - 3)^2}$$

$$= \dfrac{A}{u + 2} + \dfrac{B}{u - 3} + \dfrac{C}{(u - 3)^2}$$

$$= \dfrac{A(u - 3)^2 + B(u + 2)(u - 3) + C(u + 2)}{(u + 2)(u - 3)^2}$$

Equate the numerators:

$$1 + 2u = A(u - 3)^2 + B(u + 2)(u - 3) + C(u + 2)$$

Let $u - 3 = 0$, or $u = 3$.

$$1 + 2 \cdot 3 = 0 + 0 + C(5)$$

$$7 = 5C$$

$$\dfrac{7}{5} = C$$

Let $u + 2 = 0$, or $u = -2$.

$$1 + 2(-2) = A(-2 - 3)^2 + 0 + 0$$

$$-3 = 25A$$

$$-\dfrac{3}{25} = A$$

To find B, we equate the coefficients of u^2:

$$0 = A + B$$

Substituting $-\dfrac{3}{25}$ for A and solving for B, we get $B = \dfrac{3}{25}$.

The decomposition of $\dfrac{1 + 2u}{(u + 2)(u - 3)^2}$ is as follows:

$$-\dfrac{3}{25(u + 2)} + \dfrac{3}{25(u - 3)} + \dfrac{7}{5(u - 3)^2}$$

Substituting $\ln x$ for u we get

$$-\dfrac{3}{25(\ln x + 2)} + \dfrac{3}{25(\ln x - 3)} + \dfrac{7}{5(\ln x - 3)^2}.$$

Chapter 5 Review Exercises

1. The statement is true. See page 435 in the text.

3. The statement is true. See page 473 in the text.

5. (a)

7. (h)

9. (b)

11. (c)

13. $\begin{aligned} 5x - 3y &= -4, \quad (1) \\ 3x - y &= -4 \quad (2) \end{aligned}$

Multiply equation (2) by -3 and add.

$$\begin{aligned} 5x - 3y &= -4 \\ -9x + 3y &= 12 \\ \hline -4x &= 8 \\ x &= -2 \end{aligned}$$

Back-substitute to find y.

$$\begin{aligned} 3(-2) - y &= -4 \quad \text{Using equation (2)} \\ -6 - y &= -4 \\ -y &= 2 \\ y &= -2 \end{aligned}$$

The solution is $(-2, -2)$.

15. $\begin{aligned} x + 5y &= 12, \quad (1) \\ 5x + 25y &= 12 \quad (2) \end{aligned}$

Solve equation (1) for x.

$$x = -5y + 12$$

Substitute in equation (2) and solve for y.

$$\begin{aligned} 5(-5y + 12) + 25y &= 12 \\ -25y + 60 + 25y &= 12 \\ 60 &= 12 \end{aligned}$$

We get a false equation, so there is no solution.

17. $\begin{aligned} 2x - 4y + 3z &= -3, \quad (1) \\ -5x + 2y - z &= 7, \quad (2) \\ 3x + 2y - 2z &= 4 \quad (3) \end{aligned}$

Multiply equations (2) and (3) by 2.

$$\begin{aligned} 2x - 4y + 3z &= -3 \quad (1) \\ -10x + 4y - 2z &= 14 \quad (4) \\ 6x + 4y - 4z &= 8 \quad (5) \end{aligned}$$

Multiply equation (1) by 5 and add it to equation (4).

Multiply equation (1) by -3 and add it to equation (5).

$$\begin{aligned} 2x - 4y + 3z &= -3 \quad (1) \\ -16y + 13z &= -1 \quad (6) \\ 16y - 13z &= 17 \quad (7) \end{aligned}$$

Add equation (6) to equation (7).

$$\begin{aligned} 2x - 4y + 3z &= -3 \quad (1) \\ -16y + 13z &= -1 \quad (6) \\ 0 &= 16 \quad (8) \end{aligned}$$

Equation (8) is false, so the system of equations has no solution.

19.
$$\begin{aligned} x - y \qquad\quad &= 5, \quad (1) \\ y - z \quad\;\; &= 6, \quad (2) \\ z - w &= 7, \quad (3) \\ x \qquad\quad + w &= 8 \quad\;\; (4) \end{aligned}$$

Multiply equation (1) by -1 and add it to equation (4).
$$\begin{aligned} x - y \qquad\quad &= 5 \quad (1) \\ y - z \quad\;\; &= 6 \quad (2) \\ z - w &= 7 \quad (3) \\ y \qquad + w &= 3 \quad (5) \end{aligned}$$

Multiply equation (2) by -1 and add it to equation (5).
$$\begin{aligned} x - y \qquad\quad &= 5 \qquad (1) \\ y - z \quad\;\; &= 6 \qquad (2) \\ z - w &= 7 \qquad (3) \\ z + w &= -3 \quad (6) \end{aligned}$$

Multiply equation (3) by -1 and add it to equation (6).
$$\begin{aligned} x - y \qquad\quad &= 5 \qquad (1) \\ y - z \quad\;\; &= 6 \qquad (2) \\ z - \; w &= 7 \qquad (3) \\ 2w &= -10 \quad (7) \end{aligned}$$

Solve equation (7) for w.
$$\begin{aligned} 2w &= -10 \\ w &= -5 \end{aligned}$$

Back-substitute -5 for w in equation (3) and solve for z.
$$\begin{aligned} z - w &= 7 \\ z - (-5) &= 7 \\ z + 5 &= 7 \\ z &= 2 \end{aligned}$$

Back-substitute 2 for z in equation (2) and solve for y.
$$\begin{aligned} y - z &= 6 \\ y - 2 &= 6 \\ y &= 8 \end{aligned}$$

Back-substitute 8 for y in equation (1) and solve for x.
$$\begin{aligned} x - y &= 5 \\ x - 8 &= 5 \\ x &= 13 \end{aligned}$$

Writing the solution as (w, x, y, z), we have $(-5, 13, 8, 2)$.

21. Systems 13, 14, 15, 17, 18, and 19 each have either no solution or exactly one solution, so the equations in those systems are independent. System 16 has infinitely many solutions, so the equations in that system are dependent.

23.
$$\begin{aligned} 3x + 4y + \;\; 2z &= 3 \\ 5x - 2y - 13z &= 3 \\ 4x + 3y - \;\; 3z &= 6 \end{aligned}$$

Write the augmented matrix. We will use Gaussian elimination.

$$\left[\begin{array}{rrr|r} 3 & 4 & 2 & 3 \\ 5 & -2 & -13 & 3 \\ 4 & 3 & -3 & 6 \end{array}\right]$$

Multiply row 2 and row 3 by 3.

$$\left[\begin{array}{rrr|r} 3 & 4 & 2 & 3 \\ 15 & -6 & -39 & 9 \\ 12 & 9 & -9 & 18 \end{array}\right]$$

Multiply row 1 by -5 and add it to row 2.
Multiply row 1 by -4 and add it to row 3.

$$\left[\begin{array}{rrr|r} 3 & 4 & 2 & 3 \\ 0 & -26 & -49 & -6 \\ 0 & -7 & -17 & 6 \end{array}\right]$$

Multiply row 3 by 26.

$$\left[\begin{array}{rrr|r} 3 & 4 & 2 & 3 \\ 0 & -26 & -49 & -6 \\ 0 & -182 & -442 & 156 \end{array}\right]$$

Multiply row 2 by -7 and add it to row 3.

$$\left[\begin{array}{rrr|r} 3 & 4 & 2 & 3 \\ 0 & -26 & -49 & -6 \\ 0 & 0 & -99 & 198 \end{array}\right]$$

Multiply row 1 by $\dfrac{1}{3}$, row 2 by $-\dfrac{1}{26}$, and row 3 by $-\dfrac{1}{99}$.

$$\left[\begin{array}{ccc|c} 1 & \dfrac{4}{3} & \dfrac{2}{3} & 1 \\[2mm] 0 & 1 & \dfrac{49}{26} & \dfrac{3}{13} \\[2mm] 0 & 0 & 1 & -2 \end{array}\right]$$

$$\begin{aligned} x + \frac{4}{3}y + \frac{2}{3}z &= 1 \quad\; (1) \\ y + \frac{49}{26}z &= \frac{3}{13} \quad (2) \\ z &= -2 \end{aligned}$$

Back-substitute in equation (2) and solve for y.
$$\begin{aligned} y + \frac{49}{26}(-2) &= \frac{3}{13} \\ y - \frac{49}{13} &= \frac{3}{13} \\ y &= \frac{52}{13} = 4 \end{aligned}$$

Back-substitute in equation (1) and solve for x.
$$\begin{aligned} x + \frac{4}{3}(4) + \frac{2}{3}(-2) &= 1 \\ x + \frac{16}{3} - \frac{4}{3} &= 1 \\ x &= -3 \end{aligned}$$

The solution is $(-3, 4, -2)$.

25.
$$-w + x + y + z = -2,$$
$$-3w - 2x + 3y + 2z = 10,$$
$$2w + 3x + 2y - z = -12,$$
$$2w + 4x - y + z = 1$$

Write the augmented matrix. We will use Gauss-Jordan elimination.

$$\begin{bmatrix} 1 & 1 & 1 & 1 & | & -2 \\ -3 & -2 & 3 & 2 & | & 10 \\ 2 & 3 & 2 & -1 & | & -12 \\ 2 & 4 & -1 & 1 & | & 1 \end{bmatrix}$$

Multiply row 1 by 3 and add it to row 2.

Multiply row 1 by -2 and add it to row 3.

Multiply row 1 by -2 and add it to row 4.

$$\begin{bmatrix} 1 & 1 & 1 & 1 & | & -2 \\ 0 & 1 & 6 & 5 & | & 4 \\ 0 & 1 & 0 & -3 & | & -8 \\ 0 & 2 & -3 & -1 & | & 5 \end{bmatrix}$$

Multiply row 2 by -1 and add it to row 1.

Multiply row 2 by -1 and add it to row 3.

Multiply row 2 by -2 and add it to row 4.

$$\begin{bmatrix} 1 & 0 & -5 & -4 & | & -6 \\ 0 & 1 & 6 & 5 & | & 4 \\ 0 & 0 & -6 & -8 & | & -12 \\ 0 & 0 & -15 & -11 & | & -3 \end{bmatrix}$$

Multiply row 1 by 3.

Multiply row 3 by $-\dfrac{1}{2}$.

$$\begin{bmatrix} 3 & 0 & -15 & -12 & | & -18 \\ 0 & 1 & 6 & 5 & | & 4 \\ 0 & 0 & 3 & 4 & | & 6 \\ 0 & 0 & -15 & -11 & | & -3 \end{bmatrix}$$

Multiply row 3 by 5 and add it to row 1.

Multiply row 3 by -2 and add it to row 2.

Multiply row 3 by 5 and add it to row 4.

$$\begin{bmatrix} 3 & 0 & 0 & 8 & | & 12 \\ 0 & 1 & 0 & -3 & | & -8 \\ 0 & 0 & 3 & 4 & | & 6 \\ 0 & 0 & 0 & 9 & | & 27 \end{bmatrix}$$

Multiply row 4 by $\dfrac{1}{9}$.

$$\begin{bmatrix} 3 & 0 & 0 & 8 & | & 12 \\ 0 & 1 & 0 & -3 & | & -8 \\ 0 & 0 & 3 & 4 & | & 6 \\ 0 & 0 & 0 & 1 & | & 3 \end{bmatrix}$$

Multiply row 4 by -8 and add it to row 1.

Multiply row 4 by 3 and add it to row 2.

Multiply row 4 by -4 and add it to row 3.

$$\begin{bmatrix} 3 & 0 & 0 & 0 & | & -12 \\ 0 & 1 & 0 & 0 & | & 1 \\ 0 & 0 & 3 & 0 & | & -6 \\ 0 & 0 & 0 & 1 & | & 3 \end{bmatrix}$$

Multiply rows 1 and 3 by $\dfrac{1}{3}$.

$$\begin{bmatrix} 1 & 0 & 0 & 0 & | & -4 \\ 0 & 1 & 0 & 0 & | & 1 \\ 0 & 0 & 1 & 0 & | & -2 \\ 0 & 0 & 0 & 1 & | & 3 \end{bmatrix}$$

The solution is $(-4, 1, -2, 3)$.

27. **Familiarize.** Let $x =$ the amount invested at 3% and $y =$ the amount invested at 3.5%. Then the interest from the investments is $3\%x$ and $3.5\%y$, or $0.03x$ and $0.035y$.

Translate.

The total investment is $5000.
$$x + y = 5000$$

The total interest is $167.
$$0.03x + 0.035y = 167$$

We have a system of equations.
$$x + y = 5000,$$
$$0.03x + 0.035y = 167$$

Multiplying the second equation by 1000 to clear the decimals, we have:
$$x + y = 5000, \qquad (1)$$
$$30x + 35y = 167,000. \qquad (2)$$

Carry out. We begin by multiplying equation (1) by -30 and adding.
$$-30x - 30y = -150,000$$
$$\underline{30x + 35y = 167,000}$$
$$5y = 17,000$$
$$y = 3400$$

Back-substitute to find x.
$$x + 3400 = 5000 \quad \text{Using equation (1)}$$
$$x = 1600$$

Check. The total investment is $1600 + $3400, or $5000. The total interest is 0.03($1600) + 0.035($3400), or $48 + $119, or $167. The solution checks.

State. $1600 was invested at 3% and $3400 was invested at 3.5%.

29. **Familiarize.** Let x, y, and z represent the scores on the first, second, and third tests, respectively.

Translate.

The total score on the three tests is 226.
$$x + y + z = 226$$

The sum of the scores on the first and second tests exceeds the score on the third test by 62.

$$x + y = z + 62$$

The first score exceeds the second by 6.

$$x = y + 6$$

We have a system of equations.

$$x + y + z = 226,$$
$$x + y = z + 62,$$
$$x = y + 6$$

or

$$x + y + z = 226,$$
$$x + y - z = 62,$$
$$x - y \quad = 6$$

Carry out. Solving the system of equations, we get $(75, 69, 82)$.

Check. The sum of the scores is $75 + 69 + 82$, or 226. The sum of the scores on the first two tests is $75 + 69$, or 144. This exceeds the score on the third test, 82, by 62. The score on the first test, 75, exceeds the score on the second test, 69, by 6. The solution checks.

State. The scores on the first, second, and third tests were 75, 69, and 82, respectively.

31. $\mathbf{A} + \mathbf{B} = \begin{bmatrix} 1 & -1 & 0 \\ 2 & 3 & -2 \\ -2 & 0 & 1 \end{bmatrix} + \begin{bmatrix} -1 & 0 & 6 \\ 1 & -2 & 0 \\ 0 & 1 & -3 \end{bmatrix}$

$$= \begin{bmatrix} 1 + (-1) & -1 + 0 & 0 + 6 \\ 2 + 1 & 3 + (-2) & -2 + 0 \\ -2 + 0 & 0 + 1 & 1 + (-3) \end{bmatrix}$$

$$= \begin{bmatrix} 0 & -1 & 6 \\ 3 & 1 & -2 \\ -2 & 1 & -2 \end{bmatrix}$$

33. $-\mathbf{A} = -1 \begin{bmatrix} 1 & -1 & 0 \\ 2 & 3 & -2 \\ -2 & 0 & 1 \end{bmatrix} = \begin{bmatrix} -1 & 1 & 0 \\ -2 & -3 & 2 \\ 2 & 0 & -1 \end{bmatrix}$

35. $\mathbf{A}$ and $\mathbf{B}$ do not have the same order, so it is not possible to find $\mathbf{A} + \mathbf{B}$.

37. $\mathbf{BA} = \begin{bmatrix} -1 & 0 & 6 \\ 1 & -2 & 0 \\ 0 & 1 & -3 \end{bmatrix} \cdot \begin{bmatrix} 1 & -1 & 0 \\ 2 & 3 & -2 \\ -2 & 0 & 1 \end{bmatrix}$

$$= \begin{bmatrix} -1 + 0 - 12 & 1 + 0 + 0 & 0 + 0 + 6 \\ 1 - 4 + 0 & -1 - 6 + 0 & 0 + 4 + 0 \\ 0 + 2 + 6 & 0 + 3 + 0 & 0 - 2 - 3 \end{bmatrix}$$

$$= \begin{bmatrix} -13 & 1 & 6 \\ -3 & -7 & 4 \\ 8 & 3 & -5 \end{bmatrix}$$

39. a) $\mathbf{M} = \begin{bmatrix} 46.1 & 5.9 & 10.1 & 8.5 & 11.4 \\ 54.6 & 4.6 & 9.6 & 7.6 & 10.6 \\ 48.9 & 5.5 & 12.7 & 9.4 & 9.3 \\ 51.3 & 4.8 & 11.3 & 6.9 & 12.7 \end{bmatrix}$

b) $\mathbf{N} = \begin{bmatrix} 32 & 19 & 43 & 38 \end{bmatrix}$

c) $\mathbf{NM} = \begin{bmatrix} 6564.7 & 695.1 & 1481.1 & 1082.8 & 1448.7 \end{bmatrix}$

d) The entries of $\mathbf{NM}$ represent the total cost, in cents, for each item for the day's meal.

41. $\mathbf{A} = \begin{bmatrix} 0 & 0 & 3 \\ 0 & -2 & 0 \\ 4 & 0 & 0 \end{bmatrix}$

Write the augmented matrix.

$$\left[\begin{array}{ccc|ccc} 0 & 0 & 3 & 1 & 0 & 0 \\ 0 & -2 & 0 & 0 & 1 & 0 \\ 4 & 0 & 0 & 0 & 0 & 1 \end{array} \right]$$

Interchange rows 1 and 3.

$$\left[\begin{array}{ccc|ccc} 4 & 0 & 0 & 0 & 0 & 1 \\ 0 & -2 & 0 & 0 & 1 & 0 \\ 0 & 0 & 3 & 1 & 0 & 0 \end{array} \right]$$

Multiply row 1 by $\frac{1}{4}$, row 2 by $-\frac{1}{2}$, and row 3 by $\frac{1}{3}$.

$$\left[\begin{array}{ccc|ccc} 1 & 0 & 0 & 0 & 0 & \frac{1}{4} \\ 0 & 1 & 0 & 0 & -\frac{1}{2} & 0 \\ 0 & 0 & 1 & \frac{1}{3} & 0 & 0 \end{array} \right]$$

$$\mathbf{A}^{-1} = \begin{bmatrix} 0 & 0 & \frac{1}{4} \\ 0 & -\frac{1}{2} & 0 \\ \frac{1}{3} & 0 & 0 \end{bmatrix}$$

43. $3x - 2y + 4z = 13,$
$\quad x + 5y - 3z = 7,$
$\quad 2x - 3y + 7z = -8$

Write the coefficients on the left in a matrix. Then write the product of that matrix and the column matrix containing the variables, and set the result equal to the column matrix containing the constants on the right.

$$\begin{bmatrix} 3 & -2 & 4 \\ 1 & 5 & -3 \\ 2 & -3 & 7 \end{bmatrix} \begin{bmatrix} x \\ y \\ z \end{bmatrix} = \begin{bmatrix} 13 \\ 7 \\ -8 \end{bmatrix}$$

45. $5x - y + 2z = 17,$
$\quad 3x + 2y - 3z = -16,$
$\quad 4x - 3y - z = 5$

Write an equivalent matrix equation, $\mathbf{AX} = \mathbf{B}$.

$$\begin{bmatrix} 5 & -1 & 2 \\ 3 & 2 & -3 \\ 4 & -3 & -1 \end{bmatrix} \begin{bmatrix} x \\ y \\ z \end{bmatrix} = \begin{bmatrix} 17 \\ -16 \\ 5 \end{bmatrix}$$

Then,

$$\mathbf{X} = \mathbf{A}^{-1}\mathbf{B} = \begin{bmatrix} \dfrac{11}{80} & \dfrac{7}{80} & \dfrac{1}{80} \\ \dfrac{9}{80} & \dfrac{13}{80} & -\dfrac{21}{80} \\ \dfrac{17}{80} & -\dfrac{11}{80} & -\dfrac{13}{80} \end{bmatrix} \begin{bmatrix} 17 \\ -16 \\ 5 \end{bmatrix} = \begin{bmatrix} 1 \\ -2 \\ 5 \end{bmatrix}$$

The solution is $(1, -2, 5)$.

47. $\begin{vmatrix} 1 & -2 \\ 3 & 4 \end{vmatrix} = 1 \cdot 4 - 3(-2) = 4 + 6 = 10$

49. We will expand across the first row.

$$\begin{vmatrix} 2 & -1 & 1 \\ 1 & 2 & -1 \\ 3 & 4 & -3 \end{vmatrix}$$

$$= 2(-1)^{1+1} \begin{vmatrix} 2 & -1 \\ 4 & -3 \end{vmatrix} + (-1)(-1)^{1+2} \begin{vmatrix} 1 & -1 \\ 3 & -3 \end{vmatrix} +$$

$$1(-1)^{1+3} \begin{vmatrix} 1 & 2 \\ 3 & 4 \end{vmatrix}$$

$$= 2 \cdot 1[2(-3) - 4(-1)] + (-1)(-1)[1(-3) - 3(-1)] +$$
$$1 \cdot 1[1(4) - 3(2)]$$
$$= 2(-2) + 1(0) + 1(-2)$$
$$= -6$$

51. $5x - 2y = 19,$
$\quad 7x + 3y = 15$

$$D = \begin{vmatrix} 5 & -2 \\ 7 & 3 \end{vmatrix} = 5(3) - 7(-2) = 29$$

$$D_x = \begin{vmatrix} 19 & -2 \\ 15 & 3 \end{vmatrix} = 19(3) - 15(-2) = 87$$

$$D_y = \begin{vmatrix} 5 & 19 \\ 7 & 15 \end{vmatrix} = 5(15) - 7(19) = -58$$

$$x = \frac{D_x}{D} = \frac{87}{29} = 3$$

$$y = \frac{D_y}{D} = \frac{-58}{29} = -2$$

The solution is $(3, -2)$.

53. $3x - 2y + z = 5,$
$\quad 4x - 5y - z = -1,$
$\quad 3x + 2y - z = 4$

$$D = \begin{vmatrix} 3 & -2 & 1 \\ 4 & -5 & -1 \\ 3 & 2 & -1 \end{vmatrix} = 42$$

$$D_x = \begin{vmatrix} 5 & -2 & 1 \\ -1 & -5 & -1 \\ 4 & 2 & -1 \end{vmatrix} = 63$$

$$D_y = \begin{vmatrix} 3 & 5 & 1 \\ 4 & -1 & -1 \\ 3 & 4 & -1 \end{vmatrix} = 39$$

$$D_z = \begin{vmatrix} 3 & -2 & 5 \\ 4 & -5 & -1 \\ 3 & 2 & 4 \end{vmatrix} = 99$$

$$x = \frac{D_x}{D} = \frac{63}{42} = \frac{3}{2}$$

$$y = \frac{D_y}{D} = \frac{39}{42} = \frac{13}{14}$$

$$z = \frac{D_z}{D} = \frac{99}{42} = \frac{33}{14}$$

The solution is $\left(\dfrac{3}{2}, \dfrac{13}{14}, \dfrac{33}{14} \right)$.

55.

57.

59. Let $x =$ the number of questions answered from group A and $y =$ the number of questions answered from group B. Find the maximum value of $S = 7x + 12y$ subject to

$$x + y \geq 8,$$
$$8x + 10y \leq 80,$$
$$x \geq 0,$$
$$y \geq 0$$

Graph the system of inequalities and find the vertices, $(0, 8)$, $(8, 0)$ and $(10, 0)$. Evaluate $S = 7x + 12y$ at each vertex.

$(0, 8): \ 7(0) + 12(8) = 96$
$(8, 0): \ 7(8) + 12(0) = 56$
$(10, 0): \ 7(10) + 12(0) = 70$

The maximum score of 96 occurs when 0 questions from group A and 8 questions from group B are answered correctly.

61. $\dfrac{-8x + 23}{2x^2 + 5x - 12} = \dfrac{-8x + 23}{(2x - 3)(x + 4)}$

$$= \frac{A}{2x - 3} + \frac{B}{x + 4}$$

$$= \frac{A(x + 4) + B(2x - 3)}{(2x - 3)(x + 4)}$$

Equate the numerators.

$$-8x + 23 = A(x + 4) + B(2x - 3)$$

Let $x = \dfrac{3}{2}$: $-8\left(\dfrac{3}{2}\right) + 23 = A\left(\dfrac{3}{2} + 4\right) + 0$

$$-12 + 23 = \dfrac{11}{2}A$$

$$11 = \dfrac{11}{2}A$$

$$2 = A$$

Let $x = -4$: $-8(-4) + 23 = 0 + B[2(-4) - 3]$

$$32 + 23 = -11B$$

$$55 = -11B$$

$$-5 = B$$

The decomposition is $\dfrac{2}{2x - 3} - \dfrac{5}{x + 4}$.

63. Interchanging columns of a matrix is not a row-equivalent operation, so answer A is correct. (See page 463 in the text.)

65. In general, $(\mathbf{AB})^2 \neq \mathbf{A}^2\mathbf{B}^2$. $(\mathbf{AB})^2 = \mathbf{ABAB}$ and $\mathbf{A}^2\mathbf{B}^2 = \mathbf{AABB}$. Since matrix multiplication is not commutative, $\mathbf{BA} \neq \mathbf{AB}$, so $(\mathbf{AB})^2 \neq \mathbf{A}^2\mathbf{B}^2$.

67. $\dfrac{2}{3x} + \dfrac{4}{5y} = 8$,

$\dfrac{5}{4x} - \dfrac{3}{2y} = -6$

Let $a = \dfrac{1}{x}$ and $b = \dfrac{1}{y}$. Then we have:

$$\dfrac{2}{3}a + \dfrac{4}{5}b = 8,$$

$$\dfrac{5}{4}a - \dfrac{3}{2}b = -6$$

Multiply the first equation by 15 and the second by 4 to clear the fractions.

$$10a + 12b = 120, \quad (1)$$

$$5a - 6b = -24 \quad (2)$$

Multiply equation (2) by 2 and add.

$$10a + 12b = 120$$
$$\underline{10a - 12b = -48}$$
$$20a \qquad = 72$$
$$a = \dfrac{18}{5}$$

Back-substitute to find b.

$$10 \cdot \dfrac{18}{5} + 12b = 120 \quad \text{Using equation (1)}$$

$$36 + 12b = 120$$

$$12b = 84$$

$$b = 7$$

Now find x and y.

$$a = \dfrac{1}{x} \qquad\qquad b = \dfrac{1}{y}$$

$$\dfrac{18}{5} = \dfrac{1}{x} \qquad\qquad 7 = \dfrac{1}{y}$$

$$x = \dfrac{5}{18} \qquad\qquad y = \dfrac{1}{7}$$

The solution is $\left(\dfrac{5}{18}, \dfrac{1}{7}\right)$.

69.

$|x| - |y| \le 1$

Chapter 5 Test

1. $3x + 2y = 1$, (1)

$2x - y = -11$ (2)

Multiply equation (2) by 2 and add.

$$3x + 2y = 1$$
$$\underline{4x - 2y = -22}$$
$$7x \qquad = -21$$
$$x = -3$$

Back-substitute to find y.

$$2(-3) - y = -11 \quad \text{Using equation (2).}$$

$$-6 - y = -11$$

$$-y = -5$$

$$y = 5$$

The solution is $(-3, 5)$. Since the system of equations has exactly one solution, it is consistent and independent.

2. $2x - y = 3$, (1)

$2y = 4x - 6$ (2)

Solve equation (1) for y.

$$y = 2x - 3$$

Subsitute in equation (2) and solve for x.

$$2(2x - 3) = 4x - 6$$

$$4x - 6 = 4x - 6$$

$$0 = 0$$

The equation $0 = 0$ is true for all values of x and y. Thus the system of equations has infinitely many solutions. Solving either equation for y, we get $y = 2x - 3$, so the solutions are ordered pairs of the form $(x, 2x - 3)$. Equivalently, if we solve either equation for x, we get $x = \dfrac{y + 3}{2}$, so the solutions can also be expressed as $\left(\dfrac{y + 3}{2}, y\right)$. Since there are infinitely many solutions, the system of equations is consistent and dependent.

3. $x - y = 4$, (1)

$3y = 3x - 8$ (2)

Solve equation (1) for x.

$$x = y + 4$$

Subsitute in equation (2) and solve for y.

$$3y = 3(y + 4) - 8$$
$$3y = 3y + 12 - 8$$
$$0 = 4$$

We get a false equation so there is no solution. Since there is no solution the system of equations is inconsistent and independent.

4. $2x - 3y = 8,$ (1)
$5x - 2y = 9$ (2)

Multiply equation (1) by 5 and equation (2) by -2 and add.

$$10x - 15y = 40$$
$$\underline{-10x + 4y = -18}$$
$$ - 11y = 22$$
$$y = -2$$

Back-substitute to find x.

$$2x - 3(-2) = 8$$
$$2x + 6 = 8$$
$$2x = 2$$
$$x = 1$$

The solution is $(1, -2)$. Since the system of equations has exactly one solution, it is consistent and independent.

5. $4x + 2y + z = 4,$ (1)
$3x - y + 5z = 4,$ (2)
$5x + 3y - 3z = -2$ (3)

Multiply equations (2) and (3) by 4.

$$4x + 2y + z = 4 \quad (1)$$
$$12x - 4y + 20z = 16 \quad (4)$$
$$20x + 12y - 12z = -8 \quad (5)$$

Multiply equation (1) by -3 and add it to equation (4).
Multiply equation (1) by -5 and add it to equation (5).

$$4x + 2y + z = 4 \quad (1)$$
$$-10y + 17z = 4 \quad (6)$$
$$2y - 17z = -28 \quad (7)$$

Interchange equations (6) and (7).

$$4x + 2y + z = 4 \quad (1)$$
$$2y - 17z = -28 \quad (7)$$
$$-10y + 17z = 4 \quad (6)$$

Multiply equation (7) by 5 and add it to equation (6).

$$4x + 2y + z = 4 \quad (1)$$
$$2y - 17z = -28 \quad (7)$$
$$-68z = -136 \quad (8)$$

Solve equation (8) for z.

$$-68z = -136$$
$$z = 2$$

Back-substitute 2 for z in equation (7) and solve for y.

$$2y - 17 \cdot 2 = -28$$
$$2y - 34 = -28$$
$$2y = 6$$
$$y = 3$$

Back-substitute 3 for y and 2 for z in equation (1) and solve for x.

$$4x + 2 \cdot 3 + 2 = 4$$
$$4x + 8 = 4$$
$$4x = -4$$
$$x = -1$$

The solution is $(-1, 3, 2)$.

6. *Familiarize*. Let x and y represent the number of student and nonstudent tickets sold, respectively. Then the receipts from the student tickets were $3x$ and the receipts from the nonstudent tickets were $5y$.

Translate. One equation comes from the fact that 750 tickets were sold.

$$x + y = 750$$

A second equation comes from the fact that the total receipts were $3066.

$$3x + 5y = 3066$$

Carry out. We solve the system of equations.

$$x + y = 750, \quad (1)$$
$$3x + 5y = 3066 \quad (2)$$

Multiply equation (1) by -3 and add.

$$-3x - 3y = -2250$$
$$\underline{3x + 5y = 3066}$$
$$2y = 816$$
$$y = 408$$

Substitute 408 for y in equation (1) and solve for x.

$$x + 408 = 750$$
$$x = 342$$

Check. The number of tickets sold was $342 + 408$, or 750. The total receipts were $\$3 \cdot 342 + \$5 \cdot 408 = \$1026 + \$2040 = \$3066$. The solution checks.

State. 342 student tickets and 408 nonstudent tickets were sold.

7. *Familiarize*. Let x, y, and z represent the number of orders that can be processed per day by Tricia, Maria, and Antonio, respectively.

Translate.

Tricia, Maria, and Antonio can process 352 orders per day.

$$x + y + z = 352$$

Tricia and Maria together can process 224 orders per day.

$$x + y = 224$$

Tricia and Antonio together can process 248 orders per day.

$$x + z = 248$$

We have a system of equations:

$$x + y + z = 352,$$
$$x + y = 224,$$
$$x + z = 248.$$

Carry out. Solving the system of equations, we get $(120, 104, 128)$.

Check. Tricia, Maria, and Antonio can process $120+104+128$, or 352, orders per day. Together, Tricia and Maria can process $120 + 104$, or 224, orders per day. Together, Tricia and Antonio can process $120 + 128$, or 248, orders per day. The solution checks.

State. Tricia can process 120 orders per day, Maria can process 104 orders per day, and Antonio can process 128 orders per day.

8. $\mathbf{B} + \mathbf{C} = \begin{bmatrix} -5 & 1 \\ -2 & 4 \end{bmatrix} + \begin{bmatrix} 3 & -4 \\ -1 & 0 \end{bmatrix}$

$= \begin{bmatrix} -5+3 & 1+(-4) \\ -2+(-1) & 4+0 \end{bmatrix}$

$= \begin{bmatrix} -2 & -3 \\ -3 & 4 \end{bmatrix}$

9. $\mathbf{A}$ and $\mathbf{C}$ do not have the same order, so it is not possible to find $\mathbf{A} - \mathbf{C}$.

10. $\mathbf{CB} = \begin{bmatrix} 3 & -4 \\ -1 & 0 \end{bmatrix} \begin{bmatrix} -5 & 1 \\ -2 & 4 \end{bmatrix}$

$= \begin{bmatrix} 3(-5)+(-4)(-2) & 3(1)+(-4)(4) \\ -1(-5)+0(-2) & -1(1)+0(4) \end{bmatrix}$

$= \begin{bmatrix} -7 & -13 \\ 5 & -1 \end{bmatrix}$

11. The product $\mathbf{AB}$ is not defined because the number of columns of $\mathbf{A}$, 3, is not equal to the number of rows of $\mathbf{B}$, 2.

12. $2\mathbf{A} = 2\begin{bmatrix} 1 & -1 & 3 \\ -2 & 5 & 2 \end{bmatrix} = \begin{bmatrix} 2 & -2 & 6 \\ -4 & 10 & 4 \end{bmatrix}$

13. $\mathbf{C} = \begin{bmatrix} 3 & -4 \\ -1 & 0 \end{bmatrix}$

Write the augmented matrix.

$\begin{bmatrix} 3 & -4 & | & 1 & 0 \\ -1 & 0 & | & 0 & 1 \end{bmatrix}$

Interchange rows.

$\begin{bmatrix} -1 & 0 & | & 0 & 1 \\ 3 & -4 & | & 1 & 0 \end{bmatrix}$

Multiply row 1 by 3 and add it to row 2.

$\begin{bmatrix} -1 & 0 & | & 0 & 1 \\ 0 & -4 & | & 1 & 3 \end{bmatrix}$

Multiply row 1 by -1 and row 2 by $-\dfrac{1}{4}$.

$\begin{bmatrix} 1 & 0 & | & 0 & -1 \\ 0 & 1 & | & -\dfrac{1}{4} & -\dfrac{3}{4} \end{bmatrix}$

$\mathbf{C}^{-1} = \begin{bmatrix} 0 & -1 \\ -\dfrac{1}{4} & -\dfrac{3}{4} \end{bmatrix}$

14. a) $\mathbf{M} = \begin{bmatrix} 49 & 10 & 13 \\ 43 & 12 & 11 \\ 51 & 8 & 12 \end{bmatrix}$

b) $\mathbf{N} = \begin{bmatrix} 26 & 18 & 23 \end{bmatrix}$

c) $\mathbf{NM} = \begin{bmatrix} 3221 & 660 & 812 \end{bmatrix}$

d) The entries of $\mathbf{NM}$ represent the total cost, in cents, for each type of menu item served on the given day.

15. $\begin{bmatrix} 3 & -4 & 2 \\ 2 & 3 & 1 \\ 1 & -5 & -3 \end{bmatrix} \begin{bmatrix} x \\ y \\ z \end{bmatrix} = \begin{bmatrix} -8 \\ 7 \\ 3 \end{bmatrix}$

16. $3x + 2y + 6z = 2,$

$\quad x + \ y + 2z = 1,$

$\quad 2x + 2y + 5z = 3$

Write an equivalent matrix equation, $\mathbf{AX} = \mathbf{B}$.

$\begin{bmatrix} 3 & 2 & 6 \\ 1 & 1 & 2 \\ 2 & 2 & 5 \end{bmatrix} \begin{bmatrix} x \\ y \\ z \end{bmatrix} = \begin{bmatrix} 2 \\ 1 \\ 3 \end{bmatrix}$

Then,

$\mathbf{X} = \mathbf{A}^{-1}\mathbf{B} = \begin{bmatrix} 1 & 2 & -2 \\ -1 & 3 & 0 \\ 0 & -2 & 1 \end{bmatrix} \begin{bmatrix} 2 \\ 1 \\ 3 \end{bmatrix} = \begin{bmatrix} -2 \\ 1 \\ 1 \end{bmatrix}$

The solution is $(-2, 1, 1)$.

17. $\begin{vmatrix} 3 & -5 \\ 8 & 7 \end{vmatrix} = 3 \cdot 7 - 8(-5) = 21 + 40 = 61$

18. We will expand across the first row.

$\begin{vmatrix} 2 & -1 & 4 \\ -3 & 1 & -2 \\ 5 & 3 & -1 \end{vmatrix}$

$= 2(-1)^{1+1}\begin{vmatrix} 1 & -2 \\ 3 & -1 \end{vmatrix} + (-1)(-1)^{1+2}\begin{vmatrix} -3 & -2 \\ 5 & -1 \end{vmatrix} +$

$\quad 4(-1)^{1+3}\begin{vmatrix} -3 & 1 \\ 5 & 3 \end{vmatrix}$

$= 2 \cdot 1[1(-1) - 3(-2)] + (-1)(-1)[-3(-1) - 5(-2)] +$

$\quad 4 \cdot 1[-3(3) - 5(1)]$

$= 2(5) + 1(13) + 4(-14)$

$= -33$

19. $5x + 2y = -1,$

$7x + 6y = 1$

$$D = \begin{vmatrix} 5 & 2 \\ 7 & 6 \end{vmatrix} = 5(6) - 7(2) = 16$$

$$D_x = \begin{vmatrix} -1 & 2 \\ 1 & 6 \end{vmatrix} = -1(6) - (1)(2) = -8$$

$$D_y = \begin{vmatrix} 5 & -1 \\ 7 & 1 \end{vmatrix} = 5(1) - 7(-1) = 12$$

$$x = \frac{D_x}{D} = \frac{-8}{16} = -\frac{1}{2}$$

$$y = \frac{D_y}{D} = \frac{12}{16} = \frac{3}{4}$$

The solution is $\left(-\dfrac{1}{2}, \dfrac{3}{4} \right)$.

20.

21. Graph the system of inequalities and find the vertices, $(1, 0)$, $\left(1, \dfrac{5}{3} \right)$, $(3, 3)$, and $(6, 0)$.

Evaluate $Q = 2x + 3y$ at each vertex.

$(1, 0):\ 2(1) + 3(0) = 2$ Minimum

$\left(1, \dfrac{5}{3} \right):\ 2(1) + 3\left(\dfrac{5}{3} \right) = 7$

$(3, 3):\ 2(3) + 3(3) = 15$ Maximum

$(6, 0):\ 2(6) + 3(0) = 12$

22. Let $x =$ the number of pound cakes prepared and $y =$ the number of carrot cakes. Find the maximum value of $P = 3x + 4y$ subject to

$x + y \le 100,$

$x \ge 25,$

$y \ge 15$

Graph the system of inequalities and find the vertices, $(25, 15)$, $(25, 75)$, and $(85, 15)$.

Evaluate $P = 3x + 4y$ at each vertex.

$(25, 15):\ P = 3(25) + 4(15) = 135$

$(25, 75):\ P = 3(25) + 4(75) = 375$

$(85, 15):\ P = 3(85) + 4(15) = 315$

The maximum profit of \$375 occurs when 25 pound cakes and 75 carrot cakes are prepared.

23. $\dfrac{3x - 11}{x^2 + 2x - 3} = \dfrac{3x - 11}{(x - 1)(x + 3)}$

$$= \frac{A}{x - 1} + \frac{B}{x + 3}$$

$$= \frac{A(x + 3) + B(x - 1)}{(x - 1)(x + 3)}$$

Equate the numerators.

$3x - 11 = A(x + 3) + B(x - 1)$

Let $x = -3:\ 3(-3) - 11 = 0 + B(-3 - 1)$

$$-20 = -4B$$

$$5 = B$$

Let $x = 1:\ 3(1) - 11 = A(1 + 3) + 0$

$$-8 = 4A$$

$$-2 = A$$

The decomposition is $-\dfrac{2}{x - 1} + \dfrac{5}{x + 3}$.

24. Solve:

$$A(2) - B(-2) = C(2) - 8$$

$$A(-3) - B(-1) = C(1) - 8$$

$$A(4) - B(2) = C(9) - 8$$

or

$$2A + 2B - 2C = -8$$

$$-3A + B - C = -8$$

$$4A - 2B - 9C = -8$$

The solution is $(1, -3, 2)$, so $A = 1$, $B = -3$, and $C = 2$.

Chapter 6
Conic Sections

1. Graph (f) is the graph of $x^2 = 8y$.

3. Graph (b) is the graph of $(y-2)^2 = -3(x+4)$.

5. Graph (d) is the graph of $13x^2 - 8y - 9 = 0$.

7. $x^2 = 20y$

$x^2 = 4 \cdot 5 \cdot y$ Writing $x^2 = 4py$

Vertex: $(0,0)$

Focus: $(0,5)$ $[(0,p)]$

Directrix: $y = -5$ $(y = -p)$

9. $y^2 = -6x$

$y^2 = 4\left(-\dfrac{3}{2}\right)x$ Writing $y^2 = 4px$

Vertex: $(0,0)$

Focus: $\left(-\dfrac{3}{2},0\right)$ $[(p,0)]$

Directrix: $x = -\left(-\dfrac{3}{2}\right) = \dfrac{3}{2}$ $(x = -p)$

11. $x^2 - 4y = 0$

$x^2 = 4y$

$x^2 = 4 \cdot 1 \cdot y$ Writing $x^2 = 4py$

Vertex: $(0,0)$

Focus: $(0,1)$ $[(0,p)]$

Directrix: $y = -1$ $(y = -p)$

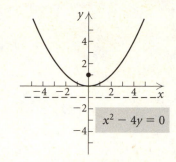

13. $x = 2y^2$

$y^2 = \dfrac{1}{2}x$

$y^2 = 4 \cdot \dfrac{1}{8} \cdot x$ Writing $y^2 = 4px$

Vertex: $(0,0)$

Focus: $\left(\dfrac{1}{8},0\right)$

Directrix: $x = -\dfrac{1}{8}$

15. Since the directrix, $x = -4$, is a vertical line, the equation is of the form $(y-k)^2 = 4p(x-h)$. The focus, $(4,0)$, is on the x-axis so the line of symmetry is the x-axis and $p = 4$. The vertex, (h,k), is the point on the x-axis midway between the directrix and the focus. Thus, it is $(0,0)$. We have

$$(y-k)^2 = 4p(x-h)$$
$$(y-0)^2 = 4 \cdot 4(x-0) \quad \text{Substituting}$$
$$y^2 = 16x.$$

17. Since the directrix, $y = \pi$, is a horizontal line, the equation is of the form $(x-h)^2 = 4p(y-k)$. The focus, $(0,-\pi)$, is on the y-axis so the line of symmetry is the y-axis and $p = -\pi$. The vertex (h,k) is the point on the y-axis midway between the directrix and the focus. Thus, it is $(0,0)$. We have

$$(x-h)^2 = 4p(y-k)$$
$$(x-0)^2 = 4(-\pi)(y-0) \quad \text{Substituting}$$
$$x^2 = -4\pi y$$

19. Since the directrix, $x = -4$, is a vertical line, the equation is of the form $(y - k)^2 = 4p(x - h)$. The focus, $(3, 2)$, is on the horizontal line $y = 2$, so the line of symmetry is $y = 2$. The vertex is the point on the line $y = 2$ that is midway between the directrix and the focus. That is, it is the midpoint of the segment from $(-4, 2)$ to $(3, 2)$: $\left(\dfrac{-4+3}{2}, \dfrac{2+2}{2}\right)$, or $\left(-\dfrac{1}{2}, 2\right)$. Then $h = -\dfrac{1}{2}$ and the directrix is $x = h - p$, so we have

$$x = h - p$$
$$-4 = -\frac{1}{2} - p$$
$$-\frac{7}{2} = -p$$
$$\frac{7}{2} = p.$$

Now we find the equation of the parabola.
$$(y - k)^2 = 4p(x - h)$$
$$(y - 2)^2 = 4\left(\frac{7}{2}\right)\left[x - \left(-\frac{1}{2}\right)\right]$$
$$(y - 2)^2 = 14\left(x + \frac{1}{2}\right)$$

21.
$$(x + 2)^2 = -6(y - 1)$$
$$[x - (-2)]^2 = 4\left(-\frac{3}{2}\right)(y - 1) \quad [(x-h)^2 = 4p(y-k)]$$

Vertex: $(-2, 1)$ $\qquad [(h, k)]$

Focus: $\left(-2, 1 + \left(-\dfrac{3}{2}\right)\right)$, or $\left(-2, -\dfrac{1}{2}\right)$
$\qquad\qquad [(h, k+p)]$

Directrix: $y = 1 - \left(-\dfrac{3}{2}\right) = \dfrac{5}{2}$ $\quad (y = k - p)$

$$(x + 2)^2 = -6(y - 1)$$

23. $x^2 + 2x + 2y + 7 = 0$
$$x^2 + 2x = -2y - 7$$
$$(x^2 + 2x + 1) = -2y - 7 + 1 = -2y - 6$$
$$(x + 1)^2 = -2(y + 3)$$
$$[x - (-1)]^2 = 4\left(-\frac{1}{2}\right)[y - (-3)]$$
$$[(x - h)^2 = 4p(y - k)]$$

Vertex: $(-1, -3)$ $\qquad [(h, k)]$

Focus: $\left(-1, -3 + \left(-\dfrac{1}{2}\right)\right)$, or $\left(-1, -\dfrac{7}{2}\right)$
$\qquad\qquad [(h, k+p)]$

Directrix: $y = -3 - \left(-\dfrac{1}{2}\right) = -\dfrac{5}{2}$ $\quad (y = k - p)$

$$x^2 + 2x + 2y + 7 = 0$$

25. $x^2 - y - 2 = 0$
$$x^2 = y + 2$$
$$(x - 0)^2 = 4 \cdot \frac{1}{4} \cdot [y - (-2)]$$
$$[(x - h)^2 = 4p(y - k)]$$

Vertex: $(0, -2)$ $\qquad [(h, k)]$

Focus: $\left(0, -2 + \dfrac{1}{4}\right)$, or $\left(0, -\dfrac{7}{4}\right)$ $[(h, k + p)]$

Directrix: $y = -2 - \dfrac{1}{4} = -\dfrac{9}{4}$ $\quad (y = k - p)$

$$x^2 - y - 2 = 0$$

27.
$$y = x^2 + 4x + 3$$
$$y - 3 = x^2 + 4x$$
$$y - 3 + 4 = x^2 + 4x + 4$$
$$y + 1 = (x + 2)^2$$
$$4 \cdot \frac{1}{4} \cdot [y - (-1)] = [x - (-2)]^2$$
$$[(x - h)^2 = 4p(y - k)]$$

Vertex: $(-2, -1)$ $\qquad [(h, k)]$

Focus: $\left(-2, -1 + \dfrac{1}{4}\right)$, or $\left(-2, -\dfrac{3}{4}\right)$ $[(h, k + p)]$

Directrix: $y = -1 - \dfrac{1}{4} = -\dfrac{5}{4}$ $\quad (y = k - p)$

$$y = x^2 + 4x + 3$$

29. $y^2 - y - x + 6 = 0$

$$y^2 - y = x - 6$$

$$y^2 - y + \frac{1}{4} = x - 6 + \frac{1}{4}$$

$$\left(y - \frac{1}{2}\right)^2 = x - \frac{23}{4}$$

$$\left(y - \frac{1}{2}\right)^2 = 4 \cdot \frac{1}{4}\left(x - \frac{23}{4}\right)$$

$$[(y - k)^2 = 4p(x - h)]$$

Vertex: $\left(\dfrac{23}{4}, \dfrac{1}{2}\right)$ $[(h, k)]$

Focus: $\left(\dfrac{23}{4} + \dfrac{1}{4}, \dfrac{1}{2}\right)$, or $\left(6, \dfrac{1}{2}\right)$ $[(h + p, k)]$

Directrix: $x = \dfrac{23}{4} - \dfrac{1}{4} = \dfrac{22}{4}$ or $\dfrac{11}{2}$ $(x = h - p)$

$$y^2 - y - x + 6 = 0$$

31. a) The vertex is $(0, 0)$. The focus is $(4, 0)$, so $p = 4$. The parabola has a horizontal axis of symmetry so the equation is of the form $y^2 = 4px$. We have

$$y^2 = 4px$$

$$y^2 = 4 \cdot 4 \cdot x$$

$$y^2 = 16x$$

b) We make a drawing.

The depth of the satellite dish at the vertex is x where $\left(x, \dfrac{15}{2}\right)$ is a point on the parabola.

$$y^2 = 16x$$

$$\left(\frac{15}{2}\right)^2 = 16x \qquad \text{Substituting } \frac{15}{2} \text{ for } y$$

$$\frac{225}{4} = 16x$$

$$\frac{225}{64} = x, \text{ or}$$

$$3\frac{33}{64} = x$$

The depth of the satellite dish at the vertex is $3\dfrac{33}{64}$ ft.

33. We position a coordinate system with the origin at the vertex and the x-axis on the parabola's axis of symmetry.

The parabola is of the form $y^2 = 4px$ and a point on the parabola is $\left(1.5, \dfrac{4}{2}\right)$, or $(1.5, 2)$.

$$y^2 = 4px$$

$$2^2 = 4 \cdot p \cdot (1.5) \quad \text{Substituting}$$

$$4 = 6p$$

$$\frac{4}{6} = p, \text{ or}$$

$$\frac{2}{3} = p$$

Since the focus is at $(p, 0)$, or $\left(\dfrac{2}{3}, 0\right)$, the focus is $\dfrac{2}{3}$ ft, or 8 in., from the vertex.

35. Vertex: $(0.867, 0.348)$

Focus: $(0.867, -0.190)$

Directrix: $y = 0.887$

37. Discussion and Writing

39. When we let $y = 0$ and solve for x, the only equation for which $x = \dfrac{2}{3}$ is (h), so only equation (h) has x-intercept $\left(\dfrac{2}{3}, 0\right)$.

41. Note that equation (g) is equivalent to $y = 2x - \dfrac{7}{4}$ and equation (h) is equivalent to $y = -\dfrac{1}{2}x + \dfrac{1}{3}$. When we look at the equations in the form $y = mx + b$, we see that $m > 0$ for (a), (b), (f), and (g) so these equations have positive slope, or slant up front left to right.

43. When we look at the equations in the form $y = mx + b$ (See Exercise 41.), only (b) has $m = \dfrac{1}{3}$ so only (b) has slope $\dfrac{1}{3}$.

45. Parallel lines have the same slope and different y-intercepts. When we look at the equations in the form $y = mx + b$ (See Exercise 41.), we see that (a) and (g) represent parallel lines.

47. A parabola with a vertical axis of symmetry has an equation of the type $(x - h)^2 = 4p(y - k)$.

Solve for p substituting $(-1, 2)$ for (h, k) and $(-3, 1)$ for (x, y).

$$[-3 - (-1)]^2 = 4p(1 - 2)$$
$$4 = -4p$$
$$-1 = p$$

The equation of the parabola is

$$[x - (-1)]^2 = 4(-1)(y - 2), \text{ or}$$
$$(x + 1)^2 = -4(y - 2).$$

49. Position a coordinate system as shown below with the y-axis on the parabola's axis of symmetry.

The equation of the parabola is of the form $(x - h)^2 = 4p(y - k)$. Substitute 100 for x, 50 for y, 0 for h, and 10 for k and solve for p.

$$(x - h)^2 = 4p(y - k)$$
$$(100 - 0)^2 = 4p(50 - 10)$$
$$10,000 = 160p$$
$$\frac{250}{4} = p$$

Then the equation is

$$x^2 = 4\left(\frac{250}{4}\right)(y - 10), \text{ or}$$
$$x^2 = 250(y - 10).$$

To find the lengths of the vertical cables, find y when $x = 0$, 20, 40, 60, 80, and 100.

When $x = 0$: $0^2 = 250(y - 10)$
$$0 = y - 10$$
$$10 = y$$

When $x = 20$: $20^2 = 250(y - 10)$
$$400 = 250(y - 10)$$
$$1.6 = y - 10$$
$$11.6 = y$$

When $x = 40$: $40^2 = 250(y - 10)$
$$1600 = 250(y - 10)$$
$$6.4 = y - 10$$
$$16.4 = y$$

When $x = 60$: $60^2 = 250(y - 10)$
$$3600 = 250(y - 10)$$
$$14.4 = y - 10$$
$$24.4 = y$$

When $x = 80$: $80^2 = 250(y - 10)$
$$6400 = 250(y - 10)$$
$$25.6 = y - 10$$
$$35.6 = y$$

When $x = 100$, we know from the given information that $y = 50$.

The lengths of the vertical cables are 10 ft, 11.6 ft, 16.4 ft, 24.4 ft, 35.6 ft, and 50 ft.

Exercise Set 6.2

1. Graph (b) is the graph of $x^2 + y^2 = 5$.

3. Graph (d) is the graph of $x^2 + y^2 - 6x + 2y = 6$.

5. Graph (a) is the graph of $x^2 + y^2 - 5x + 3y = 0$.

7. Complete the square twice.
$$x^2 + y^2 - 14x + 4y = 11$$
$$x^2 - 14x + y^2 + 4y = 11$$
$$x^2 - 14x + 49 + y^2 + 4y + 4 = 11 + 49 + 4$$
$$(x - 7)^2 + (y + 2)^2 = 64$$
$$(x - 7)^2 + [y - (-2)]^2 = 8^2$$

Center: $(7, -2)$

Radius: 8

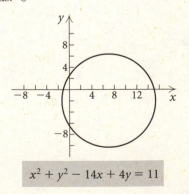

$$x^2 + y^2 - 14x + 4y = 11$$

9. Complete the square twice.

$$x^2 + y^2 + 6x - 2y = 6$$
$$x^2 + 6x + y^2 - 2y = 6$$
$$x^2 + 6x + 9 + y^2 - 2y + 1 = 6 + 9 + 1$$
$$(x + 3)^2 + (y - 1)^2 = 16$$
$$[x - (-3)]^2 + (y - 1)^2 = 4^2$$

Center: $(-3, 1)$

Radius: 4

$$x^2 + y^2 + 6x - 2y = 6$$

11. Complete the square twice.

$$x^2 + y^2 + 4x - 6y - 12 = 0$$
$$x^2 + 4x + y^2 - 6y = 12$$
$$x^2 + 4x + 4 + y^2 - 6y + 9 = 12 + 4 + 9$$
$$(x + 2)^2 + (y - 3)^2 = 25$$
$$[x - (-2)]^2 + (y - 3)^2 = 5^2$$

Center: $(-2, 3)$

Radius: 5

$$x^2 + y^2 + 4x - 6y - 12 = 0$$

13. Complete the square twice.

$$x^2 + y^2 - 6x - 8y + 16 = 0$$
$$x^2 - 6x + y^2 - 8y = -16$$
$$x^2 - 6x + 9 + y^2 - 8y + 16 = -16 + 9 + 16$$
$$(x - 3)^2 + (y - 4)^2 = 9$$
$$(x - 3)^2 + (y - 4)^2 = 3^2$$

Center: $(3, 4)$

Radius: 3

$$x^2 + y^2 - 6x - 8y + 16 = 0$$

15. Complete the square twice.

$$x^2 + y^2 + 6x - 10y = 0$$
$$x^2 + 6x + y^2 - 10y = 0$$
$$x^2 + 6x + 9 + y^2 - 10y + 25 = 0 + 9 + 25$$
$$(x + 3)^2 + (y - 5)^2 = 34$$
$$[x - (-3)]^2 + (y - 5)^2 = (\sqrt{34})^2$$

Center: $(-3, 5)$

Radius: $\sqrt{34}$

$$x^2 + y^2 + 6x - 10y = 0$$

17. Complete the square twice.

$$x^2 + y^2 - 9x = 7 - 4y$$
$$x^2 - 9x + y^2 + 4y = 7$$
$$x^2 - 9x + \frac{81}{4} + y^2 + 4y + 4 = 7 + \frac{81}{4} + 4$$
$$\left(x - \frac{9}{2}\right)^2 + (y + 2)^2 = \frac{125}{4}$$
$$\left(x - \frac{9}{2}\right)^2 + [y - (-2)]^2 = \left(\frac{5\sqrt{5}}{2}\right)^2$$

Center: $\left(\frac{9}{2}, -2\right)$

Radius: $\dfrac{5\sqrt{5}}{2}$

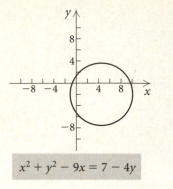

$$x^2 + y^2 - 9x = 7 - 4y$$

19. Graph (c) is the graph of $16x^2 + 4y^2 = 64$.

21. Graph (d) is the graph of $x^2 + 9y^2 - 6x + 90y = -225$.

23. $\dfrac{x^2}{4} + \dfrac{y^2}{1} = 1$

$\dfrac{x^2}{2^2} + \dfrac{y^2}{1^2} = 1$ Standard form

$a = 2, \, b = 1$

The major axis is horizontal, so the vertices are $(-2, 0)$ and $(2, 0)$. Since we know that $c^2 = a^2 - b^2$, we have $c^2 = 4 - 1 = 3$, so $c = \sqrt{3}$ and the foci are $(-\sqrt{3}, 0)$ and $(\sqrt{3}, 0)$.

To graph the ellipse, plot the vertices. Note also that since $b = 1$, the y-intercepts are $(0, -1)$ and $(0, 1)$. Plot these points as well and connect the four plotted points with a smooth curve.

$$\dfrac{x^2}{4} + \dfrac{y^2}{1} = 1$$

25. $16x^2 + 9y^2 = 144$

$\dfrac{x^2}{9} + \dfrac{y^2}{16} = 1$ Dividing by 144

$\dfrac{x^2}{3^2} + \dfrac{y^2}{4^2} = 1$ Standard form

$a = 4, \, b = 3$

The major axis is vertical, so the vertices are $(0, -4)$ and $(0, 4)$. Since $c^2 = a^2 - b^2$, we have $c^2 = 16 - 9 = 7$, so $c = \sqrt{7}$ and the foci are $(0, -\sqrt{7})$ and $(0, \sqrt{7})$.

To graph the ellipse, plot the vertices. Note also that since $b = 3$, the x-intercepts are $(-3, 0)$ and $(3, 0)$. Plot these

points as well and connect the four plotted points with a smooth curve.

$$16x^2 + 9y^2 = 144$$

27. $2x^2 + 3y^2 = 6$

$\dfrac{x^2}{3} + \dfrac{y^2}{2} = 1$

$\dfrac{x^2}{(\sqrt{3})^2} + \dfrac{y^2}{(\sqrt{2})^2} = 1$

$a = \sqrt{3}, \, b = \sqrt{2}$

The major axis is horizontal, so the vertices are $(-\sqrt{3}, 0)$ and $(\sqrt{3}, 0)$. Since $c^2 = a^2 - b^2$, we have $c^2 = 3 - 2 = 1$, so $c = 1$ and the foci are $(-1, 0)$ and $(1, 0)$.

To graph the ellipse, plot the vertices. Note also that since $b = \sqrt{2}$, the y-intercepts are $(0, -\sqrt{2})$ and $(0, \sqrt{2})$. Plot these points as well and connect the four plotted points with a smooth curve.

$$2x^2 + 3y^2 = 6$$

29. $4x^2 + 9y^2 = 1$

$\dfrac{x^2}{\frac{1}{4}} + \dfrac{y^2}{\frac{1}{9}} = 1$

$\dfrac{x^2}{\left(\frac{1}{2}\right)^2} + \dfrac{y^2}{\left(\frac{1}{3}\right)^2} = 1$

$a = \dfrac{1}{2}, \, b = \dfrac{1}{3}$

The major axis is horizontal, so the vertices are $\left(-\dfrac{1}{2}, 0\right)$ and $\left(\dfrac{1}{2}, 0\right)$. Since $c^2 = a^2 - b^2$, we have $c^2 = \dfrac{1}{4} - \dfrac{1}{9} = \dfrac{5}{36}$, so $c = \dfrac{\sqrt{5}}{6}$ and the foci are $\left(-\dfrac{\sqrt{5}}{6}, 0\right)$ and $\left(\dfrac{\sqrt{5}}{6}, 0\right)$.

To graph the ellipse, plot the vertices. Note also that since $b = \dfrac{1}{3}$, the y-intercepts are $\left(0, -\dfrac{1}{3}\right)$ and $\left(0, \dfrac{1}{3}\right)$. Plot these points as well and connect the four plotted points with a smooth curve.

$$4x^2 + 9y^2 = 1$$

31. The vertices are on the x-axis, so the major axis is horizontal. We have $a = 7$ and $c = 3$, so we can find b^2:

$$c^2 = a^2 - b^2$$
$$3^2 = 7^2 - b^2$$
$$b^2 = 49 - 9 = 40$$

Write the equation:

$$\frac{x^2}{a^2} + \frac{y^2}{b^2} = 1$$

$$\frac{x^2}{49} + \frac{y^2}{40} = 1$$

33. The vertices, $(0, -8)$ and $(0, 8)$, are on the y-axis, so the major axis is vertical and $a = 8$. Since the vertices are equidistant from the origin, the center of the ellipse is at the origin. The length of the minor axis is 10, so $b = 10/2$, or 5.

Write the equation:

$$\frac{x^2}{b^2} + \frac{y^2}{a^2} = 1$$

$$\frac{x^2}{5^2} + \frac{y^2}{8^2} = 1$$

$$\frac{x^2}{25} + \frac{y^2}{64} = 1$$

35. The foci, $(-2, 0)$ and $(2, 0)$ are on the x-axis, so the major axis is horizontal and $c = 2$. Since the foci are equidistant from the origin, the center of the ellipse is at the origin. The length of the major axis is 6, so $a = 6/2$, or 3. Now we find b^2:

$$c^2 = a^2 - b^2$$
$$2^2 = 3^2 - b^2$$
$$4 = 9 - b^2$$
$$b^2 = 5$$

Write the equation:

$$\frac{x^2}{a^2} + \frac{y^2}{b^2} = 1$$

$$\frac{x^2}{9} + \frac{y^2}{5} = 1$$

37.
$$\frac{(x-1)^2}{9} + \frac{(y-2)^2}{4} = 1$$

$$\frac{(x-1)^2}{3^2} + \frac{(y-2)^2}{2^2} = 1 \quad \text{Standard form}$$

The center is $(1, 2)$. Note that $a = 3$ and $b = 2$. The major axis is horizontal so the vertices are 3 units left and right of the center:

$(1-3, 2)$ and $(1+3, 2)$, or $(-2, 2)$ and $(4, 2)$.

We know that $c^2 = a^2 - b^2$, so $c^2 = 9 - 4 = 5$ and $c = \sqrt{5}$. Then the foci are $\sqrt{5}$ units left and right of the center:

$$(1 - \sqrt{5}, 2) \text{ and } (1 + \sqrt{5}, 2).$$

To graph the ellipse, plot the vertices. Since $b = 2$, two other points on the graph are 2 units below and above the center:

$(1, 2-2)$ and $(1, 2+2)$ or $(1, 0)$ and $(1, 4)$

Plot these points also and connect the four plotted points with a smooth curve.

$$\frac{(x-1)^2}{9} + \frac{(y-2)^2}{4} = 1$$

39.
$$\frac{(x+3)^2}{25} + \frac{(y-5)^2}{36} = 1$$

$$\frac{[x-(-3)]^2}{5^2} + \frac{(y-5)^2}{6^2} = 1 \quad \text{Standard form}$$

The center is $(-3, 5)$. Note that $a = 6$ and $b = 5$. The major axis is vertical so the vertices are 6 units below and above the center:

$(-3, 5-6)$ and $(-3, 5+6)$, or $(-3, -1)$ and $(-3, 11)$.

We know that $c^2 = a^2 - b^2$, so $c^2 = 36 - 25 = 11$ and $c = \sqrt{11}$. Then the foci are $\sqrt{11}$ units below and above the vertex:

$$(-3, 5 - \sqrt{11}) \text{ and } (-3, 5 + \sqrt{11}).$$

To graph the ellipse, plot the vertices. Since $b = 5$, two other points on the graph are 5 units left and right of the center:

$(-3-5, 5)$ and $(-3+5, 5)$, or $(-8, 5)$ and $(2, 5)$

Plot these points also and connect the four plotted points with a smooth curve.

$$\frac{(x + 3)^2}{25} + \frac{(y - 5)^2}{36} = 1$$

41. $3(x + 2)^2 + 4(y - 1)^2 = 192$

$$\frac{(x + 2)^2}{64} + \frac{(y - 1)^2}{48} = 1 \qquad \text{Dividing by 192}$$

$$\frac{[x - (-2)]^2}{8^2} + \frac{(y - 1)^2}{(\sqrt{48})^2} = 1 \qquad \text{Standard form}$$

The center is $(-2, 1)$. Note that $a = 8$ and $b = \sqrt{48}$, or $4\sqrt{3}$. The major axis is horizontal so the vertices are 8 units left and right of the center:

$(-2 - 8, 1)$ and $(-2 + 8, 1)$, or $(-10, 1)$ and $(6, 1)$.

We know that $c^2 = a^2 - b^2$, so $c^2 = 64 - 48 = 16$ and $c = 4$. Then the foci are 4 units left and right of the center:

$(-2 - 4, 1)$ and $(-2 + 4, 1)$ or $(-6, 1)$ and $(2, 1)$.

To graph the ellipse, plot the vertices. Since $b = 4\sqrt{3} \approx 6.928$, two other points on the graph are about 6.928 units below and above the center:

$(-2, 1 - 6.928)$ and $(-2, 1 + 6.928)$, or

$(-2, -5.928)$ and $(-2, 7.928)$.

Plot these points also and connect the four plotted points with a smooth curve.

$$3(x + 2)^2 + 4(y - 1)^2 = 192$$

43. Begin by completing the square twice.

$$4x^2 + 9y^2 - 16x + 18y - 11 = 0$$

$$4x^2 - 16x + 9y^2 + 18y = 11$$

$$4(x^2 - 4x) + 9(y^2 + 2y) = 11$$

$$4(x^2 - 4x + 4) + 9(y^2 + 2y + 1) = 11 + 4 \cdot 4 + 9 \cdot 1$$

$$4(x - 2)^2 + 9(y + 1)^2 = 36$$

$$\frac{(x - 2)^2}{9} + \frac{(y + 1)^2}{4} = 1$$

$$\frac{(x - 2)^2}{3^2} + \frac{[y - (-1)]^2}{2^2} = 1$$

The center is $(2, -1)$. Note that $a = 3$ and $b = 2$. The major axis is horizontal so the vertices are 3 units left and right of the center:

$(2 - 3, -1)$ and $(2 + 3, -1)$, or $(-1, -1)$ and $(5, -1)$.

We know that $c^2 = a^2 - b^2$, so $c^2 = 9 - 4 = 5$ and $c = \sqrt{5}$. Then the foci are $\sqrt{5}$ units left and right of the center:

$(2 - \sqrt{5}, -1)$ and $(2 + \sqrt{5}, -1)$.

To graph the ellipse, plot the vertices. Since $b = 2$, two other points on the graph are 2 units below and above the center:

$(2, -1 - 2)$ and $(2, -1 + 2)$, or $(2, -3)$ and $(2, 1)$.

Plot these points also and connect the four plotted points with a smooth curve.

$$4x^2 + 9y^2 - 16x + 18y - 11 = 0$$

45. Begin by completing the square twice.

$$4x^2 + y^2 - 8x - 2y + 1 = 0$$

$$4x^2 - 8x + y^2 - 2y = -1$$

$$4(x^2 - 2x) + y^2 - 2y = -1$$

$$4(x^2 - 2x + 1) + y^2 - 2y + 1 = -1 + 4 \cdot 1 + 1$$

$$4(x - 1)^2 + (y - 1)^2 = 4$$

$$\frac{(x - 1)^2}{1} + \frac{(y - 1)^2}{4} = 1$$

$$\frac{(x - 1)^2}{1^2} + \frac{(y - 1)^2}{2^2} = 1$$

The center is $(1, 1)$. Note that $a = 2$ and $b = 1$. The major axis is vertical so the vertices are 2 units below and above the center:

$(1, 1 - 2)$ and $(1, 1 + 2)$, or $(1, -1)$ and $(1, 3)$.

We know that $c^2 = a^2 - b^2$, so $c^2 = 4 - 1 = 3$ and $c = \sqrt{3}$. Then the foci are $\sqrt{3}$ units below and above the center:

$(1, 1 - \sqrt{3})$ and $(1, 1 + \sqrt{3})$.

To graph the ellipse, plot the vertices. Since $b = 1$, two other points on the graph are 1 unit left and right of the center:

$(1 - 1, 1)$ and $(1 + 1, 1)$ or $(0, 1)$ and $(2, 1)$.

Plot these points also and connect the four plotted points with a smooth curve.

$$4x^2 + y^2 - 8x - 2y + 1 = 0$$

47. The ellipse in Example 4 is flatter than the one in Example 2, so the ellipse in Example 2 has the smaller eccentricity.

We compute the eccentricities: In Example 2, $c = 3$ and $a = 5$, so $e = c/a = 3/5 = 0.6$. In Example 4, $c = 2\sqrt{3}$ and $a = 4$, so $e = c/a = 2\sqrt{3}/4 \approx 0.866$. These computations confirm that the ellipse in Example 2 has the smaller eccentricity.

49. Since the vertices, $(0, -4)$ and $(0, 4)$ are on the y-axis and are equidistant from the origin, we know that the major axis of the ellipse is vertical, its center is at the origin, and $a = 4$. Use the information that $e = 1/4$ to find c:

$$e = \frac{c}{a}$$

$$\frac{1}{4} = \frac{c}{4} \quad \text{Substituting}$$

$$c = 1$$

Now $c^2 = a^2 - b^2$, so we can find b^2:

$$1^2 = 4^2 - b^2$$

$$1 = 16 - b^2$$

$$b^2 = 15$$

Write the equation of the ellipse:

$$\frac{x^2}{b^2} + \frac{y^2}{a^2} = 1$$

$$\frac{x^2}{15} + \frac{y^2}{16} = 1$$

51. From the figure in the text we see that the center of the ellipse is $(0, 0)$, the major axis is horizontal, the vertices are $(-50, 0)$ and $(50, 0)$, and one y-intercept is $(0, 12)$. Then $a = 50$ and $b = 12$. The equation is

$$\frac{x^2}{a^2} + \frac{y^2}{b^2} = 1$$

$$\frac{x^2}{50^2} + \frac{y}{12^2} = 1$$

$$\frac{x^2}{2500} + \frac{y^2}{144} = 1.$$

53. Position a coordinate system as shown below where 1 unit $= 10^7$ mi.

The length of the major axis is $9.3 + 9.1$, or 18.4. Then the distance from the center of the ellipse (the origin) to V is $18.4/2$, or 9.2. Since the distance from the sun to V is 9.1, the distance from the sun to the center is $9.2 - 9.1$, or 0.1. Then the distance from the sun to the other focus is twice this distance:

$$2(0.1 \times 10^7 \text{ mi}) = 0.2 \times 10^7 \text{ mi}$$

$$= 2 \times 10^6 \text{ mi}$$

55. Center: $(2.003, -1.005)$

Vertices: $(-1.017, -1.005)$, $(5.023, -1.005)$

57. Discussion and Writing

59. midpoint

61. y-intercept

63. remainder

65. parabola

67. The center of the ellipse is the midpoint of the segment connecting the vertices:

$$\left(\frac{3 + 3}{2}, \frac{-4 + 6}{2} \right), \text{ or } (3, 1).$$

Now a is the distance from the origin to a vertex. We use the vertex $(3, 6)$.

$$a = \sqrt{(3 - 3)^2 + (6 - 1)^2} = 5$$

Also b is one-half the length of the minor axis.

$$b = \frac{\sqrt{(5 - 1)^2 + (1 - 1)^2}}{2} = \frac{4}{2} = 2$$

The vertices lie on the vertical line $x = 3$, so the major axis is vertical. We write the equation of the ellipse.

$$\frac{(x - h)^2}{b^2} + \frac{(y - k)^2}{a^2} = 1$$

$$\frac{(x - 3)^2}{4} + \frac{(y - 1)^2}{25} = 1$$

69. The center is the midpoint of the segment connecting the vertices:

$$\left(\frac{-3 + 3}{2}, \frac{0 + 0}{2} \right), \text{ or } (0, 0).$$

Then $a = 3$ and since the vertices are on the x-axis, the major axis is horizontal. The equation is of the form $\frac{x^2}{a^2} + \frac{y^2}{b^2} = 1$.

Substitute 3 for a, 2 for x, and $\frac{22}{3}$ for y and solve for b^2.

$$\frac{4}{9}+\frac{\frac{484}{9}}{b^2}=1$$

$$\frac{4}{9}+\frac{484}{9b^2}=1$$

$$4b^2+484=9b^2$$

$$484=5b^2$$

$$\frac{484}{5}=b^2$$

Then the equation is $\frac{x^2}{9}+\frac{y^2}{484/5}=1$.

71. Position a coordinate system as shown.

The equation of the ellipse is

$$\frac{x^2}{25^2}+\frac{y^2}{14^2}=1$$

$$\frac{x^2}{625}+\frac{y^2}{196}=1.$$

A point 6 ft from the riverbank corresponds to $(25-6,0)$, or $(19,0)$ or to $(-25+6,0)$, or $(-19,0)$. Substitute either 19 or -19 for x and solve for y, the clearance.

$$\frac{19^2}{625}+\frac{y^2}{196}=1$$

$$\frac{y^2}{196}=1-\frac{361}{625}$$

$$y^2=196\left(1-\frac{361}{625}\right)$$

$$y\approx9.1$$

The clearance 6 ft from the riverbank is about 9.1 ft.

Exercise Set 6.3

1. Graph (b) is the graph of $\frac{x^2}{25}-\frac{y^2}{9}=1$.

3. Graph (c) is the graph of $\frac{(y-1)^2}{16}-\frac{(x+3)^2}{1}=1$.

5. Graph (a) is the graph of $25x^2-16y^2=400$.

7. The vertices are equidistant from the origin and are on the y-axis, so the center is at the origin and the transverse axis is vertical. Since $c^2=a^2+b^2$, we have $5^2=3^2+b^2$ so $b^2=16$.

The equation is of the form $\frac{y^2}{a^2}-\frac{x^2}{b^2}=1$, so we have

$$\frac{y^2}{9}-\frac{x^2}{16}=1.$$

9. The asymptotes pass through the origin, so the center is the origin. The given vertex is on the x-axis, so the transverse axis is horizontal. Since $\frac{b}{a}x=\frac{3}{2}x$ and $a=2$, we have $b=3$. The equation is of the form $\frac{x^2}{a^2}-\frac{y^2}{b^2}=1$, so we have $\frac{x^2}{2^2}-\frac{y^2}{3^2}=1$, or $\frac{x^2}{4}-\frac{y^2}{9}=1$.

11. $\frac{x^2}{4}-\frac{y^2}{4}=1$

$$\frac{x^2}{2^2}-\frac{y^2}{2^2}=1 \quad\text{Standard form}$$

The center is $(0,0)$; $a=2$ and $b=2$. The transverse axis is horizontal so the vertices are $(-2,0)$ and $(2,0)$. Since $c^2=a^2+b^2$, we have $c^2=4+4=8$ and $c=\sqrt{8}$, or $2\sqrt{2}$. Then the foci are $(-2\sqrt{2},0)$ and $(2\sqrt{2},0)$.

Find the asymptotes:

$$y=\frac{b}{a}x \quad\text{and}\quad y=-\frac{b}{a}x$$

$$y=\frac{2}{2}x \quad\text{and}\quad y=-\frac{2}{2}x$$

$$y=x \quad\text{and}\quad y=-x$$

To draw the graph sketch the asymptotes, plot the vertices, and draw the branches of the hyperbola outward from the vertices toward the asymptotes.

$$\frac{x^2}{4}-\frac{y^2}{4}=1$$

13. $\frac{(x-2)^2}{9}-\frac{(y+5)^2}{1}=1$

$$\frac{(x-2)^2}{3^2}-\frac{[y-(-5)]^2}{1^2}=1 \quad\text{Standard form}$$

The center is $(2,-5)$; $a=3$ and $b=1$. The transverse axis is horizontal, so the vertices are 3 units left and right of the center:

$(2-3,-5)$ and $(2+3,-5)$, or $(-1,-5)$ and $(5,-5)$.

Since $c^2=a^2+b^2$, we have $c^2=9+1=10$ and $c=\sqrt{10}$. Then the foci are $\sqrt{10}$ units left and right of the center:

$$(2-\sqrt{10},-5) \text{ and } (2+\sqrt{10},-5).$$

Find the asymptotes:

$$y - k = \frac{b}{a}(x - h) \quad \text{and} \quad y - k = -\frac{b}{a}(x - h)$$

$$y - (-5) = \frac{1}{3}(x - 2) \quad \text{and} \quad y - (-5) = -\frac{1}{3}(x - 2)$$

$$y + 5 = \frac{1}{3}(x - 2) \quad \text{and} \quad y + 5 = -\frac{1}{3}(x - 2), \text{ or}$$

$$y = \frac{1}{3}x - \frac{17}{3} \quad \text{and} \quad y = -\frac{1}{3}x - \frac{13}{3}$$

Sketch the asymptotes, plot the vertices, and draw the graph.

$$\frac{(x - 2)^2}{9} - \frac{(y + 5)^2}{1} = 1$$

15.
$$\frac{(y + 3)^2}{4} - \frac{(x + 1)^2}{16} = 1$$

$$\frac{[y - (-3)]^2}{2^2} - \frac{[x - (-1)]^2}{4^2} = 1 \quad \text{Standard form}$$

The center is $(-1, -3)$; $a = 2$ and $b = 4$. The transverse axis is vertical, so the vertices are 2 units below and above the center:

$(-1, -3 - 2)$ and $(1, -3 + 2)$, or $(-1, -5)$ and $(-1, -1)$.

Since $c^2 = a^2 + b^2$, we have $c^2 = 4 + 16 = 20$ and $c = \sqrt{20}$, or $2\sqrt{5}$. Then the foci are $2\sqrt{5}$ units below and above of the center:

$(-1, -3 - 2\sqrt{5})$ and $(-1, -3 + 2\sqrt{5})$.

Find the asymptotes:

$$y - k = \frac{a}{b}(x - h) \quad \text{and} \quad y - k = -\frac{a}{b}(x - h)$$

$$y - (-3) = \frac{2}{4}(x - (-1)) \quad \text{and} \quad y - (-3) = -\frac{2}{4}(x - (-1))$$

$$y + 3 = \frac{1}{2}(x + 1) \quad \text{and} \quad y + 3 = -\frac{1}{2}(x + 1), \text{ or}$$

$$y = \frac{1}{2}x - \frac{5}{2} \quad \text{and} \quad y = -\frac{1}{2}x - \frac{7}{2}$$

Sketch the asymptotes, plot the vertices, and draw the graph.

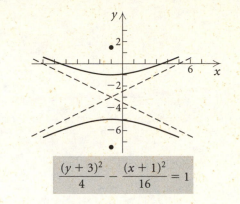

$$\frac{(y + 3)^2}{4} - \frac{(x + 1)^2}{16} = 1$$

17. $x^2 - 4y^2 = 4$

$$\frac{x^2}{4} - \frac{y^2}{1} = 1$$

$$\frac{x^2}{2^2} - \frac{y^2}{1^2} = 1 \quad \text{Standard form}$$

The center is $(0, 0)$; $a = 2$ and $b = 1$. The transverse axis is horizontal, so the vertices are $(-2, 0)$ and $(2, 0)$. Since $c^2 = a^2 + b^2$, we have $c^2 = 4 + 1 = 5$ and $c = \sqrt{5}$. Then the foci are $(-\sqrt{5}, 0)$ and $(\sqrt{5}, 0)$.

Find the asymptotes:

$$y = \frac{b}{a}x \quad \text{and} \quad y = -\frac{b}{a}x$$

$$y = \frac{1}{2}x \quad \text{and} \quad y = -\frac{1}{2}x$$

Sketch the asymptotes, plot the vertices, and draw the graph.

$$x^2 - 4y^2 = 4$$

19. $9y^2 - x^2 = 81$

$$\frac{y^2}{9} - \frac{x^2}{81} = 1$$

$$\frac{y^2}{3^2} - \frac{x^2}{9^2} = 1 \quad \text{Standard form}$$

The center is $(0, 0)$; $a = 3$ and $b = 9$. The transverse axis is vertical, so the vertices are $(0, -3)$ and $(0, 3)$. Since $c^2 = a^2 + b^2$, we have $c^2 = 9 + 81 = 90$ and $c = \sqrt{90}$, or $3\sqrt{10}$. Then the foci are $(0, -3\sqrt{10})$ and $(0, 3\sqrt{10})$.

Find the asymptotes:

$$y = \frac{a}{b}x \text{ and } y = -\frac{a}{b}x$$

$$y = \frac{3}{9}x \text{ and } y = -\frac{3}{9}x$$

$$y = \frac{1}{3}x \text{ and } y = -\frac{1}{3}x$$

Sketch the asymptotes, plot the vertices, and draw the graph.

$$9y^2 - x^2 = 81$$

21.
$$x^2 - y^2 = 2$$
$$\frac{x^2}{2} - \frac{y^2}{2} = 1$$

$$\frac{x^2}{(\sqrt{2})^2} - \frac{y^2}{(\sqrt{2})^2} = 1 \quad \text{Standard form}$$

The center is $(0,0)$; $a = \sqrt{2}$ and $b = \sqrt{2}$. The transverse axis is horizontal, so the vertices are $(-\sqrt{2}, 0)$ and $(\sqrt{2}, 0)$. Since $c^2 = a^2 + b^2$, we have $c^2 = 2 + 2 = 4$ and $c = 2$. Then the foci are $(-2, 0)$ and $(2, 0)$.

Find the asymptotes:

$$y = \frac{b}{a}x \quad \text{and} \quad y = -\frac{b}{a}x$$

$$y = \frac{\sqrt{2}}{\sqrt{2}}x \quad \text{and} \quad y = -\frac{\sqrt{2}}{\sqrt{2}}x$$

$$y = x \quad \text{and} \quad y = -x$$

Sketch the asymptotes, plot the vertices, and draw the graph.

$$x^2 - y^2 = 2$$

23.
$$y^2 - x^2 = \frac{1}{4}$$

$$\frac{y^2}{1/4} - \frac{x^2}{1/4} = 1$$

$$\frac{y^2}{(1/2)^2} - \frac{x^2}{(1/2)^2} = 1 \quad \text{Standard form}$$

The center is $(0,0)$; $a = \frac{1}{2}$ and $b = \frac{1}{2}$. The transverse axis is vertical, so the vertices are $\left(0, -\frac{1}{2}\right)$ and $\left(0, \frac{1}{2}\right)$. Since $c^2 = a^2 + b^2$, we have $c^2 = \frac{1}{4} + \frac{1}{4} = \frac{1}{2}$ and $c = \sqrt{\frac{1}{2}}$, or $\frac{\sqrt{2}}{2}$. Then the foci are $\left(0, -\frac{\sqrt{2}}{2}\right)$ and $\left(0, \frac{\sqrt{2}}{2}\right)$.

Find the asymptotes:

$$y = \frac{a}{b}x \quad \text{and} \quad y = -\frac{a}{b}x$$

$$y = \frac{1/2}{1/2}x \quad \text{and} \quad y = -\frac{1/2}{1/2}x$$

$$y = x \quad \text{and} \quad y = -x$$

Sketch the asymptotes, plot the vertices, and draw the graph.

$$y^2 - x^2 = \frac{1}{4}$$

25. Begin by completing the square twice.

$$x^2 - y^2 - 2x - 4y - 4 = 0$$
$$(x^2 - 2x) - (y^2 + 4y) = 4$$
$$(x^2 - 2x + 1) - (y^2 + 4y + 4) = 4 + 1 - 1 \cdot 4$$
$$(x - 1)^2 - (y + 2)^2 = 1$$
$$\frac{(x-1)^2}{1^2} - \frac{[y-(-2)]^2}{1^2} = 1 \quad \text{Standard form}$$

The center is $(1, -2)$; $a = 1$ and $b = 1$. The transverse axis is horizontal, so the vertices are 1 unit left and right of the center:

$(1 - 1, -2)$ and $(1 + 1, -2)$ or $(0, -2)$ and $(2, -2)$

Since $c^2 = a^2 + b^2$, we have $c^2 = 1 + 1 = 2$ and $c = \sqrt{2}$. Then the foci are $\sqrt{2}$ units left and right of the center:

$$(1 - \sqrt{2}, -2) \text{ and } (1 + \sqrt{2}, -2).$$

Find the asymptotes:

$$y - k = \frac{b}{a}(x - h) \text{ and } \quad y - k = -\frac{b}{a}(x - h)$$

$$y - (-2) = \frac{1}{1}(x - 1) \text{ and } y - (-2) = -\frac{1}{1}(x - 1)$$

$$y + 2 = x - 1 \quad \text{and} \quad y + 2 = -(x - 1), \text{ or}$$

$$y = x - 3 \quad \text{and} \quad y = -x - 1$$

Sketch the asymptotes, plot the vertices, and draw the graph.

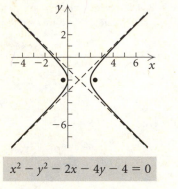

$$x^2 - y^2 - 2x - 4y - 4 = 0$$

27. Begin by completing the square twice.

$$36x^2 - y^2 - 24x + 6y - 41 = 0$$

$$(36x^2 - 24x) - (y^2 - 6y) = 41$$

$$36\left(x^2 - \frac{2}{3}x\right) - (y^2 - 6y) = 41$$

$$36\left(x^2 - \frac{2}{3}x + \frac{1}{9}\right) - (y^2 - 6y + 9) = 41 + 36 \cdot \frac{1}{9} - 1 \cdot 9$$

$$36\left(x - \frac{1}{3}\right)^2 - (y - 3)^2 = 36$$

$$\frac{\left(x - \frac{1}{3}\right)^2}{1} - \frac{(y - 3)^2}{36} = 1$$

$$\frac{\left(x - \frac{1}{3}\right)^2}{1^2} - \frac{(y - 3)^2}{6^2} = 1 \quad \text{Standard}$$
$$\text{form}$$

The center is $\left(\frac{1}{3}, 3\right)$; $a = 1$ and $b = 6$. The transverse axis is horizontal, so the vertices are 1 unit left and right of the center:

$$\left(\frac{1}{3} - 1, 3\right) \text{ and } \left(\frac{1}{3} + 1, 3\right) \text{ or } \left(-\frac{2}{3}, 3\right) \text{ and}$$

$$\left(\frac{4}{3}, 3\right).$$

Since $c^2 = a^2 + b^2$, we have $c^2 = 1 + 36 = 37$ and $c = \sqrt{37}$. Then the foci are $\sqrt{37}$ units left and right of the center:

$$\left(\frac{1}{3} - \sqrt{37}, 3\right) \text{ and } \left(\frac{1}{3} + \sqrt{37}, 3\right).$$

Find the asymptotes:

$$y - k = \frac{b}{a}(x - h) \quad \text{and} \quad y - k = -\frac{b}{a}(x - h)$$

$$y - 3 = \frac{6}{1}\left(x - \frac{1}{3}\right) \quad \text{and} \quad y - 3 = -\frac{6}{1}\left(x - \frac{1}{3}\right)$$

$$y - 3 = 6\left(x - \frac{1}{3}\right) \quad \text{and} \quad y - 3 = -6\left(x - \frac{1}{3}\right), \text{ or}$$

$$y = 6x + 1 \quad \text{and} \quad y = -6x + 5$$

Sketch the asymptotes, plot the vertices, and draw the graph.

$$36x^2 - y^2 - 24x + 6y - 41 = 0$$

29. Begin by completing the square twice.

$$9y^2 - 4x^2 - 18y + 24x - 63 = 0$$

$$9(y^2 - 2y) - 4(x^2 - 6x) = 63$$

$$9(y^2 - 2y + 1) - 4(x^2 - 6x + 9) = 63 + 9 \cdot 1 - 4 \cdot 9$$

$$9(y - 1)^2 - 4(x - 3)^2 = 36$$

$$\frac{(y - 1)^2}{4} - \frac{(x - 3)^2}{9} = 1$$

$$\frac{(y - 1)^2}{2^2} - \frac{(x - 3)^2}{3^2} = 1 \quad \text{Standard}$$
$$\text{form}$$

The center is $(3, 1)$; $a = 2$ and $b = 3$. The transverse axis is vertical, so the vertices are 2 units below and above the center:

$(3, 1 - 2)$ and $(3, 1 + 2)$, or $(3, -1)$ and $(3, 3)$.

Since $c^2 = a^2 + b^2$, we have $c^2 = 4 + 9 = 13$ and $c = \sqrt{13}$. Then the foci are $\sqrt{13}$ units below and above the center:

$(3, 1 - \sqrt{13})$ and $(3, 1 + \sqrt{13})$.

Find the asymptotes:

$$y - k = \frac{a}{b}(x - h) \text{ and } y - k = -\frac{a}{b}(x - h)$$

$$y - 1 = \frac{2}{3}(x - 3) \text{ and } y - 1 = -\frac{2}{3}(x - 3), \text{ or}$$

$$y = \frac{2}{3}x - 1 \quad \text{and} \quad y = -\frac{2}{3}x + 3$$

Sketch the asymptotes, plot the vertices, and draw the graph.

$$9y^2 - 4x^2 - 18y + 24x - 63 = 0$$

31. Begin by completing the square twice.

$$x^2 - y^2 - 2x - 4y = 4$$
$$(x^2 - 2x + 1) - (y^2 + 4y + 4) = 4 + 1 - 4$$
$$(x - 1)^2 - (y + 2)^2 = 1$$
$$\frac{(x-1)^2}{1^2} - \frac{[y-(-2)]^2}{1^2} = 1 \qquad \text{Standard form}$$

The center is $(1, -2)$; $a = 1$ and $b = 1$. The transverse axis is horizontal, so the vertices are 1 unit left and right of the center:

$(1 - 1, -2)$ and $(1 + 1, -2)$, or $(0, -2)$ and $(2, -2)$.

Since $c^2 = a^2 + b^2$, we have $c^2 = 1 + 1 = 2$ and $c = \sqrt{2}$. Then the foci are $\sqrt{2}$ units left and right of the center:

$(1 - \sqrt{2}, -2)$ and $(1 + \sqrt{2}, -2)$.

Find the asymptotes:

$$y - k = \frac{b}{a}(x - h) \quad \text{and} \quad y - k = -\frac{b}{a}(x - h)$$

$$y - (-2) = \frac{1}{1}(x - 1) \quad \text{and} \quad y - (-2) = -\frac{1}{1}(x - 1)$$

$$y + 2 = x - 1 \qquad \text{and} \qquad y + 2 = -(x - 1), \text{ or}$$

$$y = x - 3 \qquad \text{and} \qquad y = -x - 1$$

Sketch the asymptotes, plot the vertices, and draw the graph.

$$x^2 - y^2 - 2x - 4y = 4$$

33. Begin by completing the square twice.

$$y^2 - x^2 - 6x - 8y - 29 = 0$$
$$(y^2 - 8y + 16) - (x^2 + 6x + 9) = 29 + 16 - 9$$
$$(y - 4)^2 - (x + 3)^2 = 36$$
$$\frac{(y-4)^2}{36} - \frac{(x+3)^2}{36} = 1$$
$$\frac{(y-4)^2}{6^2} - \frac{[x-(-3)]^2}{6^2} = 1 \qquad \text{Standard form}$$

The center is $(-3, 4)$; $a = 6$ and $b = 6$. The transverse axis is vertical, so the vertices are 6 units below and above the center:

$(-3, 4 - 6)$ and $(-3, 4 + 6)$, or $(-3, -2)$ and $(-3, 10)$.

Since $c^2 = a^2 + b^2$, we have $c^2 = 36 + 36 = 72$ and $c = \sqrt{72}$, or $6\sqrt{2}$. Then the foci are $6\sqrt{2}$ units below and above the center:

$$(-3, 4 - 6\sqrt{2}) \text{ and } (-3, 4 + 6\sqrt{2}).$$

Find the asymptotes:

$$y - k = \frac{a}{b}(x - h) \qquad \text{and} \quad y - k = -\frac{a}{b}(x - h)$$

$$y - 4 = \frac{6}{6}(x - (-3)) \quad \text{and} \quad y - 4 = -\frac{6}{6}(x - (-3))$$

$$y - 4 = x + 3 \qquad \text{and} \quad y - 4 = -(x + 3), \text{ or}$$

$$y = x + 7 \qquad \text{and} \qquad y = -x + 1$$

Sketch the asymptotes, plot the vertices, and draw the graph.

$$y^2 - x^2 - 6x - 8y - 29 = 0$$

35. The hyperbola in Example 3 is wider than the one in Example 2, so the hyperbola in Example 3 has the larger eccentricity.

Compute the eccentricities: In Example 2, $c = 5$ and $a = 4$, so $e = 5/4$, or 1.25. In Example 3, $c = \sqrt{5}$ and $a = 1$, so $e = \sqrt{5}/1 \approx 2.24$. These computations confirm that the hyperbola in Example 3 has the larger eccentricity.

37. The center is the midpoint of the segment connecting the vertices:

$$\left(\frac{3-3}{2}, \frac{7+7}{2}\right), \text{ or } (0, 7).$$

The vertices are on the horizontal line $y = 7$, so the transverse axis is horizontal. Since the vertices are 3 units left and right of the center, $a = 3$.

Find c:

$$e = \frac{c}{a} = \frac{5}{3}$$

$$\frac{c}{3} = \frac{5}{3} \quad \text{Substituting 3 for } a$$

$$c = 5$$

Now find b^2:

$$c^2 = a^2 + b^2$$

$$5^2 = 3^2 + b^2$$

$$16 = b^2$$

Write the equation:

$$\frac{(x-h)^2}{a^2} - \frac{(y-k)^2}{b^2} = 1$$

$$\frac{x^2}{9} - \frac{(y-7)^2}{16} = 1$$

39.

One focus is 6 units above the center of the hyperbola, so $c = 6$. One vertex is 5 units above the center, so $a = 5$. Find b^2:

$$c^2 = a^2 + b^2$$

$$6^2 = 5^2 + b^2$$

$$11 = b^2$$

Write the equation:

$$\frac{y^2}{a^2} - \frac{x^2}{b^2} = 1$$

$$\frac{y^2}{25} - \frac{x^2}{11} = 1$$

41. Center: $(-1.460, -0.957)$

Vertices: $(-2.360, -0.957)$, $(-0.560, -0.957)$

Asymptotes: $y = -1.429x - 3.043$, $y = 1.429x + 1.129$

43. Discussion and Writing

45. a) The graph of $f(x) = 2x - 3$ is shown below.

Since there is no horizontal line that crosses the graph more than once, the function is one-to-one.

b) Replace $f(x)$ with y: $y = 2x - 3$

Interchange x and y: $x = 2y - 3$

Solve for y: $x + 3 = 2y$

$$\frac{x+3}{2} = y$$

Replace y with $f^{-1}(x)$: $f^{-1}(x) = \frac{x+3}{2}$

47. a) The graph of $f(x) = \dfrac{5}{x-1}$ is shown below.

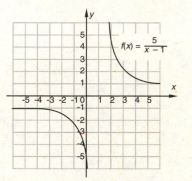

Since there is no horizontal line that crosses the graph more than once, the function is one-to-one.

b) Replace $f(x)$ with y: $y = \dfrac{5}{x-1}$

Interchange x and y: $x = \dfrac{5}{y-1}$

Solve for y: $x(y-1) = 5$

$$y - 1 = \frac{5}{x}$$

$$y = \frac{5}{x} + 1$$

Replace y with $f^{-1}(x)$: $f^{-1} = \dfrac{5}{x} + 1$, or $\dfrac{5+x}{x}$

49.
$$x + y = 5, \quad (1)$$
$$\underline{x - y = 7} \quad (2)$$
$$2x \quad\;\; = 12 \quad \text{Adding}$$
$$x = 6$$

Back-substitute in either equation (1) or (2) and solve for y. We use equation (1).

$$6 + y = 5$$

$$y = -1$$

The solution is $(6, -1)$.

51.
$$2x - 3y = 7, \quad (1)$$
$$3x + 5y = 1 \quad (2)$$

Multiply equation (1) by 5 and equation (2) by 3 and add to eliminate y.

$$10x - 15y = 35$$
$$\underline{9x + 15y = 3}$$
$$19x \qquad\; = 38$$
$$x = 2$$

Back-substitute and solve for y.

$$3 \cdot 2 + 5y = 1 \quad \text{Using equation (2)}$$
$$5y = -5$$
$$y = -1$$

The solution is $(2, -1)$.

53. The center is the midpoint of the segment connecting $(3, -8)$ and $(3, -2)$:

$$\left(\frac{3+3}{2}, \frac{-8-2}{2} \right), \text{ or } (3, -5).$$

The vertices are on the vertical line $x = 3$ and are 3 units above and below the center so the transverse axis is vertical and $a = 3$. Use the equation of an asymptote to find b:

$$y - k = \frac{a}{b}(x - h)$$

$$y + 5 = \frac{3}{b}(x - 3)$$

$$y = \frac{3}{b}x - \frac{9}{b} - 5$$

This equation corresponds to the asymptote $y = 3x - 14$, so $\frac{3}{b} = 3$ and $b = 1$.

Write the equation of the hyperbola:

$$\frac{(y-k)^2}{a^2} - \frac{(x-h)^2}{b^2} = 1$$

$$\frac{(y+5)^2}{9} - \frac{(x-3)^2}{1} = 1$$

55. S and T are the foci of the hyperbola, so $c = 300/2 = 150$.

$200 \text{ microseconds} \cdot \dfrac{0.186 \text{ mi}}{1 \text{ microsecond}} = 37.2 \text{ mi}$, the difference of the ships' distances from the foci. That is, $2a = 37.2$, so $a = 18.6$.

Find b^2:

$$c^2 = a^2 + b^2$$
$$150^2 = 18.6^2 + b^2$$
$$22,154.04 = b^2$$

Then the equation of the hyperbola is

$$\frac{x^2}{18.6^2} - \frac{y^2}{22,154.04} = 1, \text{ or } \frac{x^2}{345.96} - \frac{y^2}{22,154.04} = 1.$$

Exercise Set 6.4

1. The correct graph is (e).

3. The correct graph is (c).

5. The correct graph is (b).

7. $x^2 + y^2 = 25, \qquad (1)$

$y - x = 1 \qquad\qquad (2)$

First solve equation (2) for y.

$y = x + 1 \qquad\qquad (3)$

Then substitute $x + 1$ for y in equation (1) and solve for x.

$$x^2 + y^2 = 25$$
$$x^2 + (x+1)^2 = 25$$
$$x^2 + x^2 + 2x + 1 = 25$$
$$2x^2 + 2x - 24 = 0$$
$$x^2 + x - 12 = 0 \quad \text{Multiplying by } \frac{1}{2}$$
$$(x+4)(x-3) = 0 \quad \text{Factoring}$$
$$x + 4 = 0 \quad \text{or} \quad x - 3 = 0 \quad \begin{array}{l}\text{Principle of zero}\\ \text{products}\end{array}$$
$$x = -4 \quad \text{or} \qquad x = 3$$

Now substitute these numbers into equation (3) and solve for y.

$$y = -4 + 1 = -3$$
$$y = 3 + 1 = 4$$

The pairs $(-4, -3)$ and $(3, 4)$ check, so they are the solutions.

9. $4x^2 + 9y^2 = 36, \quad (1)$

$3y + 2x = 6 \qquad\quad (2)$

First solve equation (2) for y.

$$3y = -2x + 6$$
$$y = -\frac{2}{3}x + 2 \qquad (3)$$

Then substitute $-\frac{2}{3}x + 2$ for y in equation (1) and solve for x.

$$4x^2 + 9y^2 = 36$$
$$4x^2 + 9\left(-\frac{2}{3}x + 2 \right)^2 = 36$$
$$4x^2 + 9\left(\frac{4}{9}x^2 - \frac{8}{3}x + 4 \right) = 36$$
$$4x^2 + 4x^2 - 24x + 36 = 36$$
$$8x^2 - 24x = 0$$
$$x^2 - 3x = 0$$
$$x(x - 3) = 0$$

$x = 0 \quad \text{or} \quad x = 3$

Now substitute these numbers in equation (3) and solve for y.

$$y = -\frac{2}{3} \cdot 0 + 2 = 2$$
$$y = -\frac{2}{3} \cdot 3 + 2 = 0$$

The pairs $(0, 2)$ and $(3, 0)$ check, so they are the solutions.

11. $x^2 + y^2 = 25, \qquad (1)$

$y^2 = x + 5 \qquad\qquad (2)$

We substitute $x + 5$ for y^2 in equation (1) and solve for x.

$$x^2 + y^2 = 25$$
$$x^2 + (x+5) = 25$$
$$x^2 + x - 20 = 0$$
$$(x+5)(x-4) = 0$$

$$x + 5 = 0 \quad \text{or} \quad x - 4 = 0$$
$$x = -5 \quad \text{or} \quad x = 4$$

We substitute these numbers for x in either equation (1) or equation (2) and solve for y. Here we use equation (2).

$y^2 = -5 + 5 = 0$ and $y = 0$.

$y^2 = 4 + 5 = 9$ and $y = \pm 3$.

The pairs $(-5, 0)$, $(4, 3)$ and $(4, -3)$ check. They are the solutions.

13. $\quad x^2 + y^2 = 9, \qquad (1)$
$\qquad x^2 - y^2 = 9 \qquad\;\; (2)$

Here we use the elimination method.

$$
\begin{aligned}
x^2 + y^2 &= \;\;9 \qquad (1)\\
\underline{x^2 - y^2} &= \underline{\;\;9} \qquad (2)\\
2x^2 \quad\;\; &= 18 \qquad \text{Adding}\\
x^2 &= \;\;9\\
x &= \pm 3
\end{aligned}
$$

If $x = 3$, $x^2 = 9$, and if $x = -3$, $x^2 = 9$, so substituting 3 or -3 in equation (1) gives us

$$
\begin{aligned}
x^2 + y^2 &= 9\\
9 + y^2 &= 9\\
y^2 &= 0\\
y &= 0.
\end{aligned}
$$

The pairs $(3, 0)$ and $(-3, 0)$ check. They are the solutions.

15. $\quad y^2 - x^2 = 9 \qquad (1)$
$\qquad 2x - 3 = y \qquad (2)$

Substitute $2x - 3$ for y in equation (1) and solve for x.

$$
\begin{aligned}
y^2 - x^2 &= 9\\
(2x - 3)^2 - x^2 &= 9\\
4x^2 - 12x + 9 - x^2 &= 9\\
3x^2 - 12x &= 0\\
x^2 - 4x &= 0\\
x(x - 4) &= 0
\end{aligned}
$$

$x = 0 \quad \text{or} \quad x = 4$

Now substitute these numbers into equation (2) and solve for y.

If $x = 0$, $y = 2 \cdot 0 - 3 = -3$.

If $x = 4$, $y = 2 \cdot 4 - 3 = 5$.

The pairs $(0, -3)$ and $(4, 5)$ check. They are the solutions.

17. $\quad y^2 = x + 3, \qquad (1)$
$\qquad 2y = x + 4 \qquad (2)$

First solve equation (2) for x.

$\qquad 2y - 4 = x \qquad (3)$

Then substitute $2y - 4$ for x in equation (1) and solve for y.

$$
\begin{aligned}
y^2 &= x + 3\\
y^2 &= (2y - 4) + 3\\
y^2 &= 2y - 1\\
y^2 - 2y + 1 &= 0\\
(y - 1)(y - 1) &= 0
\end{aligned}
$$

$y - 1 = 0 \quad \text{or} \quad y - 1 = 0$
$\;\;\; y = 1 \quad \text{or} \qquad\;\; y = 1$

Now substitute 1 for y in equation (3) and solve for x.

$$
\begin{aligned}
2 \cdot 1 - 4 &= x\\
-2 &= x
\end{aligned}
$$

The pair $(-2, 1)$ checks. It is the solution.

19. $\quad x^2 + y^2 = 25, \quad (1)$
$\qquad xy = 12 \qquad\;\;\; (2)$

First we solve equation (2) for y.

$$
\begin{aligned}
xy &= 12\\
y &= \frac{12}{x}
\end{aligned}
$$

Then we substitute $\dfrac{12}{x}$ for y in equation (1) and solve for x.

$$
\begin{aligned}
x^2 + y^2 &= 25\\
x^2 + \left(\frac{12}{x}\right)^2 &= 25\\
x^2 + \frac{144}{x^2} &= 25\\
x^4 + 144 &= 25x^2 \quad \text{Multiplying by } x^2\\
x^4 - 25x^2 + 144 &= 0\\
u^2 - 25u + 144 &= 0 \qquad \text{Letting } u = x^2\\
(u - 9)(u - 16) &= 0
\end{aligned}
$$

$u = 9 \quad \text{or} \quad u = 16$

We now substitute x^2 for u and solve for x.

$\qquad x^2 = 9 \quad \text{or} \quad x^2 = 16$
$\qquad\;\; x = \pm 3 \quad \text{or} \quad\;\; x = \pm 4$

Since $y = 12/x$, if $x = 3$, $y = 4$; if $x = -3$, $y = -4$; if $x = 4$, $y = 3$; and if $x = -4$, $y = -3$. The pairs $(3, 4)$, $(-3, -4)$, $(4, 3)$, and $(-4, -3)$ check. They are the solutions.

21. $\quad x^2 + y^2 = 4, \qquad\quad (1)$
$\qquad 16x^2 + 9y^2 = 144 \quad (2)$

$$
\begin{aligned}
-9x^2 - 9y^2 &= -36 \quad \text{Multiplying (1) by } -9\\
\underline{16x^2 + 9y^2} &= \underline{\;144}\\
7x^2 \qquad\;\; &= 108 \quad \text{Adding}\\
x^2 &= \frac{108}{7}\\
x &= \pm\sqrt{\frac{108}{7}} = \pm 6\sqrt{\frac{3}{7}}\\
x &= \pm\frac{6\sqrt{21}}{7} \quad \text{Rationalizing the denominator}
\end{aligned}
$$

Substituting $\dfrac{6\sqrt{21}}{7}$ or $-\dfrac{6\sqrt{21}}{7}$ for x in equation (1) gives us

$$\frac{36 \cdot 21}{49} + y^2 = 4$$

$$y^2 = 4 - \frac{108}{7}$$

$$y^2 = -\frac{80}{7}$$

$$y = \pm\sqrt{-\frac{80}{7}} = \pm 4i\sqrt{\frac{5}{7}}$$

$$y = \pm\frac{4i\sqrt{35}}{7}. \qquad \text{Rationalizing the}$$
$$\text{denominator}$$

The pairs $\left(\dfrac{6\sqrt{21}}{7}, \dfrac{4i\sqrt{35}}{7}\right)$,

$\left(\dfrac{6\sqrt{21}}{7}, -\dfrac{4i\sqrt{35}}{7}\right)$, $\left(-\dfrac{6\sqrt{21}}{7}, \dfrac{4i\sqrt{35}}{7}\right)$, and

$\left(-\dfrac{6\sqrt{21}}{7}, -\dfrac{4i\sqrt{35}}{7}\right)$ check. They are the solutions.

23. $x^2 + 4y^2 = 25$, (1)

   $x + 2y = 7$ (2)

First solve equation (2) for x.

$x = -2y + 7$ (3)

Then substitute $-2y + 7$ for x in equation (1) and solve for y.

$$x^2 + 4y^2 = 25$$
$$(-2y + 7)^2 + 4y^2 = 25$$
$$4y^2 - 28y + 49 + 4y^2 = 25$$
$$8y^2 - 28y + 24 = 0$$
$$2y^2 - 7y + 6 = 0$$
$$(2y - 3)(y - 2) = 0$$

$$y = \frac{3}{2} \text{ or } y = 2$$

Now substitute these numbers in equation (3) and solve for x.

$$x = -2 \cdot \frac{3}{2} + 7 = 4$$
$$x = -2 \cdot 2 + 7 = 3$$

The pairs $\left(4, \dfrac{3}{2}\right)$ and $(3, 2)$ check, so they are the solutions.

25. $x^2 - xy + 3y^2 = 27$,   (1)

   $x - y = 2$ (2)

First solve equation (2) for y.

   $x - 2 = y$ (3)

Then substitute $x - 2$ for y in equation (1) and solve for x.

$$x^2 - xy + 3y^2 = 27$$
$$x^2 - x(x - 2) + 3(x - 2)^2 = 27$$
$$x^2 - x^2 + 2x + 3x^2 - 12x + 12 = 27$$
$$3x^2 - 10x - 15 = 0$$

$$x = \frac{-(-10) \pm \sqrt{(-10)^2 - 4(3)(-15)}}{2 \cdot 3}$$

$$x = \frac{10 \pm \sqrt{100 + 180}}{6} = \frac{10 \pm \sqrt{280}}{6}$$

$$x = \frac{10 \pm 2\sqrt{70}}{6} = \frac{5 \pm \sqrt{70}}{3}$$

Now substitute these numbers in equation (3) and solve for y.

$$y = \frac{5 + \sqrt{70}}{3} - 2 = \frac{-1 + \sqrt{70}}{3}$$

$$y = \frac{5 - \sqrt{70}}{3} - 2 = \frac{-1 - \sqrt{70}}{3}$$

The pairs $\left(\dfrac{5 + \sqrt{70}}{3}, \dfrac{-1 + \sqrt{70}}{3}\right)$ and

$\left(\dfrac{5 - \sqrt{70}}{3}, \dfrac{-1 - \sqrt{70}}{3}\right)$ check, so they are the solutions.

27. $x^2 + y^2 = 16$, $x^2 + y^2 = 16$,   (1)
                        or
   $y^2 - 2x^2 = 10$ $-2x^2 + y^2 = 10$ (2)

Here we use the elimination method.

$$\begin{array}{ll} 2x^2 + 2y^2 = 32 & \text{Multiplying (1) by 2} \\ \underline{-2x^2 + y^2 = 10} & \\ 3y^2 = 42 & \text{Adding} \\ y^2 = 14 & \\ y = \pm\sqrt{14} & \end{array}$$

Substituting $\sqrt{14}$ or $-\sqrt{14}$ for y in equation (1) gives us

$$x^2 + 14 = 16$$
$$x^2 = 2$$
$$x = \pm\sqrt{2}$$

The pairs $(-\sqrt{2}, -\sqrt{14})$, $(-\sqrt{2}, \sqrt{14})$, $(\sqrt{2}, -\sqrt{14})$, and $(\sqrt{2}, \sqrt{14})$ check. They are the solutions.

29. $x^2 + y^2 = 5$,   (1)

   $xy = 2$ (2)

First we solve equation (2) for y.

$$xy = 2$$
$$y = \frac{2}{x}$$

Then we substitute $\dfrac{2}{x}$ for y in equation (1) and solve for x.

$$x^2 + y^2 = 5$$
$$x^2 + \left(\frac{2}{x}\right)^2 = 5$$
$$x^2 + \frac{4}{x^2} = 5$$
$$x^4 + 4 = 5x^2 \quad \text{Multiplying by } x^2$$
$$x^4 - 5x^2 + 4 = 0$$
$$u^2 - 5u + 4 = 0 \quad \text{Letting } u = x^2$$
$$(u - 4)(u - 1) = 0$$
$$u = 4 \text{ or } u = 1$$

We now substitute x^2 for u and solve for x.

$$x^2 = 4 \quad \text{or} \quad x^2 = 1$$
$$x = \pm 2 \qquad x = \pm 1$$

Since $y = 2/x$, if $x = 2$, $y = 1$; if $x = -2$, $y = -1$; if $x = 1$, $y = 2$; and if $x = -1$, $y = -2$. The pairs $(2, 1)$, $(-2, -1)$, $(1, 2)$, and $(-1, -2)$ check. They are the solutions.

31. $3x + y = 7$ (1)
$4x^2 + 5y = 56$ (2)

First solve equation (1) for y.

$3x + y = 7$

$\quad y = 7 - 3x$ (3)

Next substitute $7 - 3x$ for y in equation (2) and solve for x.

$4x^2 + 5y = 56$

$4x^2 + 5(7 - 3x) = 56$

$4x^2 + 35 - 15x = 56$

$4x^2 - 15x - 21 = 0$

Using the quadratic formula, we find that

$$x = \frac{15 - \sqrt{561}}{8} \text{ or } x = \frac{15 + \sqrt{561}}{8}.$$

Now substitute these numbers into equation (3) and solve for y.

If $x = \dfrac{15 - \sqrt{561}}{8}$, $y = 7 - 3\left(\dfrac{15 - \sqrt{561}}{8}\right)$, or

$\dfrac{11 + 3\sqrt{561}}{8}$.

If $x = \dfrac{15 + \sqrt{561}}{8}$, $y = 7 - 3\left(\dfrac{15 + \sqrt{561}}{8}\right)$, or

$\dfrac{11 - 3\sqrt{561}}{8}$.

The pairs $\left(\dfrac{15 - \sqrt{561}}{8}, \dfrac{11 + 3\sqrt{561}}{8}\right)$ and

$\left(\dfrac{15 + \sqrt{561}}{8}, \dfrac{11 - 3\sqrt{561}}{8}\right)$ check and are the solutions.

33. $a + b = 7$,   (1)
$ab = 4$ (2)

First solve equation (1) for a.

$a = -b + 7$ (3)

Then substitute $-b + 7$ for a in equation (2) and solve for b.

$(-b + 7)b = 4$

$-b^2 + 7b = 4$

$\quad\quad 0 = b^2 - 7b + 4$

$b = \dfrac{-(-7) \pm \sqrt{(-7)^2 - 4 \cdot 1 \cdot 4}}{2 \cdot 1}$

$b = \dfrac{7 \pm \sqrt{33}}{2}$

Now substitute these numbers in equation (3) and solve for a.

$a = -\left(\dfrac{7 + \sqrt{33}}{2}\right) + 7 = \dfrac{7 - \sqrt{33}}{2}$

$a = -\left(\dfrac{7 - \sqrt{33}}{2}\right) + 7 = \dfrac{7 + \sqrt{33}}{2}$

The pairs $\left(\dfrac{7 - \sqrt{33}}{2}, \dfrac{7 + \sqrt{33}}{2}\right)$ and

$\left(\dfrac{7 + \sqrt{33}}{2}, \dfrac{7 - \sqrt{33}}{2}\right)$ check, so they are the solutions.

35. $x^2 + y^2 = 13$,   (1)
$xy = 6$ (2)

First we solve Equation (2) for y.

$xy = 6$

$y = \dfrac{6}{x}$

Then we substitute $\dfrac{6}{x}$ for y in equation (1) and solve for x.

$x^2 + y^2 = 13$

$x^2 + \left(\dfrac{6}{x}\right)^2 = 13$

$x^2 + \dfrac{36}{x^2} = 13$

$x^4 + 36 = 13x^2$ Multiplying by x^2

$x^4 - 13x^2 + 36 = 0$

$u^2 - 13u + 36 = 0$ Letting $u = x^2$

$(u - 9)(u - 4) = 0$

$u = 9$ or $u = 4$

We now substitute x^2 for u and solve for x.

$x^2 = 9$ or $x^2 = 4$

$x = \pm 3$ or $x = \pm 2$

Since $y = 6/x$, if $x = 3$, $y = 2$; if $x = -3$, $y = -2$; if $x = 2$, $y = 3$; and if $x = -2$, $y = -3$. The pairs $(3, 2)$, $(-3, -2)$, $(2, 3)$, and $(-2, -3)$ check. They are the solutions.

37. $x^2 + y^2 + 6y + 5 = 0$ (1)
$x^2 + y^2 - 2x - 8 = 0$ (2)

Using the elimination method, multiply equation (2) by -1 and add the result to equation (1).

$\quad x^2 + y^2 + 6y + 5 = 0$ (1)
$\underline{-x^2 - y^2 + 2x + 8 = 0}$ (2)
$\quad\quad\quad\quad 2x + 6y + 13 = 0$ (3)

Solve equation (3) for x.

$2x + 6y + 13 = 0$

$2x = -6y - 13$

$x = \dfrac{-6y - 13}{2}$

Substitute $\dfrac{-6y - 13}{2}$ for x in equation (1) and solve for y.

$x^2 + y^2 + 6y + 5 = 0$

$\left(\dfrac{-6y - 13}{2}\right)^2 + y^2 + 6y + 5 = 0$

$\dfrac{36y^2 + 156y + 169}{4} + y^2 + 6y + 5 = 0$

$36y^2 + 156y + 169 + 4y^2 + 24y + 20 = 0$

$\quad\quad\quad 40y^2 + 180y + 189 = 0$

Using the quadratic formula, we find that

$y = \dfrac{-45 \pm 3\sqrt{15}}{20}$. Substitute $\dfrac{-45 \pm 3\sqrt{15}}{20}$ for y in

$x = \dfrac{-6y - 13}{2}$ and solve for x.

If $y = \dfrac{-45 + 3\sqrt{15}}{20}$, then

$x = \dfrac{-6\left(\dfrac{-45 + 3\sqrt{15}}{20}\right) - 13}{2} = \dfrac{5 - 9\sqrt{15}}{20}$.

If $y = \dfrac{-45 - 3\sqrt{15}}{20}$, then

$x = \dfrac{-6\left(\dfrac{-45 - 3\sqrt{15}}{20}\right) - 13}{2} = \dfrac{5 + 9\sqrt{15}}{20}$.

The pairs $\left(\dfrac{5 + 9\sqrt{15}}{20}, \dfrac{-45 - 3\sqrt{15}}{20}\right)$ and

$\left(\dfrac{5 - 9\sqrt{15}}{20}, \dfrac{-45 + 3\sqrt{15}}{20}\right)$ check and are the solutions.

39. $2a + b = 1,$ (1)

$\quad\;\; b = 4 - a^2$ (2)

Equation (2) is already solved for b. Substitute $4 - a^2$ for b in equation (1) and solve for a.

$2a + 4 - a^2 = 1$

$\qquad 0 = a^2 - 2a - 3$

$\qquad 0 = (a - 3)(a + 1)$

$a = 3 \;$ or $\; a = -1$

Substitute these numbers in equation (2) and solve for b.

$b = 4 - 3^2 = -5$

$b = 4 - (-1)^2 = 3$

The pairs $(3, -5)$ and $(-1, 3)$ check. They are the solutions.

41. $a^2 + b^2 = 89,$ (1)

$\quad\;\; a - b = 3$ (2)

First solve equation (2) for a.

$a = b + 3$ (3)

Then substitute $b + 3$ for a in equation (1) and solve for b.

$\quad (b + 3)^2 + b^2 = 89$

$b^2 + 6b + 9 + b^2 = 89$

$\quad\; 2b^2 + 6b - 80 = 0$

$\quad\;\; b^2 + 3b - 40 = 0$

$\quad (b + 8)(b - 5) = 0$

$b = -8$ or $b = 5$

Substitute these numbers in equation (3) and solve for a.

$a = -8 + 3 = -5$

$a = 5 + 3 = 8$

The pairs $(-5, -8)$ and $(8, 5)$ check. They are the solutions.

43. $xy - y^2 = 2,$ (1)

$\quad\;\; 2xy - 3y^2 = 0$ (2)

$\quad -2xy + 2y^2 = -4$ Multiplying (1) by -2

$\underline{\quad\;\; 2xy - 3y^2 = \;\;\; 0\qquad}$

$\qquad\qquad -y^2 = -4$ Adding

$\qquad\qquad\;\;\; y^2 = \;\;\; 4$

$\qquad\qquad\;\;\;\, y = \pm 2$

We substitute for y in equation (1) and solve for x.

When $y = 2:\;\; x \cdot 2 - 2^2 = 2$

$\qquad\qquad\qquad 2x - 4 = 2$

$\qquad\qquad\qquad\quad\, 2x = 6$

$\qquad\qquad\qquad\qquad x = 3$

When $y = -2:\;\; x(-2) - (-2)^2 = 2$

$\qquad\qquad\qquad\quad -2x - 4 = 2$

$\qquad\qquad\qquad\qquad\;\; -2x = 6$

$\qquad\qquad\qquad\qquad\qquad x = -3$

The pairs $(3, 2)$ and $(-3, -2)$ check. They are the solutions.

45. $m^2 - 3mn + n^2 + 1 = 0,$ (1)

$3m^2 - mn + 3n^2 \quad\;\; = 13$ (2)

$\quad m^2 - 3mn + n^2 = -1$ (3)   Rewriting (1)

$3m^2 - mn + 3n^2 = 13$ (2)

$-3m^2 + 9mn - 3n^2 = \;\; 3$ Multiplying (3) by -3

$\underline{\;\; 3m^2 - mn + 3n^2 = 13\qquad}$

$\qquad\qquad\quad 8mn = 16$

$\qquad\qquad\qquad mn = \;\; 2$

$\qquad\qquad\qquad\;\;\; n = \dfrac{2}{m}$ (4)

Substitute $\dfrac{2}{m}$ for n in equation (1) and solve for m.

$m^2 - 3m\left(\dfrac{2}{m}\right) + \left(\dfrac{2}{m}\right)^2 + 1 = 0$

$\qquad m^2 - 6 + \dfrac{4}{m^2} + 1 = 0$

$\qquad\quad m^2 - 5 + \dfrac{4}{m^2} = 0$

$\qquad\quad m^4 - 5m^2 + 4 = 0$ Multiplying
$\qquad\qquad\qquad\qquad\qquad\qquad$ by m^2

Substitute u for m^2.

$\quad u^2 - 5u + 4 = 0$

$(u - 4)(u - 1) = 0$

$\;\; u = 4 \quad$ or $\quad u = 1$

$m^2 = 4 \quad$ or $\; m^2 = 1$

$\;\; m = \pm 2 \;\;$ or $\;\; m = \pm 1$

Substitute for m in equation (4) and solve for n.

When $m = 2,\, n = \dfrac{2}{2} = 1.$

When $m = -2,\, n = \dfrac{2}{-2} = -1.$

When $m = 1$, $n = \dfrac{2}{1} = 2$.

When $m = -1$, $n = \dfrac{2}{-1} = -2$.

The pairs $(2, 1)$, $(-2, -1)$, $(1, 2)$, and $(-1, -2)$ check. They are the solutions.

47. $x^2 + y^2 = 5$,   (1)

$x - y = 8$ (2)

First solve equation (2) for x.

$x = y + 8$ (3)

Then substitute $y + 8$ for x in equation (1) and solve for y.

$$(y + 8)^2 + y^2 = 5$$
$$y^2 + 16y + 64 + y^2 = 5$$
$$2y^2 + 16y + 59 = 0$$
$$y = \frac{-16 \pm \sqrt{(16)^2 - 4(2)(59)}}{2 \cdot 2}$$
$$y = \frac{-16 \pm \sqrt{-216}}{4}$$
$$y = \frac{-16 \pm 6i\sqrt{6}}{4}$$
$$y = -4 \pm \frac{3}{2}i\sqrt{6}$$

Now substitute these numbers in equation (3) and solve for x.

$$x = -4 + \frac{3}{2}i\sqrt{6} + 8 = 4 + \frac{3}{2}i\sqrt{6}$$
$$x = -4 - \frac{3}{2}i\sqrt{6} + 8 = 4 - \frac{3}{2}i\sqrt{6}$$

The pairs $\left(4 + \dfrac{3}{2}i\sqrt{6}, -4 + \dfrac{3}{2}i\sqrt{6}\right)$ and

$\left(4 - \dfrac{3}{2}i\sqrt{6}, -4 - \dfrac{3}{2}i\sqrt{6}\right)$ check. They are the solutions.

49. $a^2 + b^2 = 14$,   (1)

$ab = 3\sqrt{5}$ (2)

Solve equation (2) for b.

$$b = \frac{3\sqrt{5}}{a}$$

Substitute $\dfrac{3\sqrt{5}}{a}$ for b in equation (1) and solve for a.

$$a^2 + \left(\frac{3\sqrt{5}}{a}\right)^2 = 14$$
$$a^2 + \frac{45}{a^2} = 14$$
$$a^4 + 45 = 14a^2$$
$$a^4 - 14a^2 + 45 = 0$$
$$u^2 - 14u + 45 = 0 \qquad \text{Letting } u = a^2$$
$$(u - 9)(u - 5) = 0$$
$$u = 9 \quad or \quad u = 5$$
$$a^2 = 9 \quad or \quad a^2 = 5$$
$$a = \pm 3 \quad or \quad a = \pm\sqrt{5}$$

Since $b = 3\sqrt{5}/a$, if $a = 3$, $b = \sqrt{5}$; if $a = -3$, $b = -\sqrt{5}$; if $a = \sqrt{5}$, $b = 3$; and if $a = -\sqrt{5}$, $b = -3$. The pairs $(3, \sqrt{5})$, $(-3, -\sqrt{5})$, $(\sqrt{5}, 3)$, $(-\sqrt{5}, -3)$ check. They are the solutions.

51. $x^2 + y^2 = 25$,   (1)

$9x^2 + 4y^2 = 36$ (2)

$$-4x^2 - 4y^2 = -100 \qquad \text{Multiplying (1) by } -4$$

$$\frac{9x^2 + 4y^2 = 36}{5x^2 \qquad\quad = -64}$$
$$x^2 = -\frac{64}{5}$$
$$x = \pm\sqrt{\frac{-64}{5}} = \pm\frac{8i}{\sqrt{5}}$$
$$x = \pm\frac{8i\sqrt{5}}{5} \qquad \text{Rationalizing the denominator}$$

Substituting $\dfrac{8i\sqrt{5}}{5}$ or $-\dfrac{8i\sqrt{5}}{5}$ for x in equation (1) and solving for y gives us

$$-\frac{64}{5} + y^2 = 25$$
$$y^2 = \frac{189}{5}$$
$$y = \pm\sqrt{\frac{189}{5}} = \pm 3\sqrt{\frac{21}{5}}$$
$$y = \pm\frac{3\sqrt{105}}{5}. \qquad \text{Rationalizing the denominator}$$

The pairs $\left(\dfrac{8i\sqrt{5}}{5}, \dfrac{3\sqrt{105}}{5}\right)$, $\left(-\dfrac{8i\sqrt{5}}{5}, \dfrac{3\sqrt{105}}{5}\right)$,

$\left(\dfrac{8i\sqrt{5}}{5}, -\dfrac{3\sqrt{105}}{5}\right)$, and $\left(-\dfrac{8i\sqrt{5}}{5}, -\dfrac{3\sqrt{105}}{5}\right)$ check.

They are the solutions.

53. $5y^2 - x^2 = 1$,   (1)

$xy = 2$ (2)

Solve equation (2) for x.

$$x = \frac{2}{y}$$

Substitute $\dfrac{2}{y}$ for x in equation (1) and solve for y.

$$5y^2 - \left(\frac{2}{y}\right)^2 = 1$$
$$5y^2 - \frac{4}{y^2} = 1$$
$$5y^4 - 4 = y^2$$
$$5y^4 - y^2 - 4 = 0$$
$$5u^2 - u - 4 = 0 \qquad \text{Letting } u = y^2$$
$$(5u + 4)(u - 1) = 0$$

$$5u + 4 = 0 \qquad or \quad u - 1 = 0$$

$$u = -\frac{4}{5} \qquad or \qquad u = 1$$

$$y^2 = -\frac{4}{5} \qquad or \qquad y^2 = 1$$

$$y = \pm\frac{2i}{\sqrt{5}} \qquad or \qquad y = \pm 1$$

$$y = \pm\frac{2i\sqrt{5}}{5} \qquad or \qquad y = \pm 1$$

Since $x = 2/y$, if $y = \frac{2i\sqrt{5}}{5}$, $x = \dfrac{2}{\frac{2i\sqrt{5}}{5}} = \dfrac{5}{i\sqrt{5}} =$

$\dfrac{5}{i\sqrt{5}} \cdot \dfrac{-i\sqrt{5}}{-i\sqrt{5}} = -i\sqrt{5}$; if $y = -\frac{2i\sqrt{5}}{5}$,

$x = \dfrac{2}{-\frac{2i\sqrt{5}}{5}} = i\sqrt{5}$;

if $y = 1$, $x = 2/1 = 2$; if $y = -1$, $x = 2/-1 = -2$.

The pairs $\left(-i\sqrt{5}, \frac{2i\sqrt{5}}{5}\right)$, $\left(i\sqrt{5}, -\frac{2i\sqrt{5}}{5}\right)$, $(2,1)$ and $(-2,-1)$ check. They are the solutions.

55. Familiarize. We first make a drawing. We let l and w represent the length and width, respectively.

Translate. The perimeter is 28 cm.

$2l + 2w = 28$, or $l + w = 14$

Using the Pythagorean theorem we have another equation.

$l^2 + w^2 = 10^2$, or $l^2 + w^2 = 100$

Carry out. We solve the system:

$l + w = 14$,   (1)

$l^2 + w^2 = 100$ (2)

First solve equation (1) for w.

$w = 14 - l$ (3)

Then substitute $14 - l$ for w in equation (2) and solve for l.

$$l^2 + w^2 = 100$$
$$l^2 + (14 - l)^2 = 100$$
$$l^2 + 196 - 28l + l^2 = 100$$
$$2l^2 - 28l + 96 = 0$$
$$l^2 - 14l + 48 = 0$$
$$(l - 8)(l - 6) = 0$$

$l = 8$ or $l = 6$

If $l = 8$, then $w = 14 - 8$, or 6. If $l = 6$, then $w = 14 - 6$, or 8. Since the length is usually considered to be longer than the width, we have the solution $l = 8$ and $w = 6$, or $(8, 6)$.

Check. If $l = 8$ and $w = 6$, then the perimeter is $2 \cdot 8 + 2 \cdot 6$, or 28. The length of a diagonal is $\sqrt{8^2 + 6^2}$, or $\sqrt{100}$, or 10. The numbers check.

State. The length is 8 cm, and the width is 6 cm.

57. Familiarize. We first make a drawing. Let $l =$ the length and $w =$ the width of the brochure.

Translate.

    Area: $lw = 20$

    Perimeter: $2l + 2w = 18$, or $l + w = 9$

Carry out. We solve the system:

Solve the second equation for l: $l = 9 - w$

Substitute $9 - w$ for l in the first equation and solve for w.

$$(9 - w)w = 20$$
$$9w - w^2 = 20$$
$$0 = w^2 - 9w + 20$$
$$0 = (w - 5)(w - 4)$$

$w = 5$ or $w = 4$

If $w = 5$, then $l = 9 - w$, or 4. If $w = 4$, then $l = 9 - 4$, or 5. Since length is usually considered to be longer than width, we have the solution $l = 5$ and $w = 4$, or $(5, 4)$.

Check. If $l = 5$ and $w = 4$, the area is $5 \cdot 4$, or 20. The perimeter is $2 \cdot 5 + 2 \cdot 4$, or 18. The numbers check.

State. The length of the brochure is 5 in. and the width is 4 in.

59. Familiarize. We first make a drawing. Let $l =$ the length and $w =$ the width.

Translate.

Area: $lw = \sqrt{3}$ (1)

From the Pythagorean theorem: $l^2 + w^2 = 2^2$ (2)

Carry out. We solve the system of equations.

We first solve equation (1) for w.

$$lw = \sqrt{3}$$
$$w = \frac{\sqrt{3}}{l}$$

Then we substitute $\dfrac{\sqrt{3}}{l}$ for w in equation 2 and solve for l.

$$l^2 + \left(\frac{\sqrt{3}}{l}\right)^2 = 4$$

$$l^2 + \frac{3}{l^2} = 4$$

$$l^4 + 3 = 4l^2$$

$$l^4 - 4l^2 + 3 = 0$$

$$u^2 - 4u + 3 = 0 \quad \text{Letting } u = l^2$$

$$(u - 3)(u - 1) = 0$$

$$u = 3 \quad \text{or} \quad u = 1$$

We now substitute l^2 for u and solve for l.

$$l^2 = 3 \quad \text{or} \quad l^2 = 1$$

$$l = \pm\sqrt{3} \quad \text{or} \quad l = \pm 1$$

Measurements cannot be negative, so we only need to consider $l = \sqrt{3}$ and $l = 1$. Since $w = \sqrt{3}/l$, if $l = \sqrt{3}$, $w = 1$ and if $l = 1$, $w = \sqrt{3}$. Length is usually considered to be longer than width, so we have the solution $l = \sqrt{3}$ and $w = 1$, or $(\sqrt{3}, 1)$.

Check. If $l = \sqrt{3}$ and $w = 1$, the area is $\sqrt{3} \cdot 1 = \sqrt{3}$. Also $(\sqrt{3})^2 + 1^2 = 3 + 1 = 4 = 2^2$. The numbers check.

State. The length is $\sqrt{3}$ m, and the width is 1 m.

61. Familiarize. We make a drawing of the dog run. Let $l =$ the length and $w =$ the width.

Since it takes 210 yd of fencing to enclose the run, we know that the perimeter is 210 yd.

Translate.

Perimeter: $2l + 2w = 210$, or $l + w = 105$

Area: $lw = 2250$

Carry out. We solve the system:

Solve the first equation for l: $l = 105 - w$

Substitute $105 - w$ for l in the second equation and solve for w.

$$(105 - w)w = 2250$$

$$105w - w^2 = 2250$$

$$0 = w^2 - 105w + 2250$$

$$0 = (w - 30)(w - 75)$$

$$w = 30 \quad \text{or} \quad w = 75$$

If $w = 30$, then $l = 105 - 30$, or 75. If $w = 75$, then $l = 105 - 75$, or 30. Since length is usually considered to be longer than width, we have the solution $l = 75$ and $w = 30$, or $(75, 30)$.

Check. If $l = 75$ and $w = 30$, the perimeter is $2 \cdot 75 + 2 \cdot 30$, or 210. The area is $75(30)$, or 2250. The numbers check.

State. The length is 75 yd and the width is 30 yd.

63. Familiarize. We let $x =$ the length of a side of one test plot and $y =$ the length of a side of the other plot. Make a drawing.

Translate.

The sum of the areas is 832 ft².

$$x^2 + y^2 = 832$$

The difference of the areas is 320 ft².

$$x^2 - y^2 = 320$$

Carry out. We solve the system of equations.

$$\begin{aligned} x^2 + y^2 &= 832 \\ x^2 - y^2 &= 320 \\ \hline 2x^2 &= 1152 \quad \text{Adding} \\ x^2 &= 576 \\ x &= \pm 24 \end{aligned}$$

Since measurements cannot be negative, we consider only $x = 24$. Substitute 24 for x in the first equation and solve for y.

$$24^2 + y^2 = 832$$

$$576 + y^2 = 832$$

$$y^2 = 256$$

$$y = \pm 16$$

Again, we consider only the positive value, 16. The possible solution is $(24, 16)$.

Check. The areas of the test plots are 24^2, or 576, and 16^2, or 256. The sum of the areas is $576 + 256$, or 832. The difference of the areas is $576 - 256$, or 320. The values check.

State. The lengths of the test plots are 24 ft and 16 ft.

65. The correct graph is (b).

67. The correct graph is (d).

69. The correct graph is (a).

71. Graph: $x^2 + y^2 \le 16$,

$$y < x$$

The solution set of $x^2 + y^2 \le 16$ is the circle $x^2 + y^2 = 16$ and the region inside it. The solution set of $y < x$ is the half-plane below the line $y = x$. We shade the region common to the two solution sets.

To find the points of intersection of the graphs we solve the system of equations

$$x^2 + y^2 = 16,$$
$$y = x.$$

The points of intersection are $(-2\sqrt{2}, -2\sqrt{2})$ and $(2\sqrt{2}, 2\sqrt{2})$.

73. Graph: $x^2 \leq y,$

$\qquad\qquad x + y \geq 2$

The solution set of $x^2 \leq y$ is the parabola $x^2 = y$ and the region inside it. The solution set of $x + y \geq 2$ is the line $x + y = 2$ and the half-plane above the line. We shade the region common to the two solution sets.

To find the points of intersection of the graphs we solve the system of equations

$$x^2 = y,$$
$$x + y = 2.$$

The points of intersection are $(-2, 4)$ and $(1, 1)$.

75. Graph: $x^2 + y^2 \leq 25,$

$\qquad\qquad x - y > 5$

The solution set of $x^2 + y^2 \leq 25$ is the circle $x^2 + y^2 = 25$ and the region inside it. The solution set of $x - y > 5$ is the half-plane below the line $x - y = 5$. We shade the region common to the two solution sets.

To find the points of intersection of the graphs we solve the system of equations

$$x^2 + y^2 = 25,$$
$$x - y = 5.$$

The points of intersection are $(0, -5)$ and $(5, 0)$.

77. Graph: $y \geq x^2 - 3,$

$\qquad\qquad y \leq 2x$

The solution set of $y \geq x^2 - 3$ is the parabola $y = x^2 - 3$ and the region inside it. The solution set of $y \leq 2x$ is the line $y = 2x$ and the half-plane below it. We shade the region common to the two solution sets.

To find the points of intersection of the graphs we solve the system of equations

$$y = x^2 - 3,$$
$$y = 2x.$$

The points of intersection are $(-1, -2)$ and $(3, 6)$.

79. Graph: $y \geq x^2,$

$\qquad\qquad y < x + 2$

The solution set of $y \geq x^2$ is the parabola $y = x^2$ and the region inside it. The solution set of $y < x + 2$ is the half-plane below the line $y = x + 2$. We shade the region common to the two solution sets.

To find the points of intersection of the graphs we solve the system of equations

$$y = x^2,$$
$$y = x + 2.$$

The points of intersection are $(-1, 1)$ and $(2, 4)$

81. Find the points of intersection of $y_1 = \ln x + 2$ and $y_2 = x^2$. They are $(1.564, 2.448)$ and $(0.138, 0.019)$.

83. Find the point of intersection of $y_1 = e^x - 1$ and $y_2 = -3x + 4$. It is $(0.871, 1.388)$.

85. Find the points of intersection of $y_1 = e^x$ and $y_2 = x + 2$. They are $(1.146, 3.146)$ and $(-1.841, 0.159)$.

87. Graph $y_1 = \sqrt{19,380,510.36 - x^2}$,

$y_2 = -\sqrt{19,380,510.36 - x^2}$, and

$y_3 = 27,941.25x/6.125$ and find the points of intersection. They are $(0.965, 4402.33)$ and $(-0.965, -4402.33)$.

89. Graph $y_1 = \sqrt{\dfrac{14.5x^2 - 64.5}{13.5}}$,

$y_2 = -\sqrt{\dfrac{14.5x^2 - 64.5}{13.5}}$, and $y_3 = (5.5x - 12.3)/6.3$ and find the points of intersection. They are $(2.112, -0.109)$ and $(-13.041, -13.337)$.

91. Graph $y_1 = \sqrt{\dfrac{56,548 - 0.319x^2}{2688.7}}$,

$y_2 = -\sqrt{\dfrac{56,548 - 0.319x^2}{2688.7}}$,

$y_3 = \sqrt{\dfrac{0.306x^2 - 43,452}{2688.7}}$,

and $y_4 = -\sqrt{\dfrac{0.306x^2 - 43,452}{2688.7}}$ and find the points of intersection. They are $(400, 1.431)$, $(-400, 1.431)$, $(400, -1.431)$, and $(-400, -1.431)$.

93. Discussion and Writing

95. $2^{3x} = 64$

$2^{3x} = 2^6$

$3x = 6$

$x = 2$

The solution is 2.

97. $\log_3 x = 4$

$x = 3^4$

$x = 81$

The solution is 81.

99. $(x - h)^2 + (y - k)^2 = r^2$

If $(2, 4)$ is a point on the circle, then
$(2 - h)^2 + (4 - k)^2 = r^2$.
If $(3, 3)$ is a point on the circle, then
$(3 - h)^2 + (3 - k)^2 = r^2$.
Thus
$$(2 - h)^2 + (4 - k)^2 = (3 - h)^2 + (3 - k)^2$$
$$4 - 4h + h^2 + 16 - 8k + k^2 =$$
$$9 - 6h + h^2 + 9 - 6k + k^2$$
$$-4h - 8k + 20 = -6h - 6k + 18$$
$$2h - 2k = -2$$
$$h - k = -1$$
If the center (h, k) is on the line $3x - y = 3$, then $3h - k = 3$. Solving the system
$h - k = -1$,
$3h - k = 3$
we find that $(h, k) = (2, 3)$.
Find r^2, substituting $(2, 3)$ for (h, k) and $(2, 4)$ for (x, y). We could also use $(3, 3)$ for (x, y).
$$(x - h)^2 + (y - k)^2 = r^2$$
$$(2 - 2)^2 + (4 - 3)^2 = r^2$$
$$0 + 1 = r^2$$
$$1 = r^2$$

The equation of the circle is $(x - 2)^2 + (y - 3)^2 = 1$.

101. The equation of the ellipse is of the form
$\dfrac{x^2}{a^2} + \dfrac{y^2}{b^2} = 1$. Substitute $\left(1, \dfrac{\sqrt{3}}{2}\right)$ and $\left(\sqrt{3}, \dfrac{1}{2}\right)$ for (x, y) to get two equations.

$$\frac{1^2}{a^2} + \frac{\left(\dfrac{\sqrt{3}}{2}\right)^2}{b^2} = 1, \ or \ \frac{1}{a^2} + \frac{3}{4b^2} = 1$$

$$\frac{(\sqrt{3})^2}{a^2} + \frac{\left(\dfrac{1}{2}\right)^2}{b^2} = 1, \ or \ \frac{3}{a^2} + \frac{1}{4b^2} = 1$$

Substitute u for $\dfrac{1}{a^2}$ and v for $\dfrac{1}{b^2}$.

$$u + \frac{3}{4}v = 1, \qquad\qquad 4u + 3v = 4,$$
$$or$$
$$3u + \frac{1}{4}v = 1 \qquad\qquad 12u + v = 4$$

Solving for u and v, we get $u = \dfrac{1}{4}$, $v = 1$. Then
$u = \dfrac{1}{a^2} = \dfrac{1}{4}$, so $a^2 = 4$; $v = \dfrac{1}{b^2} = 1$, so $b^2 = 1$.
Then the equation of the ellipse is
$$\frac{x^2}{4} + \frac{y^2}{1} = 1, \ or \ \frac{x^2}{4} + y^2 = 1.$$

103. $(x - h)^2 + (y - k)^2 = r^2$ Standard form

Substitute $(4, 6)$, $(-6, 2)$, and $(1, -3)$ for (x, y).
$$(4 - h)^2 + (6 - k)^2 = r^2 \qquad (1)$$
$$(-6 - h)^2 + (2 - k)^2 = r^2 \qquad (2)$$
$$(1 - h)^2 + (-3 - k)^2 = r^2 \qquad (3)$$
Thus
$$(4 - h)^2 + (6 - k)^2 = (-6 - h)^2 + (2 - k)^2, \ or$$
$$5h + 2k = 3$$
and
$$(4 - h)^2 + (6 - k)^2 = (1 - h)^2 + (-3 - k)^2, \ or$$
$$h + 3k = 7.$$
We solve the system
$5h + 2k = 3$,
$h + 3k = 7.$

Solving we get $h = -\dfrac{5}{13}$ and $k = \dfrac{32}{13}$. Substituting these values in equation (1), (2), or (3), we find that $r^2 = \dfrac{5365}{169}$.
The equation of the circle is
$$\left(x + \frac{5}{13}\right)^2 + \left(y - \frac{32}{13}\right)^2 = \frac{5365}{169}.$$

105. See the answer section in the text.

107. *Familiarize.* Let x and y represent the numbers.
Translate.

The square of a certain number exceeds twice the square of another number by $\dfrac{1}{8}$.

$$x^2 = 2y^2 + \frac{1}{8}$$

The sum of the squares is $\frac{5}{16}$.

$$x^2 + y^2 = \frac{5}{16}$$

Carry out. We solve the system.

$$x^2 - 2y^2 = \frac{1}{8}, \quad (1)$$

$$x^2 + y^2 = \frac{5}{16} \quad (2)$$

$$x^2 - 2y^2 = \frac{1}{8},$$

$$\underline{2x^2 + 2y^2 = \frac{5}{8}} \quad \text{Multiplying (2) by 2}$$

$$3x^2 \qquad = \frac{6}{8}$$

$$x^2 = \frac{1}{4}$$

$$x = \pm\frac{1}{2}$$

Substitute $\pm\frac{1}{2}$ for x in (2) and solve for y.

$$\left(\pm\frac{1}{2}\right)^2 + y^2 = \frac{5}{16}$$

$$\frac{1}{4} + y^2 = \frac{5}{16}$$

$$y^2 = \frac{1}{16}$$

$$y = \pm\frac{1}{4}$$

We get $\left(\frac{1}{2}, \frac{1}{4}\right)$, $\left(-\frac{1}{2}, \frac{1}{4}\right)$, $\left(\frac{1}{2}, -\frac{1}{4}\right)$ and $\left(-\frac{1}{2}, -\frac{1}{4}\right)$.

Check. It is true that $\left(\pm\frac{1}{2}\right)^2$ exceeds twice $\left(\pm\frac{1}{4}\right)^2$ by $\frac{1}{8}$: $\frac{1}{4} = 2\left(\frac{1}{16}\right) + \frac{1}{8}$

Also $\left(\pm\frac{1}{2}\right)^2 + \left(\pm\frac{1}{4}\right)^2 = \frac{5}{16}$. The pairs check.

State. The numbers are $\frac{1}{2}$ and $\frac{1}{4}$ or $-\frac{1}{2}$ and $\frac{1}{4}$ or $\frac{1}{2}$ and $-\frac{1}{4}$ or $-\frac{1}{2}$ and $-\frac{1}{4}$.

109. See the answer section in the text.

111. $x^3 + y^3 = 72, \quad (1)$

$x + y = 6 \qquad (2)$

Solve equation (2) for y: $y = 6 - x$

Substitute for y in equation (1) and solve for x.

$$x^3 + (6 - x)^3 = 72$$

$$x^3 + 216 - 108x + 18x^2 - x^3 = 72$$

$$18x^2 - 108x + 144 = 0$$

$$x^2 - 6x + 8 = 0$$

Multiplying by $\frac{1}{18}$

$$(x - 4)(x - 2) = 0$$

$x = 4$ or $x = 2$

If $x = 4$, then $y = 6 - 4 = 2$.

If $x = 2$, then $y = 6 - 2 = 4$.

The pairs $(4, 2)$ and $(2, 4)$ check.

113. $p^2 + q^2 = 13, \quad (1)$

$$\frac{1}{pq} = -\frac{1}{6} \quad (2)$$

Solve equation (2) for p.

$$\frac{1}{q} = -\frac{p}{6}$$

$$-\frac{6}{q} = p$$

Substitute $-6/q$ for p in equation (1) and solve for q.

$$\left(-\frac{6}{q}\right)^2 + q^2 = 13$$

$$\frac{36}{q^2} + q^2 = 13$$

$$36 + q^4 = 13q^2$$

$$q^4 - 13q^2 + 36 = 0$$

$$u^2 - 13u + 36 = 0 \qquad \text{Letting } u = q^2$$

$$(u - 9)(u - 4) = 0$$

$$u = 9 \quad \text{or} \quad u = 4$$

$$x^2 = 9 \quad \text{or} \quad x^2 = 4$$

$$x = \pm 3 \text{ or} \quad x = \pm 2$$

Since $p = -6/q$, if $q = 3$, $p = -2$; if $q = -3$, $p = 2$; if $q = 2$, $p = -3$; and if $q = -2$, $p = 3$. The pairs $(-2, 3)$, $(2, -3)$, $(-3, 2)$, and $(3, -2)$ check. They are the solutions.

115. $5^{x+y} = 100,$

$3^{2x-y} = 1000$

$(x + y)\log 5 = 2, \quad$ Taking logarithms and

$(2x - y)\log 3 = 3 \quad$ simplifying

$x\log 5 + y\log 5 = 2, \quad (1)$

$2x\log 3 - y\log 3 = 3 \quad (2)$

Multiply equation (1) by $\log 3$ and equation (2) by $\log 5$ and add.

$$x\log 3 \cdot \log 5 + y\log 3 \cdot \log 5 = 2\log 3$$

$$\underline{2x\log 3 \cdot \log 5 - y\log 3 \cdot \log 5 = 3\log 5}$$

$$3x\log 3 \cdot \log 5 \qquad\qquad = 2\log 3 + 3\log 5$$

$$x = \frac{2\log 3 + 3\log 5}{3\log 3 \cdot \log 5}$$

Substitute in (1) to find y.

$$\frac{2\log 3 + 3\log 5}{3\log 3 \cdot \log 5} \cdot \log 5 + y\log 5 = 2$$

$$y\log 5 = 2 - \frac{2\log 3 + 3\log 5}{3\log 3}$$

$$y\log 5 = \frac{6\log 3 - 2\log 3 - 3\log 5}{3\log 3}$$

$$y\log 5 = \frac{4\log 3 - 3\log 5}{3\log 3}$$

$$y = \frac{4\log 3 - 3\log 5}{3\log 3 \cdot \log 5}$$

The pair $\left(\dfrac{2\log 3 + 3\log 5}{3\log 3 \cdot \log 5}, \dfrac{4\log 3 - 3\log 5}{3\log 3 \cdot \log 5} \right)$ checks. It is the solution.

Chapter 6 Review Exercises

1. $x + y^2 = 1$

   $y^2 = x - 1$

   $(y - 0)^2 = 4 \cdot \dfrac{1}{4}(x - 1)$

   This parabola has a horizontal axis of symmetry, the focus is $\left(\dfrac{5}{4}, 0 \right)$, and the directrix is $x = \dfrac{3}{4}$. Thus it opens to the left and the statement is true.

3. The statement is true. See page 546 in the text.

5. The statement is false. See Example 4 on page 561 in the text.

7. Graph (a) is the graph of $y^2 = 9 - x^2$.

9. Graph (g) is the graph of $9y^2 - 4x^2 = 36$.

11. Graph (f) is the graph of $4x^2 + y^2 - 16x - 6y = 15$.

13. Graph (c) is the graph of $\dfrac{(x+3)^2}{16} - \dfrac{(y-1)^2}{25} = 1$.

15. $y^2 = -12x$

    $y^2 = 4(-3)x$

    F: $(-3, 0)$, V: $(0, 0)$, D: $x = 3$

17. Begin by completing the square twice.

    $$16x^2 + 25y^2 - 64x + 50y - 311 = 0$$

    $$16(x^2 - 4x) + 25(y^2 + 2y) = 311$$

    $$16(x^2 - 4x + 4) + 25(y^2 + 2y + 1) = 311 + 16 \cdot 4 + 25 \cdot 1$$

    $$16(x - 2)^2 + 25(y + 1)^2 = 400$$

    $$\frac{(x-2)^2}{25} + \frac{[y - (-1)]^2}{16} = 1$$

    The center is $(2, -1)$. Note that $a = 5$ and $b = 4$. The major axis is horizontal so the vertices are 5 units left and right of the center: $(2 - 5, -1)$ and $(2 + 5, -1)$, or $(-3, -1)$ and $(7, -1)$. We know that $c^2 = a^2 - b^2 = 25 - 16 = 9$ and $c = \sqrt{9} = 3$. Then the foci are 3 units left and right of the center: $(2 - 3, -1)$ and $(2 + 3, -1)$, or $(-1, -1)$ and $(5, -1)$.

19. Begin by completing the square twice.

    $$x^2 - 2y^2 + 4x + y - \frac{1}{8} = 0$$

    $$(x^2 + 4x) - 2\left(y^2 - \frac{1}{2}y \right) = \frac{1}{8}$$

    $$(x^2 + 4x + 4) - 2\left(y^2 - \frac{1}{2}y + \frac{1}{16} \right) = \frac{1}{8} + 4 - 2 \cdot \frac{1}{16}$$

    $$(x + 2)^2 - 2\left(y - \frac{1}{4} \right)^2 = 4$$

    $$\frac{[x - (-2)]^2}{4} - \frac{\left(y - \dfrac{1}{4} \right)^2}{2} = 1$$

    The center is $\left(-2, \dfrac{1}{4} \right)$. The transverse axis is horizontal, so the vertices are 2 units left and right of the center: $\left(-2 - 2, \dfrac{1}{4} \right)$ and $\left(-2 + 2, \dfrac{1}{4} \right)$, or $\left(-4, \dfrac{1}{4} \right)$ and $\left(0, \dfrac{1}{4} \right)$. Since $c^2 = a^2 + b^2$, we have $c^2 = 4 + 2 = 6$ and $c = \sqrt{6}$. Then the foci are $\sqrt{6}$ units left and right of the center: $\left(-2 - \sqrt{6}, \dfrac{1}{4} \right)$ and $\left(-2 + \sqrt{6}, \dfrac{1}{4} \right)$.

    Find the asymptotes:

    $$y - k = \frac{b}{a}(x - h) \quad \text{and} \quad y - k = -\frac{b}{a}(x - h)$$

    $$y - \frac{1}{4} = \frac{\sqrt{2}}{2}(x + 2) \quad \text{and} \quad y - \frac{1}{4} = -\frac{\sqrt{2}}{2}(x + 2)$$

21. $x^2 - 16y = 0$, (1)

    $x^2 - y^2 = 64$ (2)

    From equation (1) we have $x^2 = 16y$. Substitute in equation (2).

    $$16y - y^2 = 64$$

    $$0 = y^2 - 16y + 64$$

    $$0 = (y - 8)^2$$

    $$0 = y - 8$$

    $$8 = y$$

    $$x^2 - (8)^2 = 64 \qquad \text{Substituting in equation (2)}$$

    $$x^2 = 128$$

    $$x = \pm\sqrt{128} = \pm 8\sqrt{2}$$

    The pairs $(-8\sqrt{2}, 8)$ and $(8\sqrt{2}, 8)$ check.

23. $\quad x^2 - y^2 = 33, \qquad (1)$

$\qquad x + y = 11 \qquad\quad (2)$

$\qquad\quad y = -x + 11$

$\qquad\quad x^2 - (-x + 11)^2 = 33 \quad$ Substituting in (1)

$\qquad\quad x^2 - (x^2 - 22x + 121) = 33$

$\qquad\quad x^2 - x^2 + 22x - 121 = 33$

$\qquad\qquad\qquad\qquad 22x = 154$

$\qquad\qquad\qquad\qquad\quad x = 7$

$y = -7 + 11 = 4$

The pair $(7, 4)$ checks.

25. $\quad x^2 - y = 3, \quad (1)$

$\qquad 2x - y = 3 \quad\ (2)$

From equation (1) we have $y = x^2 - 3$. Substitute in equation (2).

$\qquad 2x - (x^2 - 3) = 3$

$\qquad\ 2x - x^2 + 3 = 3$

$\qquad\qquad\quad 0 = x^2 - 2x$

$\qquad\qquad\quad 0 = x(x - 2)$

$x = 0 \ \ or \ \ x = 2$

$y = 0^2 - 3 = -3$

$y = 2^2 - 3 = 1$

The pairs $(0, -3)$ and $(2, 1)$ check.

27. $\quad x^2 - y^2 = 3, \quad (1)$

$\qquad\ y = x^2 - 3 \qquad (2)$

From equation (2) we have $x^2 = y + 3$. Substitute in equation (1).

$\qquad y + 3 - y^2 = 3$

$\qquad\qquad\quad 0 = y^2 - y$

$\qquad\qquad\quad 0 = y(y - 1)$

$y = 0 \ \ or \ \ y = 1$

$x^2 = 0 + 3$

$x^2 = 3$

$x = \pm\sqrt{3}$

$x^2 = 1 + 3$

$x^2 = 4$

$x = \pm 2$

The pairs $(\sqrt{3}, 0)$, $(-\sqrt{3}, 0)$, $(2, 1)$, and $(-2, 1)$ check.

29. $\quad x^2 + y^2 = 100, \qquad (1)$

$\qquad 2x^2 - 3y^2 = -120 \quad (2)$

$\qquad 3x^2 + 3y^2 = 300 \qquad$ Multiplying (1) by 3

$\qquad \underline{2x^2 - 3y^2 = -120}$

$\qquad 5x^2 \qquad\quad = 180 \qquad$ Adding

$\qquad\qquad x^2 = 36$

$\qquad\qquad\ x = \pm 6$

$(\pm 6)^2 + y^2 = 100$

$\qquad\qquad y^2 = 64$

$\qquad\qquad\ y = \pm 8$

The pairs $(6, 8)$, $(-6, 8)$, $(6, -8)$, and $(-6, -8)$ check.

31. ***Familiarize***. Let x and y represent the numbers.

Translate. The sum of the numbers is 11.

$x + y = 11$

The sum of the squares of the numbers is 65.

$x^2 + y^2 = 65$

Carry out. We solve the system of equations.

$\qquad x + y = 11, \quad (1)$

$\qquad x^2 + y^2 = 65 \quad (2)$

First we solve equation (1) for y.

$\qquad y = 11 - x$

Then substitute $11 - x$ for y in equation (2) and solve for x.

$\qquad\qquad x^2 + (11 - x)^2 = 65$

$\qquad\quad x^2 + 121 - 22x + x^2 = 65$

$\qquad\qquad\ 2x^2 - 22x + 121 = 65$

$\qquad\qquad\quad 2x^2 - 22x + 56 = 0$

$\qquad\qquad\quad\ x^2 - 11x + 28 = 0 \quad$ Dividing by 2

$\qquad\qquad\quad (x - 4)(x - 7) = 0$

$\qquad x - 4 = 0 \ \ or \ \ x - 7 = 0$

$\qquad\quad x = 4 \ \ or \qquad x = 7$

If $x = 4$, then $y = 11 - 4 = 7$.

If $x = 7$, then $y = 11 - 7 = 4$.

In either case, the possible numbers are 4 and 7.

Check. $4 + 7 = 11$ and $4^2 + 7^2 = 16 + 49 = 65$. The answer checks.

State. The numbers are 4 and 7.

33. ***Familiarize***. Let x and y represent the positive integers.

Translate. The sum of the numbers is 12.

$x + y = 12$

The sum of the reciprocals is $\dfrac{3}{8}$.

$\dfrac{1}{x} + \dfrac{1}{y} = \dfrac{3}{8}$

Carry out. We solve the system of equations.

$\qquad x + y = 12, \quad (1)$

$\qquad \dfrac{1}{x} + \dfrac{1}{y} = \dfrac{3}{8} \quad (2)$

First solve equation (1) for y.

$\qquad y = 12 - x$

Then substitute $12 - x$ for y in equation (2) and solve for x.

$$\frac{1}{x} + \frac{1}{12-x} = \frac{3}{8}, \text{ LCD is } 8x(12-x)$$

$$8x(12-x)\left(\frac{1}{x} + \frac{1}{12-x}\right) = 8x(12-x) \cdot \frac{3}{8}$$

$$8(12-x) + 8x = x(12-x) \cdot 3$$

$$96 - 8x + 8x = 36x - 3x^2$$

$$96 = 36x - 3x^2$$

$$3x^2 - 36x + 96 = 0$$

$$x^2 - 12x + 32 = 0 \quad \text{Dividing by 3}$$

$$(x-4)(x-8) = 0$$

$$x - 4 = 0 \quad or \quad x - 8 = 0$$

$$x = 4 \quad or \qquad x = 8$$

If $x = 4$, $y = 12 - 4 = 8$.

If $x = 8$, $y = 12 - 8 = 4$.

In either case, the possible numbers are 4 and 8.

Check. $4 + 8 = 12$; $\frac{1}{4} + \frac{1}{8} = \frac{2}{8} + \frac{1}{8} = \frac{3}{8}$. The answer checks.

State. The numbers are 4 and 8.

35. Familiarize. Let $x =$ the radius of the larger circle and let $y =$ the radius of the smaller circle. We will use the formula for the area of a circle, $A = \pi r^2$.

Translate. The sum of the areas is 130π ft^2.

$$\pi x^2 + \pi y^2 = 130\pi$$

The difference of the areas is 112π ft^2.

$$\pi x^2 - \pi y^2 = 112\pi$$

We have a system of equations.

$$\pi x^2 + \pi y^2 = 130\pi, \quad (1)$$
$$\pi x^2 - \pi y^2 = 112\pi \quad (2)$$

Carry out. We add.

$$\begin{array}{r} \pi x^2 + \pi y^2 = 130\pi \\ \underline{\pi x^2 - \pi y^2 = 112\pi} \\ 2\pi x^2 = 242\pi \\ x^2 = 121 \quad \text{Dividing by } 2\pi \\ x = \pm 11 \end{array}$$

Since the length of a radius cannot be negative, we consider only $x = 11$. Substitute 11 for x in equation (1) and solve for y.

$$\pi \cdot 11^2 + \pi y^2 = 130\pi$$

$$121\pi + \pi y^2 = 130\pi$$

$$\pi y^2 = 9\pi$$

$$y^2 = 9$$

$$y = \pm 3$$

Again, we consider only the positive solution.

Check. If the radii are 11 ft and 3 ft, the sum of the areas is $\pi \cdot 11^2 + \pi \cdot 3^2 = 121\pi + 9\pi = 130\pi$ ft^2. The difference of the areas is $121\pi - 9\pi = 112\pi$ ft^2. The answer checks.

State. The radius of the larger circle is 11 ft, and the radius of the smaller circle is 3 ft.

37. Graph: $x^2 + y^2 \le 16$,

$\quad\quad x + y < 4$

The solution set of $x^2 + y^2 \le 16$ is the circle $x^2 + y^2 = 16$ and the region inside it. The solution set of $x + y < 4$ is the half-plane below the line $x + y = 4$. We shade the region common to the two solution sets.

39. Graph: $x^2 + y^2 \le 9$,

$\quad\quad x \le -1$

The solution set of $x^2 + y^2 \le 9$ is the circle $x^2 + y^2 = 9$ and the region inside it. The solution set of $x \le -1$ is the line $x = -1$ and the half-plane to the left of it. We shade the region common to the two solution sets.

41. A straight line can intersect an ellipse at 0 points, 1 point, or 2 points but not at 4 points, so answer D is correct.

43. The equation of a circle can be written as

$$\frac{(x-h)^2}{a^2} + \frac{(y-k)^2}{b^2} = 1$$

where $a = b = r$, the radius of the circle. In an ellipse, $a > b$, so a circle is not a special type of ellipse.

45. Familiarize. Let x and y represent the numbers.

Translate. The product of the numbers is 4.

$$xy = 4$$

The sum of the reciprocals is $\frac{65}{56}$.

$$\frac{1}{x} + \frac{1}{y} = \frac{65}{56}$$

Carry out. We solve the system of equations.

$$xy = 4, \quad\quad (1)$$
$$\frac{1}{x} + \frac{1}{y} = \frac{65}{56} \quad (2)$$

First solve equation (1) for y.

$$y = \frac{4}{x}$$

Then substitute $\dfrac{4}{x}$ for y in equation (2) and solve for x.

$$\frac{1}{x} + \frac{1}{4/x} = \frac{65}{56}$$

$$\frac{1}{x} + \frac{x}{4} = \frac{65}{56}$$

$$56x\left(\frac{1}{x} + \frac{x}{4}\right) = 56x \cdot \frac{65}{56}$$

$$56 + 14x^2 = 65x$$

$$14x^2 - 65x + 56 = 0$$

$$(2x - 7)(7x - 8) = 0$$

$$2x - 7 = 0 \quad or \quad 7x - 8 = 0$$

$$2x = 7 \quad or \quad 7x = 8$$

$$x = \frac{7}{2} \quad or \quad x = \frac{8}{7}$$

If $x = \dfrac{7}{2}$, $y = \dfrac{4}{7/2} = 4 \cdot \dfrac{2}{7} = \dfrac{8}{7}$.

If $x = \dfrac{8}{7}$, $y = \dfrac{4}{8/7} = 4 \cdot \dfrac{7}{8} = \dfrac{7}{2}$.

In either case the possible numbers are $\dfrac{7}{2}$ and $\dfrac{8}{7}$.

Check. $\dfrac{7}{2} \cdot \dfrac{8}{7} = 4$; $\dfrac{1}{7/2} + \dfrac{1}{8/7} = \dfrac{2}{7} + \dfrac{7}{8} = \dfrac{16}{56} + \dfrac{49}{56} = \dfrac{65}{56}$.
The answer checks.

State. The numbers are $\dfrac{7}{2}$ and $\dfrac{8}{7}$.

47. A and B are the foci of the hyperbola, so $c = \dfrac{400}{2} = 200$.

300 microseconds $\cdot \dfrac{0.186 \text{ mi}}{1 \text{ microsecond}} = 55.8$ mi, the difference of the ship's distances from the foci. That is, $2a = 55.8$, so $a = 27.9$.

Find b^2.

$$c^2 = a^2 + b^2$$

$$200^2 = 27.9^2 + b^2$$

$$39,221.59 = b^2$$

Then the equation of the hyperbola is

$$\frac{x^2}{778.41} - \frac{y^2}{39,221.59} = 1.$$

Chapter 6 Test

1. Graph (c) is the graph of $4x^2 - y^2 = 4$.

2. Graph (b) is the graph of $x^2 - 2x - 3y = 5$.

3. Graph (a) is the graph of $x^2 + 4x + y^2 - 2y - 4 = 0$.

4. Graph (d) is the graph of $9x^2 + 4y^2 = 36$.

5. $x^2 = 12y$

$x^2 = 4 \cdot 3y$

$V: (0,0), F: (0,3), D: y = -3$

$x^2 = 12y$

6. $y^2 + 2y - 8x - 7 = 0$

$$y^2 + 2y = 8x + 7$$

$$y^2 + 2y + 1 = 8x + 7 + 1$$

$$(y + 1)^2 = 8x + 8$$

$$[y - (-1)]^2 = 4(2)[x - (-1)]$$

$V: (-1, -1)$

$F: (-1 + 2, -1) \quad or \quad (1, -1)$

$D: x = -1 - 2 = -3$

$y^2 + 2y - 8x - 7 = 0$

7. $(x - h)^2 = 4p(y - k)$

$$(x - 0)^2 = 4 \cdot 2(y - 0)$$

$$x^2 = 8y$$

8. Begin by completing the square twice.

$$x^2 + y^2 + 2x - 6y - 15 = 0$$

$$x^2 + 2x + y^2 - 6y = 15$$

$$(x^2 + 2x + 1) + (y^2 - 6y + 9) = 15 + 1 + 9$$

$$(x + 1)^2 + (y - 3)^2 = 25$$

$$[x - (-1)]^2 + (y - 3)^2 = 5^2$$

Center: $(-1, 3)$, radius: 5

$x^2 + y^2 + 2x - 6y - 15 = 0$

9. $9x^2 + 16y^2 = 144$

$$\frac{x^2}{16} + \frac{y^2}{9} = 1$$

$$\frac{x^2}{4^2} + \frac{y^2}{3^2} = 1$$

$a = 4, b = 3$

The center is $(0,0)$. The major axis is horizontal, so the vertices are $(-4,0)$ and $(4,0)$. Since $c^2 = a^2 - b^2$, we have $c^2 = 16 - 9 = 7$, so $c = \sqrt{7}$ and the foci are $(-\sqrt{7}, 0)$ and $(\sqrt{7}, 0)$.

$$9x^2 + 16y^2 = 144$$

10. $\dfrac{(x+1)^2}{4} + \dfrac{(y-2)^2}{9} = 1$

$\dfrac{[x-(-1)]^2}{2^2} + \dfrac{(y-2)^2}{3^2} = 1$

The center is $(-1, 2)$. Note that $a = 3$ and $b = 2$. The major axis is vertical, so the vertices are 3 units below and above the center:

$(-1, 2-3)$ and $(-1, 2+3)$ or $(-1, -1)$ and $(-1, 5)$.

We know that $c^2 = a^2 - b^2$, so $c^2 = 9 - 4 = 5$ and $c = \sqrt{5}$. Then the foci are $\sqrt{5}$ units below and above the center:

$(-1, 2 - \sqrt{5})$ and $(-1, 2 + \sqrt{5})$.

$$\dfrac{(x+1)^2}{4} + \dfrac{(y-2)^2}{9} = 1$$

11. The vertices $(0, -5)$ and $(0, 5)$ are on the y-axis, so the major axis is vertical and $a = 5$. Since the vertices are equidistant from the center, the center of the ellipse is at the origin. The length of the minor axis is 4, so $b = \dfrac{4}{2} = 2$.

The equation is $\dfrac{x^2}{4} + \dfrac{y^2}{25} = 1$.

12. $4x^2 - y^2 = 4$

$\dfrac{x^2}{1} - \dfrac{y^2}{4} = 1$

$\dfrac{x^2}{1^2} - \dfrac{y^2}{2^2} = 1$

The center is $(0,0)$; $a = 1$ and $b = 2$.

The transverse axis is horizontal, so the vertices are $(-1, 0)$ and $(1, 0)$. Since $c^2 = a^2 + b^2$, we have $c^2 = 1 + 4 = 5$ and $c = \sqrt{5}$. Then the foci are $(-\sqrt{5}, 0)$ and $(\sqrt{5}, 0)$.

Find the asymptotes:

$y = \dfrac{b}{a}x$ and $y = -\dfrac{b}{a}x$

$y = \dfrac{2}{1}x$ and $y = -\dfrac{2}{1}x$

$y = 2x$ and $y = -2x$

$$4x^2 - y^2 = 4$$

13. $\dfrac{(y-2)^2}{4} - \dfrac{(x+1)^2}{9} = 1$

$\dfrac{(y-2)^2}{2^2} - \dfrac{[x-(-1)]^2}{3^2} = 1$

The center is $(-1, 2)$; $a = 2$ and $b = 3$.

The transverse axis is vertical, so the vertices are 2 units below and above the center:

$(-1, 2-2)$ and $(-1, 2+2)$ or $(-1, 0)$ and $(-1, 4)$.

Since $c^2 = a^2 + b^2$, we have $c^2 = 4 + 9 = 13$ and $c = \sqrt{13}$. Then the foci are $\sqrt{13}$ units below and above the center:

$(-1, 2 - \sqrt{13})$ and $(-1, 2 + \sqrt{13})$.

Find the asymptotes:

$y - k = \dfrac{a}{b}(x - h)$ and $y - k = -\dfrac{a}{b}(x - h)$

$y - 2 = \dfrac{2}{3}(x - (-1))$ and $y - 2 = -\dfrac{2}{3}(x - (-1))$

$y - 2 = \dfrac{2}{3}(x + 1)$ and $y - 2 = -\dfrac{2}{3}(x + 1)$

$y = \dfrac{2}{3}x + \dfrac{8}{3}$ and $y = -\dfrac{2}{3}x + \dfrac{4}{3}$

$$\dfrac{(y-2)^2}{4} - \dfrac{(x+1)^2}{9} = 1$$

14. $2y^2 - x^2 = 18$

$$\frac{y^2}{9} - \frac{x^2}{18} = 1$$

$$\frac{y^2}{3^2} - \frac{x^2}{(3\sqrt{2})^2} = 1$$

$h = 0$, $k = 0$, $a = 3$, $b = 3\sqrt{2}$

$y - k = \dfrac{a}{b}(x - h)$ and $y - k = -\dfrac{a}{b}(x - h)$

$y - 0 = \dfrac{3}{3\sqrt{2}}(x - 0)$ and $y - 0 = -\dfrac{3}{3\sqrt{2}}(x - 0)$

$y = \dfrac{\sqrt{2}}{2}x$ and $y = -\dfrac{\sqrt{2}}{2}x$

15. The parabola is of the form $y^2 = 4px$. A point on the parabola is $\left(6, \dfrac{18}{2}\right)$, or $(6, 9)$.

$$y^2 = 4px$$
$$9^2 = 4 \cdot p \cdot 6$$
$$81 = 24p$$
$$\frac{27}{8} = p$$

Since the focus is at $(p, 0) = \left(\dfrac{27}{8}, 0\right)$, the focus is $\dfrac{27}{8}$ in. from the vertex.

16. $2x^2 - 3y^2 = -10$,   (1)

$\quad x^2 + 2y^2 = 9$ (2)

$\quad 2x^2 - 3y^2 = -10$

$\underline{-2x^2 - 4y^2 = -18}$ Multiplying (2) by -2

$\quad\quad -7y^2 = -28$ Adding

$\quad\quad\quad y^2 = 4$

$\quad\quad\quad y = \pm 2$

$x^2 + 2(\pm 2)^2 = 9$ Substituting into (2)

$\quad x^2 + 8 = 9$

$\quad\quad x^2 = 1$

$\quad\quad x = \pm 1$

The pairs $(1, 2)$, $(1, -2)$, $(-1, 2)$ and $(-1, -2)$ check.

17. $x^2 + y^2 = 13$,   (1)

$\quad x + y = 1$ (2)

First solve equation (2) for y.

$$y = 1 - x$$

Then substitute $1 - x$ for y in equation (1) and solve for x.

$$x^2 + (1 - x)^2 = 13$$
$$x^2 + 1 - 2x + x^2 = 13$$
$$2x^2 - 2x - 12 = 0$$
$$2(x^2 - x - 6) = 0$$
$$2(x - 3)(x + 2) = 0$$

$x = 3$ or $x = -2$

If $x = 3$, $y = 1 - 3 = -2$. If $x = -2$, $y = 1 - (-2) = 3$.

The pairs $(3, -2)$ and $(-2, 3)$ check.

18. $x + y = 5$,   (1)

$\quad xy = 6$ (2)

First solve equation (1) for y.

$$y = -x + 5$$

Then substitute $-x + 5$ for y in equation (2) and solve for x.

$$x(-x + 5) = 6$$
$$-x^2 + 5x - 6 = 0$$
$$-1(x^2 - 5x + 6) = 0$$
$$-1(x - 2)(x - 3) = 0$$

$x = 2$ or $x = 3$

If $x = 2$, $y = -2 + 5 = 3$. If $x = 3$, $y = -3 + 5 = 2$.

The pairs $(2, 3)$ and $(3, 2)$ check.

19. *Familiarize*. Let l and w represent the length and width of the rectangle, in feet, respectively.

Translate. The perimeter is 18 ft.

$\quad 2l + 2w = 18$ (1)

From the Pythagorean theorem, we have

$\quad l^2 + w^2 = (\sqrt{41})^2$ (2)

Carry out. We solve the system of equations. We first solve equation (1) for w.

$$2l + 2w = 18$$
$$2w = 18 - 2l$$
$$w = 9 - l$$

Then substitute $9 - l$ for w in equation (2) and solve for l.

$$l^2 + (9 - l)^2 = (\sqrt{41})^2$$
$$l^2 + 81 - 18l + l^2 = 41$$
$$2l^2 - 18l + 40 = 0$$
$$2(l^2 - 9l + 20) = 0$$
$$2(l - 4)(l - 5) = 0$$

$l = 4$ or $l = 5$

If $l = 4$, then $w = 9 - 4 = 5$. If $l = 5$, then $w = 9 - 5 = 4$. Since length is usually considered to be longer than width, we have $l = 5$ and $w = 4$.

Check. The perimeter is $2 \cdot 5 + 2 \cdot 4$, or 18 ft. The length of a diagonal is $\sqrt{5^2 + 4^2}$, or $\sqrt{41}$ ft. The solution checks.

State. The dimensions of the garden are 5 ft by 4 ft.

20. *Familiarize*. Let l and w represent the length and width of the playground, in feet, respectively.

Translate.

Perimeter: $2l + 2w = 210$ (1)

Area: $lw = 2700$ (2)

Carry out. We solve the system of equations. First solve equation (2) for w.

$$w = \frac{2700}{l}$$

Then substitute $\dfrac{2700}{l}$ for w in equation (1) and solve for l.

$$2l + 2 \cdot \frac{2700}{l} = 210$$

$$2l + \frac{5400}{l} = 210$$

$$2l^2 + 5400 = 210l \quad \text{Multiplying by } l$$

$$2l^2 - 210l + 5400 = 0$$

$$2(l^2 - 105l + 2700) = 0$$

$$2(l - 45)(l - 60) = 0$$

$l = 45 \quad or \quad l = 60$

If $l = 45$, then $w = \dfrac{2700}{45} = 60$. If $l = 60$, then $w = \dfrac{2700}{60} = 45$. Since length is usually considered to be longer than width, we have $l = 60$ and $w = 45$.

Check. Perimeter: $2 \cdot 60 + 2 \cdot 45 = 210$ ft

Area: $60 \cdot 45 = 2700$ ft^2

The solution checks.

State. The dimensions of the playground are 60 ft by 45 ft.

21. Graph: $y \geq x^2 - 4$,

$\qquad\qquad y < 2x - 1$

The solution set of $y \geq x^2 - 4$ is the parabola $y = x^2 - 4$ and the region inside it. The solution set of $y < 2x - 1$ is the half-plane below the line $y = 2x - 1$. We shade the region common to the two solution sets.

To find the points of intersection of the graphs we solve the system of equations

$\qquad y = x^2 - 4$,

$\qquad y = 2x - 1$.

The points of intersection are $(-1, -3)$ and $(3, 5)$.

22. Use the midpoint formula to find the center.

$$(h, k) = \left(\frac{1+5}{2}, \frac{1+(-3)}{2} \right) = (3, -1)$$

Use the distance formula to find the radius.

$$r = \frac{1}{2}\sqrt{(1-5)^2 + (1-(-3))^2} = \frac{1}{2}\sqrt{(-4)^2 + (4)^2} = 2\sqrt{2}$$

Write the equation of the circle.

$$(x - h)^2 + (y - k)^2 = r^2$$

$$(x - 3)^2 + [y - (-1)]^2 = (2\sqrt{2})^2$$

$$(x - 3)^2 + (y + 1)^2 = 8$$

Chapter 7

Sequences, Series, and Combinatorics

1. $a_n = 4n - 1$

$a_1 = 4 \cdot 1 - 1 = 3,$

$a_2 = 4 \cdot 2 - 1 = 7,$

$a_3 = 4 \cdot 3 - 1 = 11,$

$a_4 = 4 \cdot 4 - 1 = 15;$

$a_{10} = 4 \cdot 10 - 1 = 39;$

$a_{15} = 4 \cdot 15 - 1 = 59$

3. $a_n = \dfrac{n}{n-1}, \ n \geq 2$

The first 4 terms are $a_2, a_3, a_4,$ and a_5:

$a_2 = \dfrac{2}{2-1} = 2,$

$a_3 = \dfrac{3}{3-1} = \dfrac{3}{2},$

$a_4 = \dfrac{4}{4-1} = \dfrac{4}{3},$

$a_5 = \dfrac{5}{5-1} = \dfrac{5}{4};$

$a_{10} = \dfrac{10}{10-1} = \dfrac{10}{9};$

$a_{15} = \dfrac{15}{15-1} = \dfrac{15}{14}$

5. $a_n = \dfrac{n^2-1}{n^2+1},$

$a_1 = \dfrac{1^2-1}{1^2+1} = 0,$

$a_2 = \dfrac{2^2-1}{2^2+1} = \dfrac{3}{5},$

$a_3 = \dfrac{3^2-1}{3^2+1} = \dfrac{8}{10} = \dfrac{4}{5},$

$a_4 = \dfrac{4^2-1}{4^2+1} = \dfrac{15}{17};$

$a_{10} = \dfrac{10^2-1}{10^2+1} = \dfrac{99}{101};$

$a_{15} = \dfrac{15^2-1}{15^2+1} = \dfrac{224}{226} = \dfrac{112}{113}$

7. $a_n = (-1)^n n^2$

$a_1 = (-1)^1 1^2 = -1,$

$a_2 = (-1)^2 2^2 = 4,$

$a_3 = (-1)^3 3^2 = -9,$

$a_4 = (-1)^4 4^2 = 16;$

$a_{10} = (-1)^{10} 10^2 = 100;$

$a_{15} = (-1)^{15} 15^2 = -225$

9. $a_n = 5 + \dfrac{(-2)^{n+1}}{2^n}$

$a_1 = 5 + \dfrac{(-2)^{1+1}}{2^1} = 5 + \dfrac{4}{2} = 7,$

$a_2 = 5 + \dfrac{(-2)^{2+1}}{2^2} = 5 + \dfrac{-8}{4} = 3,$

$a_3 = 5 + \dfrac{(-2)^{3+1}}{2^3} = 5 + \dfrac{16}{8} = 7,$

$a_4 = 5 + \dfrac{(-2)^{4+1}}{2^4} = 5 + \dfrac{-32}{16} = 3;$

$a_{10} = 5 + \dfrac{(-2)^{10+1}}{2^{10}} = 5 + \dfrac{-1 \cdot 2^{11}}{2^{10}} = 3;$

$a_{15} = 5 + \dfrac{(-2)^{15+1}}{2^{15}} = 5 + \dfrac{2^{16}}{2^{15}} = 7$

11. $a_n = 5n - 6$

$a_8 = 5 \cdot 8 - 6 = 40 - 6 = 34$

13. $a_n = (2n+3)^2$

$a_6 = (2 \cdot 6 + 3)^2 = 225$

15. $a_n = 5n^2(4n - 100)$

$a_{11} = 5(11)^2(4 \cdot 11 - 100) = 5(121)(-56) =$

$\qquad -33,880$

17. $a_n = \ln e^n$

$a_{67} = \ln e^{67} = 67$

19. $2, 4, 6, 8, 10, \ldots$

These are the even integers, so the general term might be $2n$.

21. $-2, 6, -18, 54, \ldots$

We can see a pattern if we write the sequence as

$-1 \cdot 2 \cdot 1, \ 1 \cdot 2 \cdot 3, \ -1 \cdot 2 \cdot 9, \ 1 \cdot 2 \cdot 27, \ldots$

The general term might be $(-1)^n 2(3)^{n-1}$.

23. $\dfrac{2}{3}, \dfrac{3}{4}, \dfrac{4}{5}, \dfrac{5}{6}, \dfrac{6}{7}, \ldots$

These are fractions in which the denominator is 1 greater than the numerator. Also, each numerator is 1 greater than the preceding numerator. The general term might be $\dfrac{n+1}{n+2}$.

25. $1 \cdot 2, \ 2 \cdot 3, \ 3 \cdot 4, \ 4 \cdot 5, \ldots$

These are the products of pairs of consecutive natural numbers. The general term might be $n(n+1)$.

27. $0, \log 10, \log 100, \log 1000, \ldots$

We can see a pattern if we write the sequence as

$\log 1, \log 10, \log 100, \log 1000, \ldots$

The general term might be $\log 10^{n-1}$. This is equivalent to $n - 1$.

29. $1, 2, 3, 4, 5, 6, 7, \ldots$

$S_3 = 1 + 2 + 3 = 6$

$S_7 = 1 + 2 + 3 + 4 + 5 + 6 + 7 = 28$

31. $2, 4, 6, 8, \ldots$

$S_4 = 2 + 4 + 6 + 8 = 20$

$S_5 = 2 + 4 + 6 + 8 + 10 = 30$

33. $\displaystyle\sum_{k=1}^{5} \frac{1}{2k} = \frac{1}{2 \cdot 1} + \frac{1}{2 \cdot 2} + \frac{1}{2 \cdot 3} + \frac{1}{2 \cdot 4} + \frac{1}{2 \cdot 5}$

$= \frac{1}{2} + \frac{1}{4} + \frac{1}{6} + \frac{1}{8} + \frac{1}{10}$

$= \frac{60}{120} + \frac{30}{120} + \frac{20}{120} + \frac{15}{120} + \frac{12}{120}$

$= \frac{137}{120}$

35. $\displaystyle\sum_{i=0}^{6} 2^i = 2^0 + 2^1 + 2^2 + 2^3 + 2^4 + 2^5 + 2^6$

$= 1 + 2 + 4 + 8 + 16 + 32 + 64$

$= 127$

37. $\displaystyle\sum_{k=7}^{10} \ln k = \ln 7 + \ln 8 + \ln 9 + \ln 10 =$

$\ln(7 \cdot 8 \cdot 9 \cdot 10) = \ln 5040 \approx 8.5252$

39. $\displaystyle\sum_{k=1}^{8} \frac{k}{k+1} = \frac{1}{1+1} + \frac{2}{2+1} + \frac{3}{3+1} + \frac{4}{4+1} +$

$\frac{5}{5+1} + \frac{6}{6+1} + \frac{7}{7+1} + \frac{8}{8+1}$

$= \frac{1}{2} + \frac{2}{3} + \frac{3}{4} + \frac{4}{5} + \frac{5}{6} + \frac{6}{7} + \frac{7}{8} + \frac{8}{9}$

$= \frac{15,551}{2520}$

41. $\displaystyle\sum_{i=1}^{5} (-1)^i$

$= (-1)^1 + (-1)^2 + (-1)^3 + (-1)^4 + (-1)^5$

$= -1 + 1 - 1 + 1 - 1$

$= -1$

43. $\displaystyle\sum_{k=1}^{8} (-1)^{k+1} 3k$

$= (-1)^2 3 \cdot 1 + (-1)^3 3 \cdot 2 + (-1)^4 3 \cdot 3 +$

$(-1)^5 3 \cdot 4 + (-1)^6 3 \cdot 5 + (-1)^7 3 \cdot 6 +$

$(-1)^8 3 \cdot 7 + (-1)^9 3 \cdot 8$

$= 3 - 6 + 9 - 12 + 15 - 18 + 21 - 24$

$= -12$

45. $\displaystyle\sum_{k=0}^{6} \frac{2}{k^2+1} = \frac{2}{0^2+1} + \frac{2}{1^2+1} + \frac{2}{2^2+1} + \frac{2}{3^2+1} +$

$\frac{2}{4^2+1} + \frac{2}{5^2+1} + \frac{2}{6^2+1}$

$= 2 + 1 + \frac{2}{5} + \frac{2}{10} + \frac{2}{17} + \frac{2}{26} + \frac{2}{37}$

$= 2 + 1 + \frac{2}{5} + \frac{1}{5} + \frac{2}{17} + \frac{1}{13} + \frac{2}{37}$

$= \frac{157,351}{40,885}$

47. $\displaystyle\sum_{k=0}^{5} (k^2 - 2k + 3)$

$= (0^2 - 2 \cdot 0 + 3) + (1^2 - 2 \cdot 1 + 3) +$

$(2^2 - 2 \cdot 2 + 3) + (3^2 - 2 \cdot 3 + 3) +$

$(4^2 - 2 \cdot 4 + 3) + (5^2 - 2 \cdot 5 + 3)$

$= 3 + 2 + 3 + 6 + 11 + 18$

$= 43$

49. $\displaystyle\sum_{i=0}^{10} \frac{2i}{2^i + 1}$

$= \frac{2^0}{2^0+1} + \frac{2^1}{2^1+1} + \frac{2^2}{2^2+1} + \frac{2^3}{2^3+1} + \frac{2^4}{2^4+1} +$

$\frac{2^5}{2^5+1} + \frac{2^6}{2^6+1} + \frac{2^7}{2^7+1} + \frac{2^8}{2^8+1} + \frac{2^9}{2^9+1} +$

$\frac{2^{10}}{2^{10}+1}$

$= \frac{1}{2} + \frac{2}{3} + \frac{4}{5} + \frac{8}{9} + \frac{16}{17} + \frac{32}{33} + \frac{64}{65} + \frac{128}{129} +$

$\frac{256}{257} + \frac{512}{513} + \frac{1024}{1025}$

≈ 9.736

51. $5 + 10 + 15 + 20 + 25 + \ldots$

This is a sum of multiples of 5, and it is an infinite series. Sigma notation is

$\displaystyle\sum_{k=1}^{\infty} 5k.$

53. $2 - 4 + 8 - 16 + 32 - 64$

This is a sum of powers of 2 with alternating signs. Sigma notation is

$\displaystyle\sum_{k=1}^{6} (-1)^{k+1} 2k, \text{ or } \sum_{k=1}^{6} (-1)^{k-1} 2k$

55. $-\frac{1}{2} + \frac{2}{3} - \frac{3}{4} + \frac{4}{5} - \frac{5}{6} + \frac{6}{7}$

This is a sum of fractions in which the denominator is one greater than the numerator. Also, each numerator is 1 greater than the preceding numerator and the signs alternate. Sigma notation is

$\displaystyle\sum_{k=1}^{6} (-1)^k \frac{k}{k+1}.$

57. $4 - 9 + 16 - 25 + \ldots + (-1)^n n^2$

This is a sum of terms of the form $(-1)^k k^2$, beginning with $k = 2$ and continuing through $k = n$. Sigma notation is

$$\sum_{k=2}^{n} (-1)^k k^2.$$

59. $\dfrac{1}{1 \cdot 2} + \dfrac{1}{2 \cdot 3} + \dfrac{1}{3 \cdot 4} + \dfrac{1}{4 \cdot 5} + \ldots$

This is a sum of fractions in which the numerator is 1 and the denominator is a product of two consecutive integers. The larger integer in each product is the smaller integer in the succeeding product. It is an infinite series. Sigma notation is

$$\sum_{k=1}^{\infty} \frac{1}{k(k+1)}.$$

61. $a_1 = 4, \qquad a_{k+1} = 1 + \dfrac{1}{a_k}$

$a_2 = 1 + \dfrac{1}{4} = 1\dfrac{1}{4}$, or $\dfrac{5}{4}$

$a_3 = 1 + \dfrac{1}{\frac{5}{4}} = 1 + \dfrac{4}{5} = 1\dfrac{4}{5}$, or $\dfrac{9}{5}$

$a_4 = 1 + \dfrac{1}{\frac{9}{5}} = 1 + \dfrac{5}{9} = 1\dfrac{5}{9}$, or $\dfrac{14}{9}$

63. $a_1 = 6561, \qquad a_{k+1} = (-1)^k \sqrt{a_k}$

$a_2 = (-1)^1 \sqrt{6561} = -81$

$a_3 = (-1)^2 \sqrt{-81} = 9i$

$a_4 = (-1)^3 \sqrt{9i} = -3\sqrt{i}$

65. $a_1 = 2, \qquad a_{k+1} = a_k + a_{k-1}$

$a_2 = 3$

$a_3 = 3 + 2 = 5$

$a_4 = 5 + 3 = 8$

67. a) $a_1 = \$1000(1.062)^1 = \$1062$

$a_2 = \$1000(1.062)^2 \approx \$1127.84$

$a_3 = \$1000(1.062)^3 \approx \$1197.77$

$a_4 = \$1000(1.062)^4 \approx \$1272.03$

$a_5 = \$1000(1.062)^5 \approx \$1350.90$

$a_6 = \$1000(1.062)^6 \approx \$1434.65$

$a_7 = \$1000(1.062)^7 \approx \$1523.60$

$a_8 = \$1000(1.062)^8 \approx \$1618.07$

$a_9 = \$1000(1.062)^9 \approx \$1718.39$

$a_{10} = \$1000(1.062)^{10} \approx \$1824.93$

b) $a_{20} = \$1000(1.062)^{20} \approx \$3330.35$

69. Find each term by multiplying the preceding term by 2. Find 17 terms, beginning with $a_1 = 1$, since there are 16 fifteen minute periods in 4 hr.

1, 2, 4, 8, 16, 32, 64, 128, 256, 512, 1024,

2048, 4096, 8192, 16,384, 32,768, 65,536

71. $a_1 = 1$ (Given)

$a_2 = 1$ (Given)

$a_3 = a_2 + a_1 = 1 + 1 = 2$

$a_4 = a_3 + a_2 = 2 + 1 = 3$

$a_5 = a_4 + a_3 = 3 + 2 = 5$

$a_6 = a_5 + a_4 = 5 + 3 = 8$

$a_7 = a_6 + a_5 = 8 + 5 = 13$

73. See the answer section in the text.

75. See the answer section in the text.

77. a) Using a quadratic regression feature on a graphing calculator, we have $a_n = -13.21231156n^2 + 52,871.41834n - 52,893,240.07$, where a_n is in thousands.

b) $a_{1998} \approx 253$ thousand

$a_{2000} \approx 350$ thousand

$a_{2002} \approx 342$ thousand

$a_{2005} \approx 131$ thousand

79. Discussion and Writing

81. *Familiarize*. Let $y =$ the number of times Yahoo was visited in April 2006, and let $g =$ the number of times Google was visited, in millions.

Translate. The total number of visits was 228.1 million.

$$y + g = 228.1$$

Yahoo was visited 17.7 million more times than Google.

$$y = g + 17.7$$

We have a system of equations.

$$y + g = 228.1, \quad (1)$$
$$y = g + 17.7 \quad (2)$$

Carry out. We use the substitution method.

$$g + 17.7 + g = 228.1$$
$$2g + 17.7 = 228.1$$
$$2g = 210.4$$
$$g = 105.2$$

Back-substitute to find y.

$$y = 105.2 + 17.7 = 122.9$$

Check. $105.2 + 122.9 = 228.1$ million, and 122.9 million is 17.7 million more than 105.2 million. The answer checks.

State. Yahoo was visited 122.9 million times, and Google was visited 105.2 million times.

83. We complete the square twice.

$$x^2 + y^2 + 5x - 8y = 2$$

$$x^2 + 5x + y^2 - 8y = 2$$

$$x^2 + 5x + \frac{25}{4} + y^2 - 8y + 16 = 2 + \frac{25}{4} + 16$$

$$\left(x + \frac{5}{2}\right)^2 + (y - 4)^2 = \frac{97}{4}$$

$$\left[x - \left(-\frac{5}{2}\right)\right]^2 + (y - 4)^2 = \left(\frac{\sqrt{97}}{2}\right)^2$$

The center is $\left(-\dfrac{5}{2}, 4\right)$ and the radius is $\dfrac{\sqrt{97}}{2}$.

85. $a_n = i^n$

$a_1 = i$

$a_2 = i^2 = -1$

$a_3 = i^3 = -i$

$a_4 = i^4 = 1$

$a_5 = i^5 = i^4 \cdot i = i;$

$S_5 = i - 1 - i + 1 + i = i$

87. $S_n = \ln 1 + \ln 2 + \ln 3 + \cdots + \ln n$

$\quad = \ln(1 \cdot 2 \cdot 3 \cdots n)$

Exercise Set 7.2

1. 3, 8, 13, 18, . . .

$a_1 = 3$

$d = 5 \quad (8 - 3 = 5, \ 13 - 8 = 5, \ 18 - 13 = 5)$

3. 9, 5, 1, −3, . . .

$a_1 = 9$

$d = -4 \quad (5 - 9 = -4, \ 1 - 5 = -4, \ -3 - 1 = -4)$

5. $\dfrac{3}{2}, \dfrac{9}{4}, 3, \dfrac{15}{4}, \ldots$

$a_1 = \dfrac{3}{2}$

$d = \dfrac{3}{4} \quad \left(\dfrac{9}{4} - \dfrac{3}{2} = \dfrac{3}{4}, 3 - \dfrac{9}{4} = \dfrac{3}{4}\right)$

7. $a_1 = \$316$

$d = -\$3 \quad (\$313 - \$316 = -\$3,$
$\qquad \$310 - \$313 = -\$3, \ \$307 - \$310 = -\$3)$

9. 2, 6, 10, . . .

$a_1 = 2, \ d = 4, \text{ and } n = 12$

$a_n = a_1 + (n - 1)d$

$a_{12} = 2 + (12 - 1)4 = 2 + 11 \cdot 4 = 2 + 44 = 46$

11. $3, \dfrac{7}{3}, \dfrac{5}{3}, \ldots$

$a_1 = 3, \ d = -\dfrac{2}{3}, \text{ and } n = 14$

$a_n = a_1 + (n - 1)d$

$a_{14} = 3 + (14 - 1)\left(-\dfrac{2}{3}\right) = 3 - \dfrac{26}{3} = -\dfrac{17}{3}$

13. $2345.78, $2967.54, $3589.30, . . .

$a_1 = \$2345.78, \ d = \$621.76, \text{ and } n = 10$

$a_n = a_1 + (n - 1)d$

$a_{10} = \$2345.78 + (10 - 1)(\$621.76) = \$7941.62$

15. $a_1 = 0.07, \ d = 0.05$

$a_n = a_1 + (n - 1)d$

Let $a_n = 1.67$, and solve for n.

$1.67 = 0.07 + (n - 1)(0.05)$

$1.67 = 0.07 + 0.05n - 0.05$

$1.65 = 0.05n$

$33 = n$

The 33rd term is 1.67.

17. $a_1 = 3, \ d = -\dfrac{2}{3}$

$a_n = a_1 + (n - 1)d$

Let $a_n = -27$, and solve for n.

$-27 = 3 + (n - 1)\left(-\dfrac{2}{3}\right)$

$-81 = 9 + (n - 1)(-2)$

$-81 = 9 - 2n + 2$

$-92 = -2n$

$46 = n$

The 46th term is −27.

19. $a_n = a_1 + (n - 1)d$

$33 = a_1 + (8 - 1)4 \quad$ Substituting 33 for a_8,
$\qquad\qquad\qquad\qquad$ 8 for n, and 4 for d

$33 = a_1 + 28$

$5 = a_1$

(Note that this procedure is equivalent to subtracting d from a_8 seven times to get a_1: $33 - 7(4) = 33 - 28 = 5$)

21. $a_n = a_1 + (n - 1)d$

$-507 = 25 + (n - 1)(-14)$

$-507 = 25 - 14n + 14$

$-546 = -14n$

$39 = n$

23. $\dfrac{25}{3} + 15d = \dfrac{95}{6}$

$\qquad 15d = \dfrac{45}{6}$

$\qquad\quad d = \dfrac{1}{2}$

$a_1 = \dfrac{25}{3} - 16\left(\dfrac{1}{2}\right) = \dfrac{25}{3} - 8 = \dfrac{1}{3}$

The first five terms of the sequence are $\dfrac{1}{3}, \dfrac{5}{6}, \dfrac{4}{3}, \dfrac{11}{6}, \dfrac{7}{3}$.

25. 5 + 8 + 11 + 14 + . . .

Note that $a_1 = 5$, $d = 3$, and $n = 20$. First we find a_{20}:

$a_n = a_1 + (n - 1)d$

$a_{20} = 5 + (20 - 1)3$

$\qquad = 5 + 19 \cdot 3 = 62$

Then

$$S_n = \frac{n}{2}(a_1 + a_n)$$

$$S_{20} = \frac{20}{2}(5 + 62)$$

$$= 10(67) = 670.$$

27. The sum is $2 + 4 + 6 + \ldots + 798 + 800$. This is the sum of the arithmetic sequence for which $a_1 = 2$, $a_n = 800$, and $n = 400$.

$$S_n = \frac{n}{2}(a_1 + a_n)$$

$$S_{400} = \frac{400}{2}(2 + 800) = 200(802) = 160,400$$

29. The sum is $7 + 14 + 21 + \ldots + 91 + 98$. This is the sum of the arithmetic sequence for which $a_1 = 7$, $a_n = 98$, and $n = 14$.

$$S_n = \frac{n}{2}(a_1 + a_n)$$

$$S_{14} = \frac{14}{2}(7 + 98) = 7(105) = 735$$

31. First we find a_{20}:

$$a_n = a_1 + (n-1)d$$

$$a_{20} = 2 + (20 - 1)5$$

$$= 2 + 19 \cdot 5 = 97$$

Then

$$S_n = \frac{n}{2}(a_1 + a_n)$$

$$S_{20} = \frac{20}{2}(2 + 97)$$

$$= 10(99) = 990.$$

33. $\displaystyle\sum_{k=1}^{40} (2k + 3)$

Write a few terms of the sum:

$$5 + 7 + 9 + \ldots + 83$$

This is a series coming from an arithmetic sequence with $a_1 = 5$, $n = 40$, and $a_{40} = 83$. Then

$$S_n = \frac{n}{2}(a_1 + a_n)$$

$$S_{40} = \frac{40}{2}(5 + 83)$$

$$= 20(88) = 1760$$

35. $\displaystyle\sum_{k=0}^{19} \frac{k-3}{4}$

Write a few terms of the sum:

$$-\frac{3}{4} - \frac{1}{2} - \frac{1}{4} + 0 + \frac{1}{4} + \ldots + 4$$

Since k goes from 0 through 19, there are 20 terms. Thus, this is equivalent to a series coming from an arithmetic sequence with $a_1 = -\frac{3}{4}$, $n = 20$, and $a_{20} = 4$. Then

$$S_n = \frac{n}{2}(a_1 + a_n)$$

$$S_{20} = \frac{20}{2}\left(-\frac{3}{4} + 4\right)$$

$$= 10 \cdot \frac{13}{4} = \frac{65}{2}.$$

37. $\displaystyle\sum_{k=12}^{57} \frac{7 - 4k}{13}$

Write a few terms of the sum:

$$-\frac{41}{13} - \frac{45}{13} - \frac{49}{13} - \ldots - \frac{221}{13}$$

Since k goes from 12 through 57, there are 46 terms. Thus, this is equivalent to a series coming from an arithmetic sequence with $a_1 = -\frac{41}{13}$, $n = 46$, and $a_{46} = -\frac{221}{13}$. Then

$$S_n = \frac{n}{2}(a_1 + a_n)$$

$$S_{46} = \frac{46}{2}\left(-\frac{41}{13} - \frac{221}{13}\right)$$

$$= 23\left(-\frac{262}{13}\right) = -\frac{6026}{13}.$$

39. *Familiarize*. We go from 50 poles in a row, down to six poles in the top row, so there must be 45 rows. We want the sum $50 + 49 + 48 + \ldots + 6$. Thus we want the sum of an arithmetic sequence. We will use the formula $S_n = \frac{n}{2}(a_1 + a_n)$.

Translate. We want to find the sum of the first 45 terms of an arithmetic sequence with $a_1 = 50$ and $a_{45} = 6$.

Carry out. Substituting into the formula, we have

$$S_{45} = \frac{45}{2}(50 + 6)$$

$$= \frac{45}{2} \cdot 56 = 1260$$

Check. We can do the calculation again, or we can do the entire addition:

$$50 + 49 + 48 + \ldots + 6.$$

State. There will be 1260 poles in the pile.

41. We first find how many plants will be in the last row.

Familiarize. The sequence is 35, 31, 27, ... It is an arithmetic sequence with $a_1 = 35$ and $d = -4$. Since each row must contain a positive number of plants, we must determine how many times we can add -4 to 35 and still have a positive result.

Translate. We find the largest integer x for which $35 + x(-4) > 0$. Then we evaluate the expression $35 - 4x$ for that value of x.

Carry out. We solve the inequality.

$$35 - 4x > 0$$

$$35 > 4x$$

$$\frac{35}{4} > x$$

$$8\frac{3}{4} > x$$

The integer we are looking for is 8. Thus $35 - 4x = 35 - 4(8) = 3$.

Check. If we add -4 to 35 eight times we get 3, a positive number, but if we add -4 to 35 more than eight times we get a negative number.

State. There will be 3 plants in the last row.

Next we find how many plants there are altogether.

Familiarize. We want to find the sum $35 + 31 + 27 + \ldots + 3$. We know $a_1 = 35$ $a_n = 3$, and, since we add -4 to 35 eight times, $n = 9$. (There are 8 terms after a_1, for a total of 9 terms.) We will use the formula $S_n = \frac{n}{2}(a_1 + a_n)$.

Translate. We want to find the sum of the first 9 terms of an arithmetic sequence in which $a_1 = 35$ and $a_9 = 3$.

Carry out. Substituting into the formula, we have
$$S_9 = \frac{9}{2}(35 + 3)$$
$$= \frac{9}{2} \cdot 38 = 171$$

Check. We can check the calculations by doing them again. We could also do the entire addition:
$$35 + 31 + 27 + \ldots + 3.$$

State. There are 171 plants altogether.

43. Familiarize. We have a sequence 10, 20, 30, . . . It is an arithmetic sequence with $a_1 = 10$, $d = 10$, and $n = 31$.

Translate. We want to find $S_n = \frac{n}{2}(a_1 + a_n)$ where $a_n = a_1 + (n-1)d$, $a_1 = 10$, $d = 10$, and $n = 31$.

Carry out. First we find a_{31}.
$$a_{31} = 10 + (31 - 1)10 = 10 + 30 \cdot 10 = 310$$
Then $S_{31} = \frac{31}{2}(10 + 310) = \frac{31}{2} \cdot 320 = 4960.$

Check. We can do the calculation again, or we can do the entire addition:
$$10 + 20 + 30 + \ldots + 310.$$

State. A total of 4960¢, or \$49.60 is saved.

45. Familiarize. We have arithmetic sequence with $a_1 = 28$, $d = 4$, and $n = 20$.

Translate. We want to find $S_n = \frac{n}{2}(a_1 + a_n)$ where $a_n = a_1 + (n-1)d$, $a_1 = 28$, $d = 4$, and $n = 20$.

Carry out. First we find a_{20}.
$$a_{20} = 28 + (20 - 1)4 = 104$$
Then $S_{20} = \frac{20}{2}(28 + 104) = 10 \cdot 132 = 1320.$

Check. We can do the calculations again, or we can do the entire addition:
$$28 + 32 + 36 + \ldots + 104.$$

State. There are 1320 seats in the first balcony.

47. Yes; $d = 6 - 3 = 9 - 6 = 3n - 3(n-1) = 3$.

49. Discussion and Writing

51.
$$2x + y + 3z = 12$$
$$x - 3y - 2z = -1$$
$$5x + 2y - 4z = -4$$

We will use Gauss-Jordan elimination with matrices. First we write the augmented matrix.

$$\left[\begin{array}{ccc|c} 2 & 1 & 3 & 12 \\ 1 & -3 & -2 & -1 \\ 5 & 2 & -4 & -4 \end{array}\right]$$

Next we interchange the first two rows.

$$\left[\begin{array}{ccc|c} 1 & -3 & -2 & -1 \\ 2 & 1 & 3 & 12 \\ 5 & 2 & -4 & -4 \end{array}\right]$$

Now multiply the first row by -2 and add it to the second row. Also multiply the first row by -5 and add it to the third row.

$$\left[\begin{array}{ccc|c} 1 & -3 & -2 & -1 \\ 0 & 7 & 7 & 14 \\ 0 & 17 & 6 & 1 \end{array}\right]$$

Multiply the second row by $\frac{1}{7}$.

$$\left[\begin{array}{ccc|c} 1 & -3 & -2 & -1 \\ 0 & 1 & 1 & 2 \\ 0 & 17 & 6 & 1 \end{array}\right]$$

Multiply the second row by 3 and add it to the first row. Also multiply the second row by -17 and add it to the third row.

$$\left[\begin{array}{ccc|c} 1 & 0 & 1 & 5 \\ 0 & 1 & 1 & 2 \\ 0 & 0 & -11 & -33 \end{array}\right]$$

Multiply the third row by $-\frac{1}{11}$.

$$\left[\begin{array}{ccc|c} 1 & 0 & 1 & 5 \\ 0 & 1 & 1 & 2 \\ 0 & 0 & 1 & 3 \end{array}\right]$$

Multiply the third row by -1 and add it to the first row and also to the second row.

$$\left[\begin{array}{ccc|c} 1 & 0 & 0 & 2 \\ 0 & 1 & 0 & -1 \\ 0 & 0 & 1 & 3 \end{array}\right]$$

Now we can read the solution from the matrix. It is $(2, -1, 3)$.

53. The vertices are on the y-axis, so the transverse axis is vertical and $a = 5$. The length of the minor axis is 4, so $b = 4/2 = 2$. The equation is
$$\frac{x^2}{4} + \frac{y^2}{25} = 1.$$

55. $S_n = \dfrac{n}{2}(1 + 2n - 1) = n^2$

57. Let d = the common difference. Then $a_4 = a_2 + 2d$, or

$$10p + q = 40 - 3q + 2d$$
$$10p + 4q - 40 = 2d$$
$$5p + 2q - 20 = d.$$

Also, $a_1 = a_2 - d$, so we have

$$a_1 = 40 - 3q - (5p + 2q - 20)$$
$$= 40 - 3q - 5p - 2q + 20$$
$$= 60 - 5p - 5q.$$

59. $4, m_1, m_2, m_3, 12$

We look for $m_1, m_2,$ and m_3 such that $4, m_1, m_2, m_3, 12$ is an arithmetic sequence. In this case, $a_1 = 4$, $n = 5$, and $a_5 = 12$. First we find d:

$$a_n = a_1 + (n - 1)d$$
$$12 = 4 + (5 - 1)d$$
$$12 = 4 + 4d$$
$$8 = 4d$$
$$2 = d$$

Then we have

$$m_1 = a_1 + d = 4 + 2 = 6$$
$$m_2 = m_1 + d = 6 + 2 = 8$$
$$m_3 = m_2 + d = 8 + 2 = 10$$

61. $4, m_1, m_2, m_3, m_4, 13$

We look for $m_1, m_2, m_3,$ and m_4 such that $4, m_1, m_2, m_3, m_4, 13$ is an arithmetic sequence. In this case $a_1 = 4$, $n = 6$, and $a_6 = 13$. First we find d.

$$a_n = a_1 + (n - 1)d$$
$$13 = 4 + (6 - 1)d$$
$$9 = 5d$$
$$1\tfrac{4}{5} = d$$

Then we have

$$m_1 = a_1 + d = 4 + 1\tfrac{4}{5} = 5\tfrac{4}{5},$$
$$m_2 = m_1 + d = 5\tfrac{4}{5} + 1\tfrac{4}{5} = 6\tfrac{8}{5} = 7\tfrac{3}{5},$$
$$m_3 = m_2 + d = 7\tfrac{3}{5} + 1\tfrac{4}{5} = 8\tfrac{7}{5} = 9\tfrac{2}{5},$$
$$m_4 = m_3 + d = 9\tfrac{2}{5} + 1\tfrac{4}{5} = 10\tfrac{6}{5} = 11\tfrac{1}{5}.$$

63. $1, 1 + d, 1 + 2d, \ldots, 50$ has n terms and $S_n = 459$.

Find n:

$$459 = \dfrac{n}{2}(1 + 50)$$
$$18 = n$$

Find d:

$$50 = 1 + (18 - 1)d$$
$$\dfrac{49}{17} = d$$

The sequence has a total of 18 terms, so we insert 16 arithmetic means between 1 and 50 with $d = \dfrac{49}{17}$.

65.
$$m = p + d$$
$$\underline{m = q - d}$$
$$2m = p + q \quad \text{Adding}$$
$$m = \dfrac{p + q}{2}$$

Exercise Set 7.3

1. $2, 4, 8, 16, \ldots$

$$\dfrac{4}{2} = 2, \quad \dfrac{8}{4} = 2, \quad \dfrac{16}{8} = 2$$
$$r = 2$$

3. $1, -1, 1, -1, \ldots$

$$\dfrac{-1}{1} = -1, \quad \dfrac{1}{-1} = -1, \quad \dfrac{-1}{1} = -1$$
$$r = -1$$

5. $\dfrac{2}{3}, -\dfrac{4}{3}, \dfrac{8}{3}, -\dfrac{16}{3}, \ldots$

$$\dfrac{-\dfrac{4}{3}}{\dfrac{2}{3}} = -2, \quad \dfrac{\dfrac{8}{3}}{-\dfrac{4}{3}} = -2, \quad \dfrac{-\dfrac{16}{3}}{\dfrac{8}{3}} = -2$$
$$r = -2$$

7. $\dfrac{0.6275}{6.275} = 0.1, \quad \dfrac{0.06275}{0.6275} = 0.1$

$$r = 0.1$$

9. $\dfrac{\dfrac{5a}{2}}{5} = \dfrac{a}{2}, \quad \dfrac{\dfrac{5a^2}{4}}{\dfrac{5a}{2}} = \dfrac{a}{2}, \quad \dfrac{\dfrac{5a^3}{8}}{\dfrac{5a}{4}} = \dfrac{a}{2}$

$$r = \dfrac{a}{2}$$

11. $2, 4, 8, 16, \ldots$

$a_1 = 2$, $n = 7$, and $r = \dfrac{4}{2}$, or 2.

We use the formula $a_n = a_1 r^{n-1}$.

$$a_7 = 2(2)^{7-1} = 2 \cdot 2^6 = 2 \cdot 64 = 128$$

13. $2, 2\sqrt{3}, 6, \ldots$

$a_1 = 2$, $n = 9$, and $r = \dfrac{2\sqrt{3}}{2}$, or $\sqrt{3}$

$$a_n = a_1 r^{n-1}$$
$$a_9 = 2(\sqrt{3})^{9-1} = 2(\sqrt{3})^8 = 2 \cdot 81 = 162$$

15. $\dfrac{7}{625}, -\dfrac{7}{25}, \ldots$

$a_1 = \dfrac{7}{625}$, $n = 23$, and $r = \dfrac{-\dfrac{7}{25}}{\dfrac{7}{625}} = -25$.

$$a_n = a_1 r^{n-1}$$

$$a_{23} = \frac{7}{625}(-25)^{23-1} = \frac{7}{625}(-25)^{22}$$

$$= \frac{7}{25^2} \cdot 25^2 \cdot 25^{20} = 7(25)^{20}, \text{ or } 7(5)^{40}$$

17. $1, 3, 9, \ldots$

$a_1 = 1$ and $r = \dfrac{3}{1}$, or 3

$a_n = a_1 r^{n-1}$

$a_n = 1(3)^{n-1} = 3^{n-1}$

19. $1, -1, 1, -1, \ldots$

$a_1 = 1$ and $r = \dfrac{-1}{1} = -1$

$a_n = a_1 r^{n-1}$

$a_n = 1(-1)^{n-1} = (-1)^{n-1}$

21. $\dfrac{1}{x}, \dfrac{1}{x^2}, \dfrac{1}{x^3}, \cdots$

$a_1 = \dfrac{1}{x}$ and $r = \dfrac{\frac{1}{x^2}}{\frac{1}{x}} = \dfrac{1}{x}$

$a_n = a_1 r^{n-1}$

$a_n = \dfrac{1}{x}\left(\dfrac{1}{x}\right)^{n-1} = \dfrac{1}{x} \cdot \dfrac{1}{x^{n-1}} = \dfrac{1}{x^{1+n-1}} = \dfrac{1}{x^n}$

23. $6 + 12 + 24 + \ldots$

$a_1 = 6$, $n = 7$, and $r = \dfrac{12}{6}$, or 2

$S_n = \dfrac{a_1(1-r^n)}{1-r}$

$S_7 = \dfrac{6(1-2^7)}{1-2} = \dfrac{6(1-128)}{-1} = \dfrac{6(-127)}{-1} = 762$

25. $\dfrac{1}{18} - \dfrac{1}{6} + \dfrac{1}{2} - \ldots$

$a_1 = \dfrac{1}{18}$, $n = 9$, and $r = \dfrac{-\frac{1}{6}}{\frac{1}{18}} = -\dfrac{1}{6} \cdot \dfrac{18}{1} = -3$

$S_n = \dfrac{a_1(1-r^n)}{1-r}$

$S_9 = \dfrac{\frac{1}{18}\left[1-(-3)^9\right]}{1-(-3)} = \dfrac{\frac{1}{18}(1+19{,}683)}{4}$

$\dfrac{\frac{1}{18}(19{,}684)}{4} = \dfrac{1}{18}(19{,}684)\left(\dfrac{1}{4}\right) = \dfrac{4921}{18}$

27. $4 + 2 + 1 + \ldots$

$|r| = \left|\dfrac{2}{4}\right| = \left|\dfrac{1}{2}\right| = \dfrac{1}{2}$, and since $|r| < 1$, the series does have a sum.

$S_\infty = \dfrac{a_1}{1-r} = \dfrac{4}{1-\frac{1}{2}} = \dfrac{4}{\frac{1}{2}} = 4 \cdot \dfrac{2}{1} = 8$

29. $25 + 20 + 16 + \ldots$

$|r| = \left|\dfrac{20}{25}\right| = \left|\dfrac{4}{5}\right| = \dfrac{4}{5}$, and since $|r| < 1$, the series does have a sum.

$S_\infty = \dfrac{a_1}{1-r} = \dfrac{25}{1-\frac{4}{5}} = \dfrac{25}{\frac{1}{5}} = 25 \cdot \dfrac{5}{1} = 125$

31. $8 + 40 + 200 + \ldots$

$|r| = \left|\dfrac{40}{8}\right| = |5| = 5$, and since $|r| > 1$ the series does not have a sum.

33. $0.6 + 0.06 + 0.006 + \ldots$

$|r| = \left|\dfrac{0.06}{0.6}\right| = |0.1| = 0.1$, and since $|r| < 1$, the series does have a sum.

$S_\infty = \dfrac{a_1}{1-r} = \dfrac{0.6}{1-0.1} = \dfrac{0.6}{0.9} = \dfrac{6}{9} = \dfrac{2}{3}$

35. $\displaystyle\sum_{k=1}^{11} 15\left(\dfrac{2}{3}\right)^k$

$a_1 = 15 \cdot \dfrac{2}{3}$ or 10; $|r| = \left|\dfrac{2}{3}\right| = \dfrac{2}{3}$, $n = 11$

$S_{11} = \dfrac{10\left[1-\left(\frac{2}{3}\right)^{11}\right]}{1-\frac{2}{3}} = \dfrac{10\left[1-\frac{2048}{177{,}147}\right]}{\frac{1}{3}}$

$= 10 \cdot \dfrac{175{,}099}{177{,}147} \cdot 3$

$= \dfrac{1{,}750{,}990}{59{,}049}, \text{ or } 29\dfrac{38{,}569}{59{,}049}$

37. $\displaystyle\sum_{k=1}^{\infty} \left(\dfrac{1}{2}\right)^{k-1}$

$a_1 = 1$, $|r| = \left|\dfrac{1}{2}\right| = \dfrac{1}{2}$

$S_\infty = \dfrac{a_1}{1-r} = \dfrac{1}{1-\frac{1}{2}} = \dfrac{1}{\frac{1}{2}} = 2$

39. $\displaystyle\sum_{k=1}^{\infty} 12.5^k$

Since $|r| = 12.5 > 1$, the sum does not exist.

41. $\displaystyle\sum_{k=1}^{\infty} \$500(1.11)^{-k}$

$a_1 = \$500(1.11)^{-1}$, or $\dfrac{\$500}{1.11}$; $|r| = |1.11^{-1}| = \dfrac{1}{1.11}$

$S_\infty = \dfrac{a_1}{1-r} = \dfrac{\frac{\$500}{1.11}}{1-\frac{1}{1.11}} = \dfrac{\frac{\$500}{1.11}}{\frac{0.11}{1.11}} \approx \$4545.\overline{45}$

43. $\displaystyle\sum_{k=1}^{\infty} 16(0.1)^{k-1}$

$a_1 = 16$, $|r| = |0.1| = 0.1$

$S_\infty = \dfrac{a_1}{1-r} = \dfrac{16}{1-0.1} = \dfrac{16}{0.9} = \dfrac{160}{9}$

45. $0.131313\ldots = 0.13 + 0.0013 + 0.000013 + \ldots$

This is an infinite geometric series with $a_1 = 0.13$.

$|r| = \left|\dfrac{0.0013}{0.13}\right| = |0.01| = 0.01 < 1$, so the series has a limit.

$S_\infty = \dfrac{a_1}{1-r} = \dfrac{0.13}{1-0.01} = \dfrac{0.13}{0.99} = \dfrac{13}{99}$

47. We will find fractional notation for $0.999\overline{9}$ and then add 8.

$0.999\overline{9} = 0.9 + 0.09 + 0.009 + 0.0009 + \ldots$

This is an infinite geometric series with $a_1 = 0.9$.

$|r| = \left|\dfrac{0.09}{0.9}\right| = |0.1| = 0.1 < 1$, so the series has a limit.

$S_\infty = \dfrac{a_1}{1-r} = \dfrac{0.9}{1-0.1} = \dfrac{0.9}{0.9} = 1$

Then $8.999\overline{9} = 8 + 1 = 9$.

49. $3.4125\overline{125} = 3.4 + 0.0125\overline{125}$

We will find fractional notation for $0.0125\overline{125}$ and then add 3.4, or $\dfrac{34}{10}$, or $\dfrac{17}{5}$.

$0.0125\overline{125} = 0.0125 + 0.0000125 + \ldots$

This is an infinite geometric series with $a_1 = 0.0125$.

$|r| = \left|\dfrac{0.0000125}{0.0125}\right| = |0.001| = 0.001 < 1$, so the series has a limit.

$S_\infty = \dfrac{a_1}{1-r} = \dfrac{0.0125}{1-0.001} = \dfrac{0.0125}{0.999} = \dfrac{125}{9990}$

Then $\dfrac{17}{5} + \dfrac{125}{9990} = \dfrac{33{,}966}{9990} + \dfrac{125}{9990} = \dfrac{34{,}091}{9990}$

51. The amount earned is $\$0.01 + \$0.01(2) + \$0.01(2^2) + \$0.01(2^3) + \ldots + \$0.01(2^{27})$. This is a geometric series with $a_1 = \$0.01$, $r = 2$, and $n = 28$.

$$S_n = \dfrac{a_1(1-r^n)}{1-r}$$

$$S_{28} = \dfrac{\$0.01(1-2^{28})}{1-2} = \$2{,}684{,}354.55$$

53. a) The rebound distances form a geometric sequence:

$$0.6 \times 200, \ (0.6)^2 \times 200, \ (0.6)^3 \times 200, \ \ldots,$$

or $120, \ 0.6 \times 120, \ (0.6)^2 \times 120, \ \ldots$

The total rebound distance after 9 rebounds is the sum of the first 9 terms of this sequence.

We will use the formula $S_n = \dfrac{a_1(1-r^n)}{1-r}$ with $a_1 = 120$, $r = 0.6$, and $n = 9$.

$$S_9 = \dfrac{120[1-(0.6)^9]}{1-0.6} \approx 297$$

The bungee jumper has traveled about 297 ft upward after 9 rebounds.

b) $S_\infty = \dfrac{a_1}{1-r} = \dfrac{120}{1-0.6} = 300$ ft

55. The amount of the annuity is the geometric series

$\$1000 + \$1000(1.032) + \$1000(1.032)^2 + \ldots + \$1000(1.032)^{17}$, where $a_1 = \$1000$, $r = 1.032$, and $n = 18$.

Using the formula

$$S_n = \dfrac{a_1(1-r^n)}{1-r}$$

we have

$$S_{18} = \dfrac{\$1000[1-(1.032)^{18}]}{1-1.032}$$

$$S_{18} \approx \$23{,}841.50.$$

The amount of the annuity is $\$23{,}841.50$.

57. The amounts owed at the beginning of successive years form a geometric sequence:

$\$120{,}000, \ (1.12)\$120{,}000, \ (1.12)^2\$120{,}000, \ \ldots$

The amount to be repaid at the end of 13 years is the amount owed at the beginning of the 14th year.

Use the formula $a_n = a_1 r^{n-1}$ with $a_1 = 120{,}000$, $r = 1.12$, and $n = 14$:

$$a_{14} = 120{,}000(1.12)^{14-1}$$

$$a_{14} \approx \$523{,}619.17.$$

At the end of 13 years, $\$523{,}619.17$ will be repaid.

59. The total effect on the economy is the sum of an infinite geometric series

$\$13{,}000{,}000{,}000 + \$13{,}000{,}000{,}000(0.85) + \$13{,}000{,}000{,}000(0.85)^2 + \ldots$

with $a_1 = \$13{,}000{,}000{,}000$ and $r = 0.85$.

Using the formula

$$S_\infty = \dfrac{a_1}{1-r}$$

we have

$$S_\infty = \dfrac{\$13{,}000{,}000{,}000}{1-0.85}$$

$$S_\infty \approx \$86{,}666{,}666{,}667.$$

The total effect on the economy is $\$86{,}666{,}666{,}667$.

61. Left to the student

63. Discussion and Writing

65. $f(x) = x^2$, $g(x) = 4x + 5$

$(f \circ g)(x) = f(g(x)) = f(4x+5) = (4x+5)^2 = 16x^2 + 40x + 25$

$(g \circ f)(x) = g(f(x)) = g(x^2) = 4x^2 + 5$

67. $5^x = 35$

$\ln 5^x = \ln 35$

$x \ln 5 = \ln 35$

$x = \dfrac{\ln 35}{\ln 5}$

$x \approx 2.209$

69. See the answer section in the text.

71. a) If the sequence is arithmetic, then $a_2 - a_1 = a_3 - a_2$.

$$x + 7 - (x + 3) = 4x - 2 - (x + 7)$$
$$x = \frac{13}{3}$$

The three given terms are $\frac{13}{3} + 3 = \frac{22}{3}$,

$\frac{13}{3} + 7 = \frac{34}{3}$, and $4 \cdot \frac{13}{3} - 2 = \frac{46}{3}$.

Then $d = \frac{12}{3}$, or 4, so the fourth term is

$\frac{46}{3} + \frac{12}{3} = \frac{58}{3}$.

b) If the sequence is geometric, then $a_2/a_1 = a_3/a_2$.

$$\frac{x+7}{x+3} = \frac{4x-2}{x+7}$$
$$x = -\frac{11}{3} \text{ or } x = 5$$

For $x = -\frac{11}{3}$: The three given terms are

$-\frac{11}{3} + 3 = -\frac{2}{3}$, $-\frac{11}{3} + 7 = \frac{10}{3}$, and

$4\left(-\frac{11}{3}\right) - 2 = -\frac{50}{3}$.

Then $r = -5$, so the fourth term is

$-\frac{50}{3}(-5) = \frac{250}{3}$.

For $x = 5$: The three given terms are $5 + 3 = 8$,

$5 + 7 = 12$, and $4 \cdot 5 - 2 = 18$. Then $r = \frac{3}{2}$, so the

fourth term is $18 \cdot \frac{3}{2} = 27$.

73. $x^2 - x^3 + x^4 - x^5 + \ldots$

This is a geometric series with $a_1 = x^2$ and $r = -x$.

$$S_n = \frac{a_1(1 - r^n)}{1 - r} = \frac{x^2(1 - (-x)^n)}{1 - (-x)} = \frac{x^2(1 - (-x)^n)}{1 + x}$$

75. See the answer section in the text.

77. The length of a side of the first square is 16 cm. The length of a side of the next square is the length of the hypotenuse of a right triangle with legs 8 cm and 8 cm, or $8\sqrt{2}$ cm. The length of a side of the next square is the length of the hypotenuse of a right triangle with legs $4\sqrt{2}$ cm and $4\sqrt{2}$ cm, or 8 cm. The areas of the squares form a sequence:

$(16)^2$, $(8\sqrt{2})^2$, $(8)^2$, $\ldots$, or

256, 128, 64, $\ldots$.

This is a geometric sequence with $a_1 = 256$ and $r = \frac{1}{2}$.

We find the sum of the infinite geometric series $256 + 128 + 64 + \ldots$.

$$S_\infty = \frac{a_1}{1 - r}$$
$$S_\infty = \frac{256}{1 - \frac{1}{2}}$$
$$S_\infty = 512$$

The sum of the areas is 512 cm^2.

Exercise Set 7.4

1. $n^2 < n^3$

$1^2 < 1^3$, $2^2 < 2^3$, $3^2 < 3^3$, $4^2 < 4^3$, $5^2 < 5^3$

The first statement is false, and the others are true.

3. A polygon of n sides has $\dfrac{n(n-3)}{2}$ diagonals.

A polygon of 3 sides has $\dfrac{3(3-3)}{2}$ diagonals.

A polygon of 4 sides has $\dfrac{4(4-3)}{2}$ diagonals.

A polygon of 5 sides has $\dfrac{5(5-3)}{2}$ diagonals.

A polygon of 6 sides has $\dfrac{6(6-3)}{2}$ diagonals.

A polygon of 7 sides has $\dfrac{7(7-3)}{2}$ diagonals.

Each of these statements is true.

5. - 25. See the answer section in the text.

27. Discussion and Writing

29.
$$x + y + z = 3, \quad (1)$$
$$2x - 3y - 2z = 5, \quad (2)$$
$$3x + 2y + 2z = 8 \quad (3)$$

We will use Gaussian elimination. First multiply equation (1) by -2 and add it to equation (2). Also multiply equation (1) by -3 and add it to equation (3).

$$\begin{aligned} x + y + z &= 3 \\ -5y - 4z &= -1 \\ -y - z &= -1 \end{aligned}$$

Now multiply the last equation above by 5 to make the y-coefficient a multiple of the y-coefficient in the equation above it.

$$\begin{aligned} x + y + z &= 3 \quad (1) \\ -5y - 4z &= -1 \quad (4) \\ -5y - 5z &= -5 \quad (5) \end{aligned}$$

Multiply equation (4) by -1 and add it to equation (3).

$$\begin{aligned} x + y + z &= 3 \quad (1) \\ -5y - 4z &= -1 \quad (4) \\ -z &= -4 \quad (6) \end{aligned}$$

Now solve equation (6) for z.

$$-z = -4$$
$$z = 4$$

Back-substitute 4 for z in equation (4) and solve for y.

$$-5y - 4 \cdot 4 = -1$$
$$-5y - 16 = -1$$
$$-5y = 15$$
$$y = -3$$

Finally, back-substitute -3 for y and 4 for z in equation (1) and solve for x.

$$x - 3 + 4 = 3$$
$$x + 1 = 3$$
$$x = 2$$

The solution is $(2, -3, 4)$.

31. Familiarize. Let x, y, and z represent the amounts invested at 1.5%, 2%, and 3%, respectively.

Translate. We know that simple interest for one year was $104. This gives us one equation:

$$0.015x + 0.02y + 0.03z = 104$$

The amount invested at 2% is twice the amount invested at 1.5%:

$$y = 2x, \text{ or } -2x + y = 0$$

There is $400 more invested at 3% than at 2%:

$$z = y + 400, \text{ or } -y + z = 400$$

We have a system of equations:

$$0.015x + 0.02y + 0.03z = 104,$$
$$-2x + y = 0$$
$$-y + z = 400$$

Carry out. Solving the system of equations, we get $(800, 1600, 2000)$.

Check. Simple interest for one year would be $0.015(\$800) + 0.02(\$1600) + 0.03(\$2000)$, or $12 + \$32 + \$60$, or $104. The amount invested at 2%, $1600, is twice $800, the amount invested at 1.5%. The amount invested at 3%, $2000, is $400 more than $1600, the amount invested at 2%. The answer checks.

State. Martin invested $800 at 1.5%, $1600 at 2%, and $2000 at 3%.

33. - 41. See the answer section in the text.

Exercise Set 7.5

1. $_6P_6 = 6! = 6 \cdot 5 \cdot 4 \cdot 3 \cdot 2 \cdot 1 = 720$

3. Using formula (1), we have
$_{10}P_7 = 10 \cdot 9 \cdot 8 \cdot 7 \cdot 6 \cdot 5 \cdot 4 = 604,800.$

Using formula (2), we have
$$_{10}P_7 = \frac{10!}{(10-7)!} = \frac{10!}{3!} = \frac{10 \cdot 9 \cdot 8 \cdot 7 \cdot 6 \cdot 5 \cdot 4 \cdot 3!}{3!} =$$
$604,800.$

5. $5! = 5 \cdot 4 \cdot 3 \cdot 2 \cdot 1 = 120$

7. $0!$ is defined to be 1.

9. $\dfrac{9!}{5!} = \dfrac{9 \cdot 8 \cdot 7 \cdot 6 \cdot 5!}{5!} = 9 \cdot 8 \cdot 7 \cdot 6 = 3024$

11. $(8-3)! = 5! = 5 \cdot 4 \cdot 3 \cdot 2 \cdot 1 = 120$

13. $\dfrac{10!}{7!3!} = \dfrac{10 \cdot 9 \cdot 8 \cdot 7!}{7!3 \cdot 2 \cdot 1} = \dfrac{10 \cdot 3 \cdot 3 \cdot 4 \cdot 2}{3 \cdot 2 \cdot 1} =$
$10 \cdot 3 \cdot 4 = 120$

15. Using formula (2), we have
$$_8P_0 = \frac{8!}{(8-0)!} = \frac{8!}{8!} = 1.$$

17. Using a calculator, we find
$_{52}P_4 = 6,497,400$

19. Using formula (1), we have $_nP_3 = n(n-1)(n-2)$.

Using formula (2), we have
$$_nP_3 = \frac{n!}{(n-3)!} = \frac{n(n-1)(n-2)(n-3)!}{(n-3)!} =$$
$n(n-1)(n-2).$

21. Using formula (1), we have $_nP_1 = n$.

Using formula (2), we have
$$_nP_1 = \frac{n!}{(n-1)!} = \frac{n(n-1)!}{(n-1)!} = n.$$

23. $_6P_6 = 6! = 720$

25. $_9P_9 = 9! = 362,880$

27. $_9P_4 = 9 \cdot 8 \cdot 7 \cdot 6 = 3024$

29. Without repetition: $_5P_5 = 5! = 120$

With repetition: $5^5 = 3125$

31. There are $_5P_5$ choices for the order of the rock numbers and $_4P_4$ choices for the order of the speeches, so we have $_5P_5 \cdot _4P_4 = 5!4! = 2880.$

33. The first number can be any of the eight digits other than 0 and 1. The remaining 6 numbers can each be any of the ten digits 0 through 9. We have
$$8 \cdot 10^6 = 8,000,000$$

Accordingly, there can be 8,000,000 telephone numbers within a given area code before the area needs to be split with a new area code.

35. $a^2b^3c^4 = a \cdot a \cdot b \cdot b \cdot b \cdot c \cdot c \cdot c \cdot c$

There are 2 a's, 3 b's, and 4 c's, for a total of 9. We have
$$\frac{9!}{2! \cdot 3! \cdot 4!}$$
$$= \frac{9 \cdot 8 \cdot 7 \cdot 6 \cdot 5 \cdot 4!}{2 \cdot 1 \cdot 3 \cdot 2 \cdot 1 \cdot 4!} = \frac{9 \cdot 8 \cdot 7 \cdot 6 \cdot 5}{2 \cdot 3 \cdot 2} = 1260.$$

37. a) $_6P_5 = 6 \cdot 5 \cdot 4 \cdot 3 \cdot 2 = 720$

b) $6^5 = 7776$

c) The first letter can only be D. The other four letters are chosen from A, B, C, E, F without repetition. We have
$$1 \cdot _5P_4 = 1 \cdot 5 \cdot 4 \cdot 3 \cdot 2 = 120.$$

d) The first letter can only be D. The second letter can only be E. The other three letters are chosen from A, B, C, F without repetition. We have
$$1 \cdot 1 \cdot _4P_3 = 1 \cdot 1 \cdot 4 \cdot 3 \cdot 2 = 24.$$

39. a) Since repetition is allowed, each of the 5 digits can be chosen in 10 ways. The number of zip-codes possible is $10 \cdot 10 \cdot 10 \cdot 10 \cdot 10$, or 100,000.

b) Since there are 100,000 possible zip-codes, there could be 100,000 post offices.

41. a) Since repetition is allowed, each digit can be chosen in 10 ways. There can be
$10 \cdot 10 \cdot 10 \cdot 10 \cdot 10 \cdot 10 \cdot 10 \cdot 10 \cdot 10$, or $1{,}000{,}000{,}000$ social security numbers.

b) Since more than 285 million social security numbers are possible, each person can have a social security number.

43. Discussion and Writing

45.
$$x^2 + x - 6 = 0$$
$$(x+3)(x-2) = 0$$
$$x+3 = 0 \quad or \quad x-2 = 0$$
$$x = -3 \quad or \qquad x = 2$$
The solutions are -3 and 2.

47. $f(x) = x^3 - 4x^2 - 7x + 10$

We use synthetic division to find one factor of the polynomial. We try $x - 1$.

$$\underline{1|} \begin{array}{rrrr} 1 & -4 & -7 & 10 \\ & 1 & -3 & -10 \\ \hline 1 & -3 & -10 & 0 \end{array}$$

$$x^3 - 4x^2 - 7x + 10 = 0$$
$$(x-1)(x^2 - 3x - 10) = 0$$
$$(x-1)(x-5)(x+2) = 0$$
$$x-1 = 0 \quad or \quad x-5 = 0 \quad or \quad x+2 = 0$$
$$x = 1 \quad or \qquad x = 5 \quad or \qquad x = -2$$
The solutions are -2, 1, and 5.

49.
$$_nP_4 = 8 \cdot {}_{n-1}P_3$$
$$\frac{n!}{(n-4)!} = 8 \cdot \frac{(n-1)!}{(n-1-3)!}$$
$$\frac{n!}{(n-4)!} = 8 \cdot \frac{(n-1)!}{(n-4)!}$$
$$n! = 8 \cdot (n-1)! \qquad \text{Multiplying by } (n-4)!$$
$$n(n-1)! = 8 \cdot (n-1)!$$
$$n = 8 \qquad \text{Dividing by } (n-1)!$$

51.
$$_nP_4 = 8 \cdot {}_nP_3$$
$$\frac{n!}{(n-4)!} = 8 \cdot \frac{n!}{(n-3)!}$$
$$(n-3)! = 8(n-4)! \qquad \text{Multiplying by } \frac{(n-4)!(n-3)!}{n!}$$
$$(n-3)(n-4)! = 8(n-4)!$$
$$n-3 = 8 \qquad \text{Dividing by } (n-4)!$$
$$n = 11$$

53. There is one losing team per game. In order to leave one tournament winner there must be $n-1$ losers produced in $n-1$ games.

Exercise Set 7.6

1.
$$_{13}C_2 = \frac{13!}{2!(13-2)!}$$
$$= \frac{13!}{2!11!} = \frac{13 \cdot 12 \cdot 11!}{2 \cdot 1 \cdot 11!}$$
$$= \frac{13 \cdot 12}{2 \cdot 1} = \frac{13 \cdot 6 \cdot 2}{2 \cdot 1}$$
$$= 78$$

3.
$$\binom{13}{11} = \frac{13!}{11!(13-11)!}$$
$$= \frac{13!}{11!2!}$$
$$= 78 \qquad \text{(See Exercise 1.)}$$

5.
$$\binom{7}{1} = \frac{7!}{1!(7-1)!}$$
$$= \frac{7!}{1!6!} = \frac{7 \cdot 6!}{1 \cdot 6!}$$
$$= 7$$

7.
$$\frac{_5P_3}{3!} = \frac{5 \cdot 4 \cdot 3}{3!}$$
$$= \frac{5 \cdot 4 \cdot 3}{3 \cdot 2 \cdot 1} = \frac{5 \cdot 2 \cdot 2 \cdot 3}{3 \cdot 2 \cdot 1}$$
$$= 5 \cdot 2 = 10$$

9.
$$\binom{6}{0} = \frac{6!}{0!(6-0)!}$$
$$= \frac{6!}{0!6!} = \frac{6!}{6! \cdot 1}$$
$$= 1$$

11. $\binom{6}{2} = \dfrac{6 \cdot 5}{2 \cdot 1} = 15$

13. $\binom{n}{r} = \binom{n}{n-r}$, so

$$\binom{7}{0} + \binom{7}{1} + \binom{7}{2} + \binom{7}{3} + \binom{7}{4} +$$
$$\binom{7}{5} + \binom{7}{6} + \binom{7}{7}$$
$$= 2\left[\binom{7}{0} + \binom{7}{1} + \binom{7}{2} + \binom{7}{3}\right]$$
$$= 2\left[\frac{7!}{7!0!} + \frac{7!}{6!1!} + \frac{7!}{5!2!} + \frac{7!}{4!3!}\right]$$
$$= 2(1 + 7 + 21 + 35) = 2 \cdot 64 = 128$$

15. We will use form (1).
$$_{52}C_4 = \frac{52!}{4!(52-4)!}$$
$$= \frac{52 \cdot 51 \cdot 50 \cdot 49 \cdot 48!}{4 \cdot 3 \cdot 2 \cdot 1 \cdot 48!}$$
$$= \frac{52 \cdot 51 \cdot 50 \cdot 49}{4 \cdot 3 \cdot 2 \cdot 1}$$
$$= 270{,}725$$

17. We will use form (2).

$$\binom{27}{11}$$

$$= \frac{27 \cdot 26 \cdot 25 \cdot 24 \cdot 23 \cdot 22 \cdot 21 \cdot 20 \cdot 19 \cdot 18 \cdot 17}{11 \cdot 10 \cdot 9 \cdot 8 \cdot 7 \cdot 6 \cdot 5 \cdot 4 \cdot 3 \cdot 2 \cdot 1}$$

$$= 13,037,895$$

19. $\binom{n}{1} = \dfrac{n!}{1!(n-1)!} = \dfrac{n(n-1)!}{1!(n-1)!} = n$

21. $\binom{m}{m} = \dfrac{m!}{m!(m-m)!} = \dfrac{m!}{m!0!} = 1$

23. $_{23}C_4 = \dfrac{23!}{4!(23-4)!}$

$$= \frac{23!}{4!19!} = \frac{23 \cdot 22 \cdot 21 \cdot 20 \cdot 19!}{4 \cdot 3 \cdot 2 \cdot 1 \cdot 19!}$$

$$= \frac{23 \cdot 22 \cdot 21 \cdot 20}{4 \cdot 3 \cdot 2 \cdot 1} = \frac{23 \cdot 2 \cdot 11 \cdot 3 \cdot 7 \cdot 4 \cdot 5}{4 \cdot 3 \cdot 2 \cdot 1}$$

$$= 8855$$

25. $_{13}C_{10} = \dfrac{13!}{10!(13-10)!}$

$$= \frac{13!}{10!3!} = \frac{13 \cdot 12 \cdot 11 \cdot 10!}{10! \cdot 3 \cdot 2 \cdot 1}$$

$$= \frac{13 \cdot 12 \cdot 11}{3 \cdot 2 \cdot 1} = \frac{13 \cdot 3 \cdot 2 \cdot 2 \cdot 11}{3 \cdot 2 \cdot 1}$$

$$= 286$$

27. $_{10}C_7 \cdot _5C_3 = \dbinom{10}{7} \cdot \dbinom{5}{3}$ Using the fundamental counting principle

$$= \frac{10!}{7!(10-7)!} \cdot \frac{5!}{3!(5-3)!}$$

$$= \frac{10 \cdot 9 \cdot 8 \cdot 7!}{7! \cdot 3!} \cdot \frac{5 \cdot 4 \cdot 3!}{3! \cdot 2!}$$

$$= \frac{10 \cdot 9 \cdot 8}{3 \cdot 2 \cdot 1} \cdot \frac{5 \cdot 4}{2 \cdot 1} = 120 \cdot 10 = 1200$$

29. $_{52}C_5 = 2,598,960$

31. a) $_{31}P_2 = 930$

b) $31 \cdot 31 = 961$

c) $_{31}C_2 = 465$

33. Choosing k objects from a set of n objects is equivalent to not choosing the other $n - k$ objects.

35.
$$2x^2 - x = 3$$
$$2x^2 - x - 3 = 0$$
$$(2x - 3)(x + 1) = 0$$
$$2x - 3 = 0 \quad or \quad x + 1 = 0$$
$$2x = 3 \quad or \quad x = -1$$
$$x = \frac{3}{2} \quad or \quad x = -1$$

The solutions are $\dfrac{3}{2}$ and -1.

37.
$$x^3 + 3x^2 - 10x = 24$$
$$x^3 + 3x^2 - 10x - 24 = 0$$

We use synthetic division to find one factor of the polynomial on the left side of the equation. We try $x - 3$.

$$\begin{array}{r|rrrr} 3 & 1 & 3 & -10 & -24 \\ & & 3 & 18 & 24 \\ \hline & 1 & 6 & 8 & 0 \end{array}$$

Now we have:

$$(x - 3)(x^2 + 6x + 8) = 0$$
$$(x - 3)(x + 2)(x + 4) = 0$$
$$x - 3 = 0 \quad or \quad x + 2 = 0 \quad or \quad x + 4 = 0$$
$$x = 3 \quad or \quad x = -2 \quad or \quad x = -4$$

The solutions are -4, -2, and 3.

39. There are 13 diamonds, and we choose 5. We have $_{13}C_5 = 1287$.

41. Playing once: $_nC_2$

Playing twice: $2 \cdot _nC_2$

43.
$$\binom{n}{n-2} = 6$$
$$\frac{n!}{(n-(n-2))!(n-2)!} = 6$$
$$\frac{n!}{2!(n-2)!} = 6$$
$$\frac{n(n-1)(n-2)!}{2 \cdot 1 \cdot (n-2)!} = 6$$
$$\frac{n(n-1)}{2} = 6$$
$$n(n-1) = 12$$
$$n^2 - n = 12$$
$$n^2 - n - 12 = 0$$
$$(n - 4)(n + 3) = 0$$
$$n = 4 \quad or \quad n = -3$$

Only 4 checks. The solution is 4.

45.
$$\binom{n+2}{4} = 6 \cdot \binom{n}{2}$$
$$\frac{(n+2)!}{(n+2-4)!4!} = 6 \cdot \frac{n!}{(n-2)!2!}$$
$$\frac{(n+2)!}{(n-2)!4!} = 6 \cdot \frac{n!}{(n-2)!2!}$$
$$\frac{(n+2)!}{4!} = 6 \cdot \frac{n!}{2!} \quad \text{Multiplying by } (n{-}2)!$$
$$4! \cdot \frac{(n+2)!}{4!} = 4! \cdot 6 \cdot \frac{n!}{2!}$$
$$(n+2)! = 72 \cdot n!$$
$$(n+2)(n+1)n! = 72 \cdot n!$$
$$(n+2)(n+1) = 72 \quad \text{Dividing by } n!$$
$$n^2 + 3n + 2 = 72$$
$$n^2 + 3n - 70 = 0$$
$$(n + 10)(n - 7) = 0$$

$n = -10$ or $n = 7$

Only 7 checks. The solution is 7.

47. See the answer section in the text.

Exercise Set 7.7

1. Expand: $(x + 5)^4$.

We have $a = x$, $b = 5$, and $n = 4$.

Pascal's triangle method: Use the fifth row of Pascal's triangle.

$$1 \quad 4 \quad 6 \quad 4 \quad 1$$
$$(x + 5)^4$$
$$= 1 \cdot x^4 + 4 \cdot x^3 \cdot 5 + 6 \cdot x^2 \cdot 5^2 +$$
$$\quad 4 \cdot x \cdot 5^3 + 1 \cdot 5^4$$
$$= x^4 + 20x^3 + 150x^2 + 500x + 625$$

Factorial notation method:

$$(x + 5)^4$$
$$= \binom{4}{0}x^4 + \binom{4}{1}x^3 \cdot 5 + \binom{4}{2}x^2 \cdot 5^2 +$$
$$\binom{4}{3}x \cdot 5^3 + \binom{4}{4}5^4$$
$$= \frac{4!}{0!4!}x^4 + \frac{4!}{1!3!}x^3 \cdot 5 + \frac{4!}{2!2!}x^2 \cdot 5^2 +$$
$$\frac{4!}{3!1!}x \cdot 5^3 + \frac{4!}{4!0!}5^4$$
$$= x^4 + 20x^3 + 150x^2 + 500x + 625$$

3. Expand: $(x - 3)^5$.

We have $a = x$, $b = -3$, and $n = 5$.

Pascal's triangle method: Use the sixth row of Pascal's triangle.

$$1 \quad 5 \quad 10 \quad 10 \quad 5 \quad 1$$
$$(x - 3)^5$$
$$= 1 \cdot x^5 + 5x^4(-3) + 10x^3(-3)^2 + 10x^2(-3)^3 +$$
$$\quad 5x(-3)^4 + 1 \cdot (-3)^5$$
$$= x^5 - 15x^4 + 90x^3 - 270x^2 + 405x - 243$$

Factorial notation method:

$$(x - 3)^5$$
$$= \binom{5}{0}x^5 + \binom{5}{1}x^4(-3) + \binom{5}{2}x^3(-3)^2 +$$
$$\binom{5}{3}x^2(-3)^3 + \binom{5}{4}x(-3)^4 + \binom{5}{5}(-3)^5$$
$$= \frac{5!}{0!5!}x^5 + \frac{5!}{1!4!}x^4(-3) + \frac{5!}{2!3!}x^3(9) +$$
$$\frac{5!}{3!2!}x^2(-27) + \frac{5!}{4!1!}x(81) + \frac{5!}{5!0!}(-243)$$
$$= x^5 - 15x^4 + 90x^3 - 270x^2 + 405x - 243$$

5. Expand: $(x - y)^5$.

We have $a = x$, $b = -y$, and $n = 5$.

Pascal's triangle method: We use the sixth row of Pascal's triangle.

$$1 \quad 5 \quad 10 \quad 10 \quad 5 \quad 1$$
$$(x - y)^5$$
$$= 1 \cdot x^5 + 5x^4(-y) + 10x^3(-y)^2 + 10x^2(-y)^3 +$$
$$\quad 5x(-y)^4 + 1 \cdot (-y)^5$$
$$= x^5 - 5x^4y + 10x^3y^2 - 10x^2y^3 + 5xy^4 - y^5$$

Factorial notation method:

$$(x - y)^5$$
$$= \binom{5}{0}x^5 + \binom{5}{1}x^4(-y) + \binom{5}{2}x^3(-y)^2 +$$
$$\binom{5}{3}x^2(-y)^3 + \binom{5}{4}x(-y)^4 + \binom{5}{5}(-y)^5$$
$$= \frac{5!}{0!5!}x^5 + \frac{5!}{1!4!}x^4(-y) + \frac{5!}{2!3!}x^3(y^2) +$$
$$\frac{5!}{3!2!}x^2(-y^3) + \frac{5!}{4!1!}x(y^4) + \frac{5!}{5!0!}(-y^5)$$
$$= x^5 - 5x^4y + 10x^3y^2 - 10x^2y^3 + 5xy^4 - y^5$$

7. Expand: $(5x + 4y)^6$.

We have $a = 5x$, $b = 4y$, and $n = 6$.

Pascal's triangle method: Use the seventh row of Pascal's triangle.

$$1 \quad 6 \quad 15 \quad 20 \quad 15 \quad 6 \quad 1$$
$$(5x + 4y)^6$$
$$= 1 \cdot (5x)^6 + 6 \cdot (5x)^5(4y) + 15(5x)^4(4y)^2 +$$
$$\quad 20(5x)^3(4y)^3 + 15(5x)^2(4y)^4 + 6(5x)(4y)^5 +$$
$$\quad 1 \cdot (4y)^6$$
$$= 15,625x^6 + 75,000x^5y + 150,000x^4y^2 +$$
$$\quad 160,000x^3y^3 + 96,000x^2y^4 + 30,720xy^5 + 4096y^6$$

Factorial notation method:

$$(5x + 4y)^6$$
$$= \binom{6}{0}(5x)^6 + \binom{6}{1}(5x)^5(4y) +$$
$$\binom{6}{2}(5x)^4(4y)^2 + \binom{6}{3}(5x)^3(4y)^3 +$$
$$\binom{6}{4}(5x)^2(4y)^4 + \binom{6}{5}(5x)(4y)^5 + \binom{6}{6}(4y)^6$$
$$= \frac{6!}{0!6!}(15,625x^6) + \frac{6!}{1!5!}(3125x^5)(4y) +$$
$$\frac{6!}{2!4!}(625x^4)(16y^2) + \frac{6!}{3!3!}(125x^3)(64y^3) +$$
$$\frac{6!}{4!2!}(25x^2)(256y^4) + \frac{6!}{5!1!}(5x)(1024y^5) +$$
$$\frac{6!}{6!0!}(4096y^6)$$
$$= 15,625x^6 + 75,000x^5y + 150,000x^4y^2 +$$
$$\quad 160,000x^3y^3 + 96,000x^2y^4 + 30,720xy^5 +$$
$$\quad 4096y^6$$

9. Expand: $\left(2t + \dfrac{1}{t}\right)^7$.

We have $a = 2t$, $b = \dfrac{1}{t}$, and $n = 7$.

Pascal's triangle method: Use the eighth row of Pascal's triangle.

$$1 \quad 7 \quad 21 \quad 35 \quad 35 \quad 21 \quad 7 \quad 1$$

$$\left(2t + \frac{1}{t}\right)^7$$

$$= 1 \cdot (2t)^7 + 7(2t)^6\left(\frac{1}{t}\right) + 21(2t)^5\left(\frac{1}{t}\right)^2 +$$

$$35(2t)^4\left(\frac{1}{t}\right)^3 + 35(2t)^3\left(\frac{1}{t}\right)^4 + 21(2t)^2\left(\frac{1}{t}\right)^5 +$$

$$7(2t)\left(\frac{1}{t}\right)^6 + 1 \cdot \left(\frac{1}{t}\right)^7$$

$$= 128t^7 + 7 \cdot 64t^6 \cdot \frac{1}{t} + 21 \cdot 32t^5 \cdot \frac{1}{t^2} +$$

$$35 \cdot 16t^4 \cdot \frac{1}{t^3} + 35 \cdot 8t^3 \cdot \frac{1}{t^4} + 21 \cdot 4t^2 \cdot \frac{1}{t^5} +$$

$$7 \cdot 2t \cdot \frac{1}{t^6} + \frac{1}{t^7}$$

$$= 128t^7 + 448t^5 + 672t^3 + 560t + 280t^{-1} +$$

$$84t^{-3} + 14t^{-5} + t^{-7}$$

Factorial notation method:

$$\left(2t + \frac{1}{t}\right)^7$$

$$= \binom{7}{0}(2t)^7 + \binom{7}{1}(2t)^6\left(\frac{1}{t}\right) +$$

$$\binom{7}{2}(2t)^5\left(\frac{1}{t}\right)^2 + \binom{7}{3}(2t)^4\left(\frac{1}{t}\right)^3 +$$

$$\binom{7}{4}(2t)^3\left(\frac{1}{t}\right)^4 + \binom{7}{5}(2t)^2\left(\frac{1}{t}\right)^5 +$$

$$\binom{7}{6}(2t)\left(\frac{1}{t}\right)^6 + \binom{7}{7}\left(\frac{1}{t}\right)^7$$

$$= \frac{7!}{0!7!}(128t^7) + \frac{7!}{1!6!}(64t^6)\left(\frac{1}{t}\right) + \frac{7!}{2!5!}(32t^5)\left(\frac{1}{t^2}\right) +$$

$$\frac{7!}{3!4!}(16t^4)\left(\frac{1}{t^3}\right) + \frac{7!}{4!3!}(8t^3)\left(\frac{1}{t^4}\right) +$$

$$\frac{7!}{5!2!}(4t^2)\left(\frac{1}{t^5}\right) + \frac{7!}{6!1!}(2t)\left(\frac{1}{t^6}\right) + \frac{7!}{7!0!}\left(\frac{1}{t^7}\right)$$

$$= 128t^7 + 448t^5 + 672t^3 + 560t + 280t^{-1} +$$

$$84t^{-3} + 14t^{-5} + t^{-7}$$

11. Expand: $(x^2 - 1)^5$.

We have $a = x^2$, $b = -1$, and $n = 5$.

Pascal's triangle method: Use the sixth row of Pascal's triangle.

$$1 \quad 5 \quad 10 \quad 10 \quad 5 \quad 1$$

$$(x^2 - 1)^5$$

$$= 1 \cdot (x^2)^5 + 5(x^2)^4(-1) + 10(x^2)^3(-1)^2 +$$

$$10(x^2)^2(-1)^3 + 5(x^2)(-1)^4 + 1 \cdot (-1)^5$$

$$= x^{10} - 5x^8 + 10x^6 - 10x^4 + 5x^2 - 1$$

Factorial notation method:

$$(x^2 - 1)^5$$

$$= \binom{5}{0}(x^2)^5 + \binom{5}{1}(x^2)^4(-1) +$$

$$\binom{5}{2}(x^2)^3(-1)^2 + \binom{5}{3}(x^2)^2(-1)^3 +$$

$$\binom{5}{4}(x^2)(-1)^4 + \binom{5}{5}(-1)^5$$

$$= \frac{5!}{0!5!}(x^{10}) + \frac{5!}{1!4!}(x^8)(-1) + \frac{5!}{2!3!}(x^6)(1) +$$

$$\frac{5!}{3!2!}(x^4)(-1) + \frac{5!}{4!1!}(x^2)(1) + \frac{5!}{5!0!}(-1)$$

$$= x^{10} - 5x^8 + 10x^6 - 10x^4 + 5x^2 - 1$$

13. Expand: $(\sqrt{5} + t)^6$.

We have $a = \sqrt{5}$, $b = t$, and $n = 6$.

Pascal's triangle method: We use the seventh row of Pascal's triangle:

$$1 \quad 6 \quad 15 \quad 20 \quad 15 \quad 6 \quad 1$$

$$(\sqrt{5} + t)^6 = 1 \cdot (\sqrt{5})^6 + 6(\sqrt{5})^5(t) +$$

$$15(\sqrt{5})^4(t^2) + 20(\sqrt{5})^3(t^3) +$$

$$15(\sqrt{5})^2(t^4) + 6\sqrt{5}t^5 + 1 \cdot t^6$$

$$= 125 + 150\sqrt{5}\,t + 375t^2 + 100\sqrt{5}\,t^3 +$$

$$75t^4 + 6\sqrt{5}\,t^5 + t^6$$

Factorial notation method:

$$(\sqrt{5} + t)^6 = \binom{6}{0}(\sqrt{5})^6 + \binom{6}{1}(\sqrt{5})^5(t) +$$

$$\binom{6}{2}(\sqrt{5})^4(t^2) + \binom{6}{3}(\sqrt{5})^3(t^3) +$$

$$\binom{6}{4}(\sqrt{5})^2(t^4) + \binom{6}{5}(\sqrt{5})(t^5) +$$

$$\binom{6}{6}(t^6)$$

$$= \frac{6!}{0!6!}(125) + \frac{6!}{1!5!}(25\sqrt{5})t + \frac{6!}{2!4!}(25)(t^2) +$$

$$\frac{6!}{3!3!}(5\sqrt{5})(t^3) + \frac{6!}{4!2!}(5)(t^4) +$$

$$\frac{6!}{5!1!}(\sqrt{5})(t^5) + \frac{6!}{6!0!}(t^6)$$

$$= 125 + 150\sqrt{5}\,t + 375t^2 + 100\sqrt{5}\,t^3 +$$

$$75t^4 + 6\sqrt{5}\,t^5 + t^6$$

15. Expand: $\left(a - \dfrac{2}{a}\right)^9$.

We have $a = a$, $b = -\dfrac{2}{a}$, and $n = 9$.

Pascal's triangle method: Use the tenth row of Pascal's triangle.

$$1 \quad 9 \quad 36 \quad 84 \quad 126 \quad 126 \quad 84 \quad 36 \quad 9 \quad 1$$

$$\left(a - \frac{2}{a}\right)^9 = 1 \cdot a^9 + 9a^8\left(-\frac{2}{a}\right) + 36a^7\left(-\frac{2}{a}\right)^2 +$$

$$84a^6\left(-\frac{2}{a}\right)^3 + 126a^5\left(-\frac{2}{a}\right)^4 +$$

$$126a^4\left(-\frac{2}{a}\right)^5 + 84a^3\left(-\frac{2}{a}\right)^6 +$$

$$36a^2\left(-\frac{2}{a}\right)^7 + 9a\left(-\frac{2}{a}\right)^8 + 1 \cdot \left(-\frac{2}{a}\right)^9$$

$$= a^9 - 18a^7 + 144a^5 - 672a^3 + 2016a -$$

$$4032a^{-1} + 5376a^{-3} - 4608a^{-5} +$$

$$2304a^{-7} - 512a^{-9}$$

Factorial notation method:

$$\left(a - \frac{2}{a}\right)^9$$

$$= \binom{9}{0}a^9 + \binom{9}{1}a^8\left(-\frac{2}{a}\right) + \binom{9}{2}a^7\left(-\frac{2}{a}\right)^2 +$$

$$\binom{9}{3}a^6\left(-\frac{2}{a}\right)^3 + \binom{9}{4}a^5\left(-\frac{2}{a}\right)^4 +$$

$$\binom{9}{5}a^4\left(-\frac{2}{a}\right)^5 + \binom{9}{6}a^3\left(-\frac{2}{a}\right)^6 +$$

$$\binom{9}{7}a^2\left(-\frac{2}{a}\right)^7 + \binom{9}{8}a\left(-\frac{2}{a}\right)^8 +$$

$$\binom{9}{9}\left(-\frac{2}{a}\right)^9$$

$$= \frac{9!}{9!0!}a^9 + \frac{9!}{8!1!}a^8\left(-\frac{2}{a}\right) + \frac{9!}{7!2!}a^7\left(\frac{4}{a^2}\right) +$$

$$\frac{9!}{6!3!}a^6\left(-\frac{8}{a^3}\right) + \frac{9!}{5!4!}a^5\left(\frac{16}{a^4}\right) +$$

$$\frac{9!}{4!5!}a^4\left(-\frac{32}{a^5}\right) + \frac{9!}{3!6!}a^3\left(\frac{64}{a^6}\right) +$$

$$\frac{9!}{2!7!}a^2\left(-\frac{128}{a^7}\right) + \frac{9!}{1!8!}a\left(\frac{256}{a^8}\right) +$$

$$\frac{9!}{0!9!}\left(-\frac{512}{a^9}\right)$$

$$= a^9 - 9(2a^7) + 36(4a^5) - 84(8a^3) + 126(16a) -$$

$$126(32a^{-1}) + 84(64a^{-3}) - 36(128a^{-5}) +$$

$$9(256a^{-7}) - 512a^{-9}$$

$$= a^9 - 18a^7 + 144a^5 - 672a^3 + 2016a - 4032a^{-1} +$$

$$5376a^{-3} - 4608a^{-5} + 2304a^{-7} - 512a^{-9}$$

17. $(\sqrt{2} + 1)^6 - (\sqrt{2} - 1)^6$

First, expand $(\sqrt{2} + 1)^6$.

$$(\sqrt{2}+1)^6 = \binom{6}{0}(\sqrt{2})^6 + \binom{6}{1}(\sqrt{2})^5(1) +$$

$$\binom{6}{2}(\sqrt{2})^4(1)^2 + \binom{6}{3}(\sqrt{2})^3(1)^3 +$$

$$\binom{6}{4}(\sqrt{2})^2(1)^4 + \binom{6}{5}(\sqrt{2})(1)^5 +$$

$$\binom{6}{6}(1)^6$$

$$= \frac{6!}{6!0!} \cdot 8 + \frac{6!}{5!1!} \cdot 4\sqrt{2} + \frac{6!}{4!2!} \cdot 4 +$$

$$\frac{6!}{3!3!} \cdot 2\sqrt{2} + \frac{6!}{2!4!} \cdot 2 + \frac{6!}{1!5!} \cdot \sqrt{2} + \frac{6!}{0!6!}$$

$$= 8 + 24\sqrt{2} + 60 + 40\sqrt{2} + 30 + 6\sqrt{2} + 1$$

$$= 99 + 70\sqrt{2}$$

Next, expand $(\sqrt{2} - 1)^6$.

$$(\sqrt{2} - 1)^6$$

$$= \binom{6}{0}(\sqrt{2})^6 + \binom{6}{1}(\sqrt{2})^5(-1) +$$

$$\binom{6}{2}(\sqrt{2})^4(-1)^2 + \binom{6}{3}(\sqrt{2})^3(-1)^3 +$$

$$\binom{6}{4}(\sqrt{2})^2(-1)^4 + \binom{6}{5}(\sqrt{2})(-1)^5 +$$

$$\binom{6}{6}(-1)^6$$

$$= \frac{6!}{6!0!} \cdot 8 - \frac{6!}{5!1!} \cdot 4\sqrt{2} + \frac{6!}{4!2!} \cdot 4 - \frac{6!}{3!3!} \cdot 2\sqrt{2} +$$

$$\frac{6!}{2!4!} \cdot 2 - \frac{6!}{1!5!} \cdot \sqrt{2} + \frac{6!}{0!6!}$$

$$= 8 - 24\sqrt{2} + 60 - 40\sqrt{2} + 30 - 6\sqrt{2} + 1$$

$$= 99 - 70\sqrt{2}$$

$$(\sqrt{2} + 1)^6 - (\sqrt{2} - 1)^6$$

$$= (99 + 70\sqrt{2}) - (99 - 70\sqrt{2})$$

$$= 99 + 70\sqrt{2} - 99 + 70\sqrt{2}$$

$$= 140\sqrt{2}$$

19. Expand: $(x^{-2} + x^2)^4$.

We have $a = x^{-2}$, $b = x^2$, and $n = 4$.

Pascal's triangle method: Use the fifth row of Pascal's triangle.

$$1 \quad 4 \quad 6 \quad 4 \quad 1.$$

$$(x^{-2} + x^2)^4$$

$$= 1 \cdot (x^{-2})^4 + 4(x^{-2})^3(x^2) + 6(x^{-2})^2(x^2)^2 +$$

$$4(x^{-2})(x^2)^3 + 1 \cdot (x^2)^4$$

$$= x^{-8} + 4x^{-4} + 6 + 4x^4 + x^8$$

Factorial notation method:

$$(x^{-2} + x^2)^4$$

$$= \binom{4}{0}(x^{-2})^4 + \binom{4}{1}(x^{-2})^3(x^2) +$$

$$\binom{4}{2}(x^{-2})^2(x^2)^2 + \binom{4}{3}(x^{-2})(x^2)^3 +$$

$$\binom{4}{4}(x^2)^4$$

$$= \frac{4!}{4!0!}(x^{-8}) + \frac{4!}{3!1!}(x^{-6})(x^2) + \frac{4!}{2!2!}(x^{-4})(x^4) +$$

$$\frac{4!}{1!3!}(x^{-2})(x^6) + \frac{4!}{0!4!}(x^8)$$

$$= x^{-8} + 4x^{-4} + 6 + 4x^4 + x^8$$

21. Find the 3rd term of $(a + b)^7$.

First, we note that $3 = 2 + 1$, $a = a$, $b = b$, and $n = 7$. Then the 3rd term of the expansion of $(a + b)^7$ is

$$\binom{7}{2}a^{7-2}b^2, \text{ or } \frac{7!}{2!5!}a^5b^2, \text{ or } 21a^5b^2.$$

23. Find the 6th term of $(x - y)^{10}$.

First, we note that $6 = 5 + 1$, $a = x$, $b = -y$, and $n = 10$. Then the 6th term of the expansion of $(x - y)^{10}$ is

$$\binom{10}{5}x^5(-y)^5, \text{ or } -252x^5y^5.$$

25. Find the 12th term of $(a - 2)^{14}$.

First, we note that $12 = 11 + 1$, $a = a$, $b = -2$, and $n = 14$. Then the 12th term of the expansion of $(a - 2)^{14}$ is

$$\binom{14}{11}a^{14-11} \cdot (-2)^{11} = \frac{14!}{3!11!}a^3(-2048)$$

$$= 364a^3(-2048)$$

$$= -745,472a^3$$

27. Find the 5th term of $(2x^3 - \sqrt{y})^8$.

First, we note that $5 = 4 + 1$, $a = 2x^3$, $b = -\sqrt{y}$, and $n = 8$. Then the 5th term of the expansion of $(2x^3 - \sqrt{y})^8$ is

$$\binom{8}{4}(2x^3)^{8-4}(-\sqrt{y})^4$$

$$= \frac{8!}{4!4!}(2x^3)^4(-\sqrt{y})^4$$

$$= 70(16x^{12})(y^2)$$

$$= 1120x^{12}y^2$$

29. The expansion of $(2u - 3v^2)^{10}$ has 11 terms so the 6th term is the middle term. Note that $6 = 5 + 1$, $a = 2u$, $b = -3v^2$, and $n = 10$. Then the 6th term of the expansion of $(2u - 3v^2)^{10}$ is

$$\binom{10}{5}(2u)^{10-5}(-3v^2)^5$$

$$= \frac{10!}{5!5!}(2u)^5(-3v^2)^5$$

$$= 252(32u^5)(-243v^{10})$$

$$= -1,959,552u^5v^{10}$$

31. The number of subsets is 2^7, or 128.

33. The number of subsets is 2^{24}, or 16,777,216.

35. The term of highest degree of $(x^5 + 3)^4$ is the first term, or

$$\binom{4}{0}(x^5)^{4-0}3^0 = \frac{4!}{4!0!}x^{20} = x^{20}.$$

Therefore, the degree of $(x^5 + 3)^4$ is 20.

37. We use factorial notation. Note that $a = 3$, $b = i$, and $n = 5$.

$$(3 + i)^5$$

$$= \binom{5}{0}(3^5) + \binom{5}{1}(3^4)(i) + \binom{5}{2}(3^3)(i^2) +$$

$$\binom{5}{3}(3^2)(i^3) + \binom{5}{4}(3)(i^4) + \binom{5}{5}(i^5)$$

$$= \frac{5!}{0!5!}(243) + \frac{5!}{1!4!}(81)(i) + \frac{5!}{2!3!}(27)(-1) +$$

$$\frac{5!}{3!2!}(9)(-i) + \frac{5!}{4!1!}(3)(1) + \frac{5!}{5!0!}(i)$$

$$= 243 + 405i - 270 - 90i + 15 + i$$

$$= -12 + 316i$$

39. We use factorial notation. Note that $a = \sqrt{2}$, $b = -i$, and $n = 4$.

$$(\sqrt{2} - i)^4 = \binom{4}{0}(\sqrt{2})^4 + \binom{4}{1}(\sqrt{2})^3(-i) +$$

$$\binom{4}{2}(\sqrt{2})^2(-i)^2 + \binom{4}{3}(\sqrt{2})(-i)^3 +$$

$$\binom{4}{4}(-i)^4$$

$$= \frac{4!}{0!4!}(4) + \frac{4!}{1!3!}(2\sqrt{2})(-i) +$$

$$\frac{4!}{2!2!}(2)(-1) + \frac{4!}{3!1!}(\sqrt{2})(i) +$$

$$\frac{4!}{4!0!}(1)$$

$$= 4 - 8\sqrt{2}i - 12 + 4\sqrt{2}i + 1$$

$$= -7 - 4\sqrt{2}i$$

41. $(a-b)^n = \binom{n}{0}a^n(-b)^0 + \binom{n}{1}a^{n-1}(-b)^1 +$

$$\binom{n}{2}a^{n-2}(-b)^2 + \cdots +$$

$$\binom{n}{n-1}a^1(-b)^{n-1} + \binom{n}{n}a^0(-b)^n$$

$$= \binom{n}{0}(-1)^0 a^n b^0 + \binom{n}{1}(-1)^1 a^{n-1}b^1 +$$

$$\binom{n}{2}(-1)^2 a^{n-2}b^2 + \cdots +$$

$$\binom{n}{n-1}(-1)^{n-1}a^1 b^{n-1} +$$

$$\binom{n}{n}(-1)^n a^0 b^n$$

$$= \sum_{k=0}^{n}\binom{n}{k}(-1)^k a^{n-k}b^k$$

43.

$$\frac{(x+h)^n - x^n}{h}$$

$$= \frac{\binom{n}{0}x^n + \binom{n}{1}x^{n-1}h + \cdots + \binom{n}{n}h^n - x^n}{h}$$

$$= \binom{n}{1}x^{n-1} + \binom{n}{2}x^{n-2}h + \cdots + \binom{n}{n}h^{n-1}$$

$$= \sum_{k=1}^{n}\binom{n}{k}x^{n-k}h^{k-1}$$

45. Discussion and Writing

47. $(fg)(x) = f(x)g(x) = (x^2+1)(2x-3) =$
$2x^3 - 3x^2 + 2x - 3$

49. $(g \circ f)(x) = g(f(x)) = g(x^2+1) = 2(x^2+1) - 3 =$
$2x^2 + 2 - 3 = 2x^2 - 1$

51. $\sum_{k=0}^{4}\binom{4}{k}(-1)^k x^{4-k}6^k = \sum_{k=0}^{4}\binom{4}{k}x^{4-k}(-6)^k$, so

the left side of the equation is sigma notation for $(x-6)^4$.
We have:

$(x-6)^4 = 81$

$x - 6 = \pm 3$ Taking the 4th root on both
sides

$x - 6 = 3$ or $x - 6 = -3$

$x = 9$ or $x = 3$

The solutions are 9 and 3.

If we also observe that $(3i)^4 = 81$, we also find the imaginary solutions $6 \pm 3i$.

53. The expansion of $(x^2 - 6y^{3/2})^6$ has 7 terms, so the 4th term is the middle term.

$$\binom{6}{3}(x^2)^3(-6y^{3/2})^3 = \frac{6!}{3!3!}(x^6)(-216y^{9/2}) =$$

$$-4320x^6 y^{9/2}$$

55. The $(k+1)$st term of $\left(\sqrt[3]{x} - \frac{1}{\sqrt{x}}\right)^7$ is

$\binom{7}{k}(\sqrt[3]{x})^{7-k}\left(-\frac{1}{\sqrt{x}}\right)^k$. The term containing $\frac{1}{x^{1/6}}$ is the
term in which the sum of the exponents is $-1/6$. That is,

$$\left(\frac{1}{3}\right)(7-k) + \left(-\frac{1}{2}\right)(k) = -\frac{1}{6}$$

$$\frac{7}{3} - \frac{k}{3} - \frac{k}{2} = -\frac{1}{6}$$

$$-\frac{5k}{6} = -\frac{15}{6}$$

$$k = 3$$

Find the $(3+1)$st, or 4th term.

$$\binom{7}{3}(\sqrt[3]{x})^4\left(-\frac{1}{\sqrt{x}}\right)^3 = \frac{7!}{4!3!}(x^{4/3})(-x^{-3/2}) =$$

$$-35x^{-1/6}, \text{ or } -\frac{35}{x^{1/6}}.$$

57. $_{100}C_0 + {}_{100}C_1 + \cdots + {}_{100}C_{100}$ is the total number of subsets of a set with 100 members, or 2^{100}.

59. $\sum_{k=0}^{23}\binom{23}{k}(\log_a x)^{23-k}(\log_a t)^k =$

$(\log_a x + \log_a t)^{23} = [\log_a(xt)]^{23}$

61. See the answer section in the text.

Exercise Set 7.8

1. a) We use Principle P.

For 1: $P = \dfrac{18}{100}$, or 0.18

For 2: $P = \dfrac{24}{100}$, or 0.24

For 3: $P = \dfrac{23}{100}$, or 0.23

For 4: $P = \dfrac{23}{100}$, or 0.23

For 5: $P = \dfrac{12}{100}$, or 0.12

b) Opinions may vary, but it seems that people tend not to select the first or last numbers.

3. The company can expect 78% of the 15,000 pieces of advertising to be opened and read. We have:

$78\%(15,000) = 0.78(15,000) = 11,700.$

5. a) The consonants with the 5 greatest numbers of occurrences are the 5 consonants with the greatest probability of occurring. They are T, S, R, N, and L.

b) E is the vowel with the greatest number of occurrences, so E is the vowel with the greatest probability of occurring.

c) Yes

7. a) Since there are 14 equally likely ways of selecting a marble from a bag containing 4 red marbles and 10 green marbles, we have, by Principle P,

$$P(\text{selecting a red marble}) = \frac{4}{14} = \frac{2}{7}.$$

b) Since there are 14 equally likely ways of selecting a marble from a bag containing 4 red marbles and 10 green marbles, we have, by Principle P,

$$P(\text{selecting a green marble}) = \frac{10}{14} = \frac{5}{7}.$$

c) Since there are 14 equally likely ways of selecting a marble from a bag containing 4 red marbles and 10 green marbles, we have, by Principle P,

$$P(\text{selecting a purple marble}) = \frac{0}{14} = 0.$$

d) Since there are 14 equally likely ways of selecting a marble from a bag containing 4 red marbles and 10 green marbles, we have, by Principle P,

$$P(\text{selecting a red or a green marble}) =$$
$$\frac{4+10}{14} = 1.$$

9. The total number of coins is $7 + 5 + 10$, or 22 and the total number of coins to be drawn is $4 + 3 + 1$, or 8. The number of ways of selecting 8 coins from a group of 22 is $_{22}C_8$. Four dimes can be selected in $_7C_4$ ways, 3 nickels in $_5C_3$ ways, and 1 quarter in $_{10}C_1$ ways.

$$P(\text{selecting 4 dimes, 3 nickels, and 1 quarter}) =$$
$$\frac{_7C_4 \cdot {_5C_3} \cdot {_{10}C_1}}{_{22}C_8}, \text{ or } \frac{350}{31,977}$$

11. The number of ways of selecting 5 cards from a deck of 52 cards is $_{52}C_5$. Three sevens can be selected in $_4C_3$ ways and 2 kings in $_4C_2$ ways.

$$P(\text{drawing 3 sevens and 2 kings}) = \frac{_4C_3 \cdot {_4C_2}}{_{52}C_5}, \text{ or }$$
$$\frac{1}{108,290}.$$

13. The number of ways of selecting 5 cards from a deck of 52 cards is $_{52}C_5$. Since 13 of the cards are spades, then 5 spades can be drawn in $_{13}C_5$ ways

$$P(\text{drawing 5 spades}) = \frac{_{13}C_5}{_{52}C_5} = \frac{1287}{2,598,960} =$$
$$\frac{33}{66,640}$$

15. a) HHH, HHT, HTH, HTT, THH, THT, TTH, TTT

b) Three of the 8 outcomes have exactly one head. Thus, $P(\text{exactly one head}) = \frac{3}{8}$.

c) Seven of the 8 outcomes have exactly 0, 1, or 2 heads. Thus, $P(\text{at most two heads}) = \frac{7}{8}$.

d) Seven of the 8 outcomes have 1, 2, or 3 heads. Thus, $P(\text{at least one head}) = \frac{7}{8}$.

e) Three of the 8 outcomes have exactly two tails. Thus, $P(\text{exactly two tails}) = \frac{3}{8}$.

17. The roulette wheel contains 38 equally likely slots. Eighteen of the 38 slots are colored black. Thus, by Principle P,

$$P(\text{the ball falls in a black slot}) = \frac{18}{38} = \frac{9}{19}.$$

19. The roulette wheel contains 38 equally likely slots. Only 1 slot is numbered 0. Then, by Principle P,

$$P(\text{the ball falls in the 0 slot}) = \frac{1}{38}.$$

21. The roulette wheel contains 38 equally likely slots. Thirty-six of the slots are colored red or black. Then, by Principle P,

$$P(\text{the ball falls in a red or a black slot}) = \frac{36}{38} = \frac{18}{19}.$$

23. The roulette wheel contains 38 equally likely slots. Eighteen of the slots are odd-numbered. Then, by Principle P,

$$P(\text{the ball falls in a an odd-numbered slot}) =$$
$$\frac{18}{38} = \frac{9}{19}.$$

25. a), b) Answers may vary

27. Discussion and Writing

29. one-to-one

31. zero

33. inverse variation

35. geometric sequence

37. Consider a suit

A K Q J 10 9 8 7 6 5 4 3 2

A straight flush can be any of the following combinations in the same suit.

K	Q	J	10	9
Q	J	10	9	8
J	10	9	8	7
10	9	8	7	6
9	8	7	6	5
8	7	6	5	4
7	6	5	4	3
6	5	4	3	2
5	4	3	2	A

Remember a straight flush does not include A K Q J 10 which is a royal flush.

a) Since there are 9 straight flushes per suit, there are 36 straight flushes in all 4 suits.

b) Since 2,598,960, or $_{52}C_5$, poker hands can be dealt from a standard 52-card deck and 36 of those hands are straight flushes, the probability of getting a straight flush is $\frac{36}{2,598,960}$, or about 1.39×10^{-5}.

39. a) There are 13 ways to select a denomination. Then from that denomination there are $_4C_3$ ways to pick 3 of the 4 cards in that denomination. Now there are 12 ways to select any one of the remaining 12 denominations and $_4C_2$ ways to pick 2 cards from the 4 cards in that denomination. Thus the number of full houses is $(13 \cdot {_4}C_3) \cdot (12 \cdot {_4}C_2)$ or 3744.

b) $\dfrac{3744}{_{52}C_5} = \dfrac{3744}{2,598,960} \approx 0.00144$

41. a) There are 4 ways to select a suit and then $\dbinom{13}{5}$ ways to choose 5 cards from that suit. But these combinations include hands with all the cards in sequence, so we subtract the 4 royal flushes (Exercise 28) and the 36 straight flushes (Exercise 29). Thus there are $4 \cdot \dbinom{13}{5} - 4 - 36$, or 5108 flushes.

b) $\dfrac{5108}{_{52}C_5} = \dfrac{5108}{2,598,960} \approx 0.00197$

43. a) There are 10 sets of 5 consecutive cards:

A   K   Q   J   10

K   Q   J   10   9

.

.

.

5   4   3   2   A

In each of these 10 sets there are 4 ways to choose (from 4 suits) each of the 5 cards. These combinations include the 4 royal flushes and the 36 straight flushes, both of which consist of 5 cards of the *same suit* in sequence.

Thus there are $10 \cdot 4 \cdot 4 \cdot 4 \cdot 4 \cdot 4 - 4 - 36$, or 10,200 straights.

b) $\dfrac{10,200}{_{52}C_5} = \dfrac{10,200}{2,598,960} \approx 0.00392$

Chapter 7 Review Exercises

1. The statement is true. See page 576 in the text.

3. The statement is true. See page 613 in the text.

5. $a_n = (-1)^n \left(\dfrac{n^2}{n^4 + 1} \right)$

$a_1 = (-1)^1 \left(\dfrac{1^2}{1^4 + 1} \right) = -\dfrac{1}{2}$

$a_2 = (-1)^2 \left(\dfrac{2^2}{2^4 + 1} \right) = \dfrac{4}{17}$

$a_3 = (-1)^3 \left(\dfrac{3^2}{3^4 + 1} \right) = -\dfrac{9}{82}$

$a_4 = (-1)^4 \left(\dfrac{4^2}{4^4 + 1} \right) = \dfrac{16}{257}$

$a_{11} = (-1)^{11} \left(\dfrac{11^2}{11^4 + 1} \right) = -\dfrac{121}{14,642}$

$a_{23} = (-1)^{23} \left(\dfrac{23^2}{23^4 + 1} \right) = -\dfrac{529}{279,842}$

7. $\displaystyle\sum_{k=1}^{4} \dfrac{(-1)^{k+1} 3^k}{3^k - 1}$

$= \dfrac{(-1)^{1+1} 3^1}{3^1 - 1} + \dfrac{(-1)^{2+1} 3^2}{3^2 - 1} + \dfrac{(-1)^{3+1} 3^3}{3^3 - 1} + \dfrac{(-1)^{4+1} 3^4}{3^4 - 1}$

$= \dfrac{3}{2} - \dfrac{9}{8} + \dfrac{27}{26} - \dfrac{81}{80}$

$= \dfrac{417}{1040}$

9. $a_n = a_1 + (n-1)d$

$a_1 = \dfrac{3}{4}, d = \dfrac{13}{12} - \dfrac{3}{4} = \dfrac{1}{3}$, and $n = 10$

$a_{10} = \dfrac{3}{4} + (10-1)\dfrac{1}{3} = \dfrac{3}{4} + 9 \cdot \dfrac{1}{3} = \dfrac{3}{4} + 3 = \dfrac{15}{4}$

11. $a_n = a_1 + (n-1)d$

$a_{18} = 4 + (18-1)3 = 4 + 51 = 55$

$S_n = \dfrac{n}{2}(a_1 + a_n)$

$S_{18} = \dfrac{18}{2}(4 + 55) = 531$

13. $a_1 = 5, \ a_{17} = 53$

$53 = 5 + (17-1)d$

$53 = 5 + 16d$

$48 = 16d$

$3 = d$

$a_3 = 5 + (3-1)3 = 11$

15. $a_1 = -2, \ r = 2, \ a_n = -64$

$a_n = a_1 r^{n-1}$

$-64 = -2 \cdot 2^{n-1}$

$-64 = -2^n$

$-2^6 = -2^n$

$6 = n$

$S_n = \dfrac{a_1(1 - r^n)}{1 - r}$

$S_n = \dfrac{-2(1 - 2^6)}{1 - 2} = 2(1 - 64) = -126$

17. Since $|r| = \left| \dfrac{27.5}{2.5} \right| = |1.1| = 1.1 > 1$, the sum does not exist.

19. Since $|r| = \left| \dfrac{-\frac{1}{6}}{\frac{1}{2}} \right| = \left| -\dfrac{1}{3} \right| = \dfrac{1}{3} < 1$, the series has a sum.

$S_\infty = \dfrac{\frac{1}{2}}{1 - \left(-\frac{1}{3} \right)} = \dfrac{\frac{1}{2}}{\frac{4}{3}} = \dfrac{1}{2} \cdot \dfrac{3}{4} = \dfrac{3}{8}$

21. $5, m_1, m_2, m_3, m_4, 9$

We look for $m_1, m_2, m_3,$ and m_4, such that $5, m_1, m_2, m_3, m_4, 9$ is an arithmetic sequence. In this case, $a_1 = 5$, $n = 6$, and $a_6 = 9$. First we find d:

$a_n = a_1 + (n-1)d$

$9 = 5 + (6-1)d$

$4 = 5d$

$\dfrac{4}{5} = d$

Then we have:

$m_1 = a_1 + d = 5 + \dfrac{4}{5} = 5\dfrac{4}{5}$

$m_2 = m_1 + d = 5\dfrac{4}{5} + \dfrac{4}{5} = 6\dfrac{3}{5}$

$m_3 = m_2 + d = 6\dfrac{3}{5} + \dfrac{4}{5} = 7\dfrac{2}{5}$

$m_4 = m_3 + d = 7\dfrac{2}{5} + \dfrac{4}{5} = 8\dfrac{1}{5}$

23. The amount of the annuity is a geometric series

$\quad$ $\$2000 + \$2000(1.028) + \$2000(1.028)^2 + \dots$

$\quad$ $\$2000(1.028)^{17}$,

where $a_1 = \$2000$, $n = 18$, and $r = 1.028$.

Using the formula $S_n = \dfrac{a_1(1-r^n)}{1-r}$, we have

$S_{18} = \dfrac{2000\left[1-(1.028)^{18}\right]}{1-1.028} \approx \$45,993.04$.

25. The total economic effect is given by the infinite geometric series

$\quad$ $\$24,000,000,000 + \$24,000,000,000(0.73) +$

$\quad$ $\$24,000,000,000(0.73)^2 + \dots$.

Since $|r| = |0.73| = 0.73 < 1$, the series has a sum.

Using the formula $S_\infty = \dfrac{a_1}{1-r}$ with

$a_1 = \$24,000,000,000$ and $r = 0.73$, we have

$S_\infty = \dfrac{24,000,000,000}{1-0.73} \approx \$88,888,888,889$.

27. $S_n : 1 + 3 + 3^2 + \dots + 3^{n-1} = \dfrac{3^n - 1}{2}$

$S_1 : 1 = \dfrac{3^1 - 1}{2}$

$S_k : 1 + 3 + 3^2 + \dots + 3^{k-1} = \dfrac{3^k - 1}{2}$

$S_{k+1} : 1 + 3 + 3^2 + \dots + 3^{(k+1)-1} = \dfrac{3^{k+1}-1}{2}$

1. *Basis step:* $\dfrac{3^1-1}{2} = \dfrac{2}{2} = 1$ is true.

2. *Induction step:* Assume S_k. Add 3^k to both sides.

$1 + 3 + \dots + 3^{k-1} + 3^k$

$= \dfrac{3^k-1}{2} + 3^k = \dfrac{3^k-1}{2} + 3^k \cdot \dfrac{2}{2}$

$= \dfrac{3 \cdot 3^k - 1}{2} = \dfrac{3^{k+1}-1}{2}$

29. $6! = 720$

31. $\begin{pmatrix} 15 \\ 8 \end{pmatrix} = \dfrac{15!}{8!(15-8)!} = 6435$

33. $\dfrac{9!}{1!4!2!2!} = 3780$

35. a) $_6P_5 = \dfrac{6!}{(6-5)!} = 720$

b) $6^5 = 7776$

c) $_5P_4 = \dfrac{5!}{(5-4)!} = 120$

d) $_3P_2 = \dfrac{3!}{(3-2)!} = 6$

37. $(m+n)^7$

Pascal's triangle method: Use the 8th row of Pascal's triangle.

1 7 21 35 35 21 7 1

$(m+n)^7 = m^7 + 7m^6n + 21m^5n^2 + 35m^4n^3 +$

$\quad\quad 35m^3n^4 + 21m^2n^5 + 7mn^6 + n^7$

Factorial notation method:

$(m+n)^7 = \begin{pmatrix}7\\0\end{pmatrix}m^7 + \begin{pmatrix}7\\1\end{pmatrix}m^6n + \begin{pmatrix}7\\2\end{pmatrix}m^5n^2 + \begin{pmatrix}7\\3\end{pmatrix}m^4n^3 +$

$\begin{pmatrix}7\\4\end{pmatrix}m^3n^4 + \begin{pmatrix}7\\5\end{pmatrix}m^2n^5 + \begin{pmatrix}7\\6\end{pmatrix}mn^6 + \begin{pmatrix}7\\7\end{pmatrix}n^7$

$= m^7 + 7m^6n + 21m^5n^2 + 35m^4n^3 +$

$\quad\quad 35m^3n^4 + 21m^2n^5 + 7mn^6 + n^7$

39. Expand: $(x^2 - 3y)^4$

Pascal's triangle method: Use the 5th row.

1 4 6 4 1

$(x^2-3y)^4 = (x^2)^4 + 4(x^2)^3(-3y) + 6(x^2)^2(-3y)^2 +$

$\quad\quad 4(x^2)(-3y)^3 + (-3y)^4$

$= x^8 - 12x^6y + 54x^4y^2 - 108x^2y^3 + 81y^4$

Factorial notation method:

$(x^2-3y)^4 = \begin{pmatrix}4\\0\end{pmatrix}(x^2)^4 + \begin{pmatrix}4\\1\end{pmatrix}(x^2)^3(-3y) +$

$\begin{pmatrix}4\\2\end{pmatrix}(x^2)^2(-3y)^2 + \begin{pmatrix}4\\3\end{pmatrix}(x^2)(-3y)^3 +$

$\begin{pmatrix}4\\4\end{pmatrix}(-3y)^4$

$= x^8 - 12x^6y + 54x^4y^2 - 108x^2y^3 + 81y^4$

41. Expand: $(1+5i)^6$

Pascal's triangle method: Use the 7th row.

1 6 15 20 15 6 1

$(1+5i)^6 = 1^6 + 6(1)^5(5i) + 15(1)^4(5i)^2 + 20(1)^3(5i)^3 +$

$\quad\quad 15(1)^2(5i)^4 + 6(1)(5i)^5 + (5i)^6$

$= 1 + 30i - 375 - 2500i + 9375 +$

$\quad\quad 18,750i - 15,625$

$= -6624 + 16,280i$

Factorial notation method:

$$(1+5i)^6 = \binom{6}{0}1^6 + \binom{6}{1}(1)^5(5i) + \binom{6}{2}(1)^4(5i)^2 +$$

$$\binom{6}{3}(1)^3(5i)^3 + \binom{6}{4}(1)^2(5i)^4 +$$

$$\binom{6}{5}(1)(5i)^5 + \binom{6}{6}(5i)^6$$

$$= 1 + 30i - 375 - 2500i + 9375 +$$

$$18,750i - 15,625$$

$$= -6624 + 16,280i$$

43. Find 12th term of $(2a - b)^{18}$.

$$\binom{18}{11}(2a)^7(-b)^{11} = -\binom{18}{11}128a^7b^{11}$$

45. Of 52 cards, 13 are clubs.

Probability $= \dfrac{13}{52} = \dfrac{1}{4}$.

47. A: $\dfrac{86}{86 + 97 + 23} = \dfrac{86}{206} \approx 0.42$

B: $\dfrac{97}{86 + 97 + 23} = \dfrac{97}{206} \approx 0.47$

C: $\dfrac{23}{86 + 97 + 23} = \dfrac{23}{206} \approx 0.11$

49. There are 3 pairs that total 4: 1 and 3, 2 and 2, 3 and 1. There are $6 \cdot 6$, or 36, possible outcomes. Thus, we have $\dfrac{3}{36}$, or $\dfrac{1}{12}$. Answer A is correct.

51. a) $a_n = 0.9694927954n + 46.17184726$, where a_n is in millions and n is the number of years after 1980

b) In 2010, $n = 2010 - 1980 = 30$.

$a_{30} \approx 75.257$ million

53. Someone who has managed several sequences of hiring has a considerable income from the sales of the people in the lower levels. However, with a finite population, it will not be long before the salespersons in the lowest level have no one to hire and no one to sell to.

55. $\dfrac{a_{k+1}}{a_k} = r_1$, $\dfrac{b_{k+1}}{b_k} = r_2$, so $\dfrac{a_{k+1}b_{k+1}}{a_k b_k} = r_1 r_2$, a constant.

57. $f(x) = x^4 - 4x^3 - 4x^2 + 16x = x(x^3 - 4x^2 - 4x + 16)$

We know that 0 is a zero.

Consider $x^3 - 4x^2 - 4x + 16$.

Possibilities for p/q: $\pm 1, \pm 2, \pm 4, \pm 8, \pm 16$

```
-2 | 1   -4   -4    16
   |      -2   12  -16
   ----------------------
     1   -6    8|   0
```

$f(x) = x(x + 2)(x^2 - 6x + 8)$

$ = x(x + 2)(x - 2)(x - 4)$

The zeros are $-2, 0, 2,$ and 4.

59.

$$\sum_{k=0}^{10} (-1)^k \binom{10}{k}(\log x)^{10-k}(\log y)^k$$

$$= (\log x - \log y)^{10}$$

$$= \left(\log \frac{x}{y}\right)^{10}$$

61.

$$\binom{n}{n-1} = 36$$

$$\frac{n!}{(n-1)![n-(n-1)]!} = 36$$

$$\frac{n(n-1)!}{(n-1)!1!} = 36$$

$$n = 36$$

Chapter 7 Test

1. $a_n = (-1)^n(2n + 1)$

$a_{21} = (-1)^{21}[2(21) + 1]$

$\phantom{a_{21}} = -43$

2. $a_n = \dfrac{n+1}{n+2}$

$a_1 = \dfrac{1+1}{1+2} = \dfrac{2}{3}$

$a_2 = \dfrac{2+1}{2+2} = \dfrac{3}{4}$

$a_3 = \dfrac{3+1}{3+2} = \dfrac{4}{5}$

$a_4 = \dfrac{4+1}{4+2} = \dfrac{5}{6}$

$a_5 = \dfrac{5+1}{5+2} = \dfrac{6}{7}$

3. $\displaystyle\sum_{k=1}^{4} (k^2 + 1) = (1^2 + 1) + (2^2 + 1) + (3^2 + 1) + (4^2 + 1)$

$$= 2 + 5 + 10 + 7$$

$$= 34$$

4. $\displaystyle\sum_{k=1}^{6} 4k$

5. $\displaystyle\sum_{k=1}^{\infty} 2^k$

6. $a_{n+1} = 2 + \dfrac{1}{a_n}$

$a_1 = 2 + \dfrac{1}{1} = 2 + 1 = 3$

$a_2 = 2 + \dfrac{1}{3} = 2\dfrac{1}{3}$

$a_3 = 2 + \dfrac{1}{\frac{7}{3}} = 2 + \dfrac{3}{7} = 2\dfrac{3}{7}$

$a_4 = 2 + \dfrac{1}{\frac{17}{7}} = 2 + \dfrac{7}{17} = 2\dfrac{7}{17}$

7. $d = 5 - 2 = 3$

$a_n = a_1 + (n-1)d$

$a_{15} = 2 + (15-1)3 = 44$

8. $a_1 = 8$, $a_{21} = 108$, $n = 21$

$a_n = a_1 + (n-1)d$

$108 = 8 + (21-1)d$

$100 = 20d$

$5 = d$

Use $a_n = a_1 + (n-1)d$ again to find a_7.

$a_7 = 8 + (7-1)(5) = 8 + 30 = 38$

9. $a_1 = 17$, $d = 13 - 17 = -4$, $n = 20$

First find a_{20}:

$a_n = a_1 + (n-1)d$

$a_{20} = 17 + (20-1)(-4) = 17 - 76 = -59$

Now find S_{20}:

$S_n = \dfrac{n}{2}(a_1 + a_n)$

$S_{20} = \dfrac{20}{2}(17 - 59) = 10(-42) = -420$

10. $\displaystyle\sum_{k=1}^{25}(2k+1)$

$a_1 = 2 \cdot 1 + 1 = 3$

$a_{25} = 2 \cdot 25 + 1 = 51$

$S_n = \dfrac{n}{2}(a_1 + a_n)$

$S_{25} = \dfrac{25}{2}(3 + 51) = \dfrac{25}{2} \cdot 54 = 675$

11. $a_1 = 10$, $r = \dfrac{-5}{10} = -\dfrac{1}{2}$

$a_n = a_1 r^{n-1}$

$a_{11} = 10\left(-\dfrac{1}{2}\right)^{11-1} = \dfrac{5}{512}$

12. $r = 0.2$, $S_4 = 1248$

$S_n = \dfrac{a_1(1-r^n)}{1-r}$

$1248 = \dfrac{a_1(1 - 0.2^4)}{1 - 0.2}$

$1248 = \dfrac{0.9984 a_1}{0.8}$

$a_1 = 1000$

13. $\displaystyle\sum_{k=1}^{8} 2^k$

$a_1 = 2^1 = 2$, $r = 2$, $n = 8$

$S_8 = \dfrac{2(1 - 2^8)}{1 - 2} = 510$

14. $a_1 = 18$, $r = \dfrac{6}{18} = \dfrac{1}{3}$

Since $|r| = \dfrac{1}{3} < 1$, the series has a sum.

$S_\infty = \dfrac{18}{1 - \dfrac{1}{3}} = \dfrac{18}{\dfrac{2}{3}} = 18 \cdot \dfrac{3}{2} = 27$

15. $0.\overline{56} = 0.56 + 0.0056 + 0.000056 + \ldots$

$|r| = \left|\dfrac{0.0056}{0.56}\right| = |0.01| = 0.01 < 1$, so the series has a sum.

$S_\infty = \dfrac{0.56}{1 - 0.01} = \dfrac{0.56}{0.99} = \dfrac{56}{99}$

16. $a_1 = \$10,000$

$a_2 = \$10,000 \cdot 0.80 = \$8000$

$a_3 = \$8000 \cdot 0.80 = \$6400$

$a_4 = \$6400 \cdot 0.80 = \$5120$

$a_5 = \$5120 \cdot 0.80 = \$4096$

$a_6 = \$4096 \cdot 0.80 = \$3276.80$

17. We have an arithmetic sequence \$8.50, \$8.75, \$9.00, \$9.25, and so on with $d = \$0.25$. Each year there are 12/3, or 4 raises, so after 4 years the sequence will have the original hourly wage plus the $4 \cdot 4$, or 16, raises for a total of 17 terms. We use the formula $a_n = a_1 + (n-1)d$ with $a_1 = \$8.50$, $d = \$0.25$, and $n = 17$.

$a_{17} = \$8.50 + (17-1)(\$0.25) = \$8.50 + 16(\$0.25) = \$8.50 + \$4.00 = \$12.50$

At the end of 4 years Tamika's hourly wage will be \$12.50.

18. We use the formula $S_n = \dfrac{a_1(1-r^n)}{1-r}$ with $a_1 = \$2500$, $r = 1.056$, and $n = 18$.

$S_{18} = \dfrac{2500[1 - (1.056)^{18}]}{1 - 1.056} = \$74,399.77$

19. $S_n: \quad 2 + 5 + 8 + \ldots + (3n-1) = \dfrac{n(3n+1)}{2}$

$S_1: \quad 2 = \dfrac{1(3 \cdot 1 + 1)}{2}$

$S_k: \quad 2 + 5 + 8 + \ldots + (3k-1) = \dfrac{k(3k+1)}{2}$

$S_{k+1}: \quad 2 + 5 + 8 + \ldots + (3k-1) + [3(k+1)-1] = \dfrac{(k+1)[3(k+1)+1]}{2}$

1) *Basis step:* $\dfrac{1(3 \cdot 1 + 1)}{2} = \dfrac{1 \cdot 4}{2} = 2$, so S_1 is true.

2) *Induction step:*

$\underbrace{2 + 5 + 8 + \ldots + (3k-1)}_{} + [3(k+1)-1]$

$= \dfrac{k(3k+1)}{2} + [3k+3-1] \quad \text{By } S_k$

$= \dfrac{3k^2}{2} + \dfrac{k}{2} + 3k + 2$

$= \dfrac{3k^2}{2} + \dfrac{7k}{2} + 2$

$= \dfrac{3k^2 + 7k + 4}{2}$

$= \dfrac{(k+1)(3k+4)}{2}$

$= \dfrac{(k+1)[3(k+1)+1]}{2}$

20. $_{15}P_6 = \dfrac{15!}{(15-6)!} = 3,603,600$

21. $_{21}C_{10} = \dfrac{21!}{10!(21-10)!} = 352,716$

22. $\dbinom{n}{4} = \dfrac{n!}{4!(n-4)!}$

$\qquad = \dfrac{n(n-1)(n-2)(n-3)(n-4)!}{4!(n-4)!}$

$\qquad = \dfrac{n(n-1)(n-2)(n-3)}{24}$

23. $_6P_4 = \dfrac{6!}{(6-4)!} = 360$

24. a) $6^4 = 1296$

$\quad$ b) $_5P_3 = \dfrac{5!}{(5-3)!} = 60$

25. $_{28}C_4 = \dfrac{28!}{4!(28-4)!} = 20,475$

26. $_{12}C_8 \cdot _8C_4 = \dfrac{12!}{8!(12-8)!} \cdot \dfrac{8!}{4!(8-4)!} = 34,650$

27. Expand: $(x+1)^5$.

Pascal's triangle method: Use the 6th row.

1   5   10   10   5   1
$(x+1)^5 = x^5 + 5x^4 \cdot 1 + 10x^3 \cdot 1^2 + 10x^2 \cdot 1^3 + 5x \cdot 1^4 + 1^5$
$\qquad = x^5 + 5x^4 + 10x^3 + 10x^2 + 5x + 1$

Factorial notation method:

$(x+1)^5 = \dbinom{5}{0}x^5 + \dbinom{5}{1}x^4 \cdot 1 + \dbinom{5}{2}x^3 \cdot 1^2 +$

$\qquad \dbinom{5}{3}x^2 \cdot 1^3 + \dbinom{5}{4}x \cdot 1^4 + \dbinom{5}{5}1^5$

$\qquad = x^5 + 5x^4 + 10x^3 + 10x^2 + 5x + 1$

28. Find 5th term of $(x-y)^7$.

$\dbinom{7}{4}x^3(-y)^4 = 35x^3y^4$

29. $2^9 = 512$

30. $\dfrac{8}{6+8} = \dfrac{8}{14} = \dfrac{4}{7}$

31. $\dfrac{_6C_1 \cdot _5C_2 \cdot _4C_5}{_{15}C_6} = \dfrac{6 \cdot 10 \cdot 4}{5005} = \dfrac{48}{1001}$

32. $\qquad\qquad _nP_7 = 9 \cdot _nP_6$

$\qquad \dfrac{n!}{(n-7)!} = 9 \cdot \dfrac{n!}{(n-6)!}$

$\dfrac{n!}{(n-7)!} \cdot \dfrac{(n-6)!}{n!} = 9 \cdot \dfrac{n!}{(n-6)!} \cdot \dfrac{(n-6)!}{n!}$

$\qquad \dfrac{(n-6)(n-7)!}{(n-7)!} = 9$

$\qquad\qquad n - 6 = 9$

$\qquad\qquad\qquad n = 15$